D0793731

ULTRA-HIGH TEMPERATURE CERAMICS

ULTRA-HIGH TEMPERATURE CERAMICS

Materials for Extreme Environment Applications

EDITED BY

William G. Fahrenholtz

Eric J. Wuchina

William E. Lee

Yanchun Zhou

The American Ceramic Society

WILEY

Cover Images: Title background shows crystal growth steps on the surface of a polycrystalline zirconium diboride ceramic (Courtesy of D. Sciti and F. Monteverde, Institute of Science and Technology for Ceramics); Main image shows a UHTC composite being tested in the flame of an oxyacetylene torch (Courtesy of J. Binner, University of Birmingham).

Published by John Wiley & Sons, Inc., Hoboken, New Jersey.
Published simultaneously in Canada.

For general information on our other products and services or for technical support, please contact our Customer Care Department within the United States at (800) 762-2974, outside the United States at (317) 572-3993 or fax (317) 572-4002.

Wiley also publishes its books in a variety of electronic formats. Some content that appears in print may not be available in electronic formats. For more information about Wiley products, visit our web site at www.wiley.com.

Library of Congress Cataloging-in-Publication Data:

Ultra-high temperature ceramics : materials for extreme environment applications / edited by William G. Fahrenholtz, Eric. J. Wuchina, William E. Lee, Yanchun Zhou.
    pages   cm
    Includes index.
    ISBN 978-1-118-70078-5 (hardback)
1. Ceramics.   2. Ceramic materials.   3. Boron compounds.   4. Carbon compounds.   I. Fahrenholtz, William, editor.
    TP807.U48 2014
    666–dc23

2014024691

Printed in the United States of America.

10   9   8   7   6   5   4   3   2   1

The editors would like to dedicate this book to the memory of Dr. Jules Routbort who provided the impetus and support for the first UHTC meeting. Unfortunately, Jules passed away in early 2012 before the meeting was held.

CONTENTS

ACKNOWLEDGMENTS

The chapters in this book are based on presentations at *Ultra-High Temperature Ceramics: Materials for Extreme Environment Applications II* that was held on May 13–8, 2012, in Schloss Hernstein, Austria. The authors would like to thank everyone who supported the meeting financially, especially Dr. Joe Wells from the Office of Naval Research-Global (N62909-12-1-1014) and Dr. Randall Pollak from the European Office of Aerospace Research and Development (FA8655-12-1-0009) for providing funding to support the participation of graduate students and early-career professionals.

CONTRIBUTORS LIST

Davide Alfano Italian Aerospace Research Centre (CIRA), Capua (CE), Italy

Bikramjit Basu Materials Research Center, Indian Institute of Science, Bangalore, India

Twisampati Bhandari High Temperature and Energy Materials Laboratory, Department of Metallurgical Engineering and Material Science, IIT Bombay, Mumbai, India

Jon Binner School of Metallurgy and Materials, University of Birmingham, Birmingham, UK

Michael K. Cinibulk Materials and Manufacturing Directorate, Air Force Research Laboratory, Wright-Patterson AFB, OH, USA

Antonio Del Vecchio Italian Aerospace Research Centre (CIRA), Capua (CE), Italy

William G. Fahrenholtz Department of Materials Science and Engineering, Missouri University of Science and Technology, Rolla, MO, USA

George V. Franks Department of Chemical and Biomolecular Engineering, The University of Melbourne, Victoria and Defence Materials Technology Centre (DMTC), Hawthorn, Australia

Roberto Gardi Italian Aerospace Research Centre (CIRA), Capua (CE), Italy

Edoardo Giorgi Department of Materials, Imperial College London, London, UK

Brahma Raju Golla Metallurgical and Materials Engineering Department, National Institute of Technology, Warangal, India

Wei-Ming Guo School of Electromechanical Engineering, Guangzhou Higher Education Mega Center, Guangdong University of Technology, Guangzhou, China

Gregory J. K. Harrington Department of Materials Science and Engineering, Missouri University of Science and Technology, Rolla, MO, USA

Robert Harrison Department of Materials, Imperial College London, London, UK

Greg E. Hilmas Materials Science and Engineering Department, Missouri University of Science and Technology, Rolla, MO, USA

William E. Lee Centre for Advanced Structural Ceramics, Department of Materials, Imperial College London, London, UK

Zhen Li Shenyang National Laboratory for Materials Science, Institute of Metal Research, Chinese Academy of Science, Shenyang, China

Hai-Tao Liu State Key Laboratory of High Performance Ceramics and Superfine Microstructure, Shanghai Institute of Ceramics, Chinese Academy of Sciences, Shanghai, China

Ji-Xuan Liu State Key Laboratory of High Performance Ceramics and Superfine Microstructure, Shanghai Institute of Ceramics, Chinese Academy of Sciences, Shanghai, China

Alexandre Maître Centre Européen de la Céramique, University of Limoges, Limoges Cedex, France

Valentina Medri National Research Council of Italy, Institute of Science and Technology for Ceramics, Faenza, Italy

Frédéric Monteverde National Research Council of Italy, Institute of Science and Technology for Ceramics, Faenza, Italy

Amartya Mukhopadhyay High Temperature and Energy Materials Laboratory, Department of Metallurgical Engineering and Material Science, IIT Bombay, Mumbai, India

Eric W. Neuman Materials Science and Engineering Department, Missouri University of Science and Technology, Rolla, MO and Intel Corporation, Beaverton, OR, USA

De-Wei Ni Department of Energy Conservation and Storage, Technical University of Denmark, Roskilde, Denmark

Mark Opeka Naval Surface Warfare Center, West Bethesda, MD, USA

Triplicane A. Parthasarathy UES, Inc., Dayton, OH, USA

Anish Paul School of Metallurgy and Materials, University of Birmingham, Birmingham, UK

Olivier Rapaud Centre Européen de la Céramique, University of Limoges, Limoges Cedex, France

Luigi Scatteia Booz & Company B.V., Amsterdam, the Netherlands

Diletta Sciti National Research Council of Italy, Institute of Science and Technology for Ceramics, Faenza, Italy

Laura Silvestroni National Research Council of Italy, Institute of Science and Technology for Ceramics, Faenza, Italy

Carolina Tallon Department of Chemical and Biomolecular Engineering, The University of Melbourne, Victoria and Defence Materials Technology Centre (DMTC), Hawthorn, Australia

Gregory B. Thompson Department of Metallurgical and Materials Engineering, University of Alabama, Tuscaloosa, AL, USA

Bala Vaidhyanathan Department of Materials, Loughborough University, Loughborough, UK

L. J. Vandeperre Centre for Advanced Structural Ceramics, Department of Materials, Imperial College London, London, UK

J. Wang Centre for Advanced Structural Ceramics, Department of Materials, Imperial College London, London, UK

Jiemin Wang Shenyang National Laboratory for Materials Science, Institute of Metal Research, Chinese Academy of Science, Shenyang, China

Jingyang Wang Shenyang National Laboratory for Materials Science, Institute of Metal Research, Chinese Academy of Science, Shenyang, China

Xin-Gang Wang State Key Laboratory of High Performance Ceramics and Superfine Microstructure, Shanghai Institute of Ceramics, Chinese Academy of Sciences, Shanghai, China

Christopher R. Weinberger Sandia National Laboratories, Albuquerque, NM, USA; Department of Mechanical Engineering, Drexel University, Philadelphia, PA, USA

Wen-Wen Wu School of Engineering, Brown University, Providence, RI, USA

Eric J. Wuchina Naval Surface Warfare Center, Carderock Division, West Bethesda, MD, USA

Xun Zhan Shenyang National Laboratory for Materials Science, Institute of Metal Research, Chinese Academy of Science, Shenyang, China

Guo-Jun Zhang State Key Laboratory of High Performance Ceramics and Superfine Microstructure, Shanghai Institute of Ceramics, Chinese Academy of Sciences, Shanghai, China

Yanchun Zhou Science and Technology of Advanced Functional Composite Laboratory, Aerospace Research Institute of Materials and Processing Technology, Beijing, China

Ji Zou Department of Materials and Environmental Chemistry, Arrhenius Laboratory, Stockholm University, Stockholm, Sweden

1

INTRODUCTION

William G. Fahrenholtz[1], Eric J. Wuchina[2], William E. Lee[3], and
Yanchun Zhou[4]

[1]*Department of Materials Science and Engineering, Missouri University of
Science and Technology, Rolla, MO, USA*
[2]*Naval Surface Warfare Center, Carderock Division, West Bethesda, MD, USA*
[3]*Centre for Advanced Structural Ceramics, Department of Materials,
Imperial College London, London, UK*
[4]*Science and Technology of Advanced Functional Composite Laboratory, Aerospace
Research Institute of Materials and Processing Technology, Beijing, China*

1.1 BACKGROUND

The impetus for this book was the conference "Ultra-High Temperature Ceramics: Materials for Extreme Environment Applications II," which was held on May 13–18, 2012, in Hernstein, Austria. As the title implies, this was the second conference on this topic that has been organized by Engineering Conferences International (ECI). The four editors served as the co-organizers of the conference. Both the U.S. Navy Office of Naval Research Global and the U.S. Air Force European Office of Aerospace Research and Development provided funding for the conference that helped support participation of invited speakers and students. The conference brought together about 60 researchers from around the world to discuss the latest research related to this remarkable class of materials. The first conference in the series was organized by Eric Wuchina and Alida

Ultra-High Temperature Ceramics: Materials for Extreme Environment Applications, First Edition.
Edited by William G. Fahrenholtz, Eric J. Wuchina, William E. Lee, and Yanchun Zhou.
© 2014 The American Ceramic Society. Published 2014 by John Wiley & Sons, Inc.

Bellosi and was held in August 2008 at Lake Tahoe, California, United States. A third conference is planned for Australia in April 2015.

The book is our attempt to capture a snapshot of the current state of the art in ultra-high temperature ceramic (UHTC) materials. The chapters in this volume represent the key areas that were discussed in the meeting. The chapter authors are leaders in their fields from around the world, and all of the lead authors participated in the conference. Rather than a narrow focus on the latest scientific progress as would be expected in an article in a peer-reviewed journal, the chapters in this book provide a broader look at recent progress and information on the current understanding of this family of materials.

1.2 ULTRA-HIGH TEMPERATURE CERAMICS

Recent interest in UHTCs has been motivated by the search for materials that can withstand extreme environments. The extremes can include, individually or in combination, the effects of temperature, chemical reactivity, mechanical stress, radiation, and wear. Some potential applications for this class of materials include microelectronics, molten metal containment, high-temperature electrodes, and wear-resistant surfaces. However, a majority of the research has been motivated by unmet material needs for hypersonic aviation. Specifically, improved materials are needed to withstand the conditions encountered by wing leading edges and propulsion system components in hypersonic aerospace vehicles as well as the extreme conditions associated with atmospheric reentry and rocket propulsion. The combination of extreme temperature, chemically aggressive environments, and rapid heating/cooling is beyond the capabilities of current engineering ceramics. Recent interest in UHTCs has been high as indicated by a number of special journal issues [1–3] and review articles [4–9] devoted to the topic. Despite this interest, no clear criteria have been established to differentiate UHTCs from other structural ceramics.

Broadly, ceramic materials can be defined as inorganic, nonmetallic solids [10]. This definition encompasses most materials that are typically considered to be ceramics such as clay-based traditional ceramics, alumina, piezoelectric materials, and silicon carbide. However, the definition still leaves some gray areas such as glass, carbon, and intermetallic compounds. In some cases, the definition is expanded to include other characteristics. For example, Barsoum defines ceramics as "solid compounds that are formed by the application of heat, and sometimes heat and pressure, comprising at least one metal and a non-metallic elemental solid, a combination of at least two nonmetallic elemental solids, or a combination of at least two nonmetallic elemental solids and a nonmetal" [11]. In other cases, the definition involves characteristics such as melting temperature, bonding type, or electrical properties [12]. Likewise, several different definitions have been espoused for UHTCs. The three main classifications are melting temperature, ultimate use temperature, and chemical composition.

The most common definition for a UHTC is a material that melts at a temperature of 3000°C or higher. As shown in Figure 1.1, very few materials meet this criterion. For example, only three elements have melting temperature above 3000°C, W, Re, and Ta, all of which are metals. Note that carbon was not included in this group because of the complexity of its behavior, which has been reviewed in detail elsewhere [13, 14]. Interestingly, ThO_2 is the only oxide ceramic that has a melting temperature above

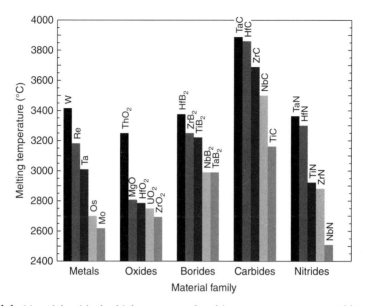

Figure 1.1. Materials with the highest reported melting temperature grouped by material family. Reprinted with permission from Ref. [6].

3000°C. Most of the materials that have melting temperatures above 3000°C are borides, carbides, and nitrides of early transition metals. Consequently, most studies on UHTCs focus on compounds such as ZrB_2, HfB_2, TaC, TaB_2, ZrC, and HfC. While this definition is probably the most common, significant uncertainty exists in melting temperatures for these compounds. For example, ZrB_2 is commonly reported to melt at 3250°C based on the phase diagrams reported by Rudy [15] and Portnoi *et al.* [16], but others report different melting temperatures including 3040°C by Glaser and Post [17] and 3517°C by Rogl and Potter [18] The discrepancies indicate that not only are temperatures difficult to measure precisely in the ultra-high temperature regime, but also that these compounds may decompose or dissociate before melting, which would also be difficult to detect in the experimental setups used for these studies. So, while melting temperature is a clearly defined metric, assessment includes uncertainty. Further, the selection of 3000°C as the criterion was arbitrary and could be set at other temperatures.

A second method that can be used to define UHTCs is the highest temperature for use in air. This practical definition fits well with the nature of engineering materials, but introduces additional questions. As with melting temperature, the selection of a use temperature to define the ultra-high temperature regime is also somewhat arbitrary. At the time of publication of this chapter, a number of choices are available for use in air at temperatures up to about 1600 °C including alumina, magnesia, silicon carbide, and silicon nitride. Hence, the minimum use temperature for the ultra-high temperature regime should be above that level. Many hypersonic applications involve higher temperatures, so 2000 °C has been cited as the cutoff of the ultra-high temperature regime [5–19]. Despite a seemingly clear definition, both use temperature and duration blur the distinction. The application-driven criterion to define UHTCs has obvious attractions as a definition for engineering materials, but this metric also has some shortcomings.

The final method that can be used to define UHTCs is chemistry, which is the least quantitative, but probably most widely used. Most UHTC compounds are borides, carbides, or nitrides of early transition metals. Hence, any compound containing a transition metal such as Zr, Hf, Ta, W, or Nb along with B, C, or N has the potential to be a UHTC.

No one definition has emerged as the way to identify UHTCs. As an example of the shortcomings of all of the three definitions described earlier, consider one of the most commonly cited UHTC compositions, ZrB_2–SiC. When compounds with melting temperatures above 3000 °C such as ZrB_2 are combined intentionally with other phases (i.e., sintering aids, grain pinning additives, oxidation-enhancing additives, etc.) or when impurities are present, the temperature at which liquid forms or melting occurs can be below 3000°C. For ZrB_2–SiC, the solidus temperature is about 2300°C due to a eutectic reaction [20]. Hence, one could argue that one of the most widely researched UHTC compositions does not meet any of the criteria defined earlier because: (i) liquid forms below 3000°C; (ii) one of the constituents (SiC) cannot be used for extended times above 1600°C; and (iii) SiC is not a boride, carbide, or nitride of an early transition metal.

Despite the uncertainty in the definition of UHTCs, a close-knit global community of researchers who focus on these materials has emerged over the past decade or so. Groups in the United States, Italy, China, Australia, Russia, and the United Kingdom, have worked competitively and collaboratively to advance our understanding of the fundamental behavior of materials that can be used in extreme environments. This book, in keeping with the spirit of the ECI conference series, takes a pragmatic approach in defining UHTCs and includes the community of researchers who focus on materials with the potential for use in extreme environments such as those associated with hypersonic flight, atmospheric reentry, and rocket propulsion. Since the first conference in the series in 2008, the community of researchers has grown to include those working on materials for nuclear applications and researchers who are investigating new methods for characterizing and testing materials under conditions that are representative of the extreme environments encountered in use.

1.3 DESCRIPTION OF CONTENTS

The sequence of the chapters in this book was selected to represent the progression of a typical experimental study. After this introductory chapter, the next chapter provides background information on previous research. The chapter focuses on historic studies on what we now consider UHTCs that were conducted in the United States in the 1950s, 1960s, and 1970s. These studies accompanied decisions that were being made about the design of the next generation of launch and reentry vehicles. The next group of chapters describes synthesis and processing. The third chapter describes synthesis of boride compounds, while the fourth focuses on calculations of fundamental bonding properties. The fifth chapter describes aqueous processing and surface chemistry. The sixth is the final chapter in the synthesis and processing section, and it describes densification and microstructure of UHTCs. The middle section of the book contains chapters describing the properties of UHTCs. This section includes contributions focused on thermomechanical properties, elevated temperature mechanical properties and deformation, and oxidation. The final section is focused on the performance of UHTC materials. The section includes a review of UHTC-based composites. Other chapters focus on TaC, UHTC nitrides, TiB_2, and UHTCs for nuclear applications. The final chapter describes the testing of UHTCs in relevant environments. That chapter is based on the conference keynote presentation.

REFERENCES

1. Joan Fuller and Michael Sacks, issue editors. Special Issue of the Journal of Materials Science, 39(19), October 2004, 5885–6066.
2. Joan Fuller, Yigal Blum, and Jochen Marschall, guest editors. Special Issue of the Journal of the American Ceramic Society, 91(5), May 2008, 1397–1502.
3. Joan Fuller, Greg Hilmas, Erica Corral, Laura Riegel, and William Fahrenholtz, guest editors. Special Issue of the Journal of the European Ceramic Society, 30(11), August 2010, 2145–2418.
4. Telle R, Sigl LS, Takagi K. Boride-based hard materials. In: Riedel R, editor. *Handbook of Ceramic Hard Materials*. Weinheim: Wiley-VCH; 2000. p 802–949.
5. Gasch MJ, Ellerby DT, Johnson SM. Ultra-High temperature ceramics. In: Bansal N, editor. *Handbook of Composites*. Boston (MA): Kluwer Academic Publishers; 2004. p 197–224.
6. Fahrenholtz WG, Hilmas GE, Talmy IG, Zaykoski JA. Refractory diborides of zirconium and hafnium. J Am Ceram Soc 2007;90 (5):1347–1364.
7. Guo S-Q. Densification of ZrB_2-based composites and their mechanical and physical properties: a review. J Eur Ceram Soc 2009;29 (6):995–1011.
8. Fahrenholtz WG, Hilmas GE. Oxidation of Ultra-High temperature transition metal diboride ceramics. Int Mater Rev 2012;57 (1):61–72.
9. Eakins E, Jayaseelan DD, Lee WE. Toward Oxidation-Resistant ZrB_2–SiC Ultra-High Temperature Ceramics. Met and Mat Trans A 2011;42A: 878–887.
10. Kingery WD, Bowen HK, Uhlmann DR. *Introduction to Ceramics*. New York: John Wiley & Sons, Inc.; 1976.
11. Barsoum MW. *Fundamentals of Ceramics*. New York: McGraw-Hill; 1997.
12. Richerson DW. *Modern Ceramic Engineering*. 2nd ed. New York: Marcel-Dekker; 1992.
13. Bundy FP. Pressure-temperature phase diagram of elemental carbon. Phys A 1989;156 (1):169–178.
14. Bundy FP, Bassett WA, Weathers MS, Hemley RJ, Mao HK, Goncharov AF. The pressure-temperature phase and transformation diagram for carbon; updated through 1994. Carbon 1996;34 (2):141–153.
15. Rudy E. Ternary phase equilibria in transition metal-boron-carbon systems: part V, compendium of phase diagram data. Technical Report AFML-TR-65-2. Wright Patterson Air Force Base (OH): Air Force Materials Laboratory; 1969.
16. Portnoi KI, Romashov VM, Vyroshina LI. Phase diagram of the zirconium-boron system. Poroshkoviaia Metallugia 1970;10 (7):68–71.
17. Glaser FW, Post B. System zirconium-boron. Trans Metallurgical Soc AIME 1953;197: 1117–1118.
18. Rogl P, Potter PE. A critical review and thermodynamic calculation of the binary system: zirconium-boron. Calphad 1988;12 (2):191–204.
19. Opeka MM, Talmy IG, Zaykoski JA. Oxidation-based materials selection for 2000°C+ hypersonic aerosurfaces: theoretical considerations and historical experience. J Mater Sci 2004;39 (13):5887–5904.
20. McHale AE, editor. *Phase Diagrams for Ceramists Volume X: Borides, Carbides, and Nitrides*. Westerville (OH): The American Ceramic Society; 1994. Figure 8672.

2

A HISTORICAL PERSPECTIVE ON RESEARCH RELATED TO ULTRA-HIGH TEMPERATURE CERAMICS*

William G. Fahrenholtz

Department of Materials Science and Engineering, Missouri University of Science and Technology, Rolla, MO, USA

2.1 ULTRA-HIGH TEMPERATURE CERAMICS

Historically, the boride and carbide ceramics that we now classify as ultra-high temperature ceramics (UHTCs) have been known by a number of different names including refractory borides or carbides [1–4], oxidation-resistant diborides [5], ceramals or cermets [6, 7], and hard metals [8]. Although the term **ultra-high temperature** was not used in any of these early reports, the aerospace community recognized the need for materials that could withstand the extreme temperatures and chemically aggressive environments. For example, reports from the late 1950s and early 1960s from the National Advisory Committee for Aeronautics (NACA) and its follow-on agency the National Aeronautics and Space Administration (NASA) described the need for rocket nozzles and thermal protection systems [2, 6, 9, 10]. In particular, Reference [10] identified heat sources associated with atmospheric reentry and rocket propulsion as well as

---

*This manuscript references a number of technical reports from projects sponsored or conducted by the U.S. Government. All of these reports have been approved for public release and are available either through the NASA Technical Report Server (http://ntrs.nasa.gov/search.jsp) or the Defense Technical Information Center (http://www.dtic.mil/dtic/).

Ultra-High Temperature Ceramics: Materials for Extreme Environment Applications, First Edition.
Edited by William G. Fahrenholtz, Eric J. Wuchina, William E. Lee, and Yanchun Zhou.
© 2014 The American Ceramic Society. Published 2014 by John Wiley & Sons, Inc.

TABLE 2.1. Selected recommendations for future research and development activities related to UHTCs from Reference [10]

Recommendation number	Recommendation
10	A systematic investigation on the application of TPS to solid propellant rocket motors to include (i) transpiration, (ii) evaporation cooling, (iii) chemical reaction cooling, (iv) heat sink, (v) radiation, (vi) film cooling, and (vii) ablation.
12	Studies on the materials behavior at extreme temperatures, and establishment of **thermal and other pertinent material properties at extremely high** as well as extremely low temperatures.
13	Development and introduction of standard test methods for the establishment of performance data for TPS.
16	Reduction of efforts in short-time high-heating-rate ablative systems for ballistic nose cones, as all present and foreseen requirements can be adequately met with achieved capabilities. Remaining efforts should be directed toward a wider variety of potential material systems.
18	Reduction of present extensive efforts of theoretical treatment of TPS in favor of technological approaches as outlined in (12) above

the needs related to leading edges, acreage, and propulsion components. The report provided a series of recommendations for future studies, and some of the pertinent recommendations are summarized in Table 2.1. Interestingly, some of the needs identified more than 50 years ago continue to be cited as priorities for current research. The Streurer report also defined heat flux and time regimes in which different types of thermal protection systems would be most effective [10]. As shown in Figure 2.1, convection and radiation cooling could be used for long-duration exposures at heat fluxes below about 40 BTU/ft^2·s (~45 W/cm^2), whereas a heat sink approach could be used for short-duration (i.e., <0.3 min) exposures for heat fluxes up to about 1000 BTU/ft^2·s (1130 W/cm^2). Either ablative materials or transpiration cooling was recommended for heat fluxes above 1000 BTU/ft^2·s (1130 W/cm^2).

Although the term UHTCs is a recent development, the term **extreme** associated with the application environments began to appear in the ceramic refractories literature in the late 1950s. Extreme was used to refer to temperatures of 2500°F (~1400°C) or higher [11, 12]. The broader term **ultra-high temperature** emerged in the 1960s and found more widespread use in Japanese articles through the 1990s [13–16]. Terms similar to **ultra-high temperature ceramics** started to appear in a series of reports and papers when the current wave of interest in these materials started in the United States in the late 1980s and early 1990s [17–21]. Reference [20] is notable as it clearly defined the ultra-high temperature regime in terms of temperature (above 3000°F or ~1650°C) as well as providing a minimum strength target (greater than ~150 MPa) for materials at ultra-high temperatures (Fig. 2.2). The NASA report [21] seems to have been particularly influential in solidifying the use of the term

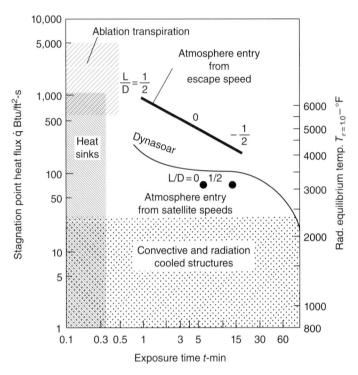

Figure 2.1. Summary of types of thermal protection systems as a function of heat flux and exposure time from Reference [10].

UHTC as it clearly identified this class of materials as candidates for hypersonic aerospace vehicles. The subsequent Sharp Hypersonic Aerothermodynamic Research Probe (SHARP) tests drew significant attention to UHTCs. In particular, the SHARP B-1 and B-2 tests as well as the related NASA reports identified UHTCs as potential candidates for the hypersonic flight environment despite failure of the UHTC strakes due to problems traced back to processing issues [22, 23]. The term UHTC and the recent resurgence of interest in research on boride and carbide UHTCs seem to have grown from these studies as well as those from a select group of other researchers [24–26].

2.2 HISTORIC RESEARCH

The synthesis of boride and carbide compounds began to draw the interest of researchers in the late 1800s and early 1900s. For the diborides, Tucker and Moody reacted elemental zirconium and boron to produce ZrB_2, although they described the compound as Zr_3B_4 [27, 28]. McKenna later reported the synthesis of ZrB_2 by carbothermal reduction according to Reaction 1 at 2000°C [29]. In contrast, reports

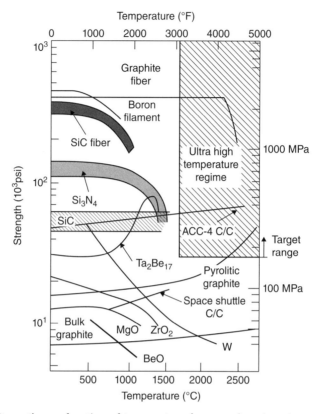

Figure 2.2. Strength as a function of temperature for several engineering materials along with a clearly defined Ultra-High temperature regime from Reference [20].

of the formation of HfB_2 were much later, presumably due to the difficulty of separating Hf from Zr [30].

$$ZrO_{2(cr)} + B_2O_{3(l)} + 5C_{(s)} \rightarrow ZrB_{2(cr)} + 5CO_{(g)} \qquad (2.1)$$

Henri Moissan was an early pioneer of research on carbide materials with over 600 scientific publications identified in a literature search.[†] Even though Acheson was the first to report a commercially viable synthesis process for SiC [31], Moissan's research on SiC [32], as well as the carbides of Mo [33], Ti [34], and Zr [35] was notable because of its breadth and depth. After the initial reports, scientific publications on borides and many of the carbides were sporadic, but progress was made in areas such as bonding [36], electrical properties [37], and electronic structure [38] through the first half of the 1900s. Although limited in scope, these early studies set the stage for the progress that would result from later studies.

[†]SciFinder Scholar, accessed June 6, 2012.

2.3 INITIAL NASA STUDIES

In the late 1950s and early 1960s, NASA (and its predecessor agency NACA) began to search for materials that could be used in the extreme environments associated with rocket propulsion and atmospheric reentry. The agency was exploring supersonic flight with vehicles such as the Bell X-1, which was the first plane to fly faster than the speed of sound, as well as a variety of concepts for hypersonic aerospace vehicles [39]. NASA conceived, studied, and tested both blunt body and lifting body designs [40]. Even in the early 1950s, NASA recognized that current materials technologies were not adequate to enable their future vehicle needs and began to search for suitable candidates [6]. For rocket motors, extreme temperatures were predicted for some applications, so the list of candidates of interest was limited to materials with melting points above 6000°F (~3300°C) including tungsten, pyrolytic graphite, hafnium carbide, and tantalum carbide [2]. In contrast, a wider variety of materials were needed for the thermal protection systems of future manned and unmanned vehicles. The agency recognized that their array of planned vehicles would present a wide variety of thermal loads based on the trajectories and other needs of specific missions [10]. For example, significant differences in heat loads were identified based on whether vehicles would land vertically as was planned for the Mercury program or horizontally as was foreseen for winged vehicles (Fig. 2.3) [41]. Because of the possible trajectories that were being considered (Fig. 2.4) and the different possible wing leading-edge radii [41], a variety of materials were considered candidates including refractory metals, ceramics, and cermets [42]. At this point, heat loads

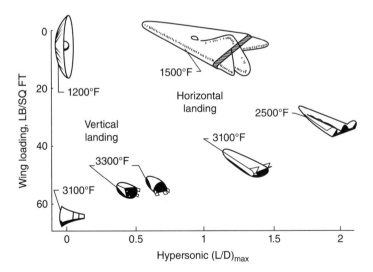

Figure 2.3. Notional temperature requirements for orbital reentry vehicles based on projected wing loading and hypersonic lift-to-drag (L/D) ratio from Reference [41].

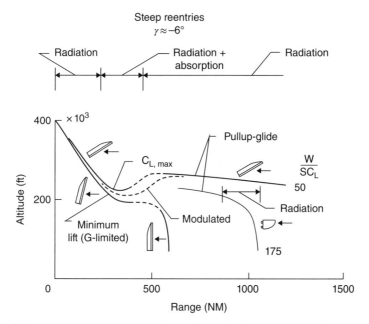

Figure 2.4. Trajectories and methods for dealing with heat loads from Reference [41].

could be predicted, and the fundamental design principles used to develop thermal protection systems for specific trajectories had been established, but neither standardized evaluation tests nor design databases populated with candidate materials were available [10]. Availability of standardized tests and design data is a problem that persist for sharp leading-edge vehicles to the present day. These early studies also motivated the U.S. Air Force to explore vehicle designs with higher cross range, which motivated interest in transition metal boride compounds. Ultimately, the blunt body Space Shuttle Orbiter was selected as the first reusable atmospheric reentry vehicle based on cost and mission flexibility [39]. However, the early studies showed the potential advantages of hypersonic vehicles with a high lift to drag ratio such as substantially higher cross range.

2.4 RESEARCH FUNDED BY THE AIR FORCE MATERIALS LABORATORY

Beginning in the early 1960s, the U.S. Air Force funded a series of studies that focused on refractory diborides and carbides as candidates for a number of potential future aerospace vehicles. For this manuscript, the research has been divided into three main categories: (1) initial thermodynamic analysis and oxidation behavior; (2) processing, properties, oxidation, and testing studies; and (3) phase equilibria research. Each of these areas is discussed in the following subsections.

2.4.1 Thermodynamic Analysis and Oxidation Behavior

Through the first half of the 1960s, the U.S. Air Force commissioned studies focused on the thermodynamic properties of refractory compounds, including the borides, carbides, and nitrides. Broad-based studies at AVCO produced fundamental thermodynamic property data for an extensive number of borides, carbides, and nitrides [43]. Data generated as part of that project continue to be cited today in references such as the NIST-JANAF tables [44]. Building on the AVCO studies, investigations at Arthur D. Little, Inc. focused on the preparation and characterization, thermodynamic data, and reaction kinetics of ZrB_2 and HfB_2 [45]. These investigations had a number of notable outcomes. The materials studied in this project were produced using zone melting techniques to remove impurities and minimize porosity. As shown in Figure 2.5, both ZrB_2 and HfB_2 exhibited parabolic oxidation kinetics over wide temperature ranges, with HfB_2 having

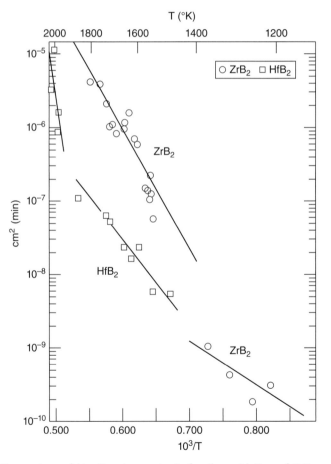

Figure 2.5. Comparison of kinetic rate constants for the oxidation of ZrB_2 and HfB_2 as a function of temperature [45].

lower overall rate constants based on lower mass gain. This research resulted in a number of scientific publications describing the oxidation of nominally pure diborides that continue to be referenced today [46, 47] This research also provided fundamental thermodynamic data including heat capacity values for ZrB_2 and HfB_2, which were more accurate than previous reports because of the higher purity of the materials examined in this study. In addition, Arthur D. Little performed early research on the oxidation of silicide compounds [48].

The U.S. Air Force also sponsored several projects that examined the thermodynamic aspects of refractory compounds. Research at the National Bureau of Standards (NBS; now the National Institute of Standards and Technology of NIST) examined the heats of formation of ZrB_2, AlB_2, and TiB_2 [49]. The NBS report is notable because of an extensive literature review of the synthesis of the borides of Al, Ce, Cr, La, Mg, La, Mo, Nb, P, Si, Sr, Ta, Th, Ti, W, U, and V. Other research examined the oxidation of carbides [50] and the vaporization of refractory compounds including ZrO_2, HfO_2, ThO_2, ZrB_2, and HfN [51]. More than 50 years later, the fundamental research sponsored by the U.S. Air Force continues to serve as the basis for understanding the thermochemical stability of borides and carbides.

2.4.2 Processing, Properties, Oxidation, and Testing

ManLabs, Inc., a small research and development company located in Cambridge, Massachusetts, was the lead contractor on a series of projects focused on boride and carbide ceramics. These projects started in the early 1960s and continued into the 1970s. A number of notable accomplishments were achieved, which resulted in the publication of a large number of highly detailed technical reports as well as a series of publications in scientific journals. While the focus was on using commercially available materials, the team provided feedback to suppliers such as improving the purity or reducing the particle size of the starting powders. In addition, several specialized pieces of equipment were produced for processing, characterizing, and testing these materials due to the extreme temperatures involved. The projects were divided into three different focus areas. In the first series, candidate materials were screened along with the evaluation of processing and characterization methods. The second series focused on measuring and understanding the properties of ZrB_2 and HfB_2 ceramics. The final series of reports focused on the evaluation of boride-, carbide-, and graphite-based materials in relevant environments. This subsection attempts to capture some of the research highlights from each series of reports in roughly the time sequence of the projects.

The first series of studies at ManLabs examined TiB_2, ZrB_2, HfB_2, NbB_2, and TaB_2 as potential candidates for hypersonic flight and atmospheric reentry applications [52]. Based on literature reports of melting temperatures and thermochemical stability, borides were identified as promising candidates. This first study focused on gathering chemical, physical, and thermodynamic property data for candidate materials from literature sources and then identifying the most promising materials for further study with some initial oxidation testing. Based on oxidation rates, HfB_2 and ZrB_2 were selected for further study. The follow-on study examined the effect of boron-to-metal ratio (B/Me) on oxidation behavior, thermal conductivity, emissivity, and electrical resistivity [53].

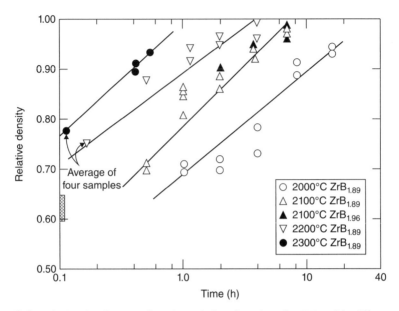

Figure 2.6. Relative density as a function of sintering time for ZrB$_2$ with different B/Me ratios [53].

This study also reported extensive evaluations of the densification behavior of ZrB$_2$ and HfB$_2$. While stoichiometric ZrB$_2$ and HfB$_2$ could not be hot pressed to relative densities of more than 98% with grain sizes of less than 50 μm, altering the B/Me ratio improved densification. As shown in Figure 2.6, ZrB$_2$ with a B/Me ratio of 1.89 could be hot pressed to nearly full density at 2200°C without exaggerated grain growth. The plot also shows that hot pressing at 2300°C resulted in a limiting density value of less than 95%, which was attributed to entrapped porosity. Further analysis of density as a function of grain size and sintering time (Fig. 2.7) was used to elucidate the densification mechanism. Densification was attributed to grain boundary diffusion through a thin liquid film where impurities were concentrated. This study also examined the oxidation behavior of the nominally pure diborides as well as diborides with additions of Si or MoSi$_2$. While detailed kinetic studies were completed as part of a project described later, this project did define regimes of behavior for protective behavior (i.e., below ~1200°C for nominally pure ZrB$_2$) and linear kinetics that are still in use today. The report ended with sections describing the thermodynamic stability and phase equilibria of the diborides that were based on literature reports and thermodynamic calculations. The most important outcome of the first two studies was that it motivated additional projects that were more focused on fundamental research on the processing, microstructure, properties, and performance of boride ceramics.

In the second series of studies, borides with carbon and silicon carbide additions were examined. As part of this project, a series of ZrB$_2$ and HfB$_2$ compositions were formulated and densified. The addition of SiC was examined to improve the oxidation resistance of the diborides while carbon was added to improve thermal shock resistance.

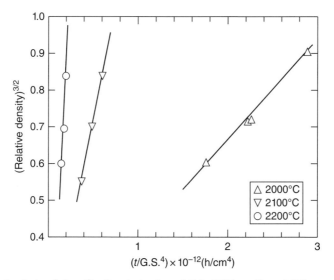

Figure 2.7. Analysis of densification behavior of ZrB$_2$ (B/Me ratio = 1.89) as a function of densification temperature where t is sintering time and GS is grain size [53].

TABLE 2.2. Compositions examined in the second series of studies by ManLabs [54]

Designation	Major phase	Additive(s)	Composition range
I	ZrB$_2$	None	
II	HfB$_2$	None	
III and IV	HfB$_2$	SiC	10 – 50 vol%
V	ZrB$_2$	SiC	10 – 50 vol%
VII	ZrB$_2$	SiC and C	14 vol% SiC, 10 vol% C
XIV	HfB$_2$	SiC and C	14 vol% SiC, 10 vol% C
XII	ZrB$_2$	C	20 vol% C
XV	HfB$_2$	C	20 vol% C

The same materials were examined through the entire series of reports, which were separated based on the focus. Some of the important compositions that were developed are summarized in Table 2.2. These reports were highly influential in later studies by narrowing the focus to ZrB$_2$ and HfB$_2$ and identifying the addition of 20 vol% SiC as the **optimal** composition based on densification, mechanical properties, and oxidation behavior.

One of the early reports examined densification behavior and characterization [54]. Up to this point, all of the borides examined by ManLabs were densified by hot pressing, but plasma spraying, pressureless sintering, and hot forging were examined as alternative densification methods. The presence of SiC was determined to be beneficial because it not only led to improved oxidation protection, but also inhibited grain growth. The results of oxidation testing were published in a series of manuscripts, including one that is used to justify SiC additions to diborides [55]. Figure 2.8 shows that the oxidation

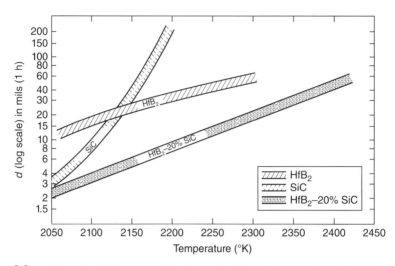

Figure 2.8. Oxide-scale thickness as a function of oxidation temperature for HfB$_2$, SiC, and HfB$_2$ containing 20 vol% SiC based on data from Reference [55].

resistance of HfB$_2$–SiC is superior to that of either HfB$_2$ or SiC alone due to the synergistic effects of the borosilicate glassy layer.

The next report in the series examined the mechanical properties of the boride compositions containing SiC and/or C [56]. For nominally pure diborides, the strength was found to increase from room temperature to a maximum at 800°C. Above 800°C, strength decreased to a minimum at 1400°C and then increased between 1400 and 1800°C (Fig. 2.9). The increase in strength up to 800°C was attributed to the relaxation of thermal residual stresses that were present after processing, but no experimental confirmation was presented to support this assertion. Above 1400°C, the increase in strength was attributed to increasing plasticity, which could act to blunt crack tips. The report also offers detailed explanations and analysis of the effects of grain size and porosity on room and elevated temperature mechanical properties. The addition of SiC was identified as beneficial to mechanical properties by increasing the elevated temperature strength (Fig. 2.10). SiC was found to reduce grain growth during densification as well as improve the strength at elevated temperatures. However, SiC was also found to increase plasticity due to grain boundary sliding, a mechanism that was proposed based on observations of deformation during mechanical testing and supported with compressive creep studies.

The last report in this series focused on the thermal, physical, electrical, and optical properties of diborides [57]. In this study, the use of additives was presented as a method for controlling properties, and extensive tables of data for heat capacity, thermal expansion, electrical resistivity, and other properties were presented (Fig. 2.11). For thermal conductivity, the importance of high relative density and minimal impurity contents was discussed.

The final series of studies at ManLabs focused on evaluating the environmental response of boride-, carbide-, and graphite-based materials. The overall goal was to

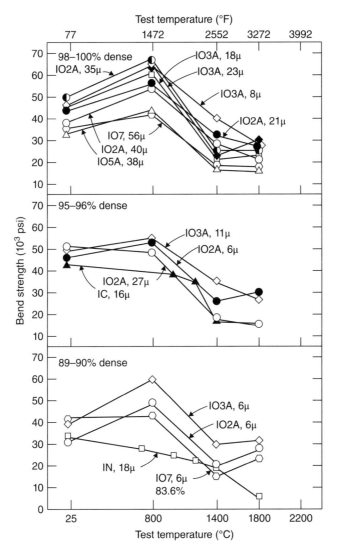

Figure 2.9. Strength as a function of temperature for nominally pure ZrB$_2$ (Material I) [56].

correlate the behavior of materials in laboratory-based tests (i.e., furnace oxidation termed the **material-centric** behavior regime in the reports) with tests that simulate hypersonic flight or atmospheric reentry (i.e., relevant conditions referred to as **environment-centric** in the reports). The final series of studies resulted in the publication of nine reports. The reports began with descriptions of facilities, progressed through tabulating experimental results, and then concluded with an attempt to correlate oxidation response in different test regimes.

The first reports from the final series described the testing methods and facilities that were utilized for the study [58, 59]. The research included conventional furnace

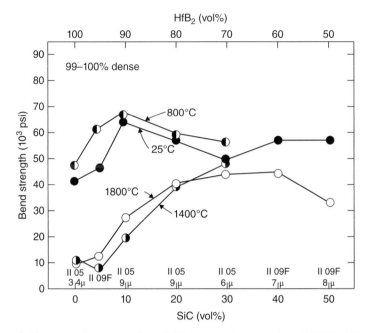

Figure 2.10. Strength as a function of SiC content in composition IV (HfB$_2$–SiC) [56].

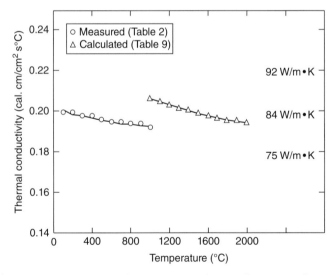

Figure 2.11. Thermal conductivity of composition I (nominally pure ZrB$_2$) as a function of temperature using two methods, cut bar for temperatures below 1000°C and thermal flash for temperature above 1000°C. The materials were nominally fully dense with a grain size of ~20 μm [57].

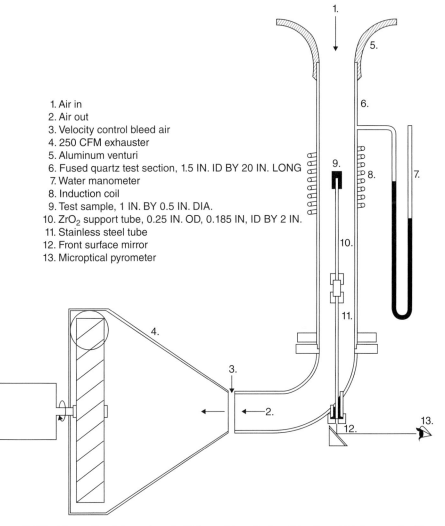

1. Air in
2. Air out
3. Velocity control bleed air
4. 250 CFM exhauster
5. Aluminum venturi
6. Fused quartz test section, 1.5 IN. ID BY 20 IN. LONG
7. Water manometer
8. Induction coil
9. Test sample, 1 IN. BY 0.5 IN. DIA.
10. ZrO$_2$ support tube, 0.25 IN. OD, 0.185 IN, ID BY 2 IN.
11. Stainless steel tube
12. Front surface mirror
13. Microptical pyrometer

Figure 2.12. Schematic of the system built for cold gas/hot wall testing at high gas velocities. The system consisted of an induction heater and high-velocity gas-handling unit [58].

oxidation (cold gas/hot wall/low velocity) along with more specialized tests including high-velocity furnace testing (cold gas/hot wall/high velocity) and simulated reentry conditions (hot gas/cold wall/high velocity). To accomplish the testing, specialized furnaces were developed for the high-velocity testing in conventional furnaces (Fig. 2.12). Arc heater facilities at AVCO and other laboratories were also utilized (Fig. 2.13). The facilities, test conditions, temperature measurements, and characterization methods are described in detail in the reports.

The second set of reports in the final series focused on experimental results [60, 61]. The materials studied included nominally pure diborides, diboride–SiC composites,

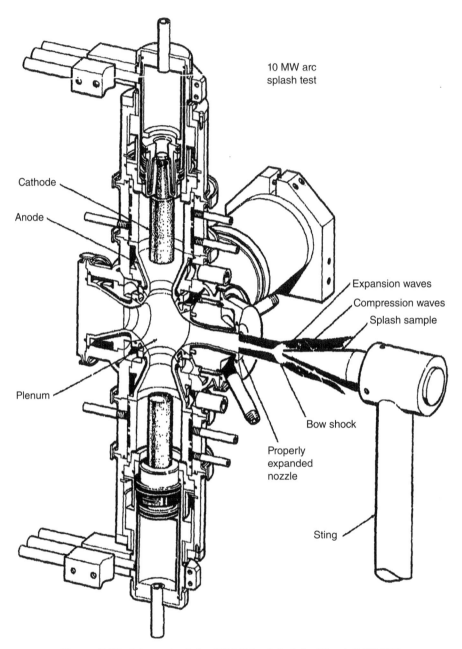

10 MW arc
splash test

Cathode

Anode

Expansion waves

Compression waves

Splash sample

Plenum

Bow shock

Properly
expanded
nozzle

Sting

Figure 2.13. Schematic of the 10 MW Arc Splash Facility at AVCO [59].

TABLE 2.3. Summary of compositions studied for the third series of ManLabs projects

Designation	Material	Major phase	Composition
A-2	HfB_2	HfB_2	Nominally pure HfB_2
A-3	ZrB_2	ZrB_2	Nominally pure ZrB_2
A-7	$HfB_2 + SiC$	HfB_2	HfB_2 plus 20 vol% SiC
A-8	$ZrB_2 + SiC$	ZrB_2	ZrB_2 plus 20 vol% SiC
A-10	$ZrB_2 + SiC + C$	ZrB_2	14 vol% SiC + 30 vol% graphite
C-11	HfC + C	HfC	Graphite (C-Hf hypereutectic)
C-12	ZrC + C	ZrC	Graphite (C-Hf hypereutectic)
D-13	JTA	Graphite	Graphite + 43 wt% ZrB_2 + 13 wt% SiC
E-14	KT-SiC	SiC	SiC + ~9 vol% Si
F-15	JT0992	Graphite	HfC + SiC
F-16	JT0981	Graphite	ZrC + SiC
H-23	$SiO_2 + W$	SiO_2	SiO_2 plus 60 wt% W
I-23	Hf-Ta-Mo	Hf-based alloy	Hf with 20 at% Ta and 2 at% Mo

carbide–carbon composites, graphite-based materials, and refractory metal-based materials (Table 2.3). For furnace testing below 3800°F (~2100°C), the lowest oxidation rates were measured for composites containing one of the diborides and 20 vol% SiC. Above that temperature, oxidation rates were similar for nominally pure diborides and the diboride–SiC composites due to the depletion of SiC from the composites. Interestingly, carbides exhibited good oxidation protection, but only above about 3450°F (~1900°C). At lower temperatures, the oxide was described as **puffy** and was not protective. Above 3450°F, the scale was more dense and provided protection to the underlying boride. In all tests, the performance of Hf compounds was slightly better than Zr compounds, presumably due to the higher refractoriness of the HfO_2 compared to ZrO_2. The temperature limits for oxidation protection for each of the candidate materials are summarized in Figure 2.14.

To complement the furnace oxidation studies, oxidation behavior was also studied in hot gas/cold wall testing. High heat fluxes were selected to be representative of the conditions that would be encountered during atmospheric reentry. The conditions were severe enough to cause graphite and tungsten to recess on the order of 7–14 in. (~18–36 cm), but resulted in recessions of less than 1 mm in the diboride and carbide materials [61]. Differences were noted in the behavior of materials in conventional furnace oxidation testing, the high-velocity gas flow described in Figure 2.12. The working hypothesis used to explain the results was that different heating methods resulted in different temperature gradients across the oxide scale, which are depicted schematically in Figure 2.15. Specimens in conventional furnace oxidation testing achieve thermal equilibrium. In contrast, the arc heater provides heat to the outside of the specimen, which results in a hotter outside and cooler inside. Likewise, induction heating produces higher temperatures inside the specimen due to generation of heat inside a specimen with an insulating external scale. The materials were tested in a cyclic manner to evaluate the reuse potential (Fig. 2.16). From the tests, the nominally pure diborides and the

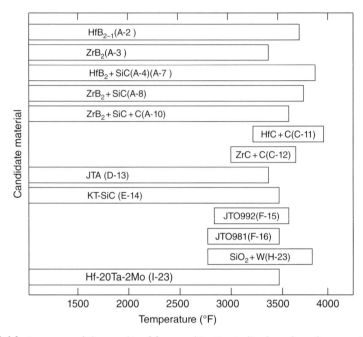

Figure 2.14. Summary of the results of furnace kinetic studies based on the recession of the parent material after 2 h. HfB$_2$ is always marginally better than the corresponding ZrB$_2$ composition. Carbides are not good at "moderate temperature, but extend temperature to higher values. JTA = graphite grade with ZrB$_2$ and Si additions. KT-SiC = Si-rich SiC (9 vol% free Si). SiO$_2$ + W = SiO$_2$ + 60 wt% W. JTO992 is C–HfC–SiC and JTO981 is C–ZrC–SiC [60].

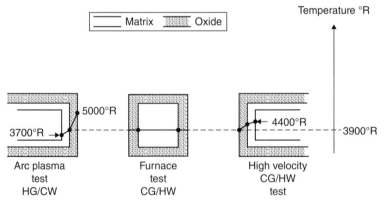

Figure 2.15. Schematic description of the temperature gradients observed across the oxide scales for specimens tested in (left) an arc plasma facility that uses hot gas, cold way, and high velocity; (center) conventional tube furnace with low velocity flow; and (right) inductively heated specimen in cold gas/hot wall/high velocity [61].

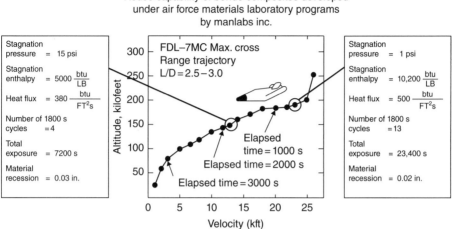

Figure 2.16. Schematic with information about a model trajectory for a lifting body reentry vehicle showing stagnation pressure and enthalpy along with other information for a vehicle with a 3-in. leading-edge radius [61].

diboride–SiC composites were identified as the only materials capable of meeting the requirements for the model trajectory. Even though it did not meet the recession criteria established for this project, the merits of JTA graphite (graphite containing a diboride and SiC) also exhibited promising behavior.

The final reports focused on correlating the behavior observed in **material-centered** testing in laboratory furnaces with behavior in **environment-centered** testing that simulated atmospheric reentry conditions [62]. The best success was achieved for describing the behavior of graphite, including the JTA graphite–boride–SiC composite material. For ceramic composites, correlations were more difficult, but some predictive capability was achieved. One of the critical aspects of the research was correlating flight conditions (i.e., altitude and velocity) with test conditions (i.e., enthalpy and heat flux) as shown in Figure 2.17. While the initial goal of correlating behavior in cold gas/hot wall tests in the laboratory to hot gas/cold wall simulations of atmospheric reentry was not met, the project still had significant success by identifying candidate materials for applications. One of the major conclusions of the study was that ZrB_2 and HfB_2 along with the composites with SiC were the only materials studied that were capable of surviving the extreme environment associated with atmospheric reentry of a sharp leading-edged vehicle.

2.4.3 Phase Equilibria

In parallel with the efforts underway at ManLabs, the U.S. Air Force also commissioned a series of studies on phase equilibria in transition metal–based systems. These projects were led by Dr. Edwin Rudy at Aerojet-General Corporation. The overall program resulted in the publication of 38 reports that were broken down into five parts: (1) 14 reports on binary systems, (2) 18 reports on ternary systems, (3) 2 reports on experimental techniques,

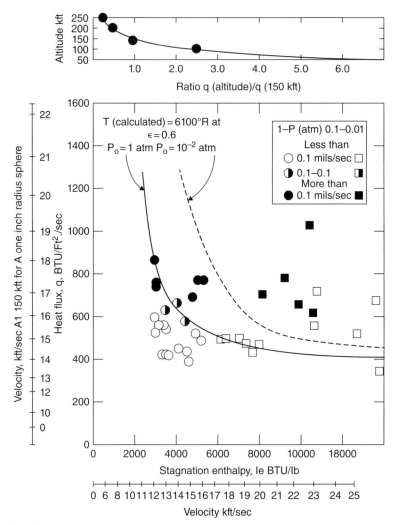

Figure 2.17. Comparison of measured recession rates (points) as a function of heat flux and total enthalpy. Lines are shown for a calculated surface temperature of 6100°F (3300°C) under two different conditions. The plot also includes a line (– - –) showing the speed and altitude of a potential reentry trajectory [62].

(4) 3 reports on thermochemical calculations, and (5) a compendium of phase diagram data. Prof. Hans Nowotny from University of Vienna, who remains one of the most prolific authors on transition metal carbide systems, acted as a consultant on the project.

The materials for these projects were produced by the reaction of elemental powders [63]. Starting with high-purity fine powders, the specimens were produced by rapid heating in graphite dies. The total cycle times ranged from 2 to 5 min with a typical heat treating temperature of 1900°C. A pressure of ~39 MPa (380 atm quoted in the report) was applied. After initial reaction, specimens were heat treated at temperatures ranging

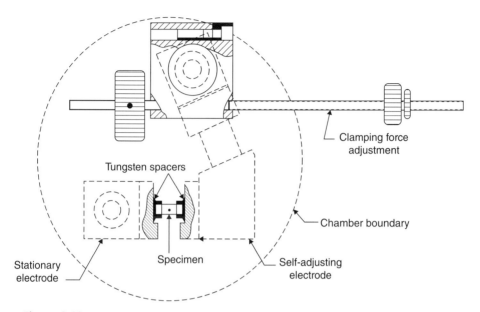

Figure 2.18. Schematic of the Pirani furnace used for melting point determinations [64].

from 1550 to 1800°C in a W-mesh element furnace. The heat treatments were conducted in sealed tantalum containers under a vacuum of about 7×10^{-3} Pa followed by quenching into a molten tin bath.

Phase diagrams were determined using a modified Pirani-style furnace to melt cylindrical specimens (Fig. 2.18) that was constructed especially for Rudy's experiments [64]. Among the possible designs, the Pirani-style furnace was selected because it allowed for controlled atmosphere, precision measurement of temperature, and elevated temperature while minimizing reactions between the furnace and the specimen. The design used a cylindrical specimen with a thinned center section that had a black body hole that allowed for temperature measurement with a micro-optical pyrometer. This design had the advantage that the hottest temperatures were attained inside the specimen, which enabled more accurate measurement of melting temperatures. To highlight the capabilities of the apparatus, one of the first systems examined was the HfC–TaC binary. A previous study [65] had concluded that a solid solution of HfC and TaC had a higher melting temperature than either end member. Rudy, in an effort to clarify this point, highlighted the capabilities of the system at Aerojet by examining melting temperature across the composition range in this system (Fig. 2.19). Rudy concluded that the melting temperature of Ta-rich compositions was nearly constant across the composition range of interest and did not show a maximum as described by Agte [65]. Interestingly, some researchers continue to disagree on this point.

The culmination of the phase equilibria studies was an immense volume of data that was compiled at the conclusion of the project [66]. The compendium is more than 700 pages in length and includes data for over 50 binary and 25 ternary systems. Not only does the volume show phase diagrams, but it also includes liquidus projects,

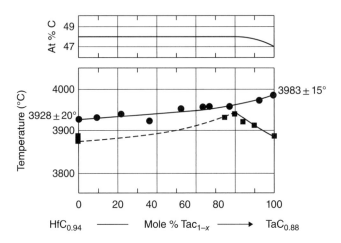

Figure 2.19. Comparison of melting temperature as a function of composition in the HfC–TaC system for data from Rudy [64] and Agte [65].

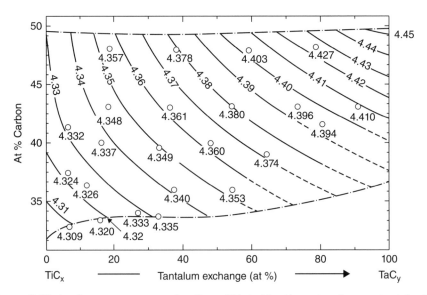

Figure 2.20. Lattice parameter as a function of Ta to Ti ratio and carbon content for the TiC–TaC system [66].

three-dimensional images of ternary systems, vertical sections, and isothermal sections in addition to other information that was collected during the project. For example, a tremendous amount of lattice parameter information was analyzed as part of the project and used to construct composition–lattice parameter maps such as the one shown in Figure 2.20. Many of the diagrams produced by Rudy *et al.*, such as the Zr–B–C ternary shown in Figure 2.21, continue to guide researchers working in these materials. These diagrams have been widely reproduced in technical papers

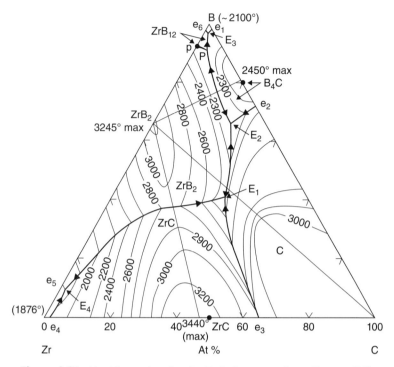

Figure 2.21. Liquidus project for the Zr–B–C ternary phase diagram [66].

and student theses. In addition, Volume X of the *Phase Diagrams for Ceramists* compilation is devoted to borides, carbides, and nitrides, and features many diagrams from Rudy's work.

2.5 SUMMARY

The progression of research on boride-based ultra-high temperature ceramic materials was reviewed. Although boride and carbide ceramics have been studied for over a century, the use of the term ultra-high temperature ceramics is a relatively recent development as this term has become popular only in the past 20 years. For the most part, boride and carbide ceramics remained a scientific curiosity from their first reports in the late 1800s and early 1900s until the 1950s. During the space race, the need for materials that could withstand the extreme environments associated with rocket propulsion, hypersonic flight, and atmospheric reentry became a national priority. Initial studies by NASA identified boride and carbide ceramics as candidates for rocket nozzles based on melting temperatures in excess of 3000°C. In the 1960s, research sponsored by the United States focused on finding materials for reusable atmospheric reentry vehicles with high lift-to-drag ratios. Projections for leading-edge temperatures in excess of 2200°C motivated a series of projects that focused on boride ceramics.

Initial screening studies confirmed that not only did these materials possess melting temperatures above 3000°C, but they had also demonstrated resistance to oxidation. Research at ManLabs consisted of three phases that (1) identified HfB_2 and ZrB_2 as the most promising candidates; (2) studied the processing, microstructure, and properties of ZrB_2, HfB_2, and their composites with SiC; and (3) evaluated the response of materials to oxidation and exposure to simulated atmospheric reentry conditions. Although the research was not able to correlate performance in laboratory oxidation studies to behavior in high-enthalpy flows, HfB_2- and ZrB_2-based ceramics were identified as the only materials capable of withstanding the combination of heat flux and surface temperature projected for atmospheric reentry. Parallel studies at Aerojet evaluated the phase equilibria in UHTC systems. An impressive number of binary and ternary phase diagrams resulted from that research, many of which continue to be widely used today.

ACKNOWLEDGMENTS

The author wishes to acknowledge funding provided by the Office of Naval Research Global office in London (grant number N62909-12-1-1014 overseen by Dr. Joe Wells) and the European Office of Aerospace Research and Development (grant number FA8655-12-1-0009 overseen by Dr. Randall Pollak) for the conference "Ultra-High Temperature Ceramics: Materials for Extreme Environment Applications II" that was held on May 13–18, 2012, in Hernstein, Austria.

REFERENCES

1. Markovich VI. *Boron and Refractory Borides*. Berlin: Springer-Verlag; 1977.
2. Jaffe HA. Development and testing of superior nozzle materials. National Aeronautics and Space Administration Final Report for Project NASw-67. Washington, DC: National Aeronautics and Space Administration; April 1961.
3. Fenter JR. Refractory diborides as engineering materials. SAMPE Q 1971;2:1–15.
4. Storms EK. *The Refractory Carbides*. New York: Academic Press; 1967.
5. Kaufman L, Clougherty EV. Investigation of Boride compounds for very high temperature applications. Air Force Materials Laboratory Report Number RTD-TDR-63-4096, Part II. Wright-Patterson Air Force Base, OH: Air Force Materials Laboratory; February 1965.
6. Freeman JW, Cross HC. Notes on heat-resistant materials in Britain from technical mission October 13 to November 30, 1950. National Advisory Committee for Aeronautics Technical Memorandum RM51D23. Washington, DC: National Advisory Committee for Aeronautics (NACA); May 14, 1951.
7. Samsonov GV. *Refractory Transition Metal Compounds; High Temperature Cermets*. New York: Academic Press; 1964.
8. Schwartzkopf P, Kieffer R, Leszynski W, Benesovsky F. *Refractory Hard Metals: Borides, Carbides, Nitrides, and Silicides*. New York: Macmillan; 1953.
9. U.S. Air Force and National Aeronautics and Space Administration Joint Conference on Manned Hypervelocity and Reentry Vehicles: A Compilation of Papers Presented; NASA-TMX-67563; April 11–12, 1960; Langley Research Center, Langley Field, VA.

10. Steurer WH, Crane RM, Gilbert LL, Hermach CA, Scala E, Zeilberger EJ, Raring RH. *Thermal Protection Systems: Report on the Aspects of Thermal Protection of Interest to NASA and the Related Materials R&D Requirements.* Washington, DC: National Aeronautics and Space Administration; February 1962.

11. Loch LD. Above 2500°F: what materials to use? Chem Eng 1958;65 (13):105–109.

12. Whittemore OJ Jr. Extreme temperature refractories. J Can Ceram Soc 1959;28:43–48.

13. Sata T. Materials for Ultra-High temperature. Oyo Butsuri 1967;36 (7):537–544.

14. Noguchi T. Synthesis of Ultrahigh-Temperature ceramic materials. Nippon Kessho Gakkaishi 1974;16 (4):288–293.

15. Sata T. Ultra-High temperature materials. Kino Zairyo 1984;4 (1):10–17.

16. Tanaka T. Development and problem of Ultra-High temperature materials. Shin Kinzoku Kogyo 1992;348:83–90.

17. Hillig WB. Prospects for Ultrahigh-Temperature ceramic composites. In: Tressler RE et al., editors. *Tailoring Multiphase and Composite Ceramics.* New York: Plenum Press; 1986. p 697–712.

18. Vedula KM. Ultra-High temperature ceramic-ceramic composites. Final Report on Project WRDC-TR-89-4089. Wright Patterson Air Force Base (OH): Air Force Materials Laboratory; October 1989.

19. Mehrotra GM. Chemical compatibility and oxidation resistance of potential matrix and reinforcement materials in ceramic composites for Ultra-High temperature applications. Final Report Number WRDC-TR-90-4127. Wright Patterson Air Force Base (OH): Air Force Systems Command; March 1, 1991.

20. Courtright EL, Graham HC, Katz AP, Kerans RJ. Ultra-High temperature assessment study–ceramic matrix composites. Final Report WL-TR-91-4061. Wright Patterson Air Force Base (OH): Wright Laboratory Materials Directorate; September 1992.

21. Rasky D, Bull J. Ultra-High temperature ceramics. NASA Report RTOP 232-01-04. Washington, DC: National Aeronautics and Space Administration; May 2, 1994.

22. Kolodziej P, Salute J, Keese DL. First flight demonstration of a sharp Ultra-High temperature ceramic nosetip. NASA Technical Report TM-112215. Washington, DC: National Aeronautics and Space Administration; December 1997.

23. Kowalski T, Buesking K, Koodziej P, Bull J. A thermostructural analysis of a diboride composite leading edge. NASA Technical Report TM-110407. Washington, DC: National Aeronautics and Space Administration; July 1996.

24. Opeka MM, Talmy IG, Wuchina EJ, Zaykoski JA, Causey SJ. Mechanical, thermal, and oxidation properties of refractory Hafnium and Zirconium compounds. J Eur Ceram Soc 1999;19 (13–14):2405–2414.

25. Talmy IG, Wuchina EJ, Zaykoski JA, Opeka MM. Ceramics in the system NbB_2-CrB_2. In: Materials Research Society Symposium Proceedings, Volume 365; November 28–December 2, 1994; Materials Research Society Meeting, Boston, MA. Materials Research Society, Pittsburgh, PA. p 81–87.

26. Zhang G-J, Deng Z-Y, Kondo N, Yang J-F, Ohji T. Reactive hot pressing of ZrB_2–SiC composites. J Am Ceram Soc 2000;83 (9):2330–2332.

27. Tucker SA, Moody HR. The preparation of a new metal Boride. Proc Chem Soc Lond 1901; 17 (238):129–130.

28. Tucker SA, Moody HR. The preparation of some new metal Borides. J Chem Soc 1902;81:14–17.

29. McKenna PM. Tantalum carbide: its relation to other hard refractory compounds. Ind Eng Chem 1936;28 (7):767–772.

30. Agte C, Moers K. Methoden zur Reindarstellung hochschmelzender Carbide, Nitride und Boride und Beschreibung einiger ihrer Eigenschaften. Z Anorg Allg Chem 1931;198 (1): 233–275.

31. Acheson EG. Article of carborundum and process of the manufacture thereof. US patent 645648. 1898.

32. Moissan H. Volatilisation of Zirconia and Silica at a high temperature and their reduction by carbon. C R Acad Sci Hebd Seances Acad Sci 1894;116:1222–1224.

33. Moissan H. Preparation of tungsten, molybdenum, and vanadium in the electric arc furnace. C R Acad Sci Hebd Seances Acad Sci 1894;116:1225–1227.

34. Moissan H. Titanium. C R Acad Sci Hebd Seances Acad Sci 1895;120:290–296.

35. Moissan H. A new zirconium carbide. C R Acad Sci Hebd Seances Acad Sci 1896; 122:651–654.

36. Samsonov GV. Continuous-discrete character of variation of the type of bonding in refractory compounds of transition metals and principles of classification of refractory compounds. In: Samsonov GV, editor. *Refractory Transition Metal Compounds; High Temperature Cermets.* Academic Press: New York; 1964. p 1–12.

37. Samsonov GV, Sinel'nikova VS. Electrical resistance of refractory compounds at high temperatures. In: Samsonov GV, editor. *Refractory Transition Metal Compounds; High Temperature Cermets.* Academic Press: New York; 1964. p 172–177.

38. Kiessling R. The borides of some transition metals. Acta Chem Scand 1995;4:209–227.

39. Heppenheimer TA. *The Space Shuttle Decision: NASA's Search for a Reusable Space Vehicle.* Washington (DC): The National Aeronautics and Space Administration; 1999.

40. Hallion RP. The NACA, NASA, and the supersonic-hypersonic frontier. In: Dick SJ, editor. *NASA's First 50 Years.* The National Aeronautics and Space Administration: Washington (DC); 2010. p 223–274.

41. Love ES. Introductory considerations of manned reentry orbital vehicles. U.S. Air Force and National Aeronautics and Space Administration Joint Conference on Manned Hypervelocity and Reentry Vehicles: A Compilation of Papers Presented; April 11–12, 1960, NASA Report TMX 67563. Langley Field (VA): Langley Research Center. p 39–54.

42. Mathauser EE. Materials for application to manned reentry vehicles. U.S. Air Force and National Aeronautics and Space Administration Joint Conference on Manned Hypervelocity and Reentry Vehicles: A Compilation of Papers Presented; April 11–12, 1960, NASA Report TMX 67563. Langley Field (VA): Langley Research Center. p 559–570.

43. Schick HL. *Thermodynamics of Certain Refractory Compounds: Thermodynamic Tables, Bibliography, and Property File.* New York: Academic Press; 1966.

44. Chase MW Jr. Zirconium diboride. In: *NIST-JANAF Thermochemical Tables, Fourth Edition, Journal of Physical and Chemical Reference Data, Monograph Number 9.* Woodbury (NY): American Chemical Society and American Institute of Physics; 1998. p 279.

45. McClaine LA. Thermodynamic and kinetic studies for a refractory materials program. Technical Report ASD-TDR-62-204. Wright-Patterson Air Force Base (OH): Air Force Materials Laboratory; April 1964.

46. Kuriakose AK, Margrave JL. The oxidation kinetics of zirconium diboride and zirconium carbide at high temperatures. J Electrochem Soc 1966;111 (7):827–831.

47. Berkowitz-Mattuck JB. High-temperature oxidation III. zirconium and hafnium diborides. J Electrochem Soc 1966;113 (9):908–914.

48. Berkowitz-Mattuck JB. Kinetics of oxidation of refractory metals and alloys at 1000°C-2000°C. ASD-TDR-62-203. Wright-Patterson Air Force Base (OH): Aeronautical Systems Division; March 1963.

49. Domalski ES, Armstrong GT. Heats of formation of metallic borides by fluorine bomb calorimetry. Technical Report APL-TDR-64-39. Wright-Patterson Air Force Base (OH): U.S. Air Force Aeropropulsion Laboratory; March 1964.

50. Janowski KR, Carnahan RD, Rossi RC. Static and dynamic oxidation of ZrC. SSD-TR-66-33. Los Angeles (CA): Ballistic Systems and Space Systems Division, Air Force Systems Command; January 1966.

51. Kibler GM, Lyon TF, Linevsky MJ, DeSantis VJ. Refractory materials research. WADD-TR-60-646. Wright-Patterson Air Force Base (OH): Air Force Materials Laboratory; August 1964.

52. Kaufman L. Investigation of boride compounds for very high temperature applications. Report Number RTD-TDR-63-4096, Part I. Wright-Patterson Air Force Base (OH): Air Force Materials Laboratory; December 1963.

53. Kaufman L, Clougherty EV. Investigation of boride compounds for very high temperature applications. Report Number RTD-TDR-63-4096, Part II. Wright-Patterson Air Force Base (OH): Air Force Materials Laboratory; February 1965.

54. Clougherty EV, Hill RJ, Rhodes WJ, Peters ET. Research and development of refractory oxidation-resistant diborides part II, volume II: processing and characterization. Technical Report AFML-TR-68-190. Wright-Patterson Air Force Base (OH): Air Force Materials Laboratory; January 1970.

55. Clougherty EV, Pober RL, Kaufman L. Synthesis of oxidation resistant metal diboride composites. Trans Metallurgical Society AIME 1968;242 (6):1077–1082.

56. Rhodes WH, Clougherty EV, Kalish D. Research and development of refractory oxidation-resistant diborides part II, volume IV: mechanical properties. Technical Report AFML-TR-68-190. Wright-Patterson Air Force Base (OH): Air Force Materials Laboratory; November 1969.

57. Clougherty EV, Wilkes KE, Tye RP. Research and development of refractory oxidation-resistant diborides part II, volume V: thermal, physical, electrical, and optical properties. Technical Report AFML-TR-68-190. Wright-Patterson Air Force Base (OH): Air Force Materials Laboratory; November 1969.

58. Kaufman L, Nesor H. Stability characterization of refractory materials under high velocity atmospheric flight conditions part II, volume II: facilities and techniques employed for cold gas/hot wall tests. Technical Report AFML-TR-69-84. Wright-Patterson Air Force Base (OH): Air Force Materials Laboratory; December 1969.

59. Kauman L, Nesor H. Stability characterization of refractory materials under high velocity atmospheric flight conditions part II, volume III: facilities and techniques employed for hot gas/gold wall tests. Technical Report AFML-TR-69-84. Wright-Patterson Air Force Base (OH): Air Force Materials Laboratory; December 1969.

60. Kaufman L, Nesor H. Stability characterization of refractory materials under high velocity atmospheric flight conditions part III, volume I: experimental results of low velocity cold gas/hot wall tests. Technical Report AFML-TR-69-84. Wright-Patterson Air Force Base (OH): Air Force Materials Laboratory; December 1969.

61. Kaufman L, Nesor H. Stability characterization of refractory materials under high velocity atmospheric flight conditions part III, volume III: experimental results of high velocity hot gas/cold wall tests. Technical Report AFML-TR-69-84. Wright-Patterson Air Force Base (OH): Air Force Materials Laboratory; February 1970.

62. Kaufman L, Nesor H, Bernstein H, Baron JR. Stability characterization of refractory materials under high velocity atmospheric flight conditions part IV, volume I: theoretical correlation of material performance with stream conditions. Technical Report AFML-TR-69-84. Wright-Patterson Air Force Base (OH): Air Force Materials Laboratory; December 1969.

63. Rudy E, Windisch S, Chang YA. Ternary phase equilibria in transition metal-boron-carbon-silicon systems: part I. related binary systems, volume I. Mo-C System. Technical Report AFML-TR-65-2. Wright-Patterson Air Force Base (OH): Air Force Materials Laboratory; January 1965.

64. Rudy E, Progulski G. Ternary phase equilibria in transition metal-boron-carbon-silicon systems: part III. Special techniques, volume II. A pirani-furnace for precision determination of melting temperatures of refractory metallic substances. Technical Report AFML-TR-65-2. Wright-Patterson Air Force Base (OH): Air Force Materials Laboratory; May 1967.

65. Agte C, Alterthum H. Systems of high-melting carbides; Contributions to the problem of carbon fusion. Z Tech Phys 1930;11:182–191.

66. Rudy E. Ternary phase equilibria in transition metal-boron-carbon-silicon systems: part V. Compendium of phase diagram data. Technical Report AFML-TR-65-2. Wright-Patterson Air Force Base (OH): Air Force Materials Laboratory; May 1969.

3

REACTIVE PROCESSES FOR DIBORIDE-BASED ULTRA-HIGH TEMPERATURE CERAMICS

Guo-Jun Zhang, Hai-Tao Liu, Wen-Wen Wu*, Ji Zou*,
De-Wei Ni*, Wei-Ming Guo*, Ji-Xuan Liu,
and Xin-Gang Wang

*State Key Laboratory of High Performance Ceramics and Superfine
Microstructure, Shanghai Institute of Ceramics, Chinese Academy of Sciences,
Shanghai, China*

3.1 INTRODUCTION

From the standpoint of material flow, the life cycle of diboride-based Ultra-High temperature ceramics (UHTCs) consists of two parts: fabrication and application (Fig. 3.1).

During fabrication, transition metal oxides (MeO$_2$, such as ZrO$_2$ and HfO$_2$) or elemental metals (Me, such as Zr and Hf) are combined with boron sources (B, B$_2$O$_3$, B$_4$C, etc.) to synthesize transition metal diboride (MeB$_2$, such as ZrB$_2$ and HfB$_2$) powders.

*Current address: School of Engineering, Brown University, Providence, RI, USA (W.-W. Wu); Department of Materials and Environmental Chemistry, Stockholm University, Stockholm, Sweden (J. Zou); Department of Energy Conservation and Storage, Technical University of Denmark, Roskilde, Denmark (D.-W. Ni); School of Electromechanical Engineering, Guangdong University of Technology, Guangzhou, China (W.-M.Guo).

Ultra-High Temperature Ceramics: Materials for Extreme Environment Applications, First Edition.
Edited by William G. Fahrenholtz, Eric J. Wuchina, William E. Lee, and Yanchun Zhou.
© 2014 The American Ceramic Society. Published 2014 by John Wiley & Sons, Inc.

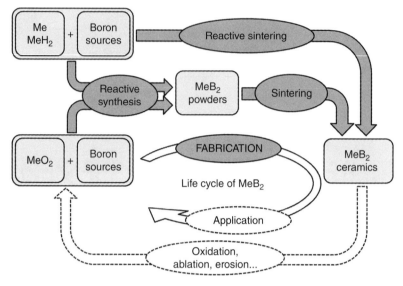

Figure 3.1. The life cycle of diboride-based ceramics.

Subsequently, MeB_2 powders are densified using methods such as pressureless sintering (PLS), hot pressing (HP), or spark plasma sintering (SPS), to obtain the final UHTC components. When using MeO_2, the synthesis of diborides includes reduction from $Me^{4+}(O^{2-})_2$ to $Me^{2+}(B^-)_2$. In contrast, when using elemental metals, oxidation from Me^0 to $Me^{2+}(B^-)_2$ is required. Powder synthesis and densification can be combined to fabricate final UHTC products in one step by processes such as reactive sintering. Reactive HP (RHP) and reactive spark plasma sintering (R-SPS) are two examples of this technique.

On the other hand, in the application process, UHTCs are used in environments that are rich in oxygen at high temperature. Some examples include sharp noses and leading edges for atmospheric reentry and hypersonic vehicles, or the corrosive environments of acids, alkalis, and molten metals. During application, especially at high temperature, MeB_2 can be oxidized under ablation conditions, leaving products of MeO_2 and gaseous B_2O_3. In some respects, the application process is the opposite of the reduction reactions used in fabrication. In view of this, the life cycle for diboride-based UHTCs can be summarized as a reduction–oxidation reaction cycle, with material flow from MeO_2 (or Me) → Fabrication → MeB_2 → Application → MeO_2.

Both the fabrication and application processes involve chemical reactions, so chemical reactions play important roles during the entire diboride-based UHTC life cycle. In this chapter, discussion of **reactive processes** is limited to the fabrication process where reduction reactions dominated, but some oxidations reactions are used to convert Me to MeB_2. In contrast, oxidation reactions that occur during application are not included. The discussion of **reactive processes** focuses on (i) synthesis of MeB_2 powders by different chemical reactions in Section 3.2, (ii) oxygen removal reactions for sintering UHTCs when using MeB_2 as starting materials in Section 3.3, and finally (iii) reactive sintering processes to fabricate UHTCs in a combined synthesis and

Fabrication process of MeB$_2$

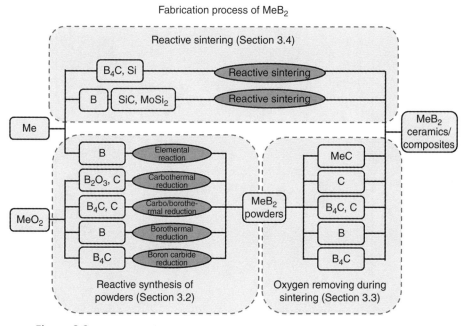

Figure 3.2. Roadmap of the fabrication process of MeB$_2$ discussed in this chapter.

densification process in Section 3.4. A roadmap for the fabrication of MeB$_2$ is shown in Figure 3.2. In addition, thermodynamic aspects of the main reactions in Figure 3.2, including the temperatures at which the reactions become favorable at standard state and under mild vacuum, are shown in Table 3.1.

3.2 REACTIVE PROCESSES FOR THE SYNTHESIS OF DIBORIDE POWDERS

Diboride powders can be synthesized by different approaches that can be divided into three main groups: (1) elemental reactions between Me and B, (2) reduction processes using MeO$_2$, and (3) chemical routes from polymeric precursors. Chemical routes used to synthesize borides include solutions, reactions with boron-containing polymers, and pre-ceramic polymers. None of these will be discussed in this chapter. This section describes elemental reactions between Me and B and reduction processes using MeO$_2$ as starting reactants.

For elemental reactions, diboride powders are synthesized from elemental metals (Me) or metal hydrides (MeH$_2$) and boron. For reduction processes, transition metal oxides, MeO$_2$, are usually used as the metal source and boron-containing materials, such as B or B$_4$C or B$_2$O$_3$, are used as the boron source. Some of the boron sources can also act as reducing agents, specifically B and B$_4$C. In some cases, reducing boron sources are combined with carbon. To prevent oxidation of the powders, reductions can be performed under vacuum or inert atmospheres. During the reduction process, two factors to consider are as follows:

TABLE 3.1. Thermodynamics of the main reactions including the temperatures at which the reactions become favorable at standard state and under mild vacuum

Reaction number in this chapter	Reaction	Type of process	Enthalpy ΔH°_{298} (kJ)	Free enthalpy ΔG°_{298} (kJ)	Minimum favorable temperature at standard state (°C)	Minimum favorable temperature with pCO=15Pa (°C)
(3.1)	$Zr+2B=ZrB_2$	Elemental reaction/reactive sintering	−322.586	−318.102	$\Delta G<0$ (R.T.—2000°C)	$\Delta G<0$ (R.T.—2000°C)
(3.2)	$ZrO_2+B_2O_3+5C=ZrB_2+5CO(g)$	Carbothermal reduction/oxygen removing	1496.907	1231.202	1140	775
(3.3)	$2ZrO_2+B_4C+3C=2ZrB_2+4CO(g)$	Carbo/borothermal reduction/oxygen removing	1175.262	961.366	1428	911
(3.10)	$3ZrO_2+10B=3ZrB_2+2B_2O_3$ (g)	Borothermal reduction/oxygen removing	661.141	522.426	1275	849
(3.5)	$7ZrO_2+5B_4C=7ZrB_2+3B_2O_3+5CO$ (g)	Boron carbide reduction/oxygen removing	1385.588	1113.223	1219	797
(3.16)	$ZrO_2+3WC=ZrC+3W+2CO$ (g)	Oxygen removing	803.069	690.083	1944	1298
(3.23)	$2Zr+B_4C+Si=2ZrB_2+SiC$	Reactive sintering	−655.075	−644.364	$\Delta G<0$ (R.T.—2000°C)	$\Delta G<0$ (R.T.—2000°C)

1. Loss and compensation of boron. In reduction processes, gaseous species, such as B_2O_3 (g), may form during the process, which results in the loss of boron source from the powder mixture. Therefore, excess boron is typically added to the raw materials to compensate for any boron loss. Evaporation of boron-containing gaseous species is affected by the synthesis conditions (e.g., gas pressure in the chamber, synthesis temperature, and heating rate), so the amount of the excess boron source is not predictable.

2. The balance between purity and grain size. The starting materials unavoidably contain impurities. In addition, if the MeO_2 used in reduction processes is not completely consumed, it will remain in the final powders as oxygen impurities. Higher synthesis temperatures can remove impurities, such as B_2O_3, which evaporate at higher temperature. On the other hand, higher temperatures lead to increased grain growth, which is contrary to the desire for finer powders. So the interplay between the purity and the grain size should be considered. One approach to balance them is to choose appropriate synthesis temperatures and holding times to prepare high purity and superfine MeB_2 powders.

This chapter discusses MeB_2 synthesis by carbothermal reduction, borothermal reduction, boron carbide reduction, and carbo/borothermal reduction. Among all of the processes for MeB_2 synthesis, these are the most popular and economical methods, so they are used most often.

3.2.1 Elemental Reactions

Elemental reactions such as those shown in Reaction 3.1 are the simplest process for MeB_2 powder synthesis and have been used for the longest time. These reactions are highly exothermic ($G_{rxn,298K} = -318\,kJ$ for ZrB_2 and $-325\,kJ$ for HfB_2), so self-propagating reactions between the precursors can be ignited. These reactions generate large quantities of heat and can promote local melting of the transition metal, which further accelerates the reaction. Elemental reactions are beneficial for self-propagating high-temperature synthesis because the high heating and cooling rates can produce high defect concentrations in as-synthesized powders, which can improve subsequent densification.

$$Me + 2B \rightarrow MeB_2 \qquad (3.1)$$

3.2.2 Reduction Processes

3.2.2.1 Carbothermal Reduction. Reaction 3.2 is a carbothermal reduction reaction that is commonly used to synthesize MeB_2 powders. However, the volatility of B_2O_3 can result in substantial loss of B_2O_3 (10–30 wt%), which requires addition of excess B_2O_3 during the synthesis process. A balance exists between formation of borides and evaporation of B_2O_3, and, under the right conditions, the rate of reduction of MeO_2 and B_2O_3 can exceed the rate of evaporation of B_2O_3. Minimizing carbon contamination is another factor to consider. For the synthesis of ZrB_2, Karasev [1] observed opposite

trends for B and C content in the final powders based on the addition of excess B_2O_3. For an excess of B_2O_3 in the range of 10–30 wt%, the reduction rate of ZrO_2 and B_2O_3 exceeded the B_2O_3 evaporation rate. This resulted in a stoichiometric powder composition with a C concentration of less than 1 wt%. When the B_2O_3 excess increased to the range of 80–200 wt%, the rate of evaporation of B_2O_3 grew, resulting in a higher concentration of C in the final powder. Increasing the excess B_2O_3 concentration further to 300–400 wt% enabled ZrB_2 to be produced with less C. Unfortunately, using large excesses of B_2O_3 is not economical. Commercially, synthesis of ZrB_2 typically uses an excess B_2O_3 content of 10–30 wt%, which results in pure ZrB_2 powders with low C contents. Although Karasev synthesized powders at 2000°C, ZrB_2 powders can be produced at lower temperatures (e.g., ~1500°C).

$$MeO_2 + B_2O_3 + 5C \rightarrow MeB_2 + 5CO(g) \qquad (3.2)$$

3.2.2.2 Carbo/Borothermal Reduction.
Another method used to synthesize MeB_2 powders uses MeO_2 as the transition metal source and B_4C as the boron source (Reaction 3.3) [2, 3], which may not occur in one step [4] based on thermodynamic calculations showing that some intermediate reactions may be more favorable (Fig. 3.3). From the thermodynamic calculations, Reactions 3.4 and 3.2 are likely intermediate steps for the overall processes described in Reaction 3.3.

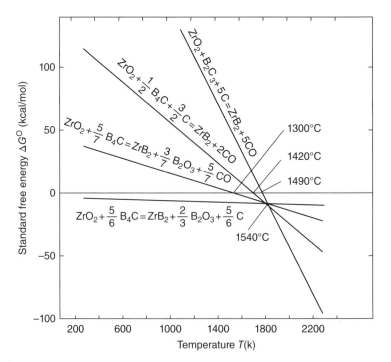

Figure 3.3. Standard free energy of reactions as a function of temperature [4].

$$2MeO_2 + B_4C + 3C \rightarrow 2MeB_2 + 4CO(g) \tag{3.3}$$

$$ZrO_2 + \frac{5}{6}B_4C \rightarrow ZrB_2 + \frac{2}{3}B_2O_3 + \frac{5}{6}C \tag{3.4}$$

The analysis summarized in Figure 3.1 suggests that Reaction 3.4 is predominant below 1540°C, while above 1540°C, Reaction 3.2 is more favorable.

Several reports have proposed another possible intermediate process as shown in Reaction 3.5 [4, 5], which may take place before Reactions 3.4 or 3.2.

$$ZrO_2 + \frac{5}{7}B_4C \rightarrow ZrB_2 + \frac{3}{7}B_2O_3 + \frac{5}{7}CO(g) \tag{3.5}$$

Some phases, such as carbon, are both a reactant and an intermediate product, indicating that intermediate reactions may take place simultaneously.

Both thermodynamic calculations and experiments demonstrate that the overall synthesis process (Reaction 3.3) goes to completion only above 1500°C. Further, the loss of B_2O_3 at the synthesis temperature affects the overall process and can result in the formation of ZrC in the final products by Reaction 3.6.

$$ZrO_2 + 3C \rightarrow ZrC + 2CO(g) \tag{3.6}$$

One method to eliminate ZrC formation is to utilize excess B_4C. The presence of C in B_4C means that the amount of carbon could be decreased. Guo et al. [5] used 20–25 wt% excess B_4C to produce nominally pure ZrB_2 after 1 h at 1650–1750°C. Even if ZrC forms as an intermediate product, it can be removed in the presence of excess B_2O_3 by Reaction 3.7 at temperatures above 1650°C.

$$ZrC + B_2O_3 + 2C \rightarrow ZrB_2 + 3CO(g) \tag{3.7}$$

Similarly, fine HfB_2 powders can also be obtained by Reaction 3.8 at 1500–1600°C for 1–2 h [6].

$$(1+x)HfO_2 + \frac{5}{7}B_4C + 5xC \rightarrow (1+x)HfB_2 + \left(\frac{3}{7}-x\right)B_2O_3 + \left(\frac{5}{7}+5x\right)CO \tag{3.8}$$

where x can vary from 0 to 3/7. When $x = 3/7$, Reaction 3.8 reduces to Reaction 3.3, meaning that Reaction 3.3 is a special case of Reaction 3.8. In practice, setting x to different values can produce high purity, fine HfB_2 powders with a quasi-columnar morphology. The best synthesis conditions seem to be for $0 \leq x \leq 1/4$ together with 0–10 wt% excess B_4C and 0–15 wt% excess carbon.

3.2.2.3 Borothermal Reduction.
Carbon is used in both carbothermal and carbo/borothermal reduction. The problem with both processes is that carbon may remain in the final products as an impurity. The use of elemental B as raw material to

synthesize MeB_2 by borothermal reduction could minimize or eliminate carbide or carbon impurities. Common borothermal processes are shown as Reactions 3.9 and 3.10 [7, 8].

$$MeO_2 + 4B \rightarrow ZrB_2 + B_2O_2(g) \tag{3.9}$$

$$ZrO_2 + \frac{10}{3}B \rightarrow ZrB_2 + \frac{2}{3}B_2O_3(g) \tag{3.10}$$

During reaction, detection of and discrimination between different boron oxides are difficult. As a result, potential vapor species are typically identified by simulation from thermodynamic calculations. For the reaction of 1 mol ZrO_2 and 4 mol B, a probable reaction path is that Reaction 3.10 would take place first, followed by the reaction of excess B (i.e., 0.67 mol based on 4 mol for Reaction 3.9 less 3.33 mol for Reaction 3.10) with $B_2O_3(l)$ to form boron-rich gaseous species such as B_2O_2 (g) and BO (g) by processes such as those described by Reactions 3.11 and 3.12.

$$2B_2O_3(l) + 2B \rightarrow 3B_2O_2(g) \tag{3.11}$$

$$B_2O_3(l) + B \rightarrow 3BO(g) \tag{3.12}$$

In contrast to carbo/borothermal reduction, B_2O_3 is not an intermediate phase, but a final product in borothermal reduction. Hence, the starting B can be consumed by reaction with B_2O_3. Experimental studies have shown that a ZrO_2/B molar ratio in the range of 3.33–4 is appropriate to synthesize ZrB_2. Although carbon impurities are avoided in this carbon-free reaction, boron oxides are formed and will be the main source of oxygen impurities. The retained B_2O_3 can be removed by washing with hot water or vaporization at 1500°C or higher with the latter thought to be more effective at producing a pure product. Fine HfB_2 powders have also been obtained by the same method [9].

With respect to the final particle size, higher temperatures lead to coarsening. Guo et al. [10] reported a two-step process that included an intermediate water washing (RWR) step that included the following: (i) borothermal reduction at approximately 1000°C to obtain a mixture of ZrB_2 powder and boron oxide, (ii) water washing to remove the oxide, and then (iii) a second reduction stage at 1550°C to remove residual oxygen. Particle coarsening was effectively restrained by the intermediate water washing process, resulting in pure, submicrometer ZrB_2 powders with low levels of oxygen impurities (Fig. 3.4).

3.2.2.4 Boron Carbide Reduction.

Boron carbide reduction is another approach to produce transition metal diborides. Using ZrB_2 as an example, Reaction 3.5 indicates that ZrO_2 will react with B_4C to form ZrB_2 plus gaseous oxides [11]. Zou et al. [11] compared ZrB_2 powders synthesized by three different approaches under vacuum at 1600°C for 1.5 h, which are summarized as Reactions 3.3, 3.5, and 3.9. The particle size (~1.0 μm) from Reaction 3.5 was larger than that produced by 3.3 (~0.85 μm), but smaller than Reaction 3.9 (~1.6 μm). The carbon impurity level for Reaction 3.5 was lower than Reaction 3.3. Further, Reaction 3.5 resulted in the lowest oxygen impurity levels, with an oxygen content of only 0.46 wt% compared with 0.51 wt% for Reaction 3.3 and

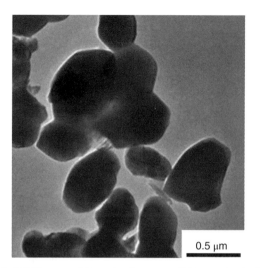

Figure 3.4. TEM image of the submicrometer ZrB_2 powder by RWR [10].

1.02 wt% for Reaction 3.9. Furthermore, Reaction 3.5 can take place at lower temperatures than Reaction 3.3. However, the composition must be controlled to avoid Reaction 3.4 instead of Reaction 3.5 since the latter produces carbon as an impurity in the as-synthesized powders.

3.2.3 Synthesis of Composite Powders

In practice, MeB_2 is usually produced with other phases, such as SiC, to form composites. The second phase can promote densification and improve properties of the composites. MeB_2-based composite powders can be prepared by reaction synthesis. One example is combustion synthesis of ZrB_2–SiC–ZrC composite powders by combustion in air according to Reaction 3.13 [12].

$$(2+x)Me + B_4C + (1-x)Si \rightarrow 2MeB_2 + (1-x)SiC + xMeC \qquad (3.13)$$

Using a mixture of Zr, B_4C, and Si as starting materials, ZrB_2-based composite powders with different ZrC contents can be synthesized by varying x. No combustion occurred under vacuum, but in air, the combustion reaction ignited easily. The heat generated from Reaction 3.13 was not sufficient for combustion under vacuum. In contrast, when reacted in air, Zr first reacted with oxygen to form ZrO_2, providing sufficient heat to ignite combustion with B_4C and Si. Although the reaction was done in air, the final composite powders were homogeneous, fine particles less than 1 µm in size with low oxygen content.

In summary, powder synthesis is an important part of the life cycle and should be designed based on the desired properties of the as-synthesized MeB_2 powders. Characteristics such as purity, particle size, and morphology affect densification and microstructure of the final ceramics. Dense UHTCs with finer grain sizes have superior mechanical properties at both room and elevated temperatures [13].

3.3 REACTIVE PROCESSES FOR OXYGEN REMOVING DURING SINTERING

The strong covalent bonding and low-volume self-diffusion coefficients of MeB_2 phases make densification difficult. Usually, diboride UHTC are prepared by densification methods, such as RHP, HP, and SPS. Additives, such as SiC, $MoSi_2$, Si_3N_4, B_4C, and C, are required for processes such as PLS. Oxygen impurities, which are always present on the surfaces of nonoxide ceramic powders, have a negative effect on the densification for diboride-based UHTCs.

While it is possible to sinter ZrB_2 to full density if the starting particle sizes are small enough, the synthesis conditions required are not economical for commercial processes. Therefore, post-synthesis particle size reduction processes are commonly used. Ball milling is an effective treatment method to obtain finer powders, which increases the driving force for densification due to higher surface area and the presence of defects induced by grinding. At the same time, oxygen impurity content increases during the milling process. Oxygen contamination takes the form of MeO_2 and B_2O_3 on the particle surfaces as oxygen has very limited solubility in the MeB_2 lattice. Sintering of MeB_2 is inhibited by the presence of B_2O_3 and MeO_2, because oxygen impurities promote coarsening mechanisms, which reduces surface area and the driving force for sintering [14].

Boron oxide can be removed from MeB_2 particle surfaces by evaporation at elevated temperatures according to Reaction 3.14.

$$B_2O_3(l) \rightarrow B_2O_3(g) \tag{3.14}$$

Zhang et al. [15] studied the vapor pressure as a function of temperature using thermodynamic calculations (Fig. 3.5), indicating that B_2O_3 can be removed under mild vacuum at elevated temperatures.

In contrast, MeO_2 impurities are more difficult to remove. Thermal treatments alone are not sufficient to vaporize MeO_2. As a result, MeO_2 impurities are typically removed using reducing additives as sintering aids. In general, the criteria for the selection of sintering aids for MeB_2 include the following: (i) must facilitate removal of MeO_2 and (ii) must form only volatile or high-melting-temperature phases [15]. In this section, some examples of reactive processes for removing oxygen during the sintering of MeB_2-based UHTCs are discussed.

3.3.1 Oxygen Removal by Reduction Using Boron/ Carbon-Containing Compounds

Because MeO_2 is one of the main oxygen impurities in MeB_2 powders, the reactions used for synthesizing MeB_2 powders from MeO_2 can also be used to remove oxygen. As a result, B, B_4C, and C are all effective sintering aids.

3.3.1.1 Boron. Boron can be used to synthesize MeB_2 powders by Reaction 3.15 [16, 17].

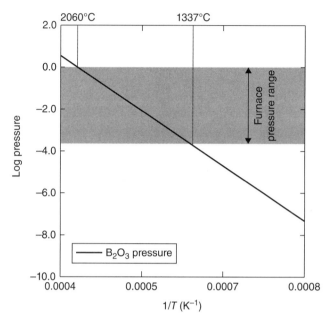

Figure 3.5. Calculated vapor pressure of B_2O_3 as a function of temperature in the pressure range maintained in the sintering furnace [15].

$$3MeO_2 + 10B \rightarrow 3MeB_2 + 2B_2O_3 \qquad (3.15)$$

Hence, B can also be used as a sintering aid to remove MeO_2 impurities. Boron is more effective for PLS processes, since B_2O_3 that is formed is more difficult to remove by evaporation during pressure-assisted sintering.

Typically, around 1 wt% boron is used to promote densification. Lower amounts (≤ 0.5 wt%) may not be enough to remove the oxide impurities, whereas excess boron (≥ 2 wt%) can form a liquid phase at above 2100°C (the melting point of boron is about 2092°C). Any boron liquid phase would promote rapid grain growth in the MeB_2, which is detrimental to densification.

3.3.1.2 Boron Carbide. Boron carbide is also a potential sintering aid for MeB_2 based on the removal of oxygen by processes similar to Reaction 3.5 [15, 18–21]. Both thermodynamic calculations and experimental results indicate that the reaction can proceed at temperatures in the ranges of 1200°C–1450°C. The addition of B_4C enables densification in MeB_2 by facilitating the removal of MeO_2 at temperatures low enough to prevent significant coarsening of the MeB_2 before densification mechanisms become active. B_4C is an ideal sintering aid for MeB_2 as it reacts with MeO_2 to form MeB_2. Furthermore, excess B_4C can pin grain growth during sintering, resulting in a finer grain size in the final ceramics.

3.3.1.3 Carbon. Carbon removed oxygen from MeB_2 particle surfaces by classic carbothermal reduction, which is also used to synthesize MeB_2 powders (Reaction 3.2)

[22, 23]. Zhu et al. [22] coated carbon onto ZrB_2 particle surfaces to densify ZrB_2. Reactions between the C coating and surface oxides on ZrB_2 particles were proposed to include carbothermal reduction by Reaction 3.2. Any loss of B_2O_3 would result in excess of ZrO_2, which could lead to ZrC formation by Reaction 3.6. Removal of oxides should minimize coarsening of ZrB_2 and consequently promote densification.

3.3.1.4 Boron Carbide and Carbon. Compared with a single additive, a combination of additives can have a synergistic effect on densification. In particular, the combination of B_4C and carbon [24–26] has been used to effectively densify ZrB_2. Reaction 3.5 is thought to be the main reaction responsible for MeO_2 removal when MeB_2 is sintered with B_4C alone. However, this reaction leads to the formation of liquid B_2O_3, which can promote grain coarsening. Analysis of Reactions 3.2 and 3.3 indicates that both ZrO_2 and liquid B_2O_3 can be removed by reaction with carbon. Thus, the combination of C and B_4C may facilitate oxide removal more effectively than either additive alone.

3.3.2 Oxygen Removing by Transition Metal Carbides

In addition to reactions with B, B_4C, and C, Chamberlain et al. discovered that WC is an effective reductant for oxygen impurities in MeB_2 [27]. Then, Zou et al. revealed that other transition metal carbides can remove oxygen [28].

3.3.2.1 Tungsten Carbide. In the presence of WC, the products with MeO_2 are MeC and W (Reaction 3.16), which are both solid phases with very high melting points. The other product, CO, can be removed as a gas.

$$MeO_2 + 3WC \rightarrow MeC + 3W + 2CO(g) \qquad (3.16)$$

Zhang et al. [15] revealed that an intermediate phase, W_2C, could be formed during sintering and also facilitate oxygen removal by Reactions 3.17 and 3.18.

$$2WC \rightarrow W_2C + C \qquad (3.17)$$

$$MeO_2 + W_2C \rightarrow MeC + 6W + 2CO(g) \qquad (3.18)$$

In addition to Reaction 3.16, Zou et al. [29, 30] revealed that another reaction with the MeB_2 matrix, the MeO_2 impurity, and WC was possible (Reaction 3.19).

$$3MeB_2 + 6WC + MeO_2 \rightarrow 4MeC + WB + 2CO(g) \qquad (3.19)$$

Experiment results show that this reaction can take place at temperature as low as 1450°C. In this case, the impurity phase, MeO_2, could also serve as a reaction promoter, which decreased the temperature at which the reaction became favorable and accelerated the reaction between WC and MeB_2 [31].

Solid solutions can form between the matrix and additives. Compared with Zr, W has a smaller covalent radius (1.57Å for Zr and 1.38Å for W) and goes into ZrB_2 crystal lattice to form a solid solution. Likewise, C (0.84Å) can substitute onto B (0.93Å) lattice positions, also forming solid solutions. Incorporation of C and W into ZrB_2 produces electron deficiencies and/or lattice vacancies. These defects increase densification rates by decreasing activation energies and increasing solid-state diffusion rates, which could be another mechanism for the enhancement of densification by WC in addition to oxygen removal [27, 32]. However, the temperatures for oxygen removal by WC are proposed to be more than 1850°C [15, 26], which are higher than other additives such as B, B_4C, and C. Finally, WC has other significant impacts on ZrB_2 properties, such as flexural strength, especially at temperatures above 1000°C.

3.3.2.2 Other Transition Metal Carbides.

Besides WC, a number of other transition metal carbides (MeC) have high melting points. Zou et al. [28, 33, 34] performed a systematic study of the effect of MeC additions on the densification behavior of ZrB_2–SiC ceramics. The transition metal carbides investigated were VC, TaC, TiC, NbC, and HfC, as well as WC.

First, thermodynamic predictions showed that MeC should react with ZrO_2; however, the reactions and the resulting products are different depending on the transition metal carbide, as shown in Figure 3.6.

WC and VC remove oxygen in the same way, as discussed earlier. Intermediate phases form after the dissociation of MeC during heating by Reaction 3.20, where x can vary between 0 and 1 and Me can be W, V, Nb, or Ta.

$$MeC \rightarrow MeC_x + (1-x)C \qquad (3.20)$$

Most of the MeO_2 can be removed below 1650°C by the successive reactions with C or Me_2C formed by reaction. For NbC and TaC, the formation of NbC_x and TaC_x becomes favorable during heating, and ZrO_2 can react with carbon, which is released by decomposition of NbC and TaC. However, subsequent reactions between residual ZrO_2 and the newly formed NbC_x or TaC_x are not favorable, namely, Reaction 3.21 is not available for Nb or Ta. As a result, some residual ZrO_2 exists in the final products.

$$3Me_2C + ZrO_2 \rightarrow 6Me + ZrC + 2CO(g) \qquad (3.21)$$

For TiC and HfC, an exchange reaction occurs between the carbide and ZrO_2 (Reaction 3.22) rather than oxygen removal.

$$MeC + ZrO_2 \rightarrow ZrC + MeO_2 \qquad (3.22)$$

Experimental results confirmed thermodynamic predictions of the ability to react with surface oxides and remove oxygen impurities in the following order:

$$WC > VC > NbC > TaC > HfC > TiC$$

The sequence is a guide for selecting transition metal carbides as sintering aids for MeB_2 ceramics based on their ability to react with and remove oxide impurities.

In summary, densification of diboride-based UHTCs requires oxygen impurity levels that are as low as possible. Oxygen removal by reaction with reducing agents can

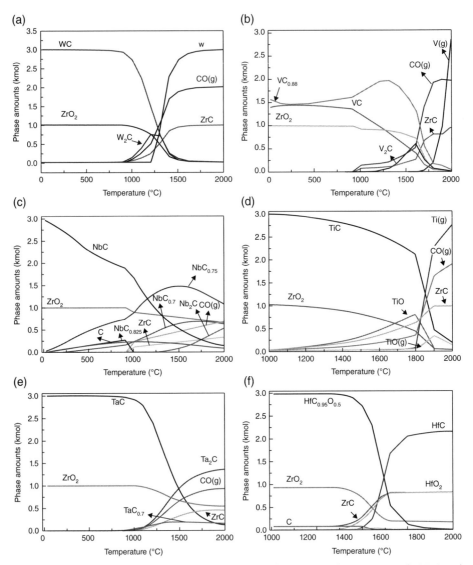

Figure 3.6. Molar content of the products calculated by reactions between 3 mole MeC and 1 mole ZrO$_2$ as a function of the temperature at a vacuum level of 5 Pa: (a) WC, (b) VC, (c) NbC, (d) TiC, (e) TaC, and (f) HfC [28].

promote densification and can have beneficial effects on the mechanical properties of the resulting ceramics. Furthermore, additives can help tailor microstructures and improve properties. An interesting example is the ZrB$_2$–SiC–WC composites prepared by Zou et al. In this study, the ZrB$_2$ starting powders were synthesized by solid-state reduction, and WC was used as a sintering aid. Nearly fully dense ceramics were obtained, and they demonstrated high fracture toughness and room temperature strength. The improved densification and properties were attributed to the presence of WC, which promoted anisotropic growth of ZrB$_2$ grains and produced a platelet morphology with

Figure 3.7. SEM images of indentation crack propagation in ZrB$_2$–SiC–WC ceramics by Zou et al. [29]

an interlocking microstructure (Fig. 3.7) [29]. Further, ZrB$_2$–SiC–WC retained a strength of at least 675 MPa at temperatures as high as 1600°C, exhibiting elastic, transgranular fracture (Fig. 3.8). This is one example of the benefits that oxygen removal has on MeB$_2$ UHTCs [35].

3.4 REACTIVE SINTERING PROCESSES

Reactive sintering, which combines synthesis and densification, can produce dense ceramics from high-purity powders in a single thermodynamically favorable *in situ* step [36–39]. Compared with conventional sintering, some advantages of reactive sintering processes and reaction sintered ceramics are as follows:

1. Reactive sintering can produce dense UHTCs at a lower temperature (usually ≤1800°C). Chemical reactions during reactive sintering are highly exothermic and thermodynamically favorable, generating enough energy and driving force for the densification of the final products under relatively low temperatures; in addition, chemical compatibility of the *in situ* formed phases and uniformity of phase distribution can also be achieved [40, 41]. Thermodynamic calculations demonstrate that most of the reactions during reactive sintering satisfy the conditions for self-sustaining combustion (T$_{adiabatic}$ ≥ 1800 K and $\Delta H°_{298}/C_{p298}$ > 2000 K). Therefore, low heating rates (e.g., ~1°C/min) and extended holding times (e.g., 360 min) at selected temperatures (e.g., 600°C) will promote reaction between the starting materials without igniting self-sustaining combustion [40, 42]. In limited cases, however, combustion reactions are beneficial for densification and can promote unique microstructure in the final products [41, 43].

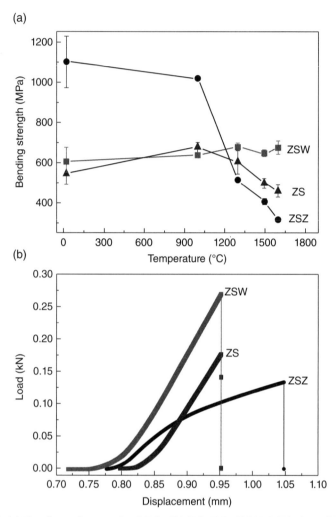

Figure 3.8. (a) The flexural strength of ZrB$_2$–SiC (ZS), ZrB$_2$–SiC–WC (ZSW), and ZrB$_2$–SiC–ZrC (ZSZ) as a function of testing temperature, (b) the load–displacement curves for ZS, ZSW, and ZSZ at 1600°C [35].

2. Reactive sintering process can produce diboride-based ceramics with aniso-tropic microstructures. Phases formed *in situ* are highly reactive due to small size and large surface areas. In addition, transient liquid phases can be formed, which promote mass transport and anisotropic grain growth. The result can be anisotropic MeB$_2$ grains in the final microstructure, either rod-like or platelet. Because MeB$_2$ ceramics have primitive hexagonal crystal structures, growth of either rodlike or platelet grains is possible due to pre-ferred growth along the *c*-axis (rods) or *a* and *b*-axes (platelets). For UHTCs, however, only a few reports have described this phenomenon. Anisotropic

grain growth mechanisms are still under investigation. Even so, this phenomenon can provide for microstructure tailoring and property improvements for UHTCs [37, 43–46].

Elemental transition metals (Me) or transition metal hydrides (MeH$_2$) are commonly used as the transition metal source (Fig. 3.1) to minimize the introduction of MeO$_2$ contamination. The boron source is elemental B or some boron-containing compounds, such as B$_4$C and BN. Moreover, because MeB$_2$-based ceramics can contain other phases, such as SiC, ZrC, and MoSi$_2$, boron-containing compounds are also C or N sources for the final products. Reduction processes that are used to synthesize MeB$_2$ powders using MeO$_2$ starting materials, as discussed in Section 3.2, are not viable approaches for reactive sintering. One reason is that most of the reduction processes are highly endothermic, and the Gibbs's free energy becomes negative at temperatures that would promote densification (e.g., >1500°C). Reduction reactions also lead to significant gas release, especially B$_2$O$_3$, which interferes with densification. In addition, the use of MeO$_2$ starting powders typically results in the retention of unreacted MeO$_2$ or other oxygen impurities in the final ceramic. As a result, most reactive sintering processes use elements or nonoxide compounds as starting materials to minimize introduction of oxygen impurities.

In this section, we discuss the fabrication of MeB$_2$-based UHTC by reactive sintering. Because some MeB$_2$ phases formed during the reactive process demonstrate anisotropic morphology, textured microstructures will be discussed.

3.4.1 Reactive Sintering from Transition Metals and Boron-Containing Compounds

3.4.1.1 MeB$_2$–SiC Ceramics. MeB$_2$–SiC ceramics have an excellent combination of properties and have received the most attention among MeB$_2$-based UHTCs. Zhang et al. [36] prepared ZrB$_2$–SiC composites by RHP using Zr, Si, and B$_4$C as starting powders under vacuum at 1900°C for 1 h, according to Reaction 3.23.

$$2Me + B_4C + Si \rightarrow 2MeB_2 + SiC \tag{3.23}$$

A mechanism for the RHP process was proposed as shown in Figure 3.9. During RHP, B and C atoms in B$_4$C will diffuse into Zr and Si, respectively, and form ZrB$_2$ and SiC *in situ*. Because the diffusion of Zr and Si atoms is slow, the as-synthesized ZrB$_2$ and SiC possess features of the starting powders. The particle size of SiC was generally <3 μm, whereas that of ZrB$_2$ was larger, in the range of 3–10 μm, compared with starting particle sizes of <10 μm for and <43 μm for Zr. The RHP process resulted in relatively small grain sizes for both ZrB$_2$ and SiC, resulting in an improvement of the mechanical properties compared with ZrB$_2$–SiC prepared by conventional HP. Subsequently, Zhao et al. also prepared ZrB$_2$–SiC by R-SPS using the same starting materials [47, 48]. After SPS, a fine homogeneous microstructure was obtained with grain sizes of <5 μm for ZrB$_2$ and <1 μm for SiC. In addition, Zimmermann et al. fabricated ZrB$_2$–SiC ceramics by RHP using ZrH$_2$, B$_4$C, and Si as given in Reaction 3.24 [49].

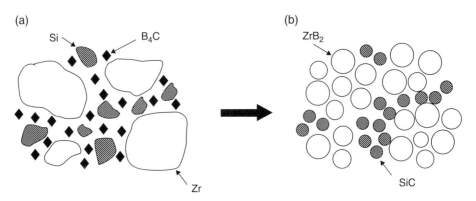

Figure 3.9. Microstructure formation mechanism of the ZrB$_2$–SiC composite in the reaction-synthesis process, depicting the transformation from (a) the powder compact to (b) the final microstructure of the composite [36].

$$2ZrH_2 + B_4C + Si \rightarrow 2ZrB_2 + SiC + 2H_2 \tag{3.24}$$

3.4.1.2 MeB$_2$–SiC–(Third Phase) Cramics.

Similar to the production of MeB$_2$–SiC composites, Zr, B$_4$C, and Si were used as starting powders to prepare ZrB$_2$–SiC–ZrC composites by RHP according to Reaction 3.25 [37].

$$(2+x)Me + B_4C + (1+x)Si \rightarrow 2MeB_2 + (1-x)SiC + xMeC \tag{3.25}$$

When $x=0$, Reaction 3.25 reduces to Reaction 3.23, which is a special case of Reaction 3.25. For Reaction 3.25, increasing the amount of Zr and reducing the amount of Si result in the formation of a third-phase ZrC, which is also a UHTC. Further investigation of Reaction 3.25 revealed that it may consist of two subreactions, Reactions 3.26 and 3.27.

$$3Zr + B_4C \rightarrow 2ZrB_2 + ZrC \tag{3.26}$$

$$2ZrC + B_4C + Si \rightarrow 2ZrB_2 + 3SiC \tag{3.27}$$

This means that reactive process could produce ZrC at relatively low temperatures (~800°C), and as the temperature increases, ZrB$_2$ could become the main phase. In addition, SiC appeared, and the amount of ZrC decreased at the same time according to Reaction 3.27. In view of these reaction steps, Wu optimized the model of the phase formation sequence during RHP of Zr, Si, and B$_4$C (Fig. 3.10). The same RHP process can also be used to prepare HfB$_2$–SiC ceramics using Hf, Si, and B$_4$C as starting materials [50].

To highlight the differences between reactive and nonreactive processes, Zhang et al. fabricated ZrB$_2$–SiC–ZrC composites in two ways: (1) from ZrB$_2$, SiC, and ZrC by HP; and (2) from Zr, B$_4$C, and Si by RHP [45, 46]. By HP, the ZrB$_2$ grains were equiaxed, while the RHP composite had a mixture of equiaxed and platelike ZrB$_2$ grains. The reactive process enabled the preparation of ZrB$_2$-based ceramics with anisotropic grains.

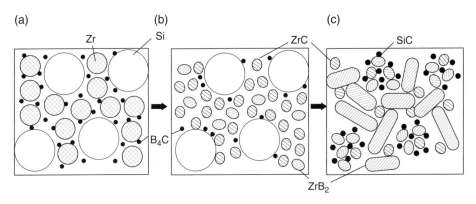

Figure 3.10. Microstructure formation mechanism of ZrB$_2$–ZrC–SiC composites in the reaction-synthesis process, depicting conversion from (a) the powder compact to (b) the intermediate state, and (c) the final microstructure [37].

In addition to ZrB$_2$–SiC and ZrB$_2$–SiC–ZrC, Wu et al. prepared ZrB$_2$–SiC–ZrN and ZrB$_2$–SiC–AlN composites by adding BN and Al to Zr, Si, and B$_4$C (Reactions 3.28 and 3.29) [51]. Similarly, ZrB$_2$–SiC–BN [52, 53] and ZrB$_2$–SiC–ZrN [54] composites were prepared using Zr, Si$_3$N$_4$, and B$_4$C batched according to Reactions 3.30 and 3.31.

$$(1+x)\,\text{Zr} + (0.5 - 0.25x)\,\text{Si} + (0.5 - 0.25x)\,\text{B}_4\text{C}$$
$$+ x\text{BN} \rightarrow \text{ZrB}_2 + (0.5 - 0.25x)\,\text{SiC} + x\text{ZrN} \tag{3.28}$$

$$\text{Zr} + (0.5 - 0.25x)\,\text{Si} + (0.5 - 0.25x)\,\text{B}_4\text{C} + x\text{BN}$$
$$+ x\text{Al} \rightarrow \text{ZrB}_2 + (0.5 - 0.25x)\,\text{SiC} + x\text{AlN} \tag{3.29}$$

$$4\text{Zr} + \text{Si}_3\text{N}_4 + 3\text{B}_4\text{C} \rightarrow 4\text{ZrB}_2 + 3\text{SiC} + 4\text{BN} \tag{3.30}$$

$$10\text{Zr} + \text{Si}_3\text{N}_4 + 3\text{B}_4\text{C} \rightarrow 6\text{ZrB}_2 + 3\text{SiC} + 4\text{ZrN} \tag{3.31}$$

Wu prepared these composites by both RHP and R-SPS. The R-SPS process formed more homogeneous and finer microstructures because of its high heating rate and short holding time, while RHP process produced coarse microstructures due to a holding time that was long enough for grain growth to proceed. The short holding time and finer microstructure of the final products in the SPS process also improved the densification behavior of the materials.

3.4.1.3 Other MeB$_2$-Based Ceramics. Johnson et al. used the directed reaction of liquid metal with B$_4$C to yield platelet-reinforced carbide matrix materials [55–57]. ZrB$_2$–ZrC$_x$–Zr were obtained by the process described by Reaction 3.32.

$$(2.2 + x)\,\text{Zr} + 0.6\text{B}_4\text{C} \rightarrow 1.2\text{ZrB}_2 + \text{ZrC}_{0.6} + x\text{Zr} \tag{3.32}$$

The composites were composed of ZrB$_2$ platelets distributed uniformly in a matrix of equiaxed ZrC$_x$ grains, while the residual Zr metal was generally situated at the grain triple points (Fig. 3.11). These platelet-reinforced ceramics exhibited an attractive

(a)   (b)

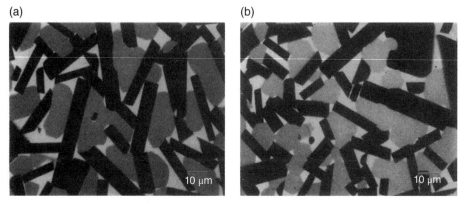

Figure 3.11. Backscattered electron images of the cross section of a typical ZrB$_2$-platelet-reinforced ZrC product taken approximately 2 mm from the top (a) and bottom (b) of a 12.7-mm-thick part. The darkest phase is ZrB$_2$, the gray phase is ZrC, and the lightest phase is zirconium metal [54].

combination of high strength (800–1030 MPa), fracture toughness (11–23 MPa·m$^{1/2}$), and thermal conductivity (50–70 W/m·K) over the temperature range of 25–600°C, which highlights the advantages of the ceramics prepared by reactive process. However, the Ultra-High temperature properties of these composites may be affected by the small amount of residual Zr metal that remained.

Zhang et al. prepared ZrB$_2$–ZrC ceramics using Zr and B$_4$C as starting materials, according to Reaction 3.33 [58].

$$3Zr + B_4C \rightarrow 2ZrB_2 + ZrC \qquad (3.33)$$

Because the diffusion of carbon was much faster than boron in zirconium, only carbon reached the center of the large Zr particles, which resulted in a microstructure with large ZrC agglomerates surrounded by fine ZrB$_2$ and ZrC particles.

Using Zr and BN as starting materials, Zhang also prepared ZrB$_2$–ZrN (Reaction 3.34). Similarly, ZrB$_2$–AlN could be prepared by adding Al to those precursors (Reaction 3.35) [58].

$$3Zr + 2BN \rightarrow ZrB_2 + 2ZrN \qquad (3.34)$$

$$Zr + 2Al + 2BN \rightarrow ZrB_2 + 2AlN \qquad (3.35)$$

The final composites had homogeneous microstructures with fine grain sizes. For the Zr–BN system, the researchers determined that the diffusion coefficient of N was smaller than that of C, but close to that of B. Accordingly, homogeneous microstructures were obtained due to the mutual constraint of grain growth for the two phases that formed, which restrained abnormal grain growth in both ZrB$_2$ and ZrN. In the Zr–Al–BN system, Al melted at temperatures as low as 660°C and the maximum solubilities of Al were 11.5 at% α-Zr and 26 at% in β-Zr. Accordingly, the redistribution of Zr and Al was remarkable in this system, producing fine, homogeneous microstructures.

3.4.2 Reactive Sintering from Transition Metals and Boron

3.4.2.1 Monolithic MeB$_2$ Ceramics.
Elemental Me or MeH$_2$ together with B can be used to prepare monolithic MeB$_2$ ceramics by Reactions 3.1 or 3.36.

$$MeH_2 + 2B \rightarrow MeB_2 + H_2 (g) \qquad (3.36)$$

Chamberlain et al. [42, 59] prepared monolithic ZrB$_2$ ceramics by RHP of Zr and B and studied the reaction mechanism. Analysis concluded that B diffused into the Zr granules to form ZrB$_2$. This behavior was observed in diffusion couple experiments in which polished Zr was heated to 1450°C in contact with B. Given this reaction path, the size and shape of the Zr precursor determined the size and shape of the resulting ZrB$_2$. Attrition milling of the precursors produced nanosized (<100 nm) Zr metal particles, which reacted with B at temperatures as low as 600°C. This process produced ZrB$_2$ with an average particle size of less than 100 nm. The nano-crystalline ZrB$_2$ exhibited significant coarsening and densification between 600 and 1450°C, which was the result of fine particle size and, possibly, a high defect concentration. Significant particle coarsening below 1650°C decreased the sinterability of ZrB$_2$. As a result, a temperature of 2100°C was required to achieve full density. Consolidation of ZrB$_2$ at 2100°C resulted in large grains (~12 μm), leading to lower strength.

3.4.2.2 MeB$_2$–SiC Ceramics.
The addition of other phases to Me and B can enable the production of MeB$_2$-based ceramics. Chamberlain et al. used Zr, B, and SiC along with small amounts of B$_4$C (0.5 wt%, to react with oxygen impurities) to prepare ZrB$_2$–SiC ceramics by RHP [40]. Samples with relative densities in excess of 95% were produced at 1650°C based on Reaction 3.37.

$$Zr + 2B + xSiC \rightarrow ZrB_2 + xSiC \qquad (3.37)$$

In this case, SiC functions not only as an important second phase that improves the microstructure and properties of the resulting ceramics, but also as an inert diluent that reduces the potential for a self-propagating reaction to ignite.

Using ZrH$_2$ as the transition metal source, together with B, SiC, and B$_4$C, Ran prepared fully densified ZrB$_2$–20 vol%SiC composites by reactive pulsed electric current sintering (R-PECS) by the process described in Reaction 3.38 [43].

$$ZrH_2 + 2B + xSiC \rightarrow ZrB_2 + xSiC + H_2 (g) \qquad (3.38)$$

The study revealed that ZrH$_2$ first decomposed into metal Zr before reacting with B to form ZrB$_2$. Since metal Zr is ductile and difficult to mill down to small particle sizes, the use of brittle ZrH$_2$ was thought to be a suitable alternative, which made it easier to obtain small starting particles. The same concept is also used to prepare TiB$_2$-based ceramics using TiH$_2$ as a precursor [60–63].

Besides enhanced densification, Ran's experiments revealed another interesting phenomenon, which was orientation of ZrB$_2$ grains. The XRD patterns indicated that the (001) peaks had higher intensity than the (100) peaks in ZrB$_2$–SiC ceramics, which was

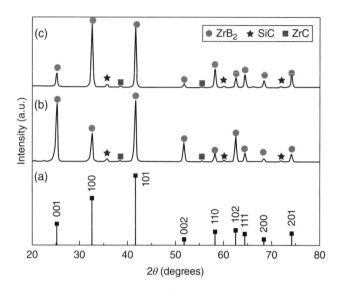

Figure 3.12. ZrB_2 JCPDS 35-0741 reference (a) and XRD patterns of cross section (b) and sintered surface (c) surface of the PECS samples [43].

different from the reference pattern of Figure 3.12. This implied that the hexagonal ZrB_2 grains had grown in a preferred direction and the mechanism of orientation of ZrB_2 grains was attributed to an anisotropic Ostwald ripening process under pressure [43].

3.4.2.3 MeB_2–$MoSi_2$ Ceramics. ZrB_2–$MoSi_2$ ceramics are another important member of the ZrB_2-based UHTC family. Using elemental Zr, B, Mo, and Si as starting materials, Wu et al. prepared ZrB_2–$MoSi_2$ ceramics via RHP, according to reaction 3.39 [44]:

$$Zr + 2B + xMo + 2xSi \rightarrow ZrB_2 + xMoSi_2 \qquad (3.39)$$

Due to the ductility of $MoSi_2$ at temperatures over 1000°C, the $MoSi_2$ grains deformed to fill the voids in the ZrB_2 skeleton, thus favoring the formation of a porosity-free material. The ZrB_2 grains had a platelet morphology, which was found only during the RHP process, not for materials prepared by PLS, HP, or SPS. The anisotropic grain growth was attributed to the *in situ* formation of ZrB_2 grains with high chemical activity. The difference in surface energy on different planes favored the elongation of ZrB_2 grains through Ostwald ripening at higher temperatures [64].

Liu et al. used RHP and subsequent hot forging to tailor the microstructure of ZrB_2–$MoSi_2$ ceramics [65]. The result was a textured and platelet-reinforced ZrB_2-based UHTC. During hot forging, the platelet ZrB_2 grains grew and simultaneously, under the applied pressure, rotated and rearranged to align along the top surface of the specimen (Fig. 3.13). The textured composites showed a remarkable improvement in mechanical properties, with flexural strengths as high as 871 MPa after hot forging, compared to 572 MPa before hot forging [65, 66].

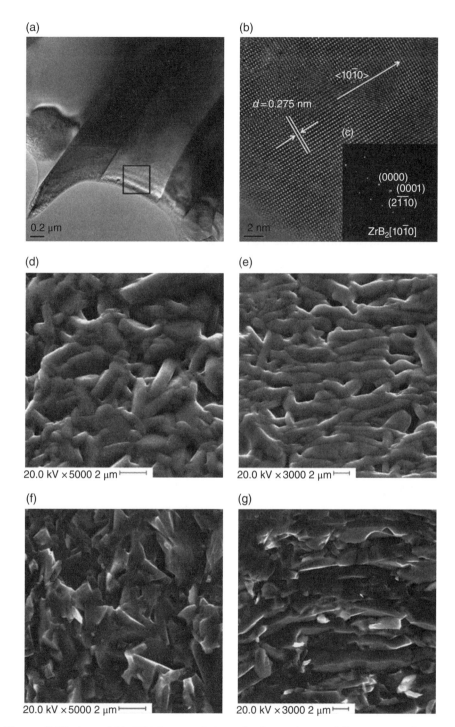

Figure 3.13. (a and b) TEM image of ZrB₂ platelet grains at different magnifications; (c) selected-area electron diffraction pattern of platelet grains of (b); (d and e) SEM images of polished surfaces before and after hot forging; (f and g) SEM images of fractured surface before and after hot forging [62].

In summary, during the MeB_2 life cycle, reactive sintering process can realize the fabrication of MeB_2-based ceramics from Me and B in one step, which simplifies the procedure compared with the two-step processes that separate synthesis and sintering densification. Relatively low densification temperatures are another advantage of reactive sintering, which are due to chemical reactions during the process and the heat generated from them. The RHP process also provides the opportunity for reduced contamination from oxygen impurities due to the removal of an oxide source to generate the MeB_2 powders. Another incomparable strong point is that reactive sintering processes provide a chance for anisotropic growth of MeB_2 grains, which is beneficial for the microstructure tailoring and property improvement. Finally, optimized microstructure and enhanced performance play important roles and increase the life of UHTCs during the application process, which is the other part of the MeB_2 life cycle.

3.5 SUMMARY

This chapter discussed reactive processes for the synthesis and densification of boride-based UHTCs as part of an overall life cycle. The properties and performances of UHTCs are dependent on the powder properties, densification processes, and microstructures, so optimization of reactive processes is important for the use of UHTCs in the high-temperature structures and other extreme environments. The discussions given earlier provide information about fabrication processes, phase formation mechanisms, microstructure evolution, and property improvements for UHTCs.

REFERENCES

1. Karasev AI. Preparation of technical zirconium diboride by the carbothermic reduction of mixtures of zirconium and boron oxides. Powder Metall Met Ceram 1973;12 (11):926–929.
2. Funke VF, Yudkovskii SI. Preparation of zirconium boride. Powder Metall Met Ceram 1964;2 (4):293–296.
3. Kuzenkova MA, Kislyi PS. Preparation of zirconium diboride. Powder Metall Met Ceram 1965;4 (12):966–969.
4. Zhao H, He Y, Jin ZZ. Preparation of zirconium boride powder. J Am Ceram Soc 1995;78 (9):2534–2536.
5. Guo WM, Zhang GJ. Reaction processes and characterization of ZrB_2 powder prepared by boro/carbothermal reduction of ZrO_2 in vacuum. J Am Ceram Soc 2009;92 (1):264–267.
6. Ni DW, Zhang GJ, Kan YM, Wang PL. Synthesis of monodispersed fine hafnium diboride powders using carbo/borothermal reduction of hafnium dioxide. J Am Ceram Soc 2008;91 (8):2709–2712.
7. Peshev P, Bliznakov G. On the borothermic preparation of titanium, zirconium and hafnium diborides. J Less Common Metals 1968;14 (1):23–32.
8. Ran SL, Van der Biest O, Vleugels J. ZrB_2 powders synthesis by borothermal reduction. J Am Ceram Soc 2010;93 (6):1586–1590.

9. Ni DW, Zhang GJ, Kan YM, Wang PL. Hot pressed HfB_2 and HfB_2-20 vol%SiC ceramics based on HfB_2 powder synthesized by borothermal reduction of HfO_2. Int J Appl Ceram Technol 2010;7 (6):830–836.

10. Guo WM, Zhang GJ. New borothermal reduction route to synthesize submicrometric ZrB_2 powders with low oxygen content. J Am Ceram Soc 2011;94 (11):3702–3705.

11. Zou J, Zhang GJ, Vleugels J, Van der Biest O. High temperature strength of hot pressed ZrB_2–20vol% SiC ceramics based on ZrB_2 starting powders prepared by different carbo/borothermal reduction routes. J Eur Ceram Soc 2013;33 (10):1609–1614.

12. Wu WW, Zhang GJ, Kan YM, Wang PL. Combustion synthesis of ZrB_2-SiC composite powders ignited in air. Mater Lett 2009;63 (16):1422–1424.

13. Zou J, Zhang GJ, Hu CF, Nishimura T, Sakka Y, Tanaka H, Vleugels J, Van der Biest O. High-temperature bending strength, internal friction and stiffness of ZrB_2-20 vol% SiC ceramics. J Eur Ceram Soc 2012;32 (10):2519–2527.

14. Baik S, Becher PF. Effect of oxygen contamination on densification of TiB2. J Am Ceram Soc 1987;70 (8):527–530.

15. Zhang SC, Hilmas GE, Fahrenholtz WG. Pressureless densification of zirconium diboride with boron carbide additions. J Am Ceram Soc 2006;89 (5):1544–1550.

16. Wang XG, Guo WM, Zhang GJ. Pressureless sintering mechanism and microstructure of ZrB_2-SiC ceramics doped with boron. Scripta Mater 2009;61 (2):177–180.

17. Guo WM, Zhang GJ, Yang ZG. Pressureless sintering of zirconium diboride ceramics with boron additive. J Am Ceram Soc 2012;95 (8):2470–2473.

18. Monteverde F. Hot pressing of hafnium diboride aided by different sinter additives. J Mater Sci 2008;43 (3):1002–1007.

19. Zhang H, Yan YJ, Huang ZR, Liu XJ, Jiang DL. Pressureless sintering of ZrB_2-SiC ceramics: the effect of B_4C content. Scripta Mater 2009;60 (7):559–562.

20. Zou J, Zhang GJ, Kan YM, Ohji T. Pressureless sintering mechanisms and mechanical properties of hafnium diboride ceramics with pre-sintering heat treatment. Scripta Mater 2010;62 (3):159–162.

21. Zou J, Zhang GJ, Kan YM. Pressureless densification and mechanical properties of hafnium diboride doped with B_4C: from solid state sintering to liquid phase sintering. J Eur Ceram Soc 2010;30 (12):2699–2705.

22. Zhu SM, Fahrenholtz WG, Hilmas GE, Zhang SC. Pressureless sintering of carbon-coated zirconium diboride powders. Mater Sci Eng A 2007;459 (1–2):167–171.

23. Guo WM, Yang ZG, Zhang GJ. Effect of carbon impurities on hot-pressed ZrB_2-SiC ceramics. J Am Ceram Soc 2011;94 (10):3241–3244.

24. Zhu S, Fahrenholtz WG, Hilmas GE, Zhang SC. Pressureless sintering of zirconium diboride using boron carbide and carbon additions. J Am Ceram Soc 2007;90:3660–3663.

25. Zhang SC, Hilmas GE, Fahrenholtz WG. Pressureless sintering of ZrB_2-SiC ceramics. J Am Ceram Soc 2008;91 (1):26–32.

26. Fahrenholtz WG, Hilmas GE, Zhang SC, Zhu S. Pressureless sintering of zirconium diboride: particle size and additive effects. J Am Ceram Soc 2008;91 (5):1398–404.

27. Chamberlain AL, Fahrenholtz WG, Hilmas GE. Pressureless sintering of zirconium diboride. J Am Ceram Soc 2006;89 (2):450–456.

28. Zou J, Zhang GJ, Sun SK, Liu HT, Kan YM, Liu JX, Xu CM. ZrO_2 removing reactions of Groups IV-VI transition metal carbides in ZrB_2 based composites. J Eur Ceram Soc 2011;31 (3):421–427.

29. Zou J, Zhang GJ, Kan YM. Formation of tough interlocking microstructure in ZrB_2-SiC-based ultrahigh-temperature ceramics by pressureless sintering. J Mater Res 2009;24 (7):2428–2434.

30. Zou J, Sun SK, Zhang GJ, Kan YM, Wang PL, Ohji T. Chemical reactions, anisotropic grain growth and sintering mechanisms of self-reinforced ZrB_2–SiC doped with WC. J Am Ceram Soc 2011;94 (5):1575–1583.

31. Ni DW, Liu JX, Zhang GJ. Pressureless sintering of HfB_2-SiC ceramics doped with WC. J Eur Ceram Soc 2012;32 (13):3627–3635.

32. Chamberlain AL, Fahrenholtz WG, Hilmas GE, Ellerby DT. High-strength zirconium diboride-based ceramics. J Am Ceram Soc 2004;87 (6):1170–1172.

33. Zou J, Zhang GJ, Kan YM, Wang PL. Pressureless densification of ZrB_2-SiC composites with vanadium carbide. Scripta Mater 2008;59 (3):309–312.

34. Zou J, Zhang GJ, Kan YM, Wang PL. Hot-pressed ZrB_2-SiC ceramics with VC addition: chemical reactions, microstructures, and mechanical properties. J Am Ceram Soc 2009;92 (12):2838–2846.

35. Zou J, Zhang GJ, Hu CF, Nishimura T, Sakka Y, Vleugels J, Biest O. Strong ZrB_2-SiC-WC ceramics at 1600 degrees C. J Am Ceram Soc 2012;95 (3):874–878.

36. Zhang GJ, Deng ZY, Kondo N, Yang JF, Ohji T. Reactive hot pressing of ZrB_2-SiC composites. J Am Ceram Soc 2000;83 (9):2330–2332.

37. Wu WW, Zhang GJ, Kan YM, Wang PL. Reactive hot pressing of ZrB_2-SiC-ZrC ultra high-temperature ceramics at 1800 degrees C. J Am Ceram Soc 2006;89 (9):2967–2969.

38. Fahrenholtz WG, Hilmas GE, Talmy IG, Zaykoski JA. Refractory diborides of zirconium and hafnium. J Am Ceram Soc 2007;90 (5):1347–1364.

39. Zhang GJ, Zou J, Ni DW, Liu HT, Kan YM. Boride ceramics: densification, microstructure tailoring and properties improvement. J Inorg Mater 2012;27 (3):225–233.

40. Chamberlain AL, Fahrenholtz WG, Hilmas GE. Low-temperature densification of zirconium diboride ceramics by reactive hot pressing. J Am Ceram Soc 2006;89:3638–3645.

41. Wu WW, Zhang GJ, Kan YM, Wang PL. Reactive hot pressing of ZrB_2-SiC-ZrC composites at 1600 degrees C. J Am Ceram Soc 2008;91 (8):2501–2508.

42. Chamberlain AL, Fahrenholtz WG, Hilmas GE. Reactive hot pressing of zirconium diboride. J Eur Ceram Soc 2009;29 (16):3401–3408.

43. Ran S, Van der Biest O, Vleugels J. ZrB_2-SiC composites prepared by reactive pulsed electric current sintering. J Eur Ceram Soc 2010;30 (12):2633–2642.

44. Wu WW, Wang Z, Zhang GJ, Kan YM, Wang PL. ZrB_2-$MoSi_2$ composites toughened by elongated ZrB2 grains via reactive hot pressing. Scripta Mater 2009;61 (3):316–319.

45. Qu Q, Zhang XH, Meng SH, Han WB, Hong CQ, Han HC. Reactive hot pressing and sintering characterization of ZrB_2-SiC-ZrC composites. Mater Sci Eng A 2008;491 (1–2):117–123.

46. Zhang XH, Qu Q, Han JC, Han WB, Hong CQ. Microstructural features and mechanical properties of ZrB_2-SiC-ZrC composites fabricated by hot pressing and reactive hot pressing. Scripta Mater 2008;59 (7):753–756.

47. Zhao Y, Wang LJ, Zhang GJ, Jiang W, Chen LD. Preparation and microstructure of a ZrB_2-SiC composite fabricated by the spark plasma sintering-reactive synthesis (SPS-RS) method. J Am Ceram Soc 2007;90 (12):4040–4042.

48. Zhao Y, Wang LJ, Zhang GJ, Jiang W, Chen LD. Effect of holding time and pressure on properties of ZrB_2-SiC composite fabricated by the spark plasma sintering reactive synthesis method. Int J Refract Method Hard Mater 2009;27 (1):177–180.

49. Zimmermann JW, Hilmas GE, Fahrenholtz WG, Monteverde F, Bellosi A. Fabrication and properties of reactively hot pressed ZrB_2-SiC ceramics. J Eur Ceram Soc 2007;27 (7): 2729–2736.

50. Monteverde F. Progress in the fabrication of ultra-high-temperature ceramics: "in situ" synthesis, microstructure and properties of a reactive hot-pressed HfB_2-SiC composite. Compos Sci Technol 2005;65 (11–12):1869–1879.

51. Wu WW, Zhang GJ, Kan YM, Wang PL, Vanmeensel K, Vleugels J, Van der Biest O. Synthesis and microstructural features of ZrB_2-SiC-based composites by reactive spark plasma sintering and reactive hot pressing. Scripta Mater 2007;57 (4):317–320.

52. Wu WW, Xiao WL, Estili M, Zhang GJ, Sakka Y. Microstructure and mechanical properties of ZrB_2-SiC-BN composites fabricated by reactive hot pressing and reactive spark plasma sintering. Scripta Mater 2013;68 (11):889–892.

53. Wu WW, Estili M, Nishimura T, Zhang GJ, Sakka Y. Machinable ZrB_2–SiC–BN composites fabricated by reactive spark plasma sintering. Mater Sci Eng A 2013;582:41–46.

54. Wu WW, Zhang GJ, Kan YM, Sakka Y. Synthesis, microstructure and mechanical properties of reactively sintered ZrB_2-SiC-ZrN composites. Ceram Int 2013;39 (6):7273–7277.

55. Johnson WB, Claar TD, Schiroky GH. Preparation and processing of platelet-reinforced ceramics by the directed reaction of zirconium with boron carbide. Ceram Eng Sci Proc 1989;10 (7-8):588–598.

56. Claar TD, Johnson WB, Andersson CA, Schiroky GH. Microsturcture and properties of platelet-reinforced ceramics formed by the directed reaction of zirconium with boron carbide. Ceram Eng Sci Proc 1989;10 (7–8):599–609.

57. Johnson WB, Nagelberg AS, Breval E. Kinetics of formation of a platelet-reinforced ceramic composite prepared by the directed reaction of zirconium with boron-carbide. J Am Ceram Soc 1991;74 (9):2093–2101.

58. Zhang GJ, Ando M, Yang JF, Ohji T, Kanzaki S. Boron carbide and nitride as reactants for in situ synthesis of boride-containing ceramic composites. J Eur Ceram Soc 2004;24 (2): 171–178.

59. Fahrenholtz WG. Reactive processing in ceramic-based systems. Int J Appl Ceram Technol 2006;3 (1):1–12.

60. Zhang GJ, Jin ZZ, Yue XM. TiB_2-Ti(C,N)-SiC composites prepared by reactive hot pressing. J Mater Sci Lett 1996;15 (1):26–28.

61. Zhang GJ, Jin ZZ, Yur XM. A multilevel ceramic composite of TiB_2-$Ti_{0.9}W_{0.1}$C-SiC prepared by in situ reactive hot pressing. Mater Lett 1996;28 (1–3):1–5.

62. Zhang GJ, Yue XM, Jin ZZ. Preparation and microstructure of TiB_2-TiC-SiC platelet-reinforced ceramics by reactive hot-pressing. J Eur Ceram Soc 1996;16 (10):1145–1148.

63. Zhang GJ, Yue XM, Jin ZZ, Dai JY. In-situ synthesized TiB_2 toughened SiC. J Eur Ceram Soc 1996;16 (4):409–412.

64. Liu HT, Wu WW, Zou J, Ni DW, Kan YM, Zhang GJ. In situ synthesis of ZrB_2-$MoSi_2$ platelet composites: reactive hot pressing process, microstructure and mechanical properties. Ceram Int 2012;38 (6):4751–4760.

65. Liu HT, Zou J, Ni DW, Wu WW, Kan YM, Zhang GJ. Textured and platelet-reinforced ZrB_2-based ultra-high-temperature ceramics. Scripta Mater 2011;65 (1):37–40.

66. Liu HT, Zou J, Ni DW, Liu JX, Zhang GJ. Anisotropy oxidation of textured ZrB_2-$MoSi_2$ ceramics. J Eur Ceram Soc 2012;32 (12):3469–3476.

4

FIRST-PRINCIPLES INVESTIGATION ON THE CHEMICAL BONDING AND INTRINSIC ELASTIC PROPERTIES OF TRANSITION METAL DIBORIDES TMB₂ (TM=Zr, Hf, Nb, Ta, AND Y)

Yanchun Zhou[1], Jiemin Wang[2], Zhen Li[2], Xun Zhan[2], and Jingyang Wang[2]

[1] *Science and Technology of Advanced Functional Composite Laboratory, Aerospace Research Institute of Materials and Processing Technology, Beijing, China*
[2] *Shenyang National Laboratory for Materials Science, Institute of Metal Research, Chinese Academy of Science, Shenyang, China*

4.1 INTRODUCTION

Transition metal borides (TMBs) are characterized by high melting points, high hardness, high strength, chemical inertness, effective wear and environmental resistance, and absence of phase changes in the solid state. These properties make them promising for applications in extreme environments such as cladding materials in

Ultra-High Temperature Ceramics: Materials for Extreme Environment Applications, First Edition.
Edited by William G. Fahrenholtz, Eric J. Wuchina, William E. Lee, and Yanchun Zhou.
© 2014 The American Ceramic Society. Published 2014 by John Wiley & Sons, Inc.

general IV nuclear reactors, nose tip and sharp leading-edge materials for hypersonic vehicles, sharp edge and hot structure component materials for scramjet engines [1–4], as well as matrix and surface coatings for ultra-high temperature ceramic (UHTC) matrix composites [5, 6].

Despite the unique combination of properties, poor thermal shock and oxidation resistance are the main barriers to their more extensive applications. Previous studies have demonstrated that poor thermal shock resistance of ZrB_2 and HfB_2 is resulted from their high Young's moduli [7], while the poor oxidation resistance is resulted from the lack of the formation of stable and protective scales on TMBs during high-temperature oxidation [7–9].

The macroscopic properties of TMBs are related to their electronic structure and bonding properties. Quantum mechanical modeling of solids through first-principles calculations based on density functional theory (DFT) has proven to be a powerful tool to obtain microscopic information that is helpful in understanding the macroscopic properties of solids. Predicting functions and properties of materials from first-principles is also helpful in scanning the properties of materials and accelerating materials development and application processes [10]. Thus, investigation of electronic structure, bonding nature, and intrinsic properties of the TMBs is essential to understand various properties of TMB_2 and to design new compositions/composites as well as to stimulate the applications of UHTCs.

The electronic structure and bonding properties of TMBs were previously investigated by several research groups [11–13]. Vajeeston *et al.* [11] calculated the electronic structure, chemical bonding, and ground-state properties of AlB_2-type transition metal diborides TMB_2 (TM = Sc, Ti, V, Cr, Mn, Fe, Y, Zr, Nb, Mo, Hf, Ta) using self-consistent tight-binding linear muffin-tin orbital method. Lawson *et al.* [12] investigated the electronic structure, elastic properties, and specific heat of ZrB_2 and HfB_2 using the plane wave DFT codes VASP [14] and ABIBIT [15]. Shein and Ivanovski [13] calculated the elastic properties of hexagonal AlB_2-like diborides of s, p, and d metals using first-principles calculations. Despite the contribution of these studies, understanding of the anisotropic chemical bonding and anisotropic mechanical properties of TMBs is still lacking. Since large-size single crystals of TMBs are not available, theoretical predications on the anisotropic mechanical properties from DFT computations could be very useful in understanding the performance and compositional design of UHTCs. In this work, the electronic structure, chemical bonding nature, and mechanical properties of some TMBs including YB_2, ZrB_2, HfB_2, NbB_2, and TaB_2 were investigated using first-principles calculations. These borides were selected because of their technological importance and because the number of valance electrons systematically changes from 3 for Y, 4 for Zr and Hf, and 5 for Nb and Ta such that a general trend on the change of electronic structure and intrinsic mechanical properties can be obtained. Anisotropic chemical bonding nature in YB_2, ZrB_2, HfB_2, NbB_2, and TaB_2 was disclosed by studying the pressure dependence of lattice parameters and bond lengths, and by analyzing their band structure, density of states (DOS), electron density difference maps, and charge density. The anisotropic chemical bonding properties resulted in high Young's moduli in x and y directions, but relatively low Young's modulus in z direction. By comparison of the Young's modulus, YB_2 is predicted to have better thermal shock resistance. By the comparison of the ratio of shear modulus to bulk modulus, G/B, TaB$_2$ is predicted to be a possible damage-tolerant UHTC.

4.2 CALCULATION METHODS

The valence states considered in the present calculations are $2s^2$ and $2p^1$ for boron, $4d^1 5s^2$ for yttrium, $4d^2 5s^2$ for zirconium, $5d^2 6s^2$ for hafnium, $4d^4 5s^1$ for niobium, and $5d^3 6s^2$ for tantalum. The first-principles calculations were performed using the CASTEP code [16], wherein the Vanderbilt-type ultrasoft pseudopotential [17] and generalized gradient approximation [18] (PBE) were employed. The plane wave basis set cutoff was 360 eV for all calculations. The special points sampling integration over the Brillouin zone was employed by using the Monkhorst Pack method with a $9 \times 9 \times 8$ special k-points mesh [19]. To investigate the anisotropic bonding nature, equilibrium crystal structures were optimized at various isotropic hydrostatic pressures from 0 to 70 GPa. Lattice parameters were modified to minimize the enthalpy and interatomic forces. The Broyden–Fletcher–Goldfarb–Shanno minimization scheme [20] was used in geometry optimization. The tolerances for geometry optimization are different on total energy within 5×10^{-6} eV/atom, maximum ionic Hellmann–Feynman force within 0.01 eV/Å, maximum ionic displacement within 5×10^{-4} Å, and maximum stress within 0.02 GPa. The calculations of the projected DOS were performed using a projection of the plane wave electronic states onto a localized linear combination of atomic orbitals basis set.

The full set of single-crystal elastic coefficients were determined from a first-principles calculation by applying a set of given homogeneous deformations with a finite value and calculating the resulting stress with respect to optimizing the internal degrees of freedoms, as implemented by Milman *et al.* [21, 22]. The criteria for convergence in optimizing atomic internal freedoms were selected as follows: difference on total energy within 5×10^{-6} eV/atom, ionic Hellmann–Feynman forces within 0.002 eV/Å, and maximum ionic displacement within 5×10^{-4} Å. The elastic stiffness was determined from a linear fit of the calculated stress as a function of strain, $c_{ij} = (\partial \sigma_i (x)/\partial \varepsilon_j)_x$. Variations in the stress tensor (σ_i) due to applied strain (ε_j) were calculated to obtain an elastic stiffness constant (c_{ij}). Both positive and negative strains were applied in the elasticity calculations. The compliance tensor S was calculated as the inverse of the stiffness tensor, $[S] = [C]^{-1}$. The polycrystalline bulk modulus B, shear modulus G, and Young's modulus E were estimated from the single-crystal elastic constants, c_{ij}, according to the Voigt, Reuss, and Hill approximation [23–25]. A detailed formula for these calculations is given in Section 3.3.

4.3 RESULTS AND DISCUSSION

4.3.1 Lattice Constants and Bond Lengths

Transition metal diborides TMB$_2$ (TM=Y, Zr, Hf, Nb, and Ta) crystallize in an AlB$_2$-type structure with the space group of P6/mmm (No. 191). Transition metal atoms are located at 1a (0,0,0) Wyckoff positions, and boron atoms are located at 2d (1/3,2/3,1/2)

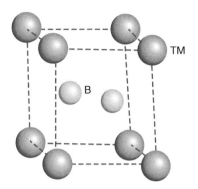

Figure 4.1. Crystal structure of TMB_2.

TABLE 4.1. Lattice constants and bond lengths of TMB_2

Lattice constants						
Compound	YB_2	ZrB_2	HfB_2	NbB_2	TaB_2	
a_{exp} (Å)	3.290	3.170	3.139	3.086	3.074	
a_{cal} (Å)	3.326	3.171	3.166	3.107	3.136	This work
	3.314	3.197	3.166	3.107	3.115	11
		3.17	3.15			12
c_{exp} (Å)	3.835	3.533	3.473	3.318	3.290	
c_{cal} (Å)	3.923	3.544	3.515	3.340	3.335	This work
	3.855	3.561	3.499	3.328	3.244	11
		3.56	3.52			12
c/a	1.179	1.117	1.110	1.075	1.063	
Bond lengths						
Compound	YB_2	ZrB_2	HfB_2	NbB_2	TaB_2	
B–B (Å)	1.920	1.831	1.828	1.794	1.811	
TM–B (Å)	2.745	2.548	2.538	2.443	2.467	

Wyckoff positions. Figure 4.1 shows the crystal structure of TMB_2. Table 4.1 lists the experimental and geometry optimized lattice constants a and c, c/a ratios and the bond lengths between B and B, and those between B and TM in TMB_2 (TM=Y, Zr, Hf, Nb, and Ta). The data listed in Table 4.1 show that GGA calculations slightly overestimate the lattice parameters as expected in any study with generalized gradient approximations. The calculated lattice constants a and c of YB_2, ZrB_2, HfB_2, NbB_2, and TaB_2 are close to the experimental lattice constants and also agree with the calculated values given in the previous work [11–13], demonstrating the reliability of the present calculations. It is also interesting to note that the lattice constants depend not only on the atomic size but also on the number of valence electrons of the transition metals. As the number of valence electrons increases from 3 for Y to 4 for Zr and Hf, and then to 5 for Nb and

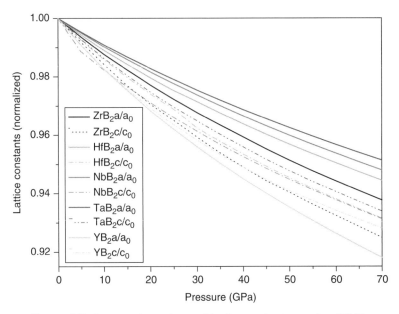

<u>Figure 4.2.</u> Pressure dependence of lattice constants *a* and *c* of TMB$_2$.

Ta, the lattice constant *c* decreases continuously; the lattice constant *a*, however, decreases from Y to Nb, but increases slightly for Ta due to its large atomic size. As a result, YB$_2$ has the largest lattice constants, ZrB$_2$ and HfB$_2$ have intermediate lattice constants, and NbB$_2$ and TaB$_2$ have the smallest lattice constants. The *c/a* ratio also decreases continuously when TM changes from Y to Zr and Hf, and then to Nb and Ta, that is, *c/a* ratio decreases with the increased number of valence electrons. The change of TM–B, B–B bond lengths shows a similar trend as the change of lattice constant *a*.

Another feature is that the TM–B bond lengths are longer than the B–B bond lengths, indicating the anisotropic chemical bonding characters in TMB$_2$ (TM=Y, Zr, Hf, Nb, and Ta). To visually illustrate the anisotropic chemical bonding nature and elastic anisotropy, the dependence of lattice constants *a* and *c*, and the B–B and TM–B bond lengths in TMB$_2$ (TM=Y, Zr, Hf, Nb, and Ta) on hydrostatic pressure was investigated. Figure 4.2 shows the pressure dependence of lattice constants for YB$_2$, ZrB$_2$, HfB$_2$, NbB$_2$, and TaB$_2$ up to 70 GPa. For all the diborides investigated in this work including YB$_2$, ZrB$_2$, HfB$_2$, NbB$_2$, and TaB$_2$, the lattice constants *a* and *c* decrease continuously with hydrostatic pressure. However, the degree of contraction is different, that is, the compressibility of the TMB$_2$ compounds is in the order YB$_2$>ZrB$_2$>HfB$_2$>NbB$_2$> TaB$_2$. In other words, the compressibility of the TMB$_2$ compounds decreases as the number of valence electrons increases from 3 for Y to 4 for Zr and Hf, and then to 5 for Nb and Ta. Careful analysis of the pressure dependence of lattice constants *a* and *c* in Figure 4.2 shows that for ZrB$_2$, HfB$_2$, NbB$_2$, and TaB$_2$, lattice constant *c* contracts more dramatically than lattice constant *a* in the pressure range examined, that is, ZrB$_2$, HfB$_2$, NbB$_2$, and TaB$_2$ are stiffer in the basal plane than in *c* direction; for YB$_2$, however, lattice constant *a* contracts more rapidly than lattice constant *c*, that is, YB$_2$ is stiffer in *c* direction

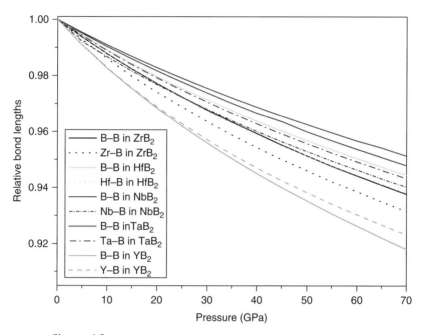

Figure 4.3. Pressure dependence of TM–B and B–B bond lengths.

than in the basal plane. This anomalous phenomenon will be explained in later sections based on the electronic structure investigation.

The strength of interatomic bonding as a function of pressure was investigated by examining changes in TM–B and B–B bond length under various pressures (Fig. 4.3). The degree of bond contraction was in the same order as that of lattice constants a and c, that is, $YB_2 > ZrB_2 > HfB_2 > NbB_2 > TaB_2$. Similar to the pressure dependence of lattice constants a and c shown in Figure 4.2, the change of TM–B and B–B bond lengths under hydrostatic pressure can also be divided into two groups. For ZrB_2, HfB_2, NbB_2, and TaB_2, the TM–B bond length contracts more rapidly than the B–B bond length, indicating that the B–B bonds in these borides are stiffer, while for YB_2, the B–B bond contracts more rapidly than the Y–B bond. The different responses of bond length to hydrostatic pressure reveal different bonding properties in TMB_2. In ZrB_2, HfB_2, NbB_2, and TaB_2, the B–B bonds are stiffer than TM–B bonds, while in YB_2, the Y–B bonds are stronger.

4.3.2 Electronic Structure and Bonding Properties

In Section 4.3.1, the pressure dependencies of lattice parameters and B–B, TM–B bond lengths were investigated, and different responses of bond lengths to hydrostatic pressure were demonstrated. These behaviors are underpinned by the electronic structure and chemical bonding nature. The electronic structure of TMB_2 (TM = Sc, Ti, Cr, Mn, Fe, Y, Zr, Nb, Mo, Hf, Ta) was studied by Vajeeston *et al.* [11] and that of ZrB_2 and HfB_2 was investigated by Lawson *et al.* [12]. However, the anisotropic chemical bonding nature was not emphasized. In this work, the electronic structure and bonding properties of TMB_2 (TM = Y, Zr, Hf, Nb, and Ta) were disclosed based on the analysis of DOS, band structure, and distribution of charge density on planes across TM, B parallel to (0001), and on (11 $\bar{2}$0) plane that is across both TM and B atoms.

To visually illustrate the chemical bonding in TMB_2, the electron density difference maps of these compounds are shown first. Electron density differences produce a density difference field that shows the changes in the electron distribution due to the formation of all the bonds in a system. The electron density difference map illustrates the charge redistribution due to chemical bonding in solids. Figure 4.4 shows the electron density difference maps on planes parallel to (0001) that are across the Zr atoms (Fig. 4.4a) and B atoms (Fig. 4.4b), and on the (11$\bar{2}$0) plane that contains both Zr and B atoms (Fig. 4.4c). In Figure 4.4a, no localized electrons were observed between neighboring Zr atoms, but triangular accumulation among groups of three Zr atoms can be seen [12], which is typical metallic bonding [31]. That is, the chemical bonding on the Zr plane is mainly metallic. In Figure 4.4b, strong localization is evident between two neighboring B atoms forming a graphitic B-plane with six-membered rings, which is an indication of σ-type covalent bonding formed by the overlapping of hybridized B-sp^2 orbitals. This B-sp^2 hybridized σ-type covalent bonding is strong and corresponds to the incompressibility of lattice constant a and B–B bond in ZrB_2. In Figure 4.4c, strong localization is obvious between the two B atoms, which is the typical π-type covalent bonding through the accumulation of p-orbitals. The valence electrons in B are $2s^2 2p^1$. The formation of σ-type bonding through overlapping of hybridized B sp^2-B sp^2

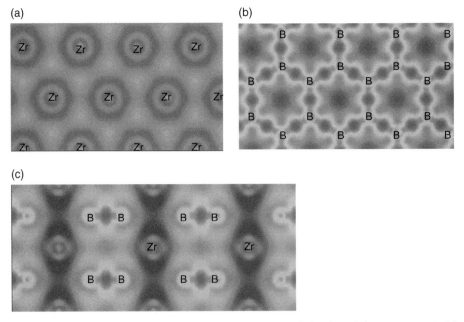

Figure 4.4. Electron density difference maps on planes parallel to (0001) that are across Zr (a) and B (b) atoms, and on (11$\bar{2}$0) that is across both Zr and B atoms (c).

orbitals and π-type bonding through overlapping of B $2p_z$-B $2p_z$ orbitals indicates charge transfer from Zr to B occurs. Thus, bonding between Zr and B is mainly ionic. However, charge accumulates between Zr and B through interactions between Zr $4d$ orbitals and B $2p$ orbitals, as can be seen in Figure 4.4c. That means besides the ionic bonding generally accepted by many researchers [11, 12], covalent bonding also contributes to the Zr–B bonding in ZrB$_2$ so that the bonding between Zr and B is ionic–covalent in nature. It is intriguing to see that the Zr–B covalent bonding is through the hybridization of Zr-$4d$ (e_g) and B-$2p_z$ orbitals, and the extension of Zr-$4d$ (e_g) orbitals in the x–y plane indicates that covalent bonding is stronger in the x-plane than in the c direction. This bonding nature is helpful in understanding the response of lattice constants and bond lengths to hydrostatic pressure shown in Figures 4.2 and 4.3, that is, strong covalent bonding in the basal plane corresponds to stiffer B–B bonds in the basal plane. These results demonstrate that all three types of bonds, that is, metallic, covalent, and ionic bonds, contribute to bonding in ZrB$_2$. The strong covalent and ionic bonding are responsible for its high modulus and strength, while the metallic bonding underpins its good electrical conductivity.

To further understand the electronic structure, band structures of TMB$_2$ (TM=Y, Zr, Hf, Nb, and Ta) were calculated and are depicted in Figure 4.5. Figure 4.5a shows band dispersion curves along some high-symmetry directions in the Brillouin zone of YB$_2$. The valance bands lying lower than the Fermi level are B $2s$, B $2p$, and Y $4d$-B $2p$, respectively. The Fermi level E$_F$ lies below the valence band maximum at the A point, indicating incomplete filling of valence bands. Then, valence and conduction bands across the Fermi level in Γ–A, A–H, and Γ–K directions of the Brillouin zone imply the presence of metallic bonding and metallic electrical conductivity of YB$_2$. The metallic electrical conductivity contributes to thermal conductivity, leading to higher thermal conductivity, which is of vital importance for UHTCs.

Figure 4.5b shows the band structure of ZrB$_2$. Compared to the band structure of YB$_2$, the Fermi level lies about 2 eV higher than in YB$_2$. The valence and conduction bands cross the Fermi level in the Γ–A and A–H directions of the Brillouin zone, but the

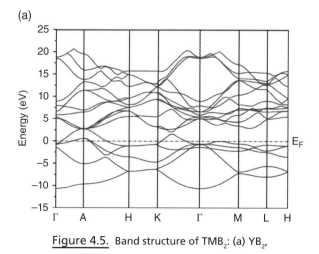

(a)

Figure 4.5. Band structure of TMB$_2$: (a) YB$_2$,

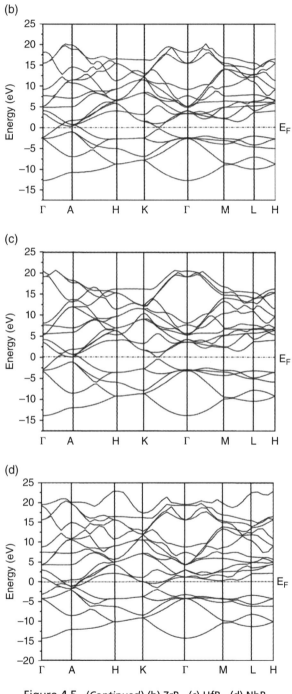

Figure 4.5. (*Continued*) (b) ZrB$_2$, (c) HfB$_2$, (d) NbB$_2$,

(e)

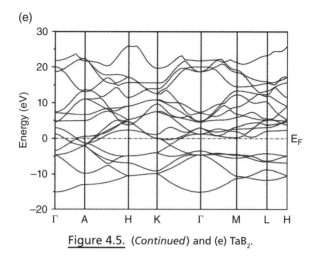

Figure 4.5. (*Continued*) and (e) TaB$_2$.

valence band in the Γ–K direction is completely filled. Metallic bonding also contribute to the bonding in ZrB$_2$, and the electrical conductivity in ZrB$_2$ is also metallic.

The band structure of HfB$_2$ is similar to that of ZrB$_2$ as is shown in Figure 4.5c, probably due to the fact that Hf has the same number of valence electrons as Zr. However, the Fermi level in HfB$_2$ is higher than in ZrB$_2$ and touches the bottom of conduction bands at the A point and in the Γ–K direction. Valence and conduction bands also cross the Fermi level in the Γ–A and A–H directions of the Brillouin zone. Since the Fermi level is at the minimum of conduction bands at A point and the conduction bands at A point are degenerated, the electron transfer from valence to conduction bands across the Fermi level in Γ–A and A–H directions might be easier, implying better metallic electrical conductivity of HfB$_2$.

In the band structure of NbB$_2$ (Fig. 4.5d), the Fermi level lies above the conduction band minimum at the A point, indicating saturation of valence electrons. Three conduction bands are across the Fermi level in the Γ–A and A–H directions, and two conduction bands cross the Γ–K direction so that the contribution from metallic bonding is higher than that in ZrB$_2$ and HfB$_2$. The band structure of TaB$_2$ is very similar to that of NbB$_2$, as is shown in Figure 4.5e. The only difference is that the Fermi level in TaB$_2$ lies slightly higher than in NbB$_2$. There are also three conduction bands across the Fermi level in the Γ–A and A–H directions and two conduction bands across the Γ–K direction.

Details of atomic bonding characteristics of TMB$_2$ (TM=Y, Zr, Hf, Nb, and Ta) can be seen from the total density of states (TDOS) and projected density of states, as well as the decomposed distribution of charge density from different covalent bonding peaks in the TDOS. Figure 4.6a shows the total and projected electronic DOS of YB$_2$. Two features are obvious in the DOS of YB$_2$: one is a peak at the Fermi level, and the other is the obvious B sp^2 hybridization. The presence of the peak at the Fermi level indicates that YB$_2$ is not stable. Strong hybridization between the B-2s and B-2p orbitals is evident because the B 2s orbitals are extended and overlapped in a large energy range with B 2p orbitals. The overlapping of hybridized B sp^2–B sp^2 orbitals ranges from −10.6 to −7.1 eV below the Fermi level, as can be seen from the distribution of charge density on (11 $\overline{2}$0) plane of

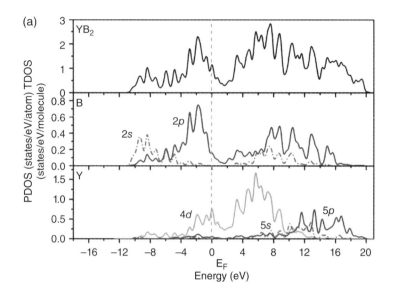

(a)

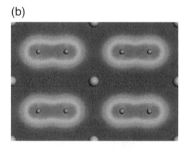

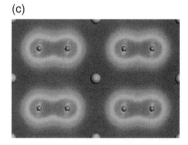

(b)

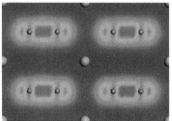

(c)

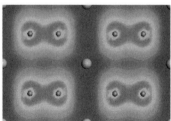

(d)

(e)

(f)

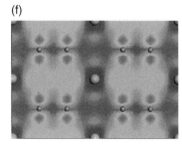

Figure 4.6. (a) Total and projected electronic density of states of YB$_2$, and decomposed charge density on (11 $\bar{2}$0) plane in the energy range from −10.6 to −7.1 eV (b), from −6.6 to −2.0 eV (c), from −4.3 to 0.45 eV (d), from −2.3 to 0.45 eV (e), and from −0.99 to 2.0 eV (f).

YB_2 (Fig. 4.6b). The peaks located from -6.6 to $-2.0\,eV$ are mainly from overlapping B sp^2–B sp^2 orbitals and with some from B $2p_z$–B $2p_z$ orbitals, as shown in Figure 4.6c. The charge density coming from states in the energy range between -4.3 and $0.44\,eV$ is shown in Figure 4.6d, which is mainly from the overlapping of B $2p_z$–B $2p_z$ orbitals with some contribution from the overlapping of B sp^2–B sp^2 orbitals. The states from -2.3 to $0.44\,eV$ are dominated by B $2p_z$–B $2p_z$ and Y-$4d(t_{2g})$–B $2p$ interactions, as shown in Figure 4.6e. In the energy range from -0.99 to $2.6\,eV$, the charge density shows $d(e_g)$ symmetry at the Y site and $2p$-like orbitals at B site, as shown in Figure 4.6f. Thus, the states near the Fermi level are dominated by the Y–Y dd interactions and B–B pp interactions.

Figure 4.7a shows the total and projected electronic DOS of ZrB_2. Besides strong hybridization between the B-$2s$ and B-$2p$ orbitals, the presence of a pseudogap is also obvious. The creation of the pseudogap is, on the one hand, ionic in origin because of charge transfer from Zr to B and, on the other hand, hybridization effects in ZrB_2. The states, which are located between -12.7 and $-7.8\,eV$ below the Fermi level, come mainly from hybridization of B sp^2–B sp^2 orbitals, as shown from the distribution of charge density in Figure 4.7b. And those at adjacent higher energy levels from -8.7 to $-3.1\,eV$ below the Fermi level are mainly from hybridization of the B sp^2–B sp^2 orbitals and some from B $2p_z$–B $2p_z$ orbitals (Fig. 4.7c). In the energy range from -6.8 to $-1.1\,eV$, charge density on the Zr site has t_{2g} symmetry, and B–B bonds have B sp^2–B sp^2 and B $2p_z$–B $2p_z$ character (Fig. 4.7d). The peaks located between -4.1 and $-1.1\,eV$ are dominated by Zr $d(t_{2g})$–B $2p_z$ and B $2p_z$–B $2p_z$ orbitals (Fig. 4.7e). The charge density in the range extending from -2.7 to $0.6\,eV$ exhibits Zr $4d(e_g)$–B $2p$ hybridizations with high accumulation in the x-axis (Fig. 4.7f). Charge density in energy range from -0.03 to $4.0\,eV$ shows e_g symmetry on Zr sites and $2p_z$ like orbitals on B sites (Fig. 4.7g). That is, states near the Fermi level are dominated by Zr–Zr dd interactions and B–B pp interactions.

The total and projected electronic DOS for HfB_2 are similar to ZrB_2 as shown in Figure 4.8a. The energy level of bonding states shifts downward, which indicates enhanced chemical bonding in HfB_2. The peaks located at -13.7 to $-9.5\,eV$ below the Fermi level are due to overlapping of hybridized B sp^2–B sp^2 orbitals, as shown in charge density distribution in Figure 4.8b. States between -9.4 and $-4.1\,eV$ are dominated by B sp^2–B sp^2 orbitals and some from B $2p_z$–B $2p_z$ orbitals, as shown in the charge density distribution in Figure 4.8c. The contribution of B $2p_z$–B $2p_z$ interactions increases in states in the energy range of -7.1 to $-1.4\,eV$ (Fig. 4.8d). Peaks located between -4.8 and $-1.4\,eV$ are dominated by Hf $d(t_{2g})$–B $2p_z$ B and $2p_z$–B $2p_z$ orbitals (Fig. 4.8e). The charge density in the range extending from -3.4 to $0.17\,eV$ exhibits Hf $5d(e_g)$–B $2p$ interactions with high accumulation in the x-axis (Fig. 4.8f). The charge density in energy range from -0.07 to $4.2\,eV$ shows e_g symmetry on Hf site and $2p_z$ like orbitals on the B site (Fig. 4.8g) so that the states near the Fermi level are dominated by the Hf–Hf dd interactions and also B–B pp interactions.

The energy level of bonding states further shifts downward for the DOS of NbB_2 as shown in Figure 4.9a. The overlapping of hybridized B sp^2–B sp^2 orbitals ranges from -14.1 to $-10.0\,eV$ below the Fermi level, as is shown in Figure 4.9b. The peaks located from -9.8 to $-4.7\,eV$ are mainly from the overlapping of B sp^2–B sp^2 and some from B $2p_z$–B $2p_z$ orbitals (Fig. 4.9c). In the energy range from -7.7 to $-2.4\,eV$, the contribution from B $2p_z$–B $2p_z$ interactions increases, and Nb-$4d(t_{2g})$–B $2p_z$ hybridization can also be

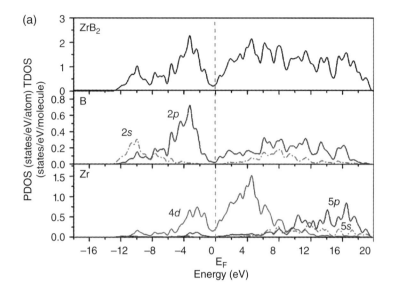

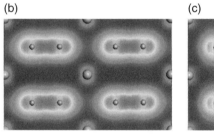
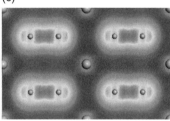
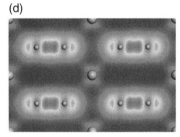

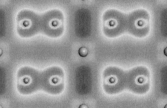

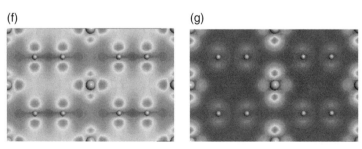

Figure 4.7. (a) Total and projected electronic density of states of ZrB$_2$, and decomposed charge density on (11$\bar{2}$0) plane in the energy range from −12.7 to −7.8 eV (b), from −8.7 to −3.1 eV (c), from −6.8 to −1.1 eV (d), from −4.1 to −1.1 eV (e), and from −2.7 to 0.6 eV (f), and from −0.03 to 4.0 eV (g).

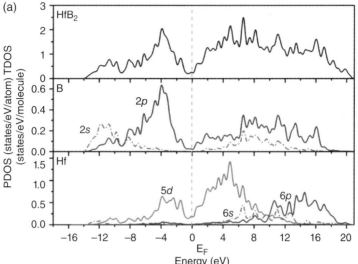

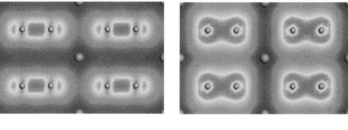

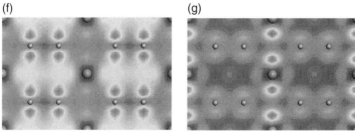

Figure 4.8. (a) Total and projected electronic density of states of HfB$_2$, and decomposed charge density on (11 $\bar{2}$0) plane in the energy range from −13.7 to −9.5 eV (b), from −9.4 to −4.1 eV (c), from −7.7 to −1.4 eV (d), from −4.8 to −1.4 eV (e), from −3.4 to 1.7 eV (f), and from −0.07 to 4.2 eV (g).

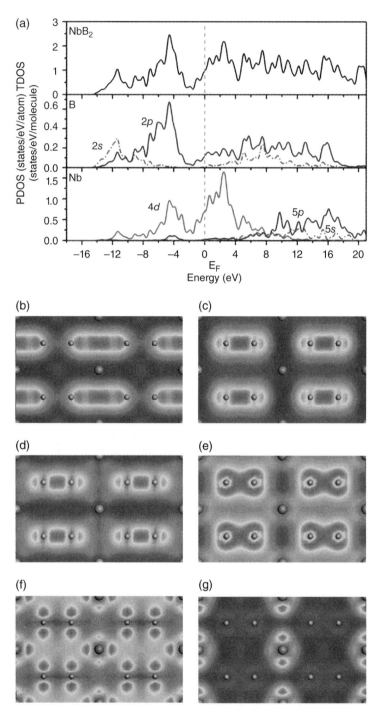

Figure 4.9. (a) Total and projected electronic density of states of NbB$_2$, and decomposed charge density on (11 $\bar{2}$0) plane in the energy range from −14.1 to −10.0 eV (b), from −9.8 to −4.7 eV (c), from −7.7 to −2.4 eV (d), from −5.5 to −2.4 eV (e), from −4.4 to −1.2 eV (f), from −1.6 to 1.9 eV (g),

(h)   (i)

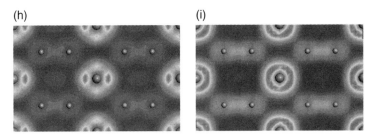

Figure 4.9. (*Continued*) from –0.7 to 3.5 eV (h), and from –0.6 to 3.8 eV (i).

seen (Fig. 4.9d). The states from –5.5 to –2.4 eV are dominated by B $2p_z$–B $2p_z$ and Nb-4$d(t_{2g})$–B $2p_z$ interactions (Fig. 4.9e) and those from –4.4 to –1.2 eV are dominated by Nb-4$d(e_g)$–B $2p_z$ interactions with high accumulation in the x-axis (Fig. 4.9f). The charge density in the energy range from –1.6 to 1.9 eV shows e_g symmetry on the Nb site and p_z-like orbitals on the B site (Fig. 4.9g). The charge density in the energy range from –0.7 to 3.5 eV shows e_g symmetry on the Nb site and sp^2-like orbitals on the B site (Fig. 4.9h), while the charge density in the energy range from –0.6 to 3.8 eV shows e_g + t_{2g} symmetry on the Nb site and sp^2-like orbitals on the B site (Fig. 4.9i). Shifts of B–B and Nd–B hybridizations to lower energy ranges indicate that B–B and Nb–B bonds in NbB$_2$ are stronger than other diborides, which is supported by the shorter B–B and Nb–B bond lengths as are given in Table 4.1. The enhanced chemical bonding in NbB$_2$ is due to the presence of one more valence electron in Nb and the increased valence electron concentration in NbB$_2$. States at the Fermi level are mainly from Nb-4d states and also from B-2p states, indicating their contribution to metallic conductivity.

Figure 4.10a shows the total and projected electronic DOS of TaB$_2$. Compared to NbB$_2$, the energy level of bonding states in TaB$_2$ further shifts downward. The overlapping of hybridized B sp^2–B sp^2 orbitals ranges from –14.9 to –10.7 eV below the Fermi level, which is shown in Figure 4.10b. The peaks located from –10.5 to –5.3 eV are mainly from overlapping of B sp^2–B sp^2 and some B $2p_z$–B $2p_z$ orbitals, as shown in Figure 4.10c. The contribution from B $2p_z$–B $2p_z$ interactions increases in energy range from –8.3 to –2.5 eV as shown in Figure 4.10d and Ta-5$d(t_{2g})$–B $2p_z$ interactions can also be seen in this energy range. The states from –6.1 to –2.5 eV are dominated by B $2p_z$–B $2p_z$ and Ta-5$d(t_{2g})$–B $2p_z$ interactions (Fig. 4.10e) and those from –4.8 to –1.3 eV are dominated by Ta-5$d(e_g)$–B $2p_z$ interactions (Fig. 4.10f). The charge density in the energy range from –1.7 to 2.1 eV shows e_g symmetry at the Ta site and p-like orbitals at the B site (Fig. 4.10g). The charge density in the energy range from –0.8 to 3.7 eV shows e_g symmetry at the Ta site and $p+sp^2$-like orbitals at the B site (Fig. 4.10h). The charge density in the energy range from –0.6 to 4.0 eV shows e_g+t_{2g} symmetry at the Ta site and $p+sp^2$-like orbitals at the B site (Fig. 4.10i). The states at the Fermi level are mainly from Ta-5d states and also from B-2p states, indicating their contribution to metallic conductivity.

The different responses of B–B and TM–B bonds to hydrostatic pressure in TMB$_2$ (TM = Y, Zr, Hf, Nb, and Ta) can be understood from atomistic bonding analysis. The TM–B hybridizations in ZrB$_2$, HfB$_2$, NbB$_2$, and TaB$_2$ are dominated by TM $d(e_g)$–B $2p_z$ interactions, while the Y–B hybridization is dominated by d (t_{2g})–B $2p_z$ interactions. As

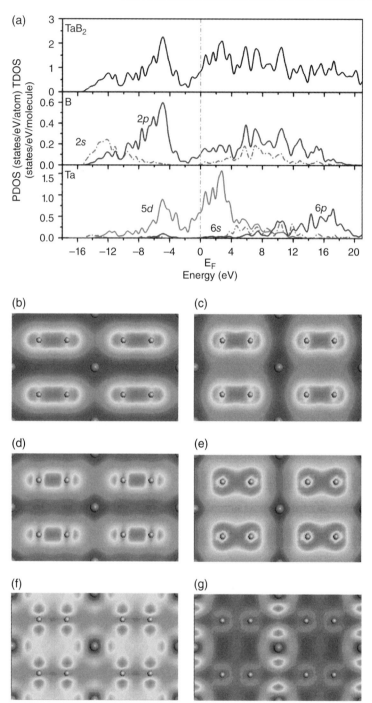

Figure 4.10. (a) Total and projected electronic density of states of TaB_2, and decomposed charge density on $(11\,\overline{2}0)$ plane in the energy range from -14.9 to $-10.7\,eV$ (b), from -10.5 to $-5.3\,eV$ (c), from -8.3 to $-2.5\,eV$ (d), from -6.1 to $-2.5\,eV$ (e), from -4.8 to $-1.3\,eV$ (f), from -1.7 to $2.1\,eV$ (g),

(h)   (i)

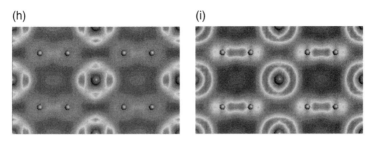

Figure 4.10. (*Continued*) from –0.8 to 3.7 eV (h), and from –0.6 to 4.0 eV (i).

a result, the lattice constant a is stiffer in ZrB_2, HfB_2, NbB_2, and TaB_2 than for YB_2. Another reason is that the Y–B pd bond in YB_2 is partially filled so that it is not as stiff as ZrB_2, HfB_2, NbB_2, and TaB_2.

4.3.3 Elastic Properties

In the last two sections, anisotropic chemical bonding nature has been revealed. Anisotropic chemical bonding can be reflected by an anisotropic Young's modulus [7, 26]. To evaluate the effect of anisotropic chemical bonding on elastic stiffness, second-order elastic constants of single crystals and elastic stiffnesses of polycrystalline TMB_2 (TM=Y, Zr, Hf, Nb, and Ta) were calculated. The second-order elastic constants were calculated using the following relations:

$$\begin{bmatrix} \sigma_1 \\ \sigma_2 \\ \sigma_3 \\ \sigma_4 \\ \sigma_5 \\ \sigma_6 \end{bmatrix} = \begin{bmatrix} c_{11} & c_{12} & c_{13} & c_{14} & c_{15} & c_{16} \\ c_{21} & c_{22} & c_{23} & c_{24} & c_{25} & c_{26} \\ c_{31} & c_{32} & c_{33} & c_{34} & c_{35} & c_{36} \\ c_{41} & c_{42} & c_{43} & c_{44} & c_{45} & c_{46} \\ c_{51} & c_{52} & c_{53} & c_{54} & c_{55} & c_{56} \\ c_{61} & c_{62} & c_{63} & c_{64} & c_{65} & c_{66} \end{bmatrix} \cdot \begin{bmatrix} \varepsilon_1 \\ \varepsilon_2 \\ \varepsilon_3 \\ 2\varepsilon_4 \\ 2\varepsilon_5 \\ 2\varepsilon_6 \end{bmatrix} \tag{4.1}$$

For hexagonal crystals, only five of the constants are independent, c_{11}, c_{12}, c_{13}, c_{33}, and c_{44}. The anisotropic Young's modulus was calculated from the compliance matrix using Equation (4.2).

$$E_{x_i} = 1/s_{ii}, i = 1, 2, 3 \tag{4.2}$$

Poisson's ratios in different directions were calculated using Equation (4.3).

$$v_{x_i x_j} = \frac{s_{ij}}{s_{ii}}, i, j = 1, 2, 3 \tag{4.3}$$

The polycrystalline bulk modulus B, shear modulus G, and Young's modulus E can be estimated from the single-crystal elastic constants, c_{ij}, according to the Voigt, Reuss, and

Hill approximations [23–25]. According to the Voigt approximation [23], the bulk and shear moduli are expressed as follows:

$$B_v = \frac{1}{9}\left(c_{11} + c_{22} + c_{33}\right) + \frac{2}{9}\left(c_{12} + c_{13} + c_{23}\right) \tag{4.4}$$

$$G_v = \frac{1}{15}\left(c_{11} + c_{22} + c_{33} - c_{12} - c_{13} - c_{23}\right) + \frac{1}{5}\left(c_{44} + c_{55} + c_{66}\right) \tag{4.5}$$

According to the Reuss approximation [24], the bulk and shear moduli are expressed by

$$B_R = \frac{1}{\left(s_{11} + s_{22} + s_{33}\right) + 2\left(s_{12} + s_{13} + s_{23}\right)} \tag{4.6}$$

$$G_R = \frac{1}{4\left(s_{11} + s_{22} + s_{33}\right) - 4\left(s_{12} + s_{13} + s_{23}\right) + 3\left(s_{44} + s_{55} + s_{66}\right)} \tag{4.7}$$

The Voigt and Reuss approximations represent the upper and lower limits of the true polycrystalline modulus. Hill proposed the arithmetic mean values of the Voigt's and Reuss's moduli expressed by [25]

$$B_H = \frac{1}{2}\left(B_R + B_V\right) \tag{4.8}$$

$$G_H = \frac{1}{2}\left(G_R + G_V\right) \tag{4.9}$$

The Young's modulus E and Poisson's ratio v can be calculated using Hill's bulk modulus B and shear modulus G using the following equations [27]:

$$E = \frac{9BG}{3B+G} \tag{4.10}$$

$$\upsilon = \frac{3B - 2G}{2\left(3B+G\right)} \tag{4.11}$$

Table 4.2 lists the second-order elastic constants c_{ij}, and bulk modulus B, shear modulus G, Poisson's ratio v of TMB_2 (TM=Y, Zr, Hf, Nb, and Ta). The computed elastic constants for ZrB_2 and HfB_2 are close to those calculated by Lawson et al. [12] and experimental values for ZrB_2 [28], demonstrating the reliability of the computed data. Analysis of the data in Table 4.2 shows that the elastic constants are consistent with bonding. For example, the elastic constants c_{11} and c_{33}, which represent stiffness against principal strains, are anisotropic, that is, c_{11} is much higher than c_{33}. As a result, the Young's moduli are also anisotropic, that is, the Young's moduli along the x and y directions E_x and E_y are higher than in z direction E_z, corresponding to stronger chemical bonding in the basal plane than in the c direction in the crystal structures of TMB_2 (TM=Y, Zr, Hf, Nb, and Ta). As the number of valence electrons increases from 3 for Y

TABLE 4.2. Second-order elastic constants (c_{ij}) and bulk modulus B, Young's modulus E, shear modulus G of TMB$_2$ (TM = Y, Zr, Hf, Nb, and Ta)

c_{ij}	Second-order elastic constants				
	YB$_2$	ZrB$_2$	HfB$_2$	NbB$_2$	TaB$_2$
c_{11} (GPa)	361	540	577	599	592
c_{33} (GPa)	316	431	448	437	426
c_{44} (GPa)	160	250	253	219	189
c_{12} (GPa)	65	56	95	107	140
c_{13} (GPa)	92	114	129	185	194
$E_x = E_y$ (GPa)	329	509	534	520	498
E_z (GPa)	276	387	398	340	323
$v_{xy} = v_{yx}$	0.116	0.051	0.107	0.055	0.102
$v_{xz} = v_{yz}$	0.257	0.252	0.260	0.400	0.408
$v_{zx} = v_{zy}$	0.216	0.191	0.193	0.262	0.265
	Elastic properties of polycrystalline materials				
B (GPa)	171	231	256	287	294
G (GPa)	145	228	230	209	189
E (GPa)	339	514	531	505	466
v	0.169	0.129	0.154	0.207	0.235
$S = 0.75 G/B$	0.636	0.740	0.674	0.546	0.482

to 4 for Zr and Hf and then to 5 for Nb and Ta, c_{11}, which represents the stiffness against principal strains in [100] direction, increases from 361 GPa for YB$_2$ to 540 GPa for ZrB$_2$ and 577 GPa for HfB$_2$, and then to 599 GPa for NbB$_2$ and 592 GPa for TaB$_2$, which is consistent with the compressibility of the lattice constant a. The elastic constant c_{33}, which represents the stiffness against principal strains in [001] direction, however, increases from 316 GPa for YB$_2$ to 431 GPa for ZrB$_2$, reaches a maximum value of 448 GPa for HfB$_2$, and then decreases to 437 GPa for NbB$_2$ and 426 GPa for TaB$_2$. The initial increase in c_{33} is due to enhanced TM d–B p bonding (TM = Y, Zr, Hf), when the number of valence electrons further increases to 5, the additional electron contributing to TM–dd interactions. Similar changes have been observed for c_{44}, which represents shear deformation resistance in (010) or (100) plane in [001] direction. Table 4.2 shows that c_{44} increases from 160 GPa for YB$_2$ to 250 GPa for ZrB$_2$, then to a maximum value of 253 GPa for HfB$_2$, and after that decreases to 219 GPa for NbB$_2$, and to 189 GPa for TaB$_2$. The value of c_{44} is closely related to the TM d–B p bonding. The initial increase in c_{44} is due to the change from partially filled pd bonding states in YB$_2$ to completely filled pd bonding states in ZrB$_2$ and HfB$_2$. Likewise, the decrease in c_{44} for NbB$_2$ and TaB$_2$ is due to the saturation of pd bonding states and the additional electrons in TM (TM = Nb, Ta) dd orbitals with t_{2g} and e_g symmetry near the Fermi level. Similar phenomena have been observed in binary and ternary carbides [32–34] that the enhanced transition metal dd bonding has a negative effect on shear resistance c_{44}. The bulk moduli of polycrystalline TMB$_2$ (TM = Y, Zr, Hf, Nb, and Ta) show similar trends. The bulk modulus B, which reflects the resistance to volume change, increases continuously when the number of valence electrons increases from 3 for Y, 4 for Zr and Hf, and 5 for Nb and Ta. The shear modulus G, however, increases from 145 GPa for YB_2 to a maximum value of 228 GPa for ZrB_2 and 230 GPa for HfB_2, and then decreases to 209 GPa for NbB_2 and 189 GPa for TaB_2. As in transition metal carbides [32–34], the enhancement of both TM d–B p and TM dd bonding is beneficial to bulk modulus B; however, the enhanced TM dd bonding has a negative effect on shear modulus G [34].

The relatively high bulk modulus B, but low shear modulus G, indicates that NbB_2 and TaB_2 are potentially damage-tolerant ceramics. In Table 4.2, the solidity index S = 0.75G/B of TMB_2 (TM = Zr, Hf, Nb, Ta, and Y), which can be used as a criterion to distinguish brittle and **ductile** materials [29]. Low–solidity index materials are more damage tolerant. For example, the solidity indices for damage-tolerant layered ternary carbides (i.e., MAX phases) are less than 0.5 [30]. Among the TMB_2 compounds, TaB_2 has the lowest solidity index (0.482). Therefore, it is expected to be the most damage tolerant among the TMB_2 investigated in the present work. The other possible damage-tolerant material is NbB_2 because it also has a low solidity index (0.546). The predicted intrinsic brittleness increases in the order $TaB_2 < NbB_2 < YB_2 < HfB_2 < ZrB_2$.

The Young's moduli of TMB_2 are strongly related to thermal shock resistance. Table 4.2 shows that the Young's moduli follows the trend $HfB_2 > ZrB_2 > NbB_2 > TaB_2 > YB_2$. Since YB_2 has the lowest Young's modulus among the investigated TMB_2, it is expected to have the best thermal shock resistance. TaB_2 has both low solidity index and Young's modulus, it is predicted to have better thermal shock resistance and damage tolerance than HfB_2 and ZrB_2.

4.4 CONCLUSION REMARKS

The chemical bonding and anisotropic elastic properties of several TMB_2 including YB_2, ZrB_2, HfB_2, NbB_2, and TaB_2 were studied. The response of lattice constants and bond lengths to hydrostatic pressure was examined as well as electronic structure, second-order elastic constants of single crystals, and elastic moduli of bulk polycrystalline materials. Lattice constants, bond lengths, and compressibility decrease when the number of valence electrons increases from 3 for Y to 4 for Zr and Hf, and then to 5 for Nb and Ta. The stiffness along the a direction is higher than that along c direction, corresponding to the stronger covalent bonding in the basal plane. DOS and band structure analysis reveal that B–B bonding is strong covalent, TM–B bonding is ionic–covalent, and TM–TM bonding is metallic. The mechanical properties are anisotropic due to the anisotropic chemical bonding nature. The Young's moduli in the x and y directions are much higher than that in the z direction. The bulk modulus increases when TM changes from Y, to Zr and Hf, and then to Nb and Ta. The shear modulus, however, increases from 145 GPa for YB_2 to 228 GPa for ZrB_2 , reaches a maximum value of 230 GPa for HfB_2, and decreases to 209 GPa for NbB_2 and 189 GPa for TaB_2. The continuous increase in bulk modulus B is due to enhancement of TM d–B p and TM dd bonding, which is beneficial to the enhanced bulk modulus. The initial increase and then decrease in shear modulus G can be explained by enhanced TM d–B p bonding, which has a positive effect on shear resistance, while excessive occupation of TM dd bonding has a negative effect on shear modulus G. The Young's moduli of TMB_2 are in the order of $HfB_2 > ZB_2 > NbB_2 > TaB_2 > YB_2$. YB_2 is predicted to have the best thermal shock resistance since it has the lowest Young's modulus. The brittleness, based on the calculated solidity index $S = 0.75 G/B$, is estimated in the order $TaB_2 < NbB_2 < YB_2 < HfB_2 < Zr B_2$. Because TaB_2 has both low solidity index and Young's modulus, it is predicted to have better thermal shock resistance and damage tolerance.

ACKNOWLEDGMENT

This work was supported by the National Outstanding Young Scientist Foundation for Y. C. Zhou under Grant No. 59925208, and the Natural Sciences Foundation of China under Grant Nos. 50672102, 50832008, 51032006, and 91226202.

REFERENCES

1. Van Wie DM, Drewry DG Jr, King DE, Hudson CM. The hypersonic environment: required operating conditions and design challenges. J Mater Sci 2004;39 (19):5915–5924.

2. Opeka MM, Talmy IG, Zaykoski JA. Oxidation-based materials selection for 2000°C+ hypersonic aerosurfaces: theoretical considerations and historical experience. J Mater Sci 2004;39 (19):5887–5904.

3. Jackson TA, Eklund DR, Fink AJ. High speed propulsion: performance advantage of advanced materials. J Mater Sci 2004;39 (19):5905–5913.

4. Wuchina E, Opeka M, Causey S, Buesking K, Spain J, Cull A, Routbort J, Guitierrez-Mora F. Designing for Ultrahigh-temperature applications: the mechanical and thermal properties of HfB_2, HfCx, and α-Hf(N). J Mater Sci 2004;39 (19):5939–5949.

5. Corral EL, Loehman RE. Ultra-high temperature ceramic coatings for oxidation protection of carbon-carbon composites. J Am Ceram Soc 2008;91 (5):1495–1502.

6. Blum YD, Marschall J, Hui D, Young S. Thick protective UHTC coatings for SiC based structures: process establishment. J Am Ceram Soc 2008;91 (5):1453–1460.

7. Fahrenholtz WG, Hilmas GE, Talmy IG, Zaykoski JA. Refractory diborides of zirconium and hafnium. J Am Ceram Soc 2007;90 (5):1347–1364.

8. Sarin P, Driemeyer PE, Haggerty RP, Kim DK, Bell JL, Apostolov ZD, Kriven WM. In situ studies of oxidation of ZrB_2 and ZrB_2-SiC composites at high temperatures. J Eur Ceram Soc 2010;30 (11):2375–2386.

9. Gangireddy S, Karlsdottir SN, Norton SJ, Tucker JC, Halloran JW. In situ microscopy observation of liquid flow, zirconia growth, and CO bubble formation during high temperature oxidation of zirconium diboride-silicon carbide. J Eur Ceram Soc 2010;30 (11):2365–2374.

10. White A. The materials genome initiative: one year on. MRS Bull 2012;37 (8):715–716.

11. Vajeeston P, Ravindran P, Ravi C, Asokamani R. Electronic structure, bonding, and ground-state properties of AlB_2-type transition-metal diborides. Phys Rev B 2001;63:045115.

12. Lawson JW Jr, Bauschlicher CW, Daw MS. Ab initio computations of electronic, mechanical, and thermal properties of ZrB_2 and HfB_2. J Am Ceram Soc 2011;94 (10):3494–3499.

13. Shein IR, Ivanovski AL. Elastic properties of mono-and polycrystalline hexagonal AlB_2 like diborides of s, p and d metals from first-principles calculations. J Phys Condens Matter 2008;20:415218.

14. Kresse G, Furthmller J. Efficient iterative schemes for ab initio total-energy calculations using a plane-wave basis set. Comput Mater Sci 1996;6:15–50.

15. Gonze X, Rignanese GM, Verstraete M, Beuken JM, Bottin F, Boulanger P, Bruneval F, Caliste D, Caracas R, Cote M, Deutsch T, Genovese L, Ghosez P, Guantomassi M, Goedecker S, Hamann DR, Hermet P, Jollet F, Jomard G, Lerous S, Mancini M, Mazevet S, Oliveira MJT, Onida G, Pouillon Y, Rangel T, Rignanese GM, Sangalli D, Shaltaf R, Torrent M, Verstraete MJ, Zerah G, Zwanzier JW. ABINI: first-principles approach of materials and nanosystem properties. Comput Phys Commun 2009;180:2582–2615.

16. Segall MD, Lindan PLD, Probert MJ, Pickard CJ, Hasnip PJ, Clarc SJ, Payne MC. First-principles simulation: ideas, illustrations and the CASTEP code. J Phys Condens Mater 2002;14:2717–2744.

17. Vanderbilt D. Soft self-consistent pseudopotential in a generalized eignevalue formalism. Phys Rev B 1990;41:7892–7895.

18. Perdew JP, Burke K, Ernzerhof M. Generalized gradient approximation made simple. Phys Rev Lett 1996;77:3865–3868.

19. Pack JD, Monkhorst HJ. Special points for brillouin-zone integrations-A reply. Phys Rev B 1977;16:1748–1749.

20. Pfrommer BG, Côté M, Louie SG, Cohen ML. Relaxation of crystals with the Quasi-Newton method. J Comput Phys 1977;131:233–240.

21. Milman V, Warren MC. Elasticity of hexagonal BeO. J Phys Condens Matter 2001;13:241–251.

22. Ravindran P, Fast L, Korzhavyi PA, Johansson B, Wills J, Eriksson O. Density functional theory for calculation of elastic properties of orthorhombic crystals: application to $TiSi_2$. J Appl Phys 1998;84:4891–4904.

23. Voigt W. *Lehrbuch der Kristallohysic*. Leipzig: Teubner; 1928.

24. Reuss A. Berechnung der Fließgrenze von Mischkristallen auf Grund der Plastizittsbedingung für Einkristalle. Z Angew Meath Mech 1929;9:49–58.

25. Hill R. The elastic behaviour of a crystalline aggregate. Proc Phys Soc A 1952;65:349–354.

26. Gilman JJ, Roberts BW. Elastic constants of TiC and TiB_2. J Appl Phys 1961;32 (7):1405.

27. Green DJ. *An Introduction to the Mechanical Properties of Ceramics*. Cambridge: Cambridge University Press; 1993. p 20–88.

28. Okamoto NL, Kusakari M, Tanaka K, Inui H, Yamaguichi M, Otani S. Temperature dependence of thermal expansion and elastic constants of single crystals of ZrB_2 and the suitability of ZrB_2 as a substrate for GaN film. J App Phys 2003;93:88–94.

29. Gilman JJ. *Electronic Basis of the Strength of Materials*. Cambridge: Cambridge University Press; 2008. p 110.

30. Wang JY, Zhou YC. Recent progress in theoretical prediction, preparation, and characterization of layered ternary transition-metal carbides. Ann Rev Mater Res 2009;39:10.01–10.29.

31. Zhou YC, Sun ZM. The electronic structure and chemical bonding in Ti_3GeC_2. Mater Chem 2000;10:343–346.

32. Wang JY, Zhou YC. Dependence of elastic stiffness on electronic band structure of nanolaminate M_2AlC (M=Ti, V, Nb, and Cr) ceramics. Phys Rev B 2001;69:214111.

33. Wang JY, Zhou YC. Ab initio elastic stiffness of nano-laminate $(M_xM'_{2-x})AlC$ (M and M'=Ti, V and Cr) solid solutions. J Phys Condens Matter 2004;16:2819–2827.

34. Jhi SH, Ihm J, Louie SG, Cohen ML. Electronic mechanism of hardness enhancement in transition-metal carbonitrides. Nature 1999;399:132–134.

5

NEAR-NET-SHAPING OF ULTRA-HIGH TEMPERATURE CERAMICS

Carolina Tallon[1,2] and George V. Franks[1,2]

[1]*Department of Chemical and Biomolecular Engineering, The University of Melbourne, Victoria, Australia*
[2]*Defence Materials Technology Centre (DMTC), Hawthorn, Australia*

5.1 INTRODUCTION

Ultra-High temperature ceramics (UHTCs) are difficult to densify for the same reason they have high melting temperatures, namely, strong bonding between atoms. As such, diffusion is typically low, and pressureless sintering (PS) of dry pressed components does not lead to ceramics with high relative density. Therefore, UHTCs are usually densified by hot pressing (HP), hot isostatic pressing (HIPing), or spark plasma sintering (SPS) [1–5]. These techniques are more expensive than PS due to the application of pressure and/or the use of consumable dies in processing. Furthermore, only simple shapes (generally limited to axially symmetric shapes of uniform cross section like cylinders or tiles) can be produced by HP and SPS. Such shapes are of limited use in most applications where complex-shaped components are required. To produce components with complex shapes, HPed or HIPed billets need extensive diamond grinding. To avoid diamond grinding, shaping components in the green stage followed by PS is preferred [6–9].

Green bodies with uniform and high particle packing density are more amenable to densification by PS than green bodies with low particle packing density [10]. High and

Ultra-High Temperature Ceramics: Materials for Extreme Environment Applications, First Edition.
Edited by William G. Fahrenholtz, Eric J. Wuchina, William E. Lee, and Yanchun Zhou.
© 2014 The American Ceramic Society. Published 2014 by John Wiley & Sons, Inc.

uniform particle packing can be obtained by consolidation of particles that interact with repulsive forces [11]. Conventional dry pressing techniques are not effective because dry systems have no mechanism to develop repulsion between particles (only van der Waals attraction is active). In contrast, wet (colloidal) processing can be effective in producing consolidated green bodies with uniform and high green density [9, 10]. Suspending ceramic particles in a liquid is a possible way to control solution chemistry and produce repulsion among particles as discussed in the next section. Well-dispersed suspensions (particles interacting via repulsion) enable a number of consolidation techniques, which produce high and uniform particle packing. These techniques are described in the third section of this chapter. Colloidal forming techniques have three desirable characteristics. First, they allow for enhanced densification without pressure and, potentially, at lower sintering temperatures. Second, they can be used to produce complex-shaped objects that need little, if any, diamond machining. Finally, they enable deagglomeration and allow for removal of flaws, which enhances reliability [9, 10].

The main driving force for sintering is the reduction in surface energy because high-energy particle surfaces are replaced by internal interfaces (such as grain boundaries) as densification proceeds. As such, fine particles (micrometer or smaller) with high surface area are more amenable to sintering than coarse particles (tens to hundreds of μm in diameter). Therefore, the use of fine particles is preferred and can produce high-density sintered components by PS.

5.2 UNDERSTANDING COLLOIDAL SYSTEMS: INTERPARTICLE FORCES

Fine particles used in colloidal powder processing have a high surface area-to-volume ratio. The behavior of fine particles is dominated by surface forces rather than volume forces (due to mass) and are called colloids. The relative influence can be understood by considering the consequences of being hit in the head with a baseball compared to having a cup of flour (same mass as the baseball) thrown on your head. The baseball has large mass so you feel the impact, but the surface adhesive forces are small so that ball drops to the ground via gravity after the impact. The particles of flour have small mass individually, so they cause little pain when they impact your head, but because the surface adhesive forces are greater than the force of gravity, the flour sticks to your hair and skin.

The mass of colloidal particles is small enough that energy from thermally induced vibrations of liquid molecules is transferred to the solid particles, which causes them to move within the liquid. This phenomenon is known as Brownian motion. In the absence of other forces, Brownian motion will maintain particles in a constantly moving randomized structure [12]. On Earth, particles also experience the force of gravity (a body force). Eventually, gravity wins, and particles settle out of suspension. However, Brownian motion can dominate for extended periods of time, which may be much longer than the processing time. Small particles, such as fine ceramics, will usually be stable against sedimentation for periods of an hour or so and up to a few days, if gently stirred or rolled.

For fine ceramic particles, surface forces dominate the behavior in suspension rather than Brownian or gravity forces. The surface forces may be either attractive or repulsive. Depending on the total of the attraction and repulsion of all the surfaces forces operating in a particular situation, either net attraction or net repulsion can result (Fig. 5.1). In the case of net repulsion, the particles remain dispersed in the suspension as individual particles and then Brownian motion usually dominates over gravity for

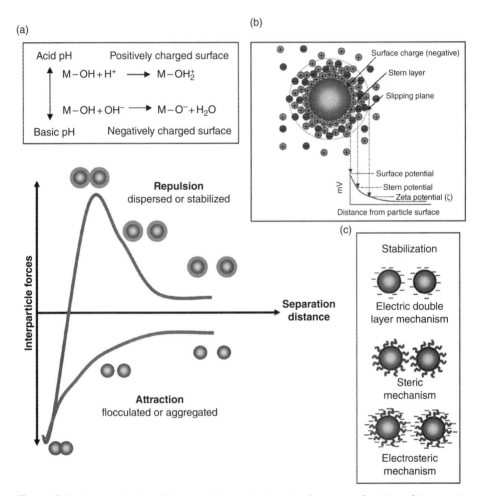

Figure 5.1. Representation of the repulsion and attraction forces as a function of interparticle distance. (a) When ceramic particles are suspended in a polar solvent, particle surfaces become charged depending on the chemistry and pH of the solution. M–OH groups represent surface hydroxyl group that reacts with acid or base. (b) The surface charge is balanced with a counterion cloud around the particle to provide electrical neutrality. Surface charge and the counterion cloud form the electrical double layer. (c) In addition to using the EDL to create repulsive forces between particles, steric and electrosteric mechanisms could be used by adding polymers or polyelectrolytes, respectively, which will adsorb on particle surfaces.

TABLE 5.1. Influence of surface forces on suspension behavior[a]

	Repulsion	Attraction
Typical behavior	Individual dispersed particles Slow settling Low viscosity High and uniform particle packing	Aggregation Rapid sedimentation High viscosity and yield stress Low and pressure-sensitive particle packing
Typical causes	EDL repulsion (high zeta potential) Low salt concentration Steric repulsion	Van der Waals attraction Low zeta potential High salt concentration Bridging polymer flocculation

[a]Refs. 12 and 13.

sufficient time to produce a suspension that is stable against sedimentation. If net attraction is the result, the particles will aggregate into larger, more massive aggregates or flocs. In these suspensions, gravity dominates, and the large heavy aggregates tend to settle out of suspension. The surface forces also control behavior including suspension rheology and particle packing as illustrated in Table 5.1.

The most technologically significant interparticle surface forces are van der Waals, electrical double layer (EDL), steric repulsion, and bridging flocculation.

The van der Waals interaction between particles is due to the sum of individual dipole moment interactions among all of the atoms in the two particles. Every atom in each particle acts as a fluctuating dipole. At every instant, a dipole is created by the separation of the nucleus of an atom (positive) and its electron cloud (negative). That dipole moment fluctuates around the atom at a very fast rate. The electric field emanating from the instantaneous fluctuating dipole moment of one atom influences the dipole moments of the other atoms in both particles. All of the atoms in both particles correlate their dipole moments resulting in the minimum energy configuration. In the case of two particles of the same material interacting across a third medium, the resulting net interaction created by the sum of all of the dipole–dipole interactions is a net attraction [12, 14]. The magnitude of the interaction depends upon the index of refraction and dielectric properties of the particles and intervening medium. Generally speaking, higher-density materials have stronger van der Waals attraction. The magnitude of the interaction can be characterized by a parameter called the Hamaker constant (A). If particles are dispersed in a fluid and no mechanism is present to create repulsion, the particles will be attracted to each other and aggregate. The van der Waals attractive force (F_{vdW}) between two spherical particles can be estimated using the following equation [1]:

$$F_{vdW} = -\frac{Ar}{12D^2} \tag{5.1}$$

where r is the particle radius, and D is the surface-to-surface separation between the particles. The negative sign indicates attraction between the particles.

When particles are suspended in aqueous solutions, they can become charged due to reaction of the surface with either hydroxide (OH⁻) or hydronium (H_3O^+) ions in solution at high and low pH, respectively (Fig. 5.1a). Surface ionization occurs with the metal hydroxyl groups on metal oxide ceramics. In the case of UHTCs, a thin layer of oxide is found on the particle surfaces, so their charging behavior in solution is expected to be similar. However, their hydrophobic nature affects suspension preparation in aqueous solutions, as observed by Leo *et al.* [17]. For the sake of discussion; it is assumed that UHTCs have at least a surface monolayer of oxide. At low pH, the surface becomes positively charged, and at high pH, the surface is negative. Each type of surface has a particular reactivity with acid and base such that the pH where the surface is neutral depends upon the surface chemistry of the material. The pH where the material is neutral is known as the isoelectric point (IEP) and is shown in Table 5.2 for a range of UHTCs. Since the overall system is electrically neutral, the surface charge on the particles is balanced by oppositely charged counterions in the solution (Fig. 5.1b). The counterions form a cloud around the particles according to the Poisson–Boltzman distribution. The surface charge and cloud of counterions are known as the Electric Double Layer (EDL) [12, 14]. When two particles with the same charge approach one another, their counterion clouds overlap. The overlapping of these ion clouds results in increased ion concentration in the gap between the two particles. The increased ion concentration produces an osmotic pressure that pushes the particles apart. This repulsive force increases as particles come closer together and is known as EDL repulsion. The magnitude of the repulsive force is related to the zeta potential (ζ) of the particle surface, which is related to the surface charge (Fig. 5.1b). The zeta potential is zero at the IEP and increases in magnitude as the pH of the suspension moves away from the IEP. The EDL force (F_{EDL}) drops off exponentially with distance between particles and can be estimated by Equation (5.2) for spherical particles (with the same zeta potential) under many conditions:

$$F_{EDL} = 2\pi\varepsilon\varepsilon_o r\kappa\zeta^2 e^{-\kappa D} \tag{5.2}$$

where ε is the dielectric constant of the solution, ε_o is the permittivity of free space, r is the radius of the particle, D is the surface-to-surface distance between the particles, and κ is the Debye screening parameter. The Debye screening parameter depends on the ionic strength (salt concentration) of the solution such that

$$\kappa = 3.29\sqrt{[c]}\left(\text{nm}^{-1}\right) \tag{5.3}$$

where c is the molar concentration of monovalent electrolyte (salt) in the solution. If a suspension contains two different types of particles with different IEPs such that one particle has negative charge and another positive charge, the EDL force will be attractive. More detail about these type of interactions is available in the colloid and surface chemistry texts [12, 14].

The van der Waals force and the EDL force are the two forces that usually dominate behavior in aqueous suspensions. Together, these forces are known as the DLVO forces (after Derjaguin, Landau, Verwey, and Overbeek). In aqueous solutions, well-dispersed suspensions can be produced by adjusting the pH of the suspension to a few pH units away from the IEP so that sufficient magnitude of zeta potential (charge) is developed that the EDL repulsion dominates over van der Waal attraction. In nonaqueous

TABLE 5.2. Isoelectric point (IEP) of most commonly used UHTCs and other relevant powders

	Powder	IEP	Supplier	Reference
Borides	ZrB_2	4	H.C. Starck, Grade B, 2.3 μm, new batch powder received July 2013	*
		4.7	H.C. Starck	[15]
		5	H.C. Starck, Grade B, 2.3 μm	[16]
		5	H.C. Starck, Grade B, 2.3 μm	[17]
		5.8	Grade ZrB_2-F, Japan New Metals Co.; 2.12 μm	[18]
	HfB_2	2.4	H.C. Starck, 11.5 μm	*
		6.4	Alfa Aesar, 2.9 μm	[19]
	TiB_2	4.5	H.C. Starck, Grade F, 2.9 μm	*
Carbides	TiC	3.3	Supplier: Zhuzhou Hard Alloy Plant, Hunan, China) 0.60 μm	[20]
		3.9	Hefei Kaier Nanotechnology Development Co., Ltd., China; 40 nm	[21]
		4	H.C. Starck, HV120, 3 μm	*
		4.3	Supplier: H.C. Starck, 1.4 μm	[22]
		4.3	Synthesized: Whiskers, 30 nm	[22]
		5	H.C. Starck	[15]
	ZrC	3.8	H.C. Starck, Grade B, 2.9 μm	*
	HfC	2.7	Micron Metals (Atlantic Equipment Engineers), HF301, -325mesh	*
	B_4C	5.5	HC Starck; Grade HD15; 2.6 μm	[17]
		6.8	HC Starck; Grade HD15; 2.6 μm	*
	SiC	3	Aldrich	[15]
		3.5	H.C. Starck, 2.1 μm	*
		4	HSC from Superior Graphite	[15]
		4.8, 4.2, 3.3	Grindwell Norton, Bangalore, India	[15]
Nitrides	TiN	2.2	T20412-20H, Sanhe Yanjiao Xinyu High Technology Company for Ceramic Materials, Beijing), 2.3 μm	[23]
		3.7	Synthesized powder, 15-30 nm	[24]
		3.9	H.C. Starck, Grade C, 2.8um	*
		4	H.C. Starck, 1.2 μm	[25]
		4.3	Supplier: H.C. Starck, 1.05 μm	[22]
Silicides	$MoSi_2$	4.3	Chida Materials Co., Ltd., China 2.5 μm	[19]

*Measured by C. Tallon for this book chapter; 100 mg/l powder suspensions were prepared in KCl 10^{-2} M; pH was adjusted with HNO_3 1 M and NaOH 1 M solutions using a Horiba pH meter. Suspensions were sonicated (Ultrasonic Horn, Misonix 4000, Farmingdale, NY) for 2 min and homogenized overnight to allow for surface equilibrium. The zeta potential measurements were carried out using Zetasizer NanoZS (Malvern, Sydney, Australia). The pH was measured just before loading the suspension in the measuring cell.

suspensions, van der Waals forces are significant, but EDL forces are not usually important because the low dielectric constant of nonpolar solvents dramatically reduces the magnitude of the EDL forces.

When soluble polymers are added to suspensions, the polymers can adsorb onto particle surfaces, which develops other forces (Fig. 5.1c). These forces, known as steric repulsion and bridging flocculation (attraction), can be important in both aqueous and non-aqueous ceramic particle suspensions. When the polymer has a relatively low molecular weight (10,000–500,000 Da) and added in concentration to completely cover the particle surface, the polymer chains form a cushion that prevents other particles from approaching, producing steric repulsion between particles [9, 26]. This mechanism is useful in developing well-dispersed suspensions in nonaqueous solvent, since the steric repulsion can overwhelm the van der Waals attraction in these systems, if the polymer type and dose are carefully selected. Steric repulsion can also be important in aqueous systems, but polyelectrolytes (charged polymers) can be used in aqueous systems so that both EDL and steric repulsion combine to aid in dispersing ceramic particles. If the polymer is only poorly soluble in the solvent, steric attraction can develop between particles. This condition is usually to be avoided in ceramic powder processing where well-dispersed suspensions are desirable. If the soluble polymer has a high molecular weight (greater than about a million dalton) and is added in concentrations such that it covers only about half of the surface of the particles, the polymer can bridge between two or more particles resulting in attraction between particles known as bridging flocculation [27]. Again, this condition is usually avoided in ceramic powder processing. Additionally, a special type of attraction could also occur between particles in water, if they are not wetted by water. This force, known as hydrophobic attraction, causes aggregation between particles that have high contact angle with water and can be a reason why some particles are not easily dispersed in water.

As illustrated in Table 5.1, the net particle interaction force has a dramatic effect on suspension behavior. Two aspects of suspension behavior, in particular, are important in near-net-shaping of UHTC components. The first aspect is suspension viscosity, which needs to be maintained as low as possible with a volume fraction of solids that is as high as possible. For most colloidal processing methods, viscosity must be kept low (except, e.g., in 3D printing of pastes), so that suspension can be poured or injected into a complex shaped die cavity for the near-net-shaping processes described in the next section. The volume fraction must be high to produce a high-density green body and enable PS. The second aspect is the particle packing during consolidation, which should be as high as possible and uniform. High-density green bodies enable PS at as low a temperature as possible, and uniformity of particle packing ensures uniform shrinkage to avoid distortion and stresses during sintering.

Suspension rheology (flow behavior) is related to several factors including particle volume fraction, particle size, and interparticle forces [12, 13, 28, 29]. Suspension viscosity increases dramatically as volume fraction increases and asymptotes to very high value as a maximum particle fraction (usually around 64 vol%) is approached. The viscosity of concentrated suspensions is shear thinning (viscosity decreases as shear rate increases) over most of the shear rate range of interest for ceramic processing. The typical suspension viscosity as a function of shear rate is shown schematically in Figure 5.2 for different particle volume fractions. For noninteracting particles or repulsive particles,

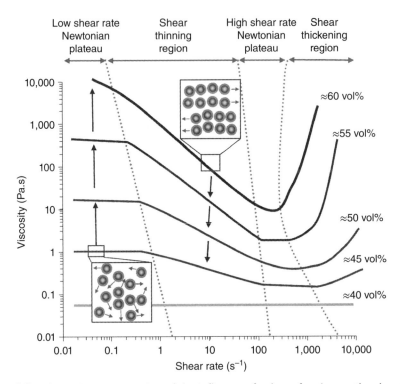

Figure 5.2. Schematic representation of the influence of volume fraction on the viscosity of suspensions as a function of shear rate. At low shear rate, Brownian motion dominates producing high-viscosity randomized structures. At high shear rates, particles line up in preferred flow structures and viscosity decreases.

shear thinning occurs because Brownian motion randomizes particle arrangement at low shear rates. As shear rate increases, particles tend to align in the flow direction to produce preferred flow structures, which minimize hydrodynamic interactions resulting in reduced viscosity as flow rate increases. At very high shear rates, and volume fractions near maximum packing, particles become jammed and viscosity increases dramatically in a phenomenon known as shear thickening. Smaller particle sizes tend to increase viscosity of both repulsive and, especially, attractive particle networks primarily because the number of particle interactions per unit volume increases with decreasing particle size. Well-dispersed suspensions of particles with repulsive interactions produce lower viscosities (at any volume fraction) than attractive particle networks. This is simply because attractive particle interactions create an additional bonding force that must be overcome before particles can rearrange and flow. As such, well-dispersed suspensions at the highest volume fraction that is pourable are the desirable conditions for suspensions for ceramic processing.

The final volume fraction (or packing) of particles during consolidation by pressure depends on the applied pressure and suspension response to applied pressure [11, 13] (Fig. 5.3). The suspension response to consolidation pressure is different for well-dispersed suspensions and attractive particle networks. Well-dispersed suspensions typically pack to

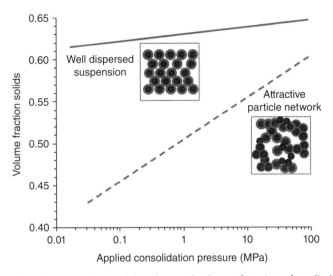

Figure 5.3. Volume fraction of consolidated green body as a function of applied pressure for a well-dispersed suspension and attractive particle networks.

near-maximum packing fraction, and density is insensitive to applied pressure. On the other hand, attractive particle networks result in lower final densities (at a particular pressure compared to dispersed suspensions), and density depends strongly on applied pressure. Pressure sensitivity (or lack thereof) is important because pressure gradients exist in complex-shaped dies during consolidation. These pressure gradients lead to density gradients in the case of attractive particle suspensions, which produce inhomogeneous green bodies. Such green bodies shrink nonuniformly during densification, tend to warp, and develop stresses, which lead to cracking and lack of shape control. Dry powder systems have no mechanism for developing repulsion between particles, so dry pressed components are particularly susceptible to green body density gradients, sintering warpage, and cracking.

One of the most important reasons for using the colloidal processing approach is that it enables the production of ceramic materials that are more reliable because of the reduced number and size of flaws in the final component [9, 30]. Soft agglomerates can be broken down by combination of physical action (mixing, milling, etc.) and chemical action (creating repulsion between particles) using the colloidal approach. Hard agglomerates and inclusions can be removed by filtration or sedimentation of the well-dispersed suspension. Removal of these heterogeneities results in more uniform and flaw-free green bodies. Once sintered, components produced by colloidal processing will have improved strength and reliability.

5.3 NEAR-NET-SHAPE COLLOIDAL PROCESSING TECHNIQUES

Near-net-shape (NNS) processing techniques are able to produce green (and sintered) pieces that resemble the final geometry of the desired component. These methods minimize (or eliminate) expensive and costly diamond machining [6, 8].

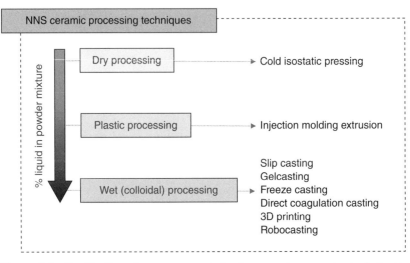

__Figure 5.4.__ Classification of green near-net-shaping ceramic processing techniques.

NNS forming techniques can be classified as dry, plastic, and wet (colloidal) processing methods [31], depending on the water (solvent) content of the initial powder mixture (Fig. 5.4). The best-known example of dry NNS processing techniques is cold isostatic pressing (CIPping), where the powder is placed into a flexible mold. The mold sits inside an oil bath, which transmits hydrostatic pressure homogeneously to produce the green body. The main disadvantage is that the use of dry powder can cause problems associated with the presence of aggregates and low green density. Plastic NNS techniques include extrusion and the use of thermoplastic injection molding using binders that melt, such as paraffin wax, thermoplastic polymeric resins, and polymer mixtures for low-pressure injection molding processes. The main drawback is the removal of large amounts of polymers/binders (typically filling the entire interstitial space among particles), which can cause cracking and undesirable porosity. Paste homogeneity can also be a challenge, depending on the particle size of the powder.

The colloidal NNS processing techniques involve the preparation of stable powder suspensions, as detailed in the previous section. The suspension is then cast into a mold with the desired geometry, and it is consolidated by different mechanisms, such as filtration (slip casting), gelation (gelcasting), freezing (freeze casting/ice templating), coagulation (direct coagulation casting), or evaporation (3D printing). The dispersion of the powder into a solvent offers the advantage of better homogeneity of the final material since aggregates are broken down. By controlling interparticle forces, packing of powder particles in the green body is more efficient, leading to a higher green density material [9, 10] and a more dense material after sintering.

Colloidal processing techniques have been used for several decades, but they are finding a second life in the application for the processing of materials that traditionally have not been prepared by these techniques, such as UHTCs. Nonoxide ceramics such as UHTCs are typically prepared by dry processing techniques in combination with HP [1, 2, 32] or, more recently, SPS [3–5], as shown in Figure 5.5. Their strong covalent

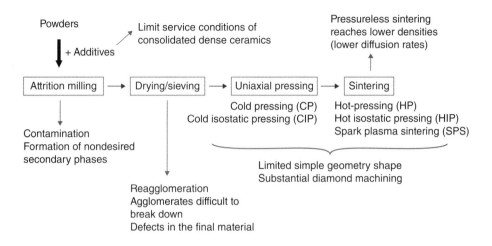

Figure 5.5. Dry processing route to prepare UHTC components. The main challenge associated with current technology for preparing UHTCs is the need for high temperature and pressure to densify the materials. This leads to the use of sintering aids, which reduces the service performance and limits shapes to very simple geometries.

bonding and low self-diffusion coefficients make the use of extremely high temperature and pressure necessary for densification. As a consequence, components are normally made in a very simple geometry (cylinders or tiles) that requires machining to achieve the final complex shape. The machining of such hard materials is time consuming and expensive. Therefore, the use of NNS techniques to produce green bodies in combination with PS is likely to be successful in producing UHTC components.

Apart from the benefits of preparing complex shapes without the need for machining, colloidal processing offers other advantages as outlined in the previous section. Dry processing could lead to the introduction of potential defects in the material, such as impurities from grinding, agglomerates after drying the milled powder, and the need of sintering aids that could affect the final material, as explained in Figure 5.5. Through colloidal processing and control of interparticle forces in suspension (Fig. 5.6), better homogenization of particles within a suspension can be achieved along with the reduction of aggregates, both of which lead to higher particle packing in green bodies and minimization of the need for sintering aids. Enhanced particle packing in the green body leads to sintering at lower temperatures, removing the need for applying pressure.

Most applications of UHTCs need large components of complex shape. Suitable colloidal processing techniques to be explored are summarized as follows:

1. Slip casting. This is the oldest and simplest colloidal processing route. A ceramic suspension up to 50 vol% solids is cast into a porous mold (made of plaster of Paris or porous plastic) [31]. Capillary pressure exerted by the empty pores of the mold drives consolidation. The solvent is extracted from the slurry, while particles are deposited as a layer on the porous mold. The process continues until the

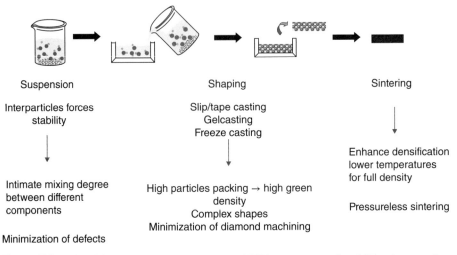

Figure 5.6. Colloidal processing route to prepare UHTC components. In addition to removing flaws related to powder aggregates and improvement in the homogeneity of the sample, the control of the interparticle forces leads to higher particle packing that reduces the temperature for sintering and the need of pressure. In addition, near-net-shaping allows the manufacturing of UHTC complex shapes.

cast body consolidates to the volume fraction determined by the applied capillary pressure and the compressibility of the sample, which depends upon the particle interaction [11, 33] (see Fig. 5.3). The body is then demolded and dried. This technique is normally recommended for components with uniform cross sections (thicknesses less than a centimeter or two) and for sizes no larger than a few hundred millimeters to ensure homogeneity. However, slip casting can be used for complex shaping, if the thickness of the plaster is sufficient to develop capillary suction pressure uniformly over the geometry. For complex shaping, a negative mold of the desired piece is made out of plaster. The slurry is cast in the cavity (Fig. 5.7a). Pressure filtration aids in the shaping of complex geometries, since it increases the particle packing density and the rate of filtration.

2. Gelcasting. The technique was popularized by researchers at Oak Ridge National Laboratory in the 1990s [34–37] and has already become a classic approach for complex shape preparation in ceramic processing. Polymerization of a monomer in solution in a highly solid-loaded suspension leads to the preparation of green bodies with outstanding mechanical strength. Gelcast bodies can be unmolded, handled, and even green machined if needed, all within a shorter consolidation time than other techniques using less than 5 wt% gelcasting additives [38, 39]. Figure 5.7b summarizes the main steps of gelcasting. The monomer and crosslinkers are dissolved in the solution prior to powder addition. After the addition of initiator and catalyst (or activator), which trigger the polymerization reaction, the suspension is cast into the desired mold. This reaction initiates the formation of microgels between monomer and crosslinkers,

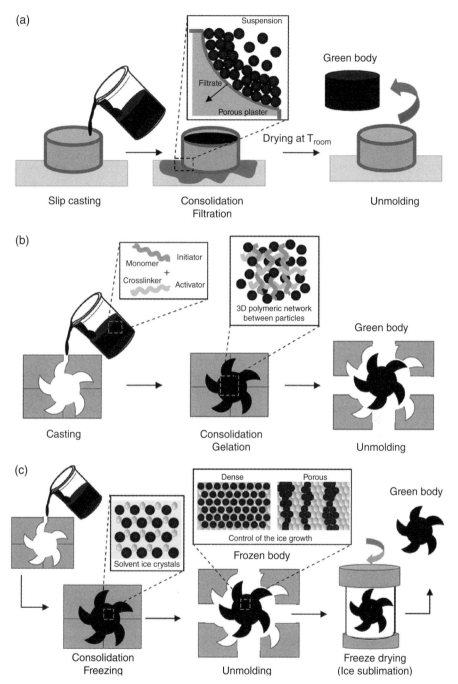

Figure 5.7. Examples of colloidal shaping techniques. (a) Slip casting: the control of suspension rheology determines particle packing; the green body is formed by the removal of solvent by filtration; (b) Gelcasting: the green body strength is determined by the amount of monomer, crosslinker, initiator (and catalyst), and the solid content of the suspension; (c) Freeze casting: the green body microstructure is designed through control of ice growth, which is controlled by solid content, freezing temperature, freezing device, and cyoprotector addition.

which eventually join into a macrogel network that holds the particles in places, keeping the same homogeneity in the green body as in the suspension. Gelcasting is very attractive when used in combination with other processing techniques to prevent cracking during drying, as in tape casting [40–42], freeze casting [43–47], or hydrolysis-assisted solidification [48]. Gelcasting has also been combined with particle stabilized foams to make porous materials [49, 50].

3. Freeze Casting. Freeze-drying is a very well-known technique for drying that was turned into a shaping technique [51–54]. The suspension is frozen inside the desired shape mold, and ice crystals formed by freezing the solvent are removed by sublimation (Fig. 5.7c). These ice crystals lock the particles in place, maintaining the homogeneity of the suspension through the green body and allowing for easy handling and unmolding the piece before the vacuum drying stage. After the sublimation, a packed (quickset) [54] or porous (ice templating) [55–57] green body is obtained, by adjusting the parameters of the method (freezing rate, solid content, cryoprotector addition) [58–60]. Setting times are very short, and drying times, depending on the size, make it an alternative to gelcasting. In most of situations, only 2 or 3 wt% dispersant is needed, minimizing the need for a binder burn-out step. Freeze casting has been used in combination with other techniques such as tape casting [61, 62], gelcasting, injection molding [54, 63], and electrophoresis [64], to aid drying and to create an extra level of porosity in materials.

4. Tape casting. In contrast with the other techniques described so far, tape casting is used to produce flexible thin films that can be stacked together to produce a multilayer structure [31]. The suspension is poured into the reservoir of a doctor blade. Upon the movement of the carrier, the slurry is spread over the sheet with a thickness determined by the gap between the blade and the carrier sheet (final tapes of thickness ranging from 10 to 1000 µm). Solvent is removed by evaporation during drying. To prevent drying cracks and provide flexibility needed for handling, a number of polymeric additives, such as dispersant, binder, and plasticizer, are added to the suspension formulation. In addition, a mixture of solvents is recommended [65, 66]. This is the main reason why tape casting has been associated recently with gelcasting and freeze casting technologies.

5.3.1 Successful Processing of UHTCs Using Colloidal Routes

The earlier section described some of the most developed colloidal NNS techniques that could be put used for processing UHTCs. Several studies have been published pursuing the idea of using colloidal processing techniques for developing UHTCs. Huang *et al.* [16] prepared aqueous suspensions of ZrB_2 for a freeze form extrusion fabrication rapid prototyping technique. Their ceramics reached relative densities of 96% by PS at 2000°C. Medri *et al.* [67] produced composites of ZrB_2–SiC–Si_3N_4 by aqueous slip casting using a commercial dispersant. The bodies were sintered at 2150°C and produced relative densities of 98%. The preparation of flexible ZrB_2 tapes was investigated using aqueous and organic suspensions of ZrB_2 to produce a thicknesses of approximately 250–280 µm with a green tensile strength of 1.21 MPa

[68, 69]. He *et al.* [70] developed aqueous gelcasting formulations using acrylamide and HEMA. Their ZrB_2–SiC green bodies contained B_4C and C as sintering aids that were sintered at 2100°C and 2 h to produce 98% relative density.

The versatility of colloidal processing techniques has led to the preparation of porous UHTCs, which opens a new range of applications where full density is not required. Landi *et al.* [71] have reported the preparation of ZrB_2 materials with aligned porosity by ice templating (freeze casting) of aqueous suspensions. By tuning the parameters controlling the ice growth, materials with 60–70% porosity and pores ranging from 30 to 180 μm were prepared. Also, Medri *et al.* [72] produced macroporous materials with porosities up to 88% using natural and synthetic foams for replica procedures, and around 70% using ovalbumin as a surfactant for foam preparation. These authors used PS at 2100 and 2150°C, respectively, which ensured enough densification of the struts to reach strengths of 67 MPa in the case of ice templating. For sacrificial templating, higher porosity and larger pore size resulted in a strength of only 4 MPa.

Despite good results, a few issues are still holding back the application of these techniques to the UHTC community's needs. Some of the ongoing issues that need to be addressed are the following:

1. Oxidation. The most important issue is the oxidation of UHTC powders. Oxidation affects the service conditions of UHTC components and can change performance from success to failure [2, 73]. The idea of putting UHTC particles into water for processing may not seem like the best way forward. However, powders supplied by manufacturers are sold in containers that allow contact with air. As a result, powders contain some oxygen impurities distributed as the corresponding oxide over the surfaces of the particles. Conventionally, the development of colloidal processing techniques has taken place using aqueous formulations. Nowadays, many different nonaqueous formulations are available, which may be more suitable for nonoxide ceramics. Compared to the as-supplied powders, no significant increase in oxygen content was measured for a common commercial ZrB_2 powder upon soaking in water or cyclohexane after 2 days of immersion and subsequent drying at 80°C for another day in air [74]. Some other compounds like HfB_2 or TaB_2 may have stronger tendencies to oxidize [2, 75], but sintering aids like carbon are commonly used to reduce the oxygen content from the starting powders to levels necessary for densification. Colloidal processing is the best option to incorporate these additives homogeneously.

2. Pressureless Sintering. If an NNS process is to be used, then densification needs to be done accordingly. PS can be used to densify materials that have very strong covalent bonds and low self-diffusion rates [2, 76]. PS is possible, if the driving force for densification is enhanced using sintering aids. Another way to improve densification is to improve particle packing or to reduce the starting particle size during synthesis [2, 76, 77] or by milling [3, 78–80].

3. Particle size. Decreasing particle size is one of the main issues for colloidal processing. The term colloidal generally refers to particle sizes below 1 μm, yet commercially available ZrB_2 powders have average particle sizes as high as approximately 20 μm. Above a few micrometers, gravity dominates particle

behavior, and stable suspensions cannot be prepared due to sedimentation. Below 1 μm, interparticle forces and Brownian motion govern the behavior of particles in suspension. This enables creation of repulsive forces to overcome attractive forces and produce a stable suspension. Grinding and milling can be used to reduce particle size to appropriate values. Lee *et al.* [18] investigated milling of ZrB_2 using a planetary mill with SiC balls and ethanol as dispersing medium. Due to the inherent properties of the powder, size was reduced from 2.3 μm initially to only 1.9 μm after milling, even after extensive milling times. More energetic milling procedures, such as attrition milling can be used, but impurities from grinding media [32] affect the purity of the powder and might compromise the final service conditions. Also grinding is an expensive operation. Synthesis of fine UHTC powders [77, 81] could address this problem although the amounts produced are typically too small to prepare large quantities of highly concentrated suspensions.

4. Particle density and shape. Finally, the density and shape of particles are also a challenge. The smallest particle size available commercially is between 1 and 3 μm. Due to the higher densities of UHTCs, sedimentation forces in suspensions are a factor to consider. Since most of the small particle size powders available commercially have been produced by grinding, the shape of the particles is angular and irregular, far from the spherical shape that is preferred for high particle packing in green bodies to allow for PS.

Next, a case study is presented whereby colloidal processing and PS of ZrB_2 demonstrate that colloidal processing is a viable option for the preparation of complex-shaped UHTC materials. In this study, bodies prepared by colloidal processing and PS are compared to those produced by dry processing and hot pressing without any sintering aids.

5.3.2 Case Study: Colloidal Processing and Pressureless Sintering of UHTCs

To investigate the feasibility of colloidal processing, zirconium diboride was selected as a reference material, since it is well studied and described in the literature. Slip casting and PS will be demonstrated to produce materials with near full density and able to survive extreme temperatures for a few minutes without melting or cracking. The details of this work are summarized here and presented in detail elsewhere [74].

5.3.2.1 Suspension Optimization through Rheological Study. For this study, ZrB_2 Grade B from H.C. Starck was used because it is used by most of the leading research groups. The powder is slightly aggregated, with a Dv_{50} of 2.3 μm, and has irregularly shaped particles. The oxygen content of the as-received powder was 0.90 wt%. Particles were suspended in water and an organic solvent (cyclohexane) for 2 days and dried at 80°C for 24 h. The oxygen content did not change from that initial value after suspension. However, to be conservative and because it produced lower-viscosity suspensions [17, 74], a nonaqueous formulation was selected to prepare the suspensions.

The first step was to study rheology to determine optimum suspension conditions. Suspensions of 20 vol% solids were prepared in cyclohexane using 1, 3, and 5 wt% of a commercial dispersant (Hypermer A70, Croda, Australia) with respect to the powder weight. This dispersant is a polymer comprised of hydroxyl and amino functional groups with a molecular weight of approximately 4.8×10^5 g/mol. The dispersant was dissolved in the solvent before adding the powder. Suspensions were sonicated for 2 min and rolled for homogenization for 6 h. Rheological behavior was studied using a rheometer (AR-G2, TA instruments) in parallel plate configuration with 40 mm diameter plates and a 700 μm gap, using shear rates from 10 to 1000 s^{-1}. The viscosity as a function of shear rate behavior (Fig. 5.8) showed that the suspension with 1 wt% of dispersant

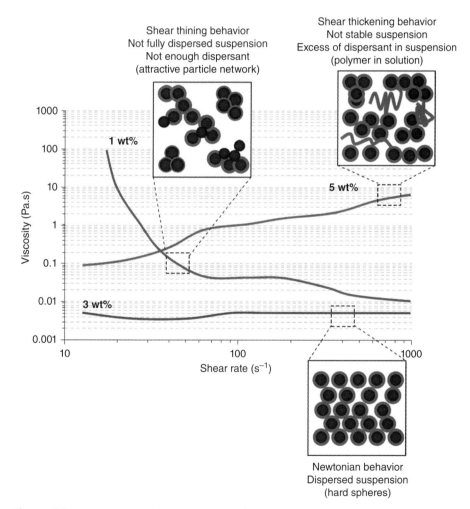

Figure 5.8. Viscosity versus shear rate curves for ZrB$_2$ suspensions prepared with different amounts of dispersant. The dispersant is Hypermer A70 and the solvent is cyclohexane. The volume fraction of solids is 50%.

exhibited shear-thinning behavior, indicating that it was not completely dispersed. With 3 wt% dispersant, the suspension was nearly Newtonian, indicating that it was well dispersed and stable. In contrast, 5 wt% dispersant was excessive, indicating that dispersant remained in solution and increased the viscosity of the suspension. Based on these observations, 3 wt% dispersant was selected as the optimum amount to prepare ZrB_2 suspensions in cyclohexane. A high solid content suspension with a low-enough viscosity to fill the desired mold was desired for producing high green density. Suspensions with up to 50 vol% solids were prepared using 3 wt% dispersant. These suspensions had appropriate viscosity for slip casting [74].

Lee *et al.* [18] prepared aqueous ZrB_2 suspensions up to 45 vol% by adjusting pH or adding PEI as a dispersant. Similarly, our group has also been working in aqueous formulations using different particle stabilization mechanisms [17]. The maximum solid contents and green densities were not as high for aqueous formulations as the ones achieved with the nonaqueous formulation because of stronger repulsion forces between particles in cyclohexane.

5.3.2.2 Green Body Formation. Slip casting was selected as the shaping technique since it is simple and widely used for all types of materials. Also, it does not require the addition of any other reagent to consolidate the component, so the effect of particle packing on sintering could be studied directly.

Suspensions prepared using the optimized conditions described earlier were ball milled with WC media in a very gentle manner for 6 h. The goal was to ensure homogenization and a good quality suspension by breaking the aggregates of the initial powder, not to reduce primary particle size. The oxygen content in the powder increased during milling, as reported by Lee *et al.* [82]. In our case, the oxygen content was 1.25% after milling compared to 0.9 wt% initially. The final oxygen content in our samples at the end of the processing was below 1 wt% for all conditions studied [74].

Suspensions were cast into lubricated metallic rings of 50 mm diameter placed on top of plaster of Paris. Specimens were covered with plastic film to minimize evaporation during casting. After 24 h, they were unmolded. Slip cast disks (50 mm diameter, 10 mm height) had an average green density of 63% of the theoretical density. No cracks were observed, and the disks had enough strength to be handled and machined. Wang *et al.* [83] reported a green density of 53% after slip casting an aqueous ZrB_2–SiC slurry of 45 vol%. The higher green density achieved in the present study was due to the stronger repulsion forces between particles in the nonaqueous solvent, which led suspensions with higher solid loadings. For comparison, the same powder was dry pressed into disks of similar dimensions applying a uniaxial load of 50 MPa. The green density of these disks was 48%. The ZrB_2 particles in our slip cast bodies were nearly close packed due to the superior quality of the suspension (colloidal processing) and the removal of powder aggregates, as compared with the dry pressed powder, which can reach only low particle packing densities due to the Van der Waals attraction between the dry particles.

5.3.2.3 Sintering. Disks produced by colloidal processing (and dry pressing) were sintered at 1900, 2000, and 2100°C in a graphite vacuum furnace set up to also run as a hot press. The dies, rams, and crucibles used were made out of graphite. Two sintering

procedures were performed at each of these temperatures: PS and HP. For PS, disks were placed in a crucible outside the die, while disks for HP were loaded inside a die. The dwell time was 1 h at each temperature, and the applied pressure in the case of HP was 40 MPa, which was applied during the dwell at the maximum temperature. Heating and cooling ramps were 10°C/min up to 1500°C and 5°C/min above that temperature.

Slip cast disks required calcination stage at 400°C for 2 h in air prior to sintering to remove any remaining organic solvent and polymeric dispersant to protect the vacuum furnace. Calcination increased oxygen content to 3 wt%, but oxygen content was reduced below 0.5 wt% after vacuum sintering in the graphite furnace. The temperature for calcination was well below 700°C, which is the temperature above which significant oxidation of ZrB_2 occurs [73, 84, 85].

Sintered samples did not show any signs of cracking for any of the selected temperatures or sintering procedures. Densification results are summarized in Figure 5.9. At 1900 and 2000°C, colloidal processing samples showed only signs of necking between particles when PS was selected, with densities around the 70%. However, at 2100°C, the disks reached 93% relative density. Although this value does not represent full densification, it is more than enough for a wide range of applications for UHTC materials [86]. This result is remarkable since the PS of borides to full density is very difficult, especially without any sintering aids as in this case. Grain coarsening can be faster than

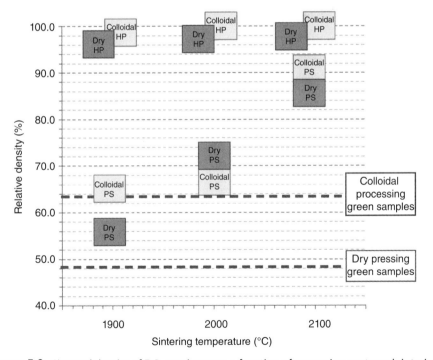

Figure 5.9. Sintered density of ZrB_2 specimens as a function of processing route and sintering temperature. Densities of the green bodies for colloidal and dry processing routes were included as dashed horizontal lines for comparison.

densification in borides, resulting in anisotropic growth and entrapped porosity [2]. Direct comparisons to other reported studies are difficult because many different starting powders have been used. However, in general, PS of ZrB_2 powder from the same supplier as used in the present work resulted in 70% relative density, after attrition milling, cold pressing, CIPping, and sintering at 2100°C [87]. Reaching full density by PS is possible with the addition of SiC [83], $MoSi_2$ [88], B_4C [80], C [69], or WC [78]. In the case of colloidal processing, the enhanced densification is related to higher particle packing in the green body.

Disks prepared by colloidal processing and HP achieved 98% relative density at 1900°C and full densification at 2000°C without any sintering aids. Densification of phase-pure ZrB_2 by HP is normally achieved at temperatures of 2000°C or higher at pressures of 20–30 MPa or at lower temperatures but higher pressures. Monteverde *et al.* [89] reported a sintered density of 87% at 1900°C and 30 MPa under vacuum for similar powder as used in the present study, while Chamberlain *et al.* [32] obtained 98% relative density at 1900°C after 32 MPa and 45 min due to the incorporation of a significant amount of WC during milling. Reactive HP resulted in full densification of ZrB_2 at lower temperatures such as 1890°C, but with the addition of 27% SiC [90]. In this context, results obtained using colloidal processing and HP in the present work have already matched and exceeded results previously reported for pure ZrB_2, due to the improved homogeneity and green density of the samples.

Compared to dry pressing and sintering under the same conditions, colloidal processing led to higher densities in both PS and HP. Higher particle packing in the green body enhanced the densification due to the close contact between particles. The as-dry pressed pellet had a relative density of only 48%, versus 63% for the colloidal counterparts. Close contact between particles enhances densification process at lower temperatures, even in the absence of pressure or sintering aids. In addition, suspension formation removed aggregates from the as-received powder, enhancing the homogeneity of the green material and, therefore, the final components [10]. To corroborate the effect of particle packing on densification, samples prepared by dry pressing were CIPped at 400 MPa prior to sintering. The additional pressing of the dry particles in the green body led to an increase of 43% in sintered density after PS at 1900°C [74]. The same procedure applied to samples prepared by colloidal processing produced only an increase of 6% in the sintered density under the same conditions, because of the higher particle packing for the colloidal processing ones.

No signs of secondary phases were found by XRD or EDS [74]. The oxygen content of sintered samples was below 0.5 wt% in all cases, which is below the content measured for as-received powders. This demonstrated the suitability of wet processing for UHTCs without compromising their performance by oxidation. Since WC grinding media were used, WC impurities could have been acted as a sintering aid as reported in the literature. Chemical analysis revealed that the gentle milling with WC media left an impurity content of only 0.09 wt% W. This percentage was well below the values reported in the literature to aid the sintering of ZrB_2. Therefore, improved particle packing was the only mechanism responsible for the enhanced densification.

Microstructures of colloidally processing samples that were sintered at 2100°C are shown in Figure 5.10. Pores were observed along with some grain pullout from the

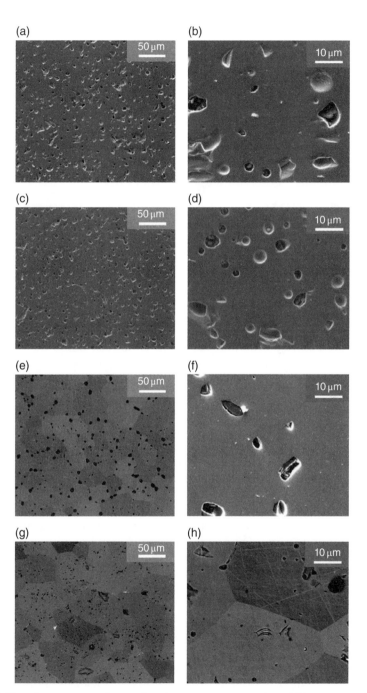

Figure 5.10. Microstructure of ZrB$_2$ after sintering at 2100°C/1 h at different magnifications. (a and b) Colloidal processing and PS; (c and d) dry processing and PS; (e and f) colloidal processing and HP; (g and h) dry pressing and HP.

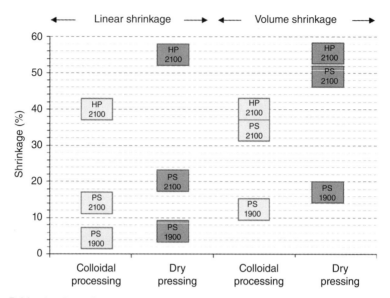

Figure 5.11. Shrinkage for ZrB$_2$ samples as a function of the processing route and type and temperature of sintering. (The linear shrinkage for PS samples showed in the graph represents the average of height and diameter of the samples tested; in the case of the HP samples, the linear shrinkage represents the average of their height only, since the diameter is constrained by die dimension.)

polishing, but the microstructures were generally homogeneous and dense. Even though these samples were prepared without any additives or secondary phases, the relative density values and microstructures indicate that colloidal processing is a promising way to produce bulk UHTCs. The density and microstructure may be suitable for some applications where UHTCs are normally used [1, 86, 91, 92].

Linear and volume shrinkage were recorded during densification (Fig. 5.11). In the case of the colloidal processing samples, there is an increase in the shrinkage observed under pressureless conditions when increasing sintering temperature, in good agreement with the final density values. The dry pressed materials showed similar trends, only with higher shrinkages due to the lower particle packing in the green bodies (48% green density after dry pressing compared to 63% green density for colloidal processing). Linear shrinkage in the axial direction measured for HP materials seems much higher than for PS, but the difference was due to geometry. In HP, the diameter is constrained so all of the shrinkage occurs in the height compared to all dimensions for PS. The volumetric shrinkage during HP was only slightly greater than for disks prepared by colloidal processing, consistent with slightly higher sintered density of the HP materials. Nevertheless, the shrinkage values for the dry pressed materials were higher than for the specimens prepared by colloidal processing. This finding is relevant because even though the improved sintered densities for disks prepared by colloidal processing samples are only slightly higher for disks prepared by dry pressing, the reduced shrinkage of the colloidal processed materials translates into better tolerance and dimension control of the final object during manufacturing. This is especially significant in the case of complex-shaped components, where the dimension control represents a very costly and time-consuming step in their fabrication.

5.3.2.4 Complex Shapes.
The systematic study described earlier confirms the viability of NNS of UHTCs by colloidal processing techniques. Using the same formulation and shaping technique, a notional leading edge was produced. A plastic mold was manufactured with the required geometry, and a negative plaster of Paris was made around it. The suspension was cast in the cavity, and after consolidation, the sample was unmolded and dried in an oven at 60°C overnight.

Figure 5.12a shows the as-molded green body, which resembles the desired leading-edge shape, including the sharp front angle. The green body had a smooth surface finish, without any additional machining. The green density was in good agreement with the values reported in the previous sections.

Leading edges were densified by PS at 2100°C/1 h under vacuum. Samples did not crack upon sintering, as seen in Figure 5.12b. A relative density of 93% was recorded for samples sintered at 2100°C. Leading edges were then tested under an oxyacetylene

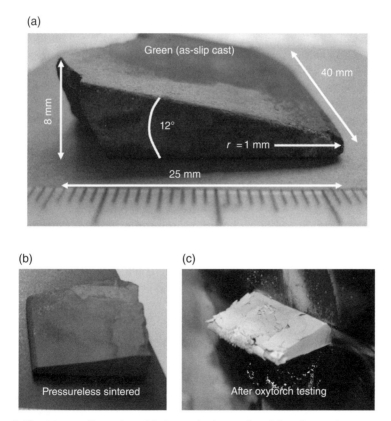

Figure 5.12. ZrB$_2$ Leading edge. (a) Green body produced by slip casting as-unmolded; (b) after pressureless sintering at 2100°C/1 h; (c) after oxytorch testing.

torch (High Temperature Materials Evaluation Rig (HoMER), developed at DSTO Fishermans Bend, VIC, Australia). Wedges were subjected to a very high heat flux for 3 min, at temperatures above 3000°C, as read by the pyrometer. The flame was located at the center of the leading edge. The leading edges survived these extreme conditions without cracking or melting (Fig. 5.12c), although the surface was oxidized [93].

5.4 SUMMARY, RECOMMENDATIONS, AND PATH FORWARD

Colloidal powder processing was an effective approach to manufacture complex-shaped UHTC components. Colloidal processing enabled enhanced reliability by controlling the interparticle forces between particles, removing flaws, and improving the particle packing in the green body. This resulted in a reduced temperature for densification and allowed for PS of as-shaped complex components.

Careful selection of solvent and processing additives is key to developing well-dispersed, low-viscosity, high-volume-fraction suspensions required for colloidal processing. To minimize the risk of oxidation of UHTC powders, nonaqueous formulations were selected. Further advances will be enabled by the development of polymers/dispersants that produce strong steric repulsion in nonaqueous solvents.

Fine powders (in the submicron-size range) are needed for colloidal processing to obtain all the benefits inherent to this approach in terms of green body homogeneity and improved reliability. In addition, fine powders also play a key role in the reduction of densification temperature due to their stronger driving force for densification.

Finally, it is expected that the development of nonaqueous gelcasting and freeze casting chemistries suitable for UHTCs will broaden the shaping capability of UHTC materials.

ACKNOWLEDGMENTS

The authors acknowledge the Defence Materials Technology Centre (DMTC) for funding and providing the support and framework to establish the collaboration to perform this work. This work was developed within the DMTC in collaboration with BAE Systems, University of Queensland, Swinburne University of Technology, Australian Nuclear Science Technology Organization (ANSTO), and Defence Science Technology Organization (DSTO). The authors would like also thank Dr. Chris Wood and Dr. Sonya Slater (Defence Science Technology Organization, DSTO, Fishermans Bend, VIC, Australia) for helping with oxytorch testing. Finally, the authors would like to especially thank Prof. Fahrenholtz for the invitation to contribute to this book.

REFERENCES

1. Levine SR, Opila EJ, Halbig MC, Kiser JD, Singh M, Salem JA. Evaluation of Ultra-High temperature ceramics for aeropropulsion use. J Eur Ceram Soc 2002;22:2757–2767.
2. Fahrenholtz WG, Hilmas GE, Talmy IG, Zaykoski JA. Refractory diborides of zirconium and hafnium. J Am Ceram Soc 2007;90 (5):1347–1364.

3. Medri V, Monteverde F, Balbo A, Bellosi A. Comparison of ZrB_2-ZrC-SiC composites fabricated by spark plasma sintering and hot-pressing. Adv Eng Mater 2005;7 (3):159–163.

4. Melendez-Martınez JJ, Dominguez-Rodrıguez A, Monteverde F, Melandri C, de Portu G. Characterisation and high temperature mechanical properties of zirconium boride-based materials. J Eur Ceram Soc 2002;22:2543–2549.

5. Snyder A, Quach D, Groza JR, Fisher T, Hodson S, Stanciu LA. Spark plasma sintering of ZrB_2–SiC–ZrC Ultra-High temperature ceramics at 1800°C. Mater Sci Eng A 2011;528: 6079–6082.

6. Sigmund WM, Bell NS, Bergstrom L. Novel powder-processing methods for advanced ceramics. J Am Ceram Soc 2000;83 (7):1557–74.

7. Tallon C, Franks GV. Recent trends in shape forming from colloidal processing: a review. J Ceram Soc Jpn 2011;119 (1387):147–160.

8. Lewis JA. Colloidal processing of ceramics. J Am Ceram Soc 2000;83 (10):2341–2359.

9. Lange FF. Colloidal processing of powder for reliable ceramics. Curr Opin Solid State Mater Sci 1998;3 (5):496–500.

10. Lange FF. Powder processing science and technology for increased reliability. J Am Ceram Soc 1989;72 (1):3–15.

11. Franks GV, Lange FF. Plastic-to-brittle transition of saturated, Alumina Powder Compacts. J Am Ceram Soc 1996;79 (12):3161–3168.

12. Hunter RJ. *Foundations of Colloid Science*. 2nd ed. New York: Oxford Press; 2001.

13. Franks GV. Colloids and fine particles. In: Rhodes M, editor. *Introduction to Particle Technology*. West Sussex: John Wiley & Sons, Ltd.; 2008. p 117–152.

14. Israelachvili JN. *Intermolecular and Surface Forces*. 2nd ed. San Diego (CA): Academic Press; 1992.

15. Komulski M. *Surface Charge and Points of Zero Charge*. Boca Raton (FL): CRC Press; 2009.

16. Huang T, Hilmas GE, Fahrenholtz WG, Leu MC. Dispersion of zirconium diboride in an aqueous, high solid paste. Int J Appl Ceram Technol 2007;4 (5):470–479.

17. Leo S, Tallon C, Franks GV. Comparison of aqueous and non-aqueous colloidal processing of difficult-to-densify ceramics. Part I: suspension rheology. J Am Ceram Soc 2013. (In press, 2014).

18. Lee S-H, Sakka Y, Kagawa Y. Dispersion behavior of ZrB_2 powder in aqueous solution. J Am Ceram Soc. 2007;90 (11):3455–3459.

19. He R, Zhang X, Han W, Hu P, Hong C. Effects of solids loading on microstructure and mechanical properties of HfB_2–20 vol.% $MoSi_2$ Ultra High temperature ceramic composites through aqueous gelcasting route. Mater Design 2013;47:35–40.

20. Zhang J-X, Jiang D-L, Tan S-H, Gui L-H, Ruan M-L. Aqueous processing of titanium carbide green sheets. J Am Ceram Soc 2001;84 (11):2537–2541.

21. Xiong J, Xiong S, Guo Z, Yang M, Chen J, Fan H. Ultrasonic dispersion of nano TiC powders aided by tween 80 addition. Ceram Int 2012;38:1815–1821.

22. Laarz E, Carlsson M, Vivien B, Johnsson M, Nygren M, Bergstrom L. Colloidal processing of Al_2O_3-based composites reinforced with TiN and TiC particulates, whiskers and nanoparticles. J Eur Ceram Soc 2001;21:1027–1035.

23. Zhang J, Duan L, Jiang D, Lin Q, Iwasa M. Dispersion of TiN in aqueous media. J Colloid Interface Sci 2005;286:209–215.

24. Wasche R, Steinborn G. Influence of the sispersants in gelcasting of nanosized TiN. J Eur Ceram Soc 1997;17:421–426.

25. Shih C-J, Hon M-H. Electrokinetic and rheological properties of aqueous TiN suspensions with ammonium salt of poly(methacrylic acid). J Eur Ceram Soc 1999;19:2773–2780.

26. Napper DH. *Polymeric Stabilization of Colloidal Dispersions*. New York: Academic Press; 1984.

27. Lamer VK, Healy TW. Adsorption-flocculation reactions of macromolecules at solid-liquid interface. Rev Pure Appl Chem 1963;13:112–133.

28. Barnes HA, Hutton JF, Walters K. *An Introduction to Rheology*. Amsterdam: Elsevier; 1989. Rheology Series, 3.

29. Pugh RJ, Bergstrom L. Rheology of concentrated suspensions. In: *Surface and Colloid Chemistry in Advanced Ceramic Processing, Surfactant Science Series*. New York: Dekker; 1994. p 193–239.

30. Pujari VK, Tracey DM, Foley MR, Paille NI, Pelletier PJ, Sales LC, Wilkens CA, Yeckley RL. Reliable ceramics for advanced heat engines. J Am Ceram Soc 1995;74 (4):86–90.

31. Reed JS. *Principles of Ceramic Processing*. 2nd ed. New York: John Wiley & Sons, Inc; 1995.

32. Chamberlain AL, Fahrenholtz WG, Hilmas GE, Ellerby DT. High-strength zirconium diboride-based ceramics. J Am Ceram Soc 2004;87 (6):1170–1172.

33. Stickland AD, Teo H-E, Franks GV and Scales PJ. Compressive strength and capillary pressure: competing properties of particulate suspensions that determine the onset of desaturation. Drying Technol 2013. DOI: 10.1080/07373937.2014.915218.

34. Janney MA, Omatete OO, Walls CA, Nunn SD, Ogle RJ, Westmoreland G. Development of low-toxicity gelcasting systems. J Am Ceram Soc 1998;81 (3):581–591.

35. Omatete OO, Janney MA, Nunn SD. Gelcasting: from laboratory development toward industrial production. J Eur Ceram Soc 1997;17 (2–3):407–413.

36. Omatete OO, Janney MA, Strehlow RA. Gelcasting—a new ceramic forming process. J Am Ceram Soc 1991;70 (10):1641.

37. Young AC, Omatete OO, Janney MA, Menchhofer PA. Gelcasting of alumina. J Am Ceram Soc 1991;74 (3):612–618.

38. Chabert F, Dunstan DE, Franks GV. Cross-linked polyvinyl alcohol as a binder for gelcasting and green machining. J Am Ceram Soc 2008;91 (10):3138–3146.

39. Tallon C, Moreno R, Nieto MI, Jach D, Rokicki G, Szafran M. Gelcasting performance of alumina aqueous suspensions with glycerol monoacrylate: a new low-toxicity acrylic monomer. J Am Ceram Soc 2007;90 (5):1386–1393.

40. J. Besida, D. E. Dunstan, J. Fawcett, C. Henderson, S. A. Khoo and G. V. Franks, Novel aqueous tape casting process. 107th Annual Meeting of the American Ceramic Society. Baltimore (MD); 2006.

41. Santanach-Carreras E, Chabert F, Dunstan DE, Franks GV. Avoiding mud cracks during drying of thin films from aqueous colloidal suspensions. J Colloid Interface Sci 2007;313: 160–168.

42. Shanti NO, Hovis DB, Seitz ME, Montgomery JK, Baskin DM, Faber KT. Ceramic laminates by gelcasting. Int J Appl Ceram Technol 2009;6 (5):593–606.

43. Lee JP, Lee KH, Song HK. Manufacture of biodegradable packaging foams from agar by freeze-drying. J Mater Sci 1997;32 (21):5825–5832.

44. Weber K, Tamandl G. Porous Al_2O_3-ceramics with uniform capillaries. Cfi-Ceram Forum Int 1998;75 (8):22–24.

45. Chen RF, Wang CA, Huang Y, Ma LG, Lin WY. Ceramics with special porous structures fabricated by freeze-gelcasting: using tert-butyl alcohol as a template. J Am Ceram Soc 2007;90 (11):3478–3484.

46. Wu HH, Li DC, Chen XJ, Sun B, Xu DY. Rapid casting of turbine blades with abnormal film cooling holes using integral ceramic casting molds. Int J Adv Manufact Technol 2010;50 (1–4):13–19.

47. Fukushima M, Nakata M, Yoshizawa Y. Fabrication and properties of ultra highly porous cordierite with oriented micrometer-sized cylindrical pores by gelation and freezing method. J Ceram Soc Jpn 2008;116 (1360):1322–1325.

48. Ganesh I, Sundararajan G, Olhero SM, Torres PMC, Ferreira JMF. A novel colloidal processing route to alumina ceramics. Ceram Int 2010;36 (4):1357–1364.

49. Chuanuwatanakul C, Tallon C, Dunstan DE, Franks GV. Controlling the microstructure of ceramic particle stabilized foams: influence of contact angle and particle aggregation. Soft Matter 2011;7:11464–11474.

50. Liu W, Xu J, Wang Y, Xu H, Xi X, Yang J. Processing and properties of porous PZT ceramics from particle-stabilized foams via gel casting. J Am Ceram Soc 2013;96 (6):1827–1831.

51. Rey L, Pirie NW, Whitman WE, Kurti N. Freezing and freeze-drying [and Discussion]. Proc R Soc Lond Biol Sci 1975;191 (1102):9–19.

52. Flosdorf EW. *Freeze-Drying. Drying by Sublimation*. New York: Reinhold Publishing Corporation; 1949.

53. Kwiatkowski A, Reszka K, Szymanski A. Preparation of corundum and steatite ceramics by the freeze-drying method. Ceram Int 1982;6 (2):79–82.

54. Novich BE, Sundback CA, Adams RW. Quickset injection molding of high-performance ceramics. In: Cima MJ, editor. *Ceramic Transactions, Forming Science and Technology for Ceramics*. Westerville (OH): American Ceramic Society; 1992. p 157–164.

55. Deville S. Freeze-casting of porous ceramics: a review of current achievements and issues. Adv Eng Mater 2008;10 (3):155–169.

56. Dogan F, Sofie SW. Microstructural control of complex-shaped ceramics processed by freeze casting. Cfi-Ceram Forum Int 2002;79 (5):E35–E38.

57. Tallon C, Moreno R, Nieto MI. Shaping of porous alumina bodies by freeze casting. Adv Appl Ceram 2009;108 (5):307–313.

58. Araki K, Halloran JW. Porous ceramic bodies with interconnected pore channels by a novel freeze casting technique. J Am Ceram Soc 2005;88 (5):1108–1114.

59. Araki K, Halloran JW. Room-temperature freeze casting for ceramics with nonaqueous sublimable vehicles in the naphthalene-camphor eutectic system. J Am Ceram Soc 2004;87 (11):2014–2019.

60. Shanti NO, Araki K, Halloran JW. Particle redistribution during dendritic solidification of particle suspensions. J Am Ceram Soc 2006;89 (8):2444–2447.

61. Sofie SW. Fabrication of functionally graded and aligned porosity in thin ceramic substrates with the novel freeze-tape-casting process. J Am Ceram Soc 2007;90 (7):2024–2031.

62. Ren LL, Zeng YP, Jiang DL. Fabrication of gradient pore TiO$_2$ sheets by a novel freeze-tape-casting process. J Am Ceram Soc 2007;90 (9):3001–3004.

63. Adams RW, Householder WB, Sundback CA. Applicability of quicksettm injection molding to intelligent processing of ceramics. Proceedings of the 15th Annual Conference on Composites and Advanced Ceramic Materials: Ceramic Engineering and Science Proceedings; January 13–16, 1991; Cocoa Beach (FL). New York: John Wiley & Sons, Inc., 2008. p 2062–2071.

64. Zhang YM, Hu LY, Han JC. Preparation of a dense/porous BiLayered ceramic by applying an electric field during freeze casting. J Am Ceram Soc 2009;92 (8):1874–1876.

65. Moreno R. The role of slip additives in tape-casting technologies: part I-solvents and dispersants. J Am Ceram Soc 1992;71 (10):1521–1531.

66. Bohnleinmauss J, Sigmund W, Wegner G, Meyer WH, Hessel F, Seitz K, Roosen A. The function of polymers in the tape casting of alumina. Adv Mater 1992;4 (2):73–81.

67. Medri V, Capiani C, Gardini D. Slip casting of ZrB_2–SiC composite aqueous suspensions. Adv Eng Mater 2010;12 (3):210–215.

68. Lu Z, Jiang D, Zhang J, Lin Q. Aqueous tape casting of zirconium diboride. J Am Ceram Soc 2009;92 (10):2212–2217.

69. Natividad SL, Marotto VR, Walker LS, Pham D, Pinc W, Corral EL. Tape casting thin, continuous, homogenous, and flexible tapes of ZrB_2. J Am Ceram Soc 2011;94 (9):2749–2753.

70. He R, Zhang X, Hu P, Liu C, Han W. Aqueous gelcasting of ZrB_2-SiC Ultra High temperature ceramics. Ceram Int 2012;38:5411–5418.

71. Landi E, Sciti D, Melandri C, Medri V. Ice templating of ZrB_2 porous architectures. J Eur Ceram Soc 2013;33:1599–1607.

72. Medri V, Mazzocchi M, Bellosi A. ZrB_2-based sponges and lightweight devices. Int J Appl Ceram Technol 2011;8 (4):815–823.

73. Fahrenholtz WG, Hilmas GE. Oxidation of Ultra-High temperature transition metal diboride ceramics. International Materials Reviews 2012;57 (1):61–72.

74. Tallon C, Chavara D, Gillen A, Riley D, Edwards L, Moricca S, Franks GV. Colloidal processing of zirconium diboride Ultra-High temperature ceramics. J Am Ceram Soc 2013;96 (8):2374–2381.

75. Silvestroni L, Guicciardi S, Melandri C, Sciti D. TaB_2-based ceramics: microstructure, mechanical properties and oxidation resistance. J Eur Ceram Soc 2012;32 (1):97–105.

76. Guo S-Q. Densification of ZrB_2-based composites and their mechanical and physical properties: a review. J Eur Ceram Soc 2009;29:995–1011.

77. Zhang Y, Li RX, Jiang YS, Zhao B, Duan HP, Li JP, Feng ZH. Morphology evolution of ZrB_2 nanoparticles synthesized by sol-gel method. J Solid State Chem 2011;184 (8):2047–2052.

78. Chamberlain AL, Fahrenholtz WG, Hilmas GE. Pressureless sintering of zirconium diboride. J Am Ceram Soc 2006;89 (2):450–456.

79. Brochu M, Gaunt BD, Boyer L, Loehman RE. Pressureless reactive sintering of ZrB_2 ceramic. J Eur Ceram Soc 2009;29:1493–1499.

80. Fahrenholtz WG, Hilmas GE, Zhang SC, Zhu S. Pressureless sintering of zirconium diboride: particle size and additive effects. J Am Ceram Soc 2008;91 (5):1398–1404.

81. Jung EY, Kim JH, Jung SH, Choi SC. Synthesis of ZrB_2 powders by carbothermal and borothermal reduction. J Alloys Compounds 2012;538:164–168.

82. Lee S-H, Sakka Y, Kagawa Y. Corrosion of ZrB_2 powder during wet processing—analysis and control. J Am Ceram Soc 2008;91 (5):1715–1717.

83. Wang X-G, Liu J-X, Kan Y-M, Zhang G-J, Wang P-L. Slip casting and pressureless sintering of ZrB_2-SiC ceramics. J Inorg Mater 2009;24 (4):831–835.

84. Fahrenholtz WG. The ZrB_2 volatility diagram. J Am Ceram Soc 2005;88 (12):3509–3512.

85. Shappirio JR, Finnegan JJ, Lux RA, Fox DC. Resistivity, oxidation kinetics and diffusion barrier properties of thin film ZrB_2, *Thin Solid Films*. I 1984;I9:23–30.

86. Glass DE. Physical challenges and limitations confronting the use of UHTCs on hypersonic vehicles. 17th AIAA Space Planes and Hypersonic Systems and Technology Conference; April 11–14, 2011; San Francisco (CA); 2011.

87. Zhu S, Fahrenholtz WG, Hilmas GE, Zhang SC. Pressureless sintering of carbon-coated zirconium diboride powders. Mater Sci Eng A 2007;459:167–171.

88. Sciti D, Monteverde F, Guicciardi S, Pezzotti G, Bellosi A. Microstructure and mechanical properties of ZrB_2–$MoSi_2$ ceramic composites produced by different sintering techniques. Mater Sci Eng A 2006;434:303–309.

89. Monteverde F, Guicciardi S, Bellosi A. Advances in microstructure and mechanical properties of zirconium diboride based ceramics. Mater Sci Eng A 2003;346:310–319.

90. Zimmermann JW, Hilmas GE, Fahrenholtz WG, Monteverde F, Bellosi A. Fabrication and properties of reactively hot-pressed ZrB_2-SiC ceramics. J Eur Ceram Soc 2007;27: 2729–2736.

91. Tallon C, Slater S, Gillen A, Wood C, Turner J. Ceramic materials for hypersonic applications. Mater Aust Mag 2011;45 (2):28–32.

92. Paul A, Jayaseelan DD, Venugopal S, Zapata-Solvas E, Binner J, Vaidhyanathan B, Heaton A, Brown P, Lee WE. UHTC composites for hypersonic applications. J Am Ceram Soc 2012;91 (1):22–28.

93. Tallon C, Slater S, Woods C, Antoniou R, Thornton J, Franks GV. High temperature testing of UHTC leading edges prepared by colloidal processing. J Am Ceram Soc 2014.

6

SINTERING AND DENSIFICATION MECHANISMS OF ULTRA-HIGH TEMPERATURE CERAMICS

Diletta Sciti, Laura Silvestroni, Valentina Medri, and Frédéric Monteverde

National Research Council of Italy, Institute of Science and Technology for Ceramics, Faenza, Italy

6.1 INTRODUCTION

Ultra-High temperature ceramics (UHTCs) are commonly referred to as a class of highly refractory nonoxide compounds characterized by melting points above 3000°C [1]. Among UHTCs, transition metal diborides MB_2 (M = Ti, Zr, Hf, Ta) occupy a prominent position of interest and, for that reason, will be treated as a focused case in the present chapter.

Sintering of UHTCs is difficult due to their strong covalent bonds and low self-diffusion rates [2]. For instance, the most common transport mechanisms for solid-state sintering of MB_2 are lattice and grain-boundary diffusion, which involve boron and the transition metal atoms. Several sintering studies conducted on MB_2 have concluded that control of surface chemistry during processing is required to achieve dense ceramics with desirable microstructures: oxide impurities at the particle surfaces enhance coarsening at temperatures below those necessary for densification by the formation of fast diffusion paths at the grain boundaries. The real surface chemistry is sensitive to powder processing techniques and often very different from what can be predicted by fundamental

Ultra-High Temperature Ceramics: Materials for Extreme Environment Applications, First Edition.
Edited by William G. Fahrenholtz, Eric J. Wuchina, William E. Lee, and Yanchun Zhou.
© 2014 The American Ceramic Society. Published 2014 by John Wiley & Sons, Inc.

surface studies in ideal conditions. For instance, an increase in the oxide content in ZrB_2 powders was observed after conventional powder processing [3, 4]. X-ray photoelectron spectroscopy studies of MB_2 [5] suggested that a number of B_xO_y, MO_y, and MB_xO_y stoichiometries may form within MB_2 ceramics. Extending the effects of these oxygen-bearing species to other Ultra-High refractory ceramics belonging to group IV–VI transition metal carbides (MC), like ZrC, HfC, or TaC, the presence of surface oxides is the most important factor hindering densification.

Sintering experiments characterized by peak processing temperatures over 2000°C and applied pressures over 30 MPa were routinely conducted to achieve full densification in monolithic UHTCs. At very high temperatures, grain coarsening becomes predominant over densification and leads to poor mechanical properties compared to ceramics with refined microstructures. Hence, different strategies to assist UHTC densification have been the focus of continuing research:

- pure and ultrafine powders synthesized via chemical or solid-state methods [6, 7];
- refinement of powder particle sizes and/or increasing the defect concentrations through mechanical activation, that is, high-energy milling [8];
- high-temperature reactive sintering of solid precursors [9];
- the use of sintering additives [10, 11].

The increase in sinterability by using very fine, pure UHTC powders is a well-known achievement and is commonly accepted for other class of materials. In the case of coarse or low-reactivity UHTC powders, mechanical activation aims to reduce particle size and increase defect concentration. Often, this procedure is difficult for UHTC powders, due to Young's modulus being above 400 GPa and fracture toughness between 3 and 4 MPa√m: the particle size can be reduced by adjusting milling time, solids loading, and speed, but most systems show practical grinding limits [12].

UHTCs have been also manufactured using high-temperature solid-state chemical reactions [9]: the potential removal of the oxides, very often confined at grain boundaries as secondary phases in the sintered microstructure, could provide higher strength at elevated temperatures. An additional advantage of this approach is motivated by the demand for reducing manufacturing costs, for instance, using cheaper elemental raw powders in place of already synthesized single-phase compounds.

Compositional engineering through addition of sintering agents is the most widely used approach [10, 11]. Additives can be ceramics or metals, and introduced in various amounts and forms (crystalline, precursor, amorphous). As a general rule of thumb, additives can either form a liquid phase at the sintering temperature (liquid-phase sintering) or facilitate removal of the oxides from the powder particle surfaces inducing densification by solid-state sintering. Usually, additives are employed in combination with pressure-assisted sintering techniques, with the primary aim of reducing the processing temperature as much as possible; however, additives and impurities that form low-temperature eutectics with UHTCs limit their high-temperature performance.

A list of sintering agents previously explored for hot pressing (HP) of MB_2 is given in Table 6.1, with an indication of sintering temperatures (T_{MAX}) that led to full

TABLE 6.1. Sintering agents and corresponding hot pressing temperature ranges (T_{MAX}) for ZrB_2 and HfB_2

Sintering agent	Content (vol%)	T_{MAX} for ZrB_2 (°C)	Ref.	T_{MAX} for HfB_2 (°C)	Ref.
Ni	<3	1850	[13]		
Si_3N_4	5	1700	[14]	1800	This work
AlN	5	1850	[15]		
$ZrSi_2$, $HfSi_2$	<10	1550	[10]	1600	[16]
$MoSi_2$	5–15	1750	[17]	1900	[18]
$TaSi_2$	5–15	1850	[19]	1900	[19]
C, B, B_4C	<5	1900–2000	[20]	1900–2000	[21]
WC	3.3	2000	[22]		
SiC	5	1900	[23]	2000	[24]

densification. Unpublished data presented here for the first time are referenced "as this work." As a general rule, corresponding temperatures for pressureless sintering (PS) cycles are 100–150°C higher. MB_2, being electrically conductive, are also especially suited for spark plasma sintering (SPS), which allows densification to occur at 100–150°C lower and in shorter times compared to conventional HP.

The purpose of this chapter is to illustrate a variety of sintering experiments applied to several different UHTCs, with special emphasis on the sintering aids used, the densification behavior during HP, and the identification of densification mechanisms. If not specified, powder mixtures were processed by wet ball milling. HP runs were conducted using induction-heated graphite dies with 15–20°C/min as typical heating rate range, under an actively pumped vacuum (1–2 mbar) at temperatures in the range 1550–1950°C and a uniaxial pressure range of 30–40 MPa.

The chapter is divided into the following sections: Sections 6.2 and 6.3 are dedicated to diborides containing metal and nitride additives, respectively. Section 6.4 is dedicated to MB_2 densified with metal disilicides. Sections 6.5 and 6.6 are devoted to carbon and carbide additions. In particular, Section 6.6 deals with the effect of SiC particles on densification: is SiC just a reinforcing phase or can be also considered a sintering additive? Section 6.7 is focused on MB_2-SiC composites containing third phases. Section 6.8 illustrates important implications of densification on high-temperature structural stability. Section 6.9 summarizes sintering and densification of carbide UHTCs.

6.2 MB_2 WITH METALS

Reactive metals like Ni [13], Mo [25], Nb [26], or Zr [27] were the first sintering additives used to densify MB_2. Densification is assisted by the presence of a liquid phase, as it exploits the melting temperature of those metals. The liquid phase causes the removal of oxide impurities from MB_2 particle surfaces and favors particle rearrangement and mass transfer mechanisms, such as surface and/or volume diffusion [13]. Typical HP temperatures were 1800, 1950, or 1900°C when Mo (5–10 vol%), Nb (25 vol%), or Zr (10–20 vol%) were used, respectively, as sintering aids. For Ni (3 vol%), a temperature of 1850°C was sufficient to achieve full density [13]. The final microstructure sometimes displays (M,Me)B$_2$ solid solutions, Me = Mo, Nb. High-temperature solubility of MB$_2$ into the liquid phase controls mass transfer rate and, therefore, the overall densification [25, 26]. Disadvantages of this processing route can be excessive grain coarsening and formation of low-melting secondary phases at grain boundaries in the final microstructure [28]. As a result, rapid degradation of flexural strength above 800°C occurs (see Section 6.8), as well as limited resistance to oxidation/corrosion. To overcome such problems, ceramic additives are preferred over metallic ones.

6.3 MB$_2$ WITH NITRIDES

Ceramic nitrides, like Si$_3$N$_4$ or AlN, improve the densification of MB$_2$, like ZrB$_2$ [14, 15] or HfB$_2$ (this work). With Si$_3$N$_4$ additions, full densification is achieved at 1700 and 1800°C for ZrB$_2$ and HfB$_2$, respectively (see Table 6.1), owing to the formation of a liquid phase that not only favors particle rearrangement, but also enhances mass transfer kinetics, according to the following sequence: (i) partial dissolution of MB$_2$, (ii) volume diffusion of the involved species, and (iii) reprecipitation on preexisting undissolved MB$_2$ particles.

The relative density versus temperature curve in Figure 6.1 exhibits the typical behavior of liquid-phase sintering: after a short stage of solid-state sintering accompanied by no measurable shrinkage, a steep increase in the densification rate occurs, due

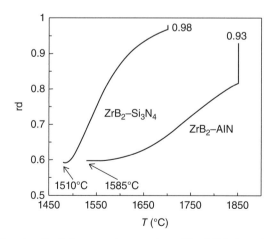

Figure 6.1. Relative density (rd) versus temperature (T) of ZrB$_2$-based compositions using 5 vol% Si$_3$N$_4$/ or AlN, as sintering aid; onset temperatures and final relative densities are also reported on each curve.

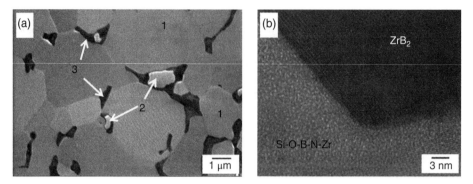

Figure 6.2. ZrB_2 ceramic sintered using 5 vol% Si_3N_4 as sintering aid: (a) SEM micrograph showing ZrB_2 (1), ZrO_2 (2), and BN (3), (b) high-resolution TEM micrograph of a grain-boundary phase.

to particle rearrangement in the presence of a liquid. The first effect of the sintering aid is to remove surface oxides from MB_2 powder particles according to:

$$Si_3N_4(s) + 2B_2O_3(l) \rightarrow 4BN(s) + 3SiO_2(l) \qquad (6.1)$$

Subsequently, the liquid formed at about 1500°C is very likely due to the reaction of Si_3N_4 remaining from Reaction 6.1 with SiO_2 and ZrB_2. During cooling, BN and ZrO_2 pockets precipitate from the liquid phase (Fig. 6.2a), but part of it is retained along the grain boundaries in the form of amorphous Si–O–B–N–Zr films (Fig. 6.2b).

AlN was also tested as a sintering aid for pure ZrB_2 [15]. In Figure 6.1, relative density versus temperature for AlN and Si_3N_4 additions is compared. AlN is much less effective than Si_3N_4, causing densification to occur at much higher temperatures (1850 vs. 1700°C) and with slower rates. In contrast to Si_3N_4, no traces of a residual liquid phase were found in the final microstructure leading to the conclusion that the main effect of AlN addition during HP was to reduce the oxygen content of the ZrB_2 particles, before significant shrinkage started taking place. In the final microstructure, Al_2O_3-based phases were found instead. Refractory nitrides such as HfN [29] and ZrN [30] have also been used but only in combination with SiC particles. In both cases, densification was accomplished at 1900°C, but without SiC and Si-based species, formation of liquid phases is very unlikely. As a matter of fact, among nitrides, Si_3N_4 was the most effective sintering aid for MB_2 promoting densification with the highest sintering rate and the lowest sintering temperature.

6.4 MB₂ WITH METAL DISILICIDES

Addition of metal disilicides ($MeSi_2$) to TiB_2 and ZrB_2 (Me = Ti, Cr, Zr, Mo, Ta, W) was first investigated by Pastor and Meyer in 1974 [31] with the purpose of improving the oxidation resistance of diborides. In Reference [31], the authors reported significant chemical interaction and solubility between ZrB_2 and several $MeSi_2$, with the exception of $TaSi_2$ and WSi_2. Studies conducted later in the 2000s have instead disclosed them as very effective additives for densifying a variety of MB_2 and through PS [32, 33], HP [17, 18], and SPS [18, 34–36]. $MeSi_2$ cannot be considered solely as sintering additives, like those illustrated in Sections 6.2 and 6.3, since their content can be varied up to 30 vol% to improve the mechanical properties and oxidation resistance, thus making the final material a composite. Table 6.2 summarizes the investigated $MeSi_2$ and related peak temperatures used during sintering experiments.

Among the disilicides, $MoSi_2$ was found to be the most versatile, allowing full densification by PS at temperatures around 1900°C [17, 18, 32–34]. Looking at Table 6.2, depending on $MoSi_2$ content from 3 to 20 vol%, temperatures for full densification range within 1850–1900°C for PS [33], 1750–1800°C for HP [17], and 1750°C for SPS [32]. Other $MeSi_2$ worth mentioning include $ZrSi_2$, $TaSi_2$, and WSi_2.

$ZrSi_2$, which has the lowest melting temperature (~1520°C) among those considered here, has been extensively used by Guo and coworkers for both PS and HP of ZrB_2 [37, 38]. $TaSi_2$ has been used as a secondary phase in combination with SiC particles, due to its ability to improve the oxidation resistance [40]. In comparison, few data are available on the densification of ZrB_2–$TaSi_2$ composites. Two studies [19, 41] indicated that densification temperatures for MB_2–$TaSi_2$ are close to that of $MoSi_2$, but PS is not possible, because $TaSi_2$ decomposes and volatilizes before significant densification starts taking place.

On the other hand, WSi_2 is the most refractory of the $MeSi_2$ compounds examined in this work (~2160°C melting point) and requires temperatures above 1900°C for good densification of MB_2. Densification curves of three ZrB_2-based systems containing 15 vol% $ZrSi_2$, $MoSi_2$, or WSi_2 are shown in Figure 6.3.

According to Pastor and Meyer [31], densification of ZrB_2 with $MeSi_2$ was successful because it was conducted at temperatures close to the $MeSi_2$ melting point. Figure 6.3 shows that the temperature range for densification is affected by the refractoriness of the added $MeSi_2$, which is lower for $ZrSi_2$ (1350–1600°C) and higher for WSi_2 (1750–1930°C). The densification curve for ZrB_2–$ZrSi_2$ shows a steep increase in the relative density as a function of temperature curve around 1450°C, that is, close to the melting point of $ZrSi_2$. For $MoSi_2$ and WSi_2, full densification is achieved at temperatures well below their melting points (2020 and 2160°C, respectively), but the slopes of the curves, and thus the densification rates, are lower.

A distinctive feature of the MB_2–$MeSi_2$ composites is the formation of solid solutions, examples of which are shown in Figure 6.4: $(Zr,Mo)B_2$, $(Zr,W)B_2$ or $(Zr,Ta)B_2$ shells grow epitaxially onto ZrB_2 cores. Multiple shells were observed in the case of HfB_2 sintered with $ZrSi_2$ (Fig. 6.4c). The core is HfB_2, the inner shell is a $(Hf,Zr)B_2$ solid solution with a Hf-to-Zr ratio around 70:30 at%, while the outer shell is a solid solution, but with Hf-to-Zr ratio around 95:5 at%.

As far as densification mechanisms are concerned, low dihedral angles of the disilicides with MB_2 suggest that $MeSi_2$ compounds have a high affinity for MB_2 ceramics. Moreover, transfer of cations from the diboride matrices into residual intergranular Si–O-based films suggests the existence of some liquid phases in which MB_2 compounds are partially soluble [42]. Although some analogies exist, densification mechanisms

TABLE 6.2. MB$_2$–MeSi$_2$ composites reached specific final relative density (rd) using different amounts of MeSi$_2$, sintering techniques (Tech), and peak temperatures (T_{MAX})

| | MB$_2$ | | | | | | | | | |
| | M = Zr | | | | | M = Hf | | | | |
MeSi$_2$	Vol%	Tech	T_{MAX} (°C)	rd (%)	Ref.	Vol%	Tech	T_{MAX} (°C)	rd (%)	Ref.
ZrSi$_2$	10–20	PS	1650	96–99	[37]					
	10–20	HP	1500	97–99	[38]					
MoSi$_2$	3	SPS	1750	98	[32]	3	SPS	1900	98	[34]
	5	PS	1900	96	[33]	–	–	–	–	–
	15	HP	1750	98	[17]	15	HP	1900	99	[18]
	20	PS	1850	99	[33]	20	PS	1950	98	[33]
TaSi$_2$	–	–	–	–	–	3	SPS	2000	96	[33]
	15	HP	1850	99	[19]	15	HP	1900	99	[19]
HfSi$_2$						5	HP	1600	99	[16]
TiSi$_2$	15	HP	1650	95–99	[39]					
WSi$_2$	20	HP	1930	99	This work					

HP, hot pressing; PS, pressureless sintering; SPS, spark plasma sintering.

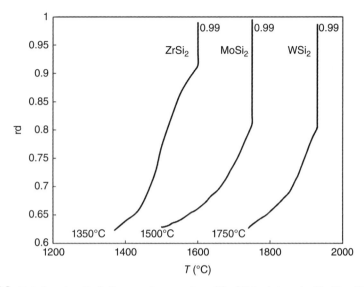

Figure 6.3. Relative density (rd) versus temperature (T) of ZrB₂ sintered with 15 vol% of ZrSi₂, MoSi₂, or WSi₂; onset temperatures and final relative densities are also reported on each curve.

can be significantly different within this group of metal disilicides. For $ZrSi_2$, a liquid phase is thought to form due to its melting and/or decomposition. For more refractory $MeSi_2$ compounds such as $MoSi_2$ or WSi_2, densification was presumed to be assisted by a transient liquid phase, due to the reaction between $MeSi_2$ and surface oxide impurities on the MB_2 particles. Effects of $MoSi_2$ or $TaSi_2$ on densification were thoroughly studied [41, 43], and several reaction paths were indicated, as suggested by the variety of crystalline and amorphous phases identified in the dense microstructures. For $MoSi_2$ additions, the presence of crystalline SiC, MoB, Mo_5SiB_2, and pockets of amorphous SiO_2 suggests the following reaction:

$$2MoSi_2(s) + B_2O_3(l) \rightarrow 2MoB(s) + 2.5Si(l) + 1.5SiO_2(l) \qquad (6.2)$$

Reaction 6.2 implies the removal of B_2O_3 from boride particle surface and formation of Si- or Si–O-based liquids. According to the Mo–Si–B phase diagram [44], liquids could form at temperatures as low as 1350°C, which is similar to the temperatures for the onset of densification reported for HP and SPS of ZrB_2. Formation of liquid promotes mass transfer mechanisms, by partial dissolution of the MB_2 matrix. The observed formation of epitaxial core–shell structures suggests a solution–reprecipitation mechanism, even if diffusion of Mo into the MB_2 lattice could also occur. During cooling, the transient liquid phase solidifies, resulting in the formation of discrete crystalline phases, such as the observed Mo_5SiB_2, and in the partial reduction of Si-based species to SiC. In the case of $TaSi_2$ additions, the starting reaction for densification could be as follows:

$$2TaSi_2(s) + 2B_2O_3(l) \rightarrow 2TaB_2(s) + Si(l) + 3SiO_2(l) \qquad (6.3)$$

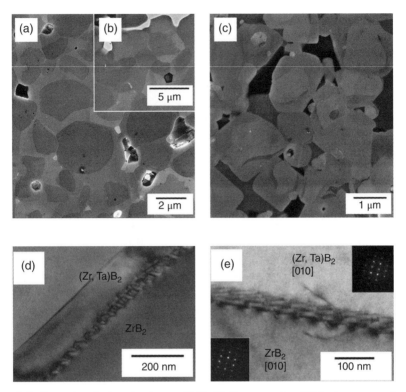

Figure 6.4. Examples of microstructures of MB_2–$MeSi_2$ composites: SEM micrographs showing (a) ZrB_2–$MoSi_2$, (b) ZrB_2–WSi_2, and (c) HfB_2–$ZrSi_2$ (d) high-resolution TEM image of the interface between core and rim in ZrB_2–$TaSi_2$ [41] and (e) the same interface showing low angle grain boundaries [41].

In contrast to the previous case, no pure Ta borides were observed in the sintered microstructures implying that the following reaction preferably takes place:

$$\left(1-x\right)MB_2\left(s\right)+xTaB_2\left(s\right)\rightarrow\left(M_{1-x}Ta_x\right)B_2\left(s\right) \qquad (6.4)$$

Indeed, the formation of solid solutions in this system was more evident than in the previous case (Fig. 6.4d and e), owing to a higher amount of host cations in the MeB_2 lattice (Ta ~ 20 at%, Mo ~ 5 at%) and to a higher difference of atomic radii between the host and the guest resulting in a more pronounced core–rim interface. The presence of intergranular residual films was dependent on the specific MB_2–MSi_2 combination and on the sintering process. Pressureless sintered ZrB_2- or HfB_2–$MoSi_2$ were almost completely free of residual phases [43], and hot-pressed TaB_2, ZrB_2, and HfB_2 with $MoSi_2$ or $TaSi_2$ showed some intergranular film, with TaB_2 being wetted the most and HfB_2 being the most refractory [41, 45].

6.5 MB$_2$ WITH CARBON OR CARBIDES

Carbon or carbides are effective sintering aids for MB$_2$ compounds at temperatures around 1900°C, even in the absence of external applied pressure [20–22, 46–50]. A series of publications describe the use of WC [49], B$_4$C + WC [50], C + WC [48, 51], and C + B$_4$C + WC [20] as sintering aids for PS of ZrB$_2$ or HfB$_2$ [21, 47, 52]. In all these studies, WC was usually introduced as a contaminant through powder milling with WC media [49], while other additives such as C or B$_4$C were deliberately added as sintering aids [20, 48, 50]. For instance, a relative density of 98% was obtained at 2150°C for attrition-milled ZrB$_2$ by Chamberlain *et al.* [49]. Zhu *et al.* [20] showed that 99% dense ZrB$_2$ can be obtained at 1900°C by adding 2 wt% B$_4$C and 1 wt% C under the form of phenolic resin, after attrition milling of ZrB$_2$ powders for 24 h with WC media. These contributions also demonstrated that improved densification was unlikely due to the reduction of particle size by attrition milling. In contrast, WC contamination (up to 8 wt% [50]) introduced by attrition milling was disclosed to play an effective role in the densification, even in the absence of any other sintering additive [49, 50]. The most important mechanism in all of these processes is the removal of surface oxides that inhibit the densification by promoting coarsening through evaporation–condensation at intermediate temperature. Oxygen contamination on the starting MB$_2$ powders takes the form of MO$_2$ (M = Zr, Hf) and B$_2$O$_3$ on the particle surface, neither of which is soluble in the boride lattice. However, B$_2$O$_3$ can be removed by evaporation beyond 1500°C under active vacuum conditions, even if some amount could remain trapped in the powder bed during the early stages of HP. On the contrary, MO$_2$ does not evaporate at the temperatures needed for the sintering of MB$_2$ matrices. These studies confirmed that C, B$_4$C, and WC reduce the oxygen-carrying chemical species present on the surfaces of the MB$_2$ powder particles (and here outlined as a mixture of MO$_2$ and B$_2$O$_3$) through the following reactions:

$$MO_2(s) + 3WC(s) \rightarrow MC(s) + 3W(s) + 2CO(g) \qquad (6.5)$$

$$7MO_2(s) + 5B_4C(s) \rightarrow 7MB_2(s) + 5CO(g) + 3B_2O_3(l) \qquad (6.6)$$

$$2MO_2(s) + B_4C(s) + 3C(s) \rightarrow 2MB_2(s) + 4CO(g) \qquad (6.7)$$

$$MO_2(s) + B_2O_3(l) + 5C(s) \rightarrow MB_2(s) + 5CO(g) \qquad (6.8)$$

$$MO_2(s) + C(s) \rightarrow MC(s) + CO(g) \qquad (6.9)$$

Efficient removal of oxides, as described in Reactions 6.5–6.9, depends on vacuum conditions. Mild vacuum with a reduced CO(g) partial pressure (P$_{CO}$) reduces the temperature at which the reactions become favorable compared to standard state conditions [20, 50]. For instance, in the case of ZrB$_2$, Reactions 6.6, 6.7, and 6.8 become favorable at 928, 1000, and 1044°C, when the P$_{CO}$ is 25 Pa [20]. After the removal of residual oxides, typical transport mechanisms for solid-state densification of MB$_2$, like lattice and grain-boundary diffusion between boride interfaces, occur. Finally, products in Reactions 6.5 and 6.9 such as W and MC may enter into solid solutions with

MB_2 matrix, and therefore they are not usually detected in the microstructure [49]. Often B_4C and WC were added in combination with SiC, and as such, they will be treated in Section 6.6.

6.6 MB_2 WITH SiC

MB_2–SiC composites were specifically designed to improve performance compared to single-phase MB_2 in terms of strength and oxidation resistance. Many studies were concerned to investigate the effect of varying SiC particulate content [30, 53–69]. Other studies were more focused on the introduction of third phases as sintering agents and their effects on densification and final properties (see Section 6.7). Recent works explored the possibility to improve the fracture toughness through the addition of SiC whiskers [70–73] or short fibers [74–76] (Table 6.3).

In this section, SiC is shown to not only to be a reinforcing phase, but also an effective sintering aid. Indeed, ZrB_2 containing ultrafine SiC powders (5, 10, 15, and 20 vol%) and no extra sintering aids, wet mixed with SiC milling media, achieved full density for all compositions after HP at 1900°C with 30–40 MPa of applied pressure [23, 53]. Reactions 6.10–6.12 were extracted from the chemical equilibria shown in Figure 6.5b.

$$3ZrO_2\left(s\right)+3B_2O_3\left(l\right)+5SiC\left(s\right)\rightarrow 3ZrB_2\left(s\right)+5SiO_2\left(s,l\right)+5CO\left(g\right) \quad (6.10)$$

$$SiO_2\left(l\right)+B_2O_3\left(l\right)\rightarrow Si-O-B\left(l\right) \quad (6.11)$$

$$Si-O-B\left(l\right)\rightarrow SiO\left(g\right)+\left(B,O\right)\left(g\right) \quad (6.12)$$

TABLE 6.3. ZrB_2–SiC_W and ZrB_2–SiC_F composites (W: whisker, F: fiber) sintered by hot pressing (HP) using different sintering aids (s.a): peak temperature (T_{MAX}), and final relative density (rd)

s.a type	s.a content	SiC content and type	T_{MAX}	rd	Sintering	Ref.
	(vol%)	(vol%)	(°C)	(%)		
YAG	3	10–30, W	1800	96–98	HP	[70]
PCS	10 wt	nd, W	1750	>99	HP	[71]
YAG	3	20, W	1950	>99	HP	[72]
ZrO_2 fiber	15–25	15–25, W	1850	>99.3	HP	[73]
Si_3N_4	5	15, F	1700	98.7	HP	[74]
Si_3N_4	5	10–30, F	1700	92–100	HP	[75]
$ZrSi_2$	10	15, F	1650	99.9	HP	[74]
$ZrSi_2$	10	20, F	1650	100	HP	[75]
$MoSi_2$	15	20, F	1900	97.6	HP	[75]
$MoSi_2$	15	15, F	1750	99	HP	[74]

nd, quantity not declared; PCS, polycarbosilane.

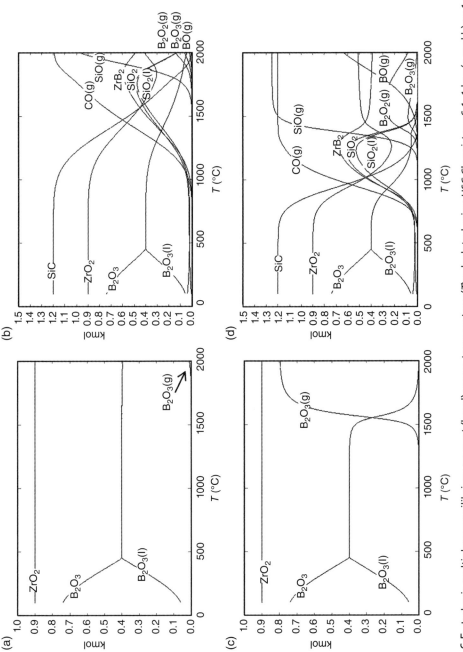

Figure 6.5. Isobaric multiphase equilibrium amount (kmol) versus temperature (T) calculated using HSC Chemistry v.6.1, 1 bar (a and b) or 1 mbar (c and d) of total equilibrium pressure; starting compositions (in kmol): 0.9 ZrO_2 + 0.8 B_2O_3 (a and c), 0.9 ZrO_2 + 0.8 B_2O_3 + 1.2 SiC + 0.05 SiO_2 (b and d).

These are considered the most important reactions driving the removal of oxide impurities from the ZrB_2 powder particle surfaces. Boron oxide exists as a liquid or vapor phase at temperature above 500°C and, according to Figure 6.5a, can be partly removed by evaporation beyond 1500°C under active vacuum conditions. Reaction 6.10 is favorable under standard conditions above approximately 900°C (see Fig. 6.5b) but switches to a thermodynamically more favorable temperature range (i.e., above ~700°C) keeping the furnace chamber vacuum level around 1 mbar: this condition facilitates the formation of gaseous CO (i.e., the removal of oxides) that can be taken away from the system via active mechanical pumping. Similar to the case of carbide additions, the presence of SiC creates favorable conditions for changing the surface chemistry of MB_2 particles, which densify significantly before grain coarsening predominates. Also, according to Reaction 6.11, limited amounts of liquid SiO_2 prevent, to some extent, the evaporation of boron oxide during the early stages of the heating cycle, react with boron oxides, and lead to the formation of a borosilicate-based liquid phase between grains, resulting in an improved densification (Fig. 6.6).

Unfortunately, the database of the thermochemical solver used (HSC for Chemistry v. 6.1) does not include a Si–O–B phase (or similar). The solid products remaining upon cooling, zirconia and silica, were mostly found trapped among SiC clusters in the sintered microstructure (Fig. 6.7), which seems to suggest that boron oxide is one of the primary inhibitors to the densification process of ZrB_2 [77].

The densification behavior of various ZrB_2–SiC compositions in Figure 6.6 implies that increasing the amount of SiC leads to shrinkage at progressively lower temperatures, that is, at 1770°C in a SiC-free ZrB_2 powder (Z0), 1710°C for 5 vol% SiC (ZS05), down to 1560°C for 20 vol% SiC (ZS20). In addition, it can be observed that, approaching 1900°C, the densification rate is higher for increasing SiC content (up to 20 vol%).

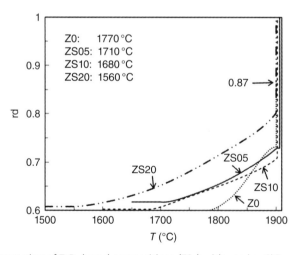

Figure 6.6. Hot pressing of ZrB_2-based compositions (ZSx) with varying SiC particulate content (x, vol%), compared to an only milled ZrB_2 powder (Z0): relative density (rd) versus temperature (T). Onset temperatures are indicated for each system.

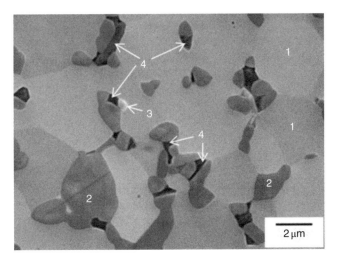

Figure 6.7. Polished section of a hot-pressed ZrB$_2$–20% SiC: ZrB$_2$- (1), SiC- (2), ZrO$_2$- (3) and SiO$_2$-based glassy pockets (4).

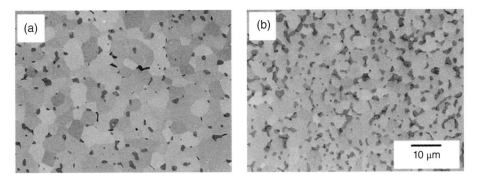

Figure 6.8. Polished sections from hot-pressed ZrB$_2$-based ceramics with 5 vol% (a) and 20 vol% SiC particulates (b).

Figure 6.8 shows the resulting microstructures of two ZrB$_2$–SiC composites, ZS05 and ZS20, where it can be clearly seen that, although both were processed at the same peak temperature of 1900°C, coarsening of the ZrB$_2$ grains was slower as a direct consequence of a larger concentration of SiC particulates at grain boundaries. SiC particles hinder grain-boundary migration during the final stage of sintering and restrain coarsening of the ZrB$_2$ matrix, increasing the driving force available for further densification. The proof is that, during the isothermal stage at 1900°C and 40 MPa of applied external pressure, a longer hold, 25 min, was required for the ZS05 composition to achieve full density compared to 15 min for ZS20. These results emphasize the double role of SiC as densification activator and grain growth controller for ZrB$_2$. On the other hand, a milled ZrB$_2$ powder sample, that is, Z0, containing no SiC particulates, exhibited a sharp drop in densification rate above 1850°C and reached a final relative density of only 87%

because most of the available driving force was consumed by coarsening of the matrix during ramping up to 1850°C.

6.7 MB$_2$–SiC COMPOSITES WITH THIRD PHASES

Table 6.4 lists a representative variety of sintering techniques and sintering aids used to produce ZrB$_2$–SiC composites. Additives range from oxides, like ZrO$_2$ [54] and Re$_2$O$_3$ (Re = La, Nd, Y, Yb [55, 56], to nitrides, like Si$_3$N$_4$ [57], AlN [58], ZrN [30]. In addition, metal disilicides, like MoSi$_2$ [59], TaSi$_2$ [60], and ZrSi$_2$ [61], and carbides, like B$_4$C [62], WC [63, 64], and VC [65], have also been studied. The primary aim was to limit peak processing temperatures and holding times, to refine microstructures, and, therefore, to improve mechanical properties. Scrolling through the series of sintering aids presented in Table 6.4, MoSi$_2$ appears one of the most effective during densification. In fact, a peak temperature of 1820°C was an appropriate condition to reach full density.

Compared to a MoSi$_2$-free ZrB$_2$–15 vol% SiC (Fig. 6.9), densification is further helped by pseudo-binary reactions involving ZrB$_2$ and transient Mo borides, similar to the MB$_2$–MeSi$_2$ system analyzed in Section 6.4. The pseudo-binary reactions gave rise to a liquid phase that assisted mass transfer through a dissolution–precipitation mechanism: the final microstructure revealed a core–rim structure of the MB$_2$ typical of the aforesaid mechanism (Fig. 6.9c).

The continuing search of third functional phases to improve both the densification process and properties of MB$_2$–SiC composites found the use of B$_4$C [62, 78] or WC

TABLE 6.4. ZrB$_2$–xSiC$_p$ composites (P: particulates) densified by HP, SPS, or PS using different sintering aids (s.a): peak temperature (T_{MAX}), and final relative density (rd)

s.a type	s.a content	x, SiC content	T_{MAX}	rd	Sintering	Ref.
	(vol%)	(vol%)	(°C)	(%)		
ZrO$_2$	5–20	10	1850	>98	HP	[54]
Re$_2$O$_3$	3	20	1900	>99	HP	[55]
Re = Yb, Nd, Y						
La$_2$O$_3$	5	20	1950	>99	SPS	[56]
Si$_3$N$_4$	5	20	1870	>98	HP	[57]
AlN	10	20	1850	>99	HP	[58]
ZrN	4.5	15	1900	>99	HP	[30]
MoSi$_2$	2	15	1820	>99	HP	[59]
TaSi$_2$	10–20	10–30	1600–1800	>99	SPS	[60]
ZrSi$_2$	4–7	17–26	2150	>98	HP	[61]
B$_4$C	2	20	2000	>99	HP	[78]
B$_4$C	2–4	20	2200	>99	PS	[62]
WC	5–10	20	2000–2200	>99	PS	[63]
WC	5	20	1900	>99	HP	[64]
VC	5	20	1900	>99	HP	[65]
LaB$_6$	5	15	1900	>99	HP	This work

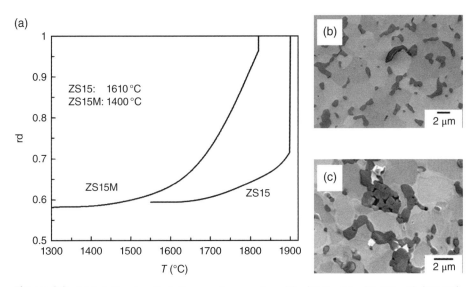

Figure 6.9. (a) Relative density (rd) versus temperature (T) of ZrB$_2$ + 15 vol% SiC with (ZS15M), or without MoSi$_2$ (ZS15): onset temperatures are indicated. Polished sections by SEM from hot-pressed ZS15 (b) and ZS15M (c) are shown.

[63, 64] very effective. Patel *et al.* [78] very recently explored possible densification mechanisms during HP of ZrB$_2$–20 vol% SiC–2 vol% B$_4$C composite. In Reference [78], the authors concluded that particle fragmentation and rearrangement were the only mechanisms operating during HP at 1700°C. However, HP at 1850°C was dominated by a plastic flow mechanism during densification, while a grain-boundary diffusion-controlled mechanism was dominant during HP at 2000°C. The benefits of using B$_4$C can be further understood by looking at the output of the equilibrium compositions in Figure 6.10, which incorporates some driving chemical reactions already considered in Section 6.6. The great improvement from the complementary addition of SiC and B$_4$C is the potential not only to fully eliminate the oxide impurities at grain boundaries, but also to obtain ZrB$_2$ as the only reaction product in the condensed state. This result can, in general, be extended to other MB$_2$ compounds like HfB$_2$. Moreover, as suggested by the thermochemical calculations, the reactions involved provide some boron, which is a powerful sintering aid for MB$_2$.

In a ZrB$_2$–15 SiC–5 B$_4$C powder mixture (vol%) recently processed by the authors of the present chapter, full densification was achieved by HP at 1900°C. No new secondary phases were identified in the hot pressed material, but a minor content of unreacted boron carbide remained. This carbide was added to react with and remove oxides during HP, as illustrated in Figure 6.10. The internal pressure of the furnace chamber continuously pumped well below ambient pressure helped chemical reactions between oxide impurities and reactive agents (i.e., SiC and B$_4$C) to occur at 450–500°C lower compared to conditions active during PS (see Fig. 6.10b). Using the latter technique, peak temperatures have to be increased, with prolonged isothermal holds, to achieve equivalent relative densities [62].

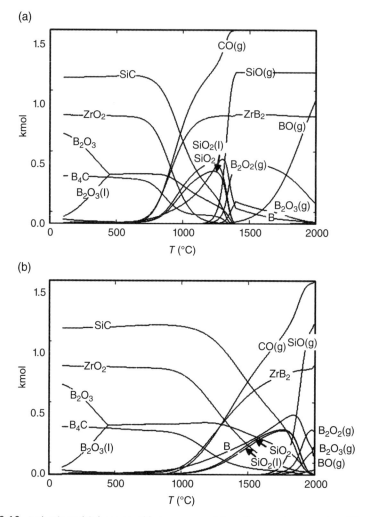

Figure 6.10. Isobaric multiphase equilibrium amount (kmol) versus temperature (T) calculated using HSC Chemistry v.6.1, 1 mbar (a) or 1 bar (b) of total equilibrium pressure, and 0.9 $ZrO_2 + 0.8\ B_2O_3 + 1.2\ SiC + 0.4\ B_4C + 0.05\ SiO_2$ of starting composition (in kmol).

On the other hand, WC as third phase is attracting great attention thanks to its ability to improve the performance of MB_2–SiC-based composites in terms of high-temperature flexure strength [64, 66] and oxidation resistance [79]. In this respect, Zhang *et al.* [79] recently highlighted the beneficial role of WC in relation to its ability to sinter porous external MO_2 scales when MB_2–SiC materials, exposed to severe environments, loose protection of the external silica-based glasses. The composition (in vol%) ZrB_2–15 SiC–5 WC, experimented by the present authors and here labeled ZSWC, was hot pressed at 1900°C and reached a final relative density of about 95% (Fig. 6.11). Despite the presence of residual porosity, the potential of this system was

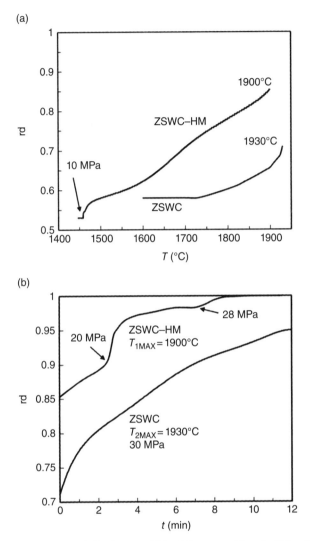

Figure 6.11. Relative density (rd) of ZrB$_2$–15SiC–xWC composition (vol%) using high-energy milled (x=2 for ZSWC–HM up to $T_{1\,MAX}$=1900°C) or only mixed ZrB$_2$ (x=5 for ZSWC up to $T_{2\,MAX}$=1930°C): (a) rd versus temperature (T), (b) rd versus time (t) at $T_{1\,MAX}$ and $T_{2\,MAX}$. For ZSWC, uniaxial pressure was applied at the constant value of 30 MPa from RT, while in three steps for ZSWC–HM.

disclosed by the measurement of residual flexure strength (measured at 25°C) after 24 h of exposure in static air at 1400°C. This composite lost only about 28% of its room temperature strength, decreasing from 540 to 390 MPa. Improvement in the densification efficiency is thus strongly recommended to exploit its potential.

Recent advances in powder processing offer practical solutions through high-energy grinding methods like planetary or shaker mill, under dry or wet conditions. Ortiz *et al.* [8]

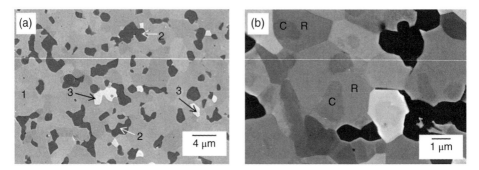

Figure 6.12. Polished regions by SEM of hot-pressed ZrB_2–15SiC–2WC composite (see ZSWC–HM in Fig. 6.11): (a) overall microstructure with ZrB_2 (1), SiC (2), (W,Zr)B (3) and (b) magnification of ZrB_2 cores (C) and $(Zr,W)B_2$ rim (R).

highlighted the outstanding opportunity that microstructure refinement down to ultrafine scale offers for speeding up sintering kinetics of refractory compounds like ZrB_2. To verify the benefits of this approach, ZrB_2 powder was first comminuted in wet conditions using a zirconia planetary mill and WC beads (this work). Then, in the composition here labeled ZSWC–HM, 15 vol% SiC and 2 vol% WC were added to the comminuted ZrB_2, and wet ball milled for several hours. Densification began as soon as the uniaxial pressure was applied at 1450°C and the powder compact continued to densify as it was heated to the final HP temperature (Fig. 6.11). Compared to the conventionally wet ball-milled powder mixture (ZSWC), the composition containing a ZrB_2 powder treated with a high-energy planetary mill (ZSWC–HM) achieved full density by HP at 1900°C (Fig. 6.11).

Analysis by X-ray diffraction and scanning electron microscopy (Fig. 6.12) showed the formation of $(Zr,W)B_2$ solid solution rims around ZrB_2 cores, and of (W,Zr)B and (Zr,W)C as secondary phases. No W silicides were observed. Compared to the ZSWC composition, activated sintering in the ZSWC–HM composition can be explained in terms of a synergic effect of a shorter diffusion mean free path of Zr and B along a greater density of grain boundaries and vacancies available for faster diffusion.

6.8 EFFECTS OF SINTERING AIDS ON HIGH-TEMPERATURE STABILITY

Proper additions of reactive sintering agents may improve the performance of UHTCs through grain refinement. However, these same additives may be deleterious for high-temperature properties due to residual secondary phases, especially for liquid-phase sintering additives. Load–displacement (L–d) curves recorded during flexure strength tests at high temperature may provide insight into this issue. Pure ZrB_2 has L–d linear curves up to at least 1200°C [80] while addition of 5 vol% of Si_3N_4 leads to a strong deviation from linearity [57]. Figure 6.13a shows that the L–d curves of various MB_2–$MeSi_2$ systems maintain linearity up to 1500°C, while the simultaneous addition of

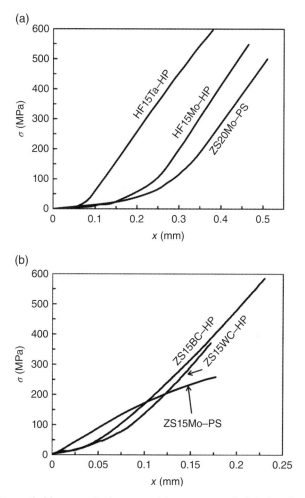

Figure 6.13. Strength (σ) versus displacement (x) curves recorded during 4-pt flexure test at 1500°C in air using different ceramics sintered by hot pressing (HP) or pressureless sintering (PS): (a) HfB_2–15 vol% $TaSi_2$ (HF15Ta–HP), HfB_2–15 vol% $MoSi_2$ (HF15Mo–HP), ZrB_2–20 vol% $MoSi_2$ (Z20Mo–PS); (b) ZrB_2–15 vol% SiC (ZS15) with B_4C, WC, or $MoSi_2$ as sinter additive.

2 vol% $MoSi_2$ and 15 vol% SiC leads to some departure from linearity (Fig. 6.13b). In this context, some measurable deviation from linearity is considered by the authors of the present chapter to be an indicator of loss in refractoriness and structural stability.

From these results, it can be inferred that, with no optimization of sintering conditions leading to the effective removal of residual oxides, addition of SiC may leave residual silica-based amorphous phases upon cooling. This issue can partly be fixed with WC or B_4C additions to ZrB_2-based systems, as shown in Figure 6.13b: even in the presence of a significant amount of SiC, linearity is retained up to 1500°C.

Flexure strength values at high temperature seem to be closely related to the type of sintering aid used, as shown in Figures 6.13 and 6.14. Almost all results presented have

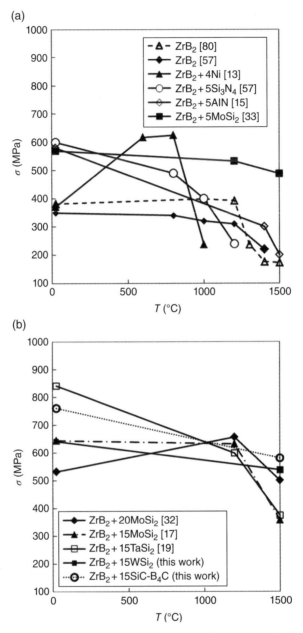

Figure 6.14. ZrB$_2$-based composites hot-pressed with different MeSi$_2$ (Me=Mo, Ta, W, Zr): average flexural strength data (σ) measured at different temperatures (T). (a) MeSi$_2$<5 vol% [13, 15, 33, 57, 80], (b) MeSi$_2$>10 vol% [17, 19, 32]. For the sake of comparison, the composite containing SiC and B$_4$C is also reported, as one of the most refractory compositions.

been obtained at ISTEC in 4-pt bending configuration in air up to 1500°C. Figure 6.13b compares different ZrB_2–SiC systems containing $MoSi_2$, WC, and B_4C as sintering aids. On one hand, the ZrB_2–SiC system with $MoSi_2$ as sintering additive (255 MPa of average flexure strength at 1500°C) retains 28% of its room temperature flexural strength. On the other hand, the ZrB_2–SiC systems with B_4C and WC as sintering aids (average flexure strengths of 590 and 360 MPa at 1500°C) maintain their RT flexure strength more efficiently: 80% for B_4C additions and 70% for WC additions. Figure 6.14a shows the effect of sintering additive in SiC-free materials, with additive content not exceeding 5 vol%. The most refractory composite is that obtained with $MoSi_2$ by PS [33], while metals and nitrides lead to reduced performance at high temperature. PS for MB_2–$MoSi_2$ is indeed the best densification method for the elimination of SiO_2-based pockets and getting non-wetted grain boundaries. The slower sintering rate compared to HP and highly reducing atmosphere favored the volatilization of silica to Si–O and C–O species, before complete pore closure occurred. The best performance of this system could also rely on the formation of refractory phases at the triple junctions [45]. Systems comprising HfB_2-3, 6, 9 vol% $MoSi_2$ were tested at 1500°C both in air and in argon, and flexure strength values were systematically lower for tests in air than in argon, implying that oxidation negatively affected strength [35]. Figure 6.14b shows the effect of different silicide additions, in the case of amounts higher than 10 vol%. Again, the most refractory composites are those HP with WSi_2 (560 MPa) or PS with $MoSi_2$ [32]. These strength values, along with those of the earlier-mentioned ZrB_2–SiC–B_4C system (see Fig. 6.14b), are among the highest found in the recent available literature for UHTCs.

6.9 TRANSITION METAL CARBIDES

Transition MC with the formula MC_x have a wide composition range for the stability of the fcc phase, enabling a great variability of stoichiometric compositions resulting in different intrinsic properties [81]. The sintering problems affecting the family of MB_2 compounds also apply to the carbides, owing to analogous highly covalent bonds. Monolithic carbides of Zr, Hf, and Ta have been densified to almost full density with pressure-assisted techniques, HP or SPS, at temperature above 2100°C and with a pressure of 70–100 MPa [82–84]. In these conditions, the resulting microstructures consist of coarse grains on the order of 10–20 μm and intragranular trapped porosity, leading to poor mechanical properties. To refine the microstructures, different strategies can be pursued: the use of ultrafine powders [85], or the addition of sintering additives, similar to the case of borides (see Table 6.5).

Table 6.5 summarizes the sintering additives and conditions adopted to densify the carbides. Three main categories can be identified: metals [86–88], transition metal silicides ($MoSi_2$, $TaSi_2$, and $ZrSi_2$) [83–85, 89–93], and carbon and carbides [94–98]. The addition of refractory metals, such as Mo, Nb, or Zr, eventually resulting in a cermet, promotes densification through the formation of a melt. The excess metallic phase at the grain boundaries results in poor corrosion resistance and decreased strength at high temperature [87–89]. In better cases, once the solubility limit of the metal into the carbide has been achieved, formation of solid solutions or other refractory carbides may occur [88].

TABLE 6.5. Literature overview of sintering techniques (Tech), parameters, and additives adopted for sintering of MC-composites

Sintering additives	MC														
	M=Zr					M=Hf					M=Ta				
	Vol%	Tech	T_{MAX}	rd	Ref.	Vol%	Tech	T_{MAX}	rd	Ref.	Vol%	Tech	T_{MAX}	rd	Ref.
Mo, Nb, Zr, V, Ta	7–17	HP	1900	98	[86–88]										
$MoSi_2$	5–20	PS	1950	83–97	[89]	5–20	PS	1950	97–99	[89]	5–20	PS	1950	93–98	[89]
	15	HP	1900	97	[90]	5–15	HP	1900	99	[85, 91]	5–15	HP	1850	96	[85, 91]
	1–9	SPS	1750–1950	90–99.9	[83]	1–9	SPS	1750–1900	88–99	[84]					
$TaSi_2$	15	HP	1700	99	[92]	15	HP	1760	99	[93]	10	PS	1850–1950	80–84	This work
											15	HP	1750	97	[93]
$ZrSi_2$						10	HP	1750	90	This work	10	HP	1700	96	This work
B_4C											9–45	PS	1800	72–99	[94]
											5–10	HP	2100–2300	98	[95]
											5	SPS	1850	99	[96]
C	5	HP	1900	97	[97]										
VC											2	PS	2300	96	[98]

Based on the experience on metal diborides, a wide-ranging scenario has been conceived for carbides sintered with silicides [89–93]. In the case of metal silicides, the major effect is the formation of a transient liquid phase upon partial dissociation of the silicide and its reaction with oxides covering the starting powders. Both $TaSi_2$ and $ZrSi_2$ allow densification to occur at lower temperature than $MoSi_2$ (1700–1750°C vs. 1900°C, respectively) [90–93], owing to their faster decomposition and formation of liquid Si. An example of densification behavior is shown in Figure 6.15, comparing the behavior of pure HfC with HfC–15 vol% $TaSi_2$. The reducing environment of the hot press, plus an additional carbon source coming from the carbide, strongly favors the silicide decomposition, which is translated in quick formation of liquid phase and enhanced matter transfer.

Additional proof of the dissociation of the silicides and the formation of liquid phase is the presence of Si–O–C-based compounds at the silicide–carbide interface, Figure 6.16a. Analogous to MB_2 sintering, oxide impurities strongly limit densification, probably owing to an increased viscosity of the melt.

The formation of core–shell microstructure has also been observed in transition MC sintered with silicides, as the examples in Figure 6.16b and c show. The mechanism of development of such morphologies has been discussed earlier as a result of the dissociation of the additive and formation of a Si-based melt, dissolution of the matrix, and epitaxial reprecipitation on the original undissolved grains. In general, the clear dislocation boundary at the core–shell interface in $TaSi_2$-doped composites was not observed in $MoSi_2$-doped ones [99, 100]. This can be explained in terms of a higher content of Ta in the solid solution, that is, approximately 20 at% Ta versus 5 at% Mo.

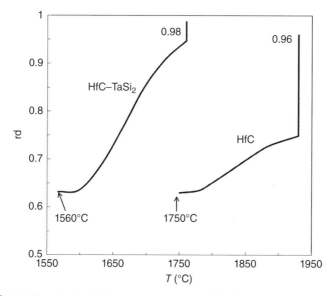

Figure 6.15. Relative density (rd) versus temperature (T) of pure HfC and HfC–15 vol% $TaSi_2$; onset temperatures and final relative densities are also reported on each curve.

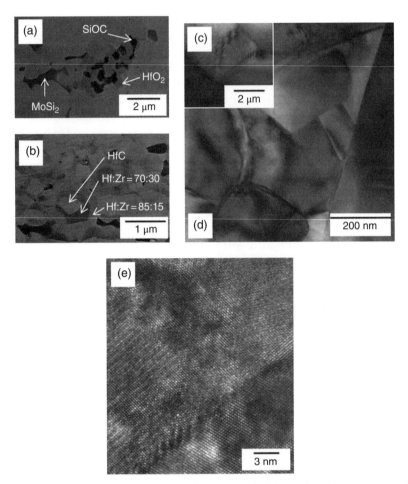

Figure 6.16. Examples of microstructures of MC sintered with MeSi$_2$: SEM micrographs showing (a) HfC–MoSi$_2$ showing the cleaning effect of the silicide trapping HfO$_2$ and reduced to SiOC, (b) HfC–ZrSi$_2$ showing multiple core–shell grains, (c) TEM image showing dislocation between core and shell in TaC–TaSi$_2$, (d) TEM image of a complex triple point in HfC–TaSi$_2$, (e) HR–TEM showing clean interfaces in ZrC–TaSi$_2$.

Actually, both Ta and Mo have smaller atomic radii than Zr and Hf. In fact, Mo has the smallest radius (Zr: 0.160 nm, Hf: 0.156 nm, Ta: 0.143 nm, Mo: 0.136 nm), so the substitution of Mo should be more favorable than Ta. Therefore, it appears that the greater supply of Ta, which is related to the lower stability of TaSi$_2$ than MoSi$_2$, may be an important factor. The higher content of Ta in the solid solution led to more pronounced differences in lattice parameters and coefficients of thermal expansion between core and shell, resulting in 45° grain boundaries in carbides (Fig. 6.16c) and low angle grain boundaries in borides (Fig. 6.4d and e). A direct consequence of the presence of dislocations and subgrains is the enhancement of impurity or cation diffusion through the dislocation core, which might continue even in the last stage of sintering. Diffusion through these discontinuous paths is six orders of magnitude faster than in the bulk [100]. This additional phenomenon might enhance the mass transfer as well and improve the densification of TaSi$_2$-containing materials over MoSi$_2$-containing materials [90]. The liquid phase at the sintering temperature presumably contains Mo/Ta/Zr–Si–C–O, and non-wetted interfaces were found for any silicide added to ZrC, HfC, or TaC [92, 93, 99]. It can be hypothesized that the liquid formed in the carbides at around 1400°C had a high viscosity and tended to precipitate in discrete pockets leaving clean grain boundaries (Fig. 6.16d and e) as opposed to the wetted interfaces observed in borides (ZrB$_2$/HfB$_2$ + TaSi$_2$, TaB$_2$ + MoSi$_2$), probably owing to a lower viscosity of the glass due to the presence of boron.

As for the effect of carbon on the densification of MC, studies demonstrated that introduction of reducing agents (carbon black, B$_4$C, or other transition MC) can remove oxygen impurities present in the starting ceramic powders [94, 98]. However, coarse grains, on the order of 20 μm, were obtained after the introduction of a small percentage of B$_4$C and/or carbon owing to an enhanced mass transfer, probably boosted by the change in M:C ratio [95]. It is reported that the coefficient of diffusion inside the MC$_x$ notably varies with carbon stoichiometry, in particular, grain growth is enhanced in the substoichiometric carbide rather than in the stoichiometric carbide [81, 95]. Reaction products originating from TaC and B$_4$C were identified as TaB$_2$ and graphite, which effectively enhanced densification and minimized grain growth. These findings originated a series of mixed carbide composites containing SiC or combination of carbides with the boride of the relevant transition metal, both classes possessing refined microstructure and improved oxidation resistance [94, 95, 98, 101, 102].

6.10 CONCLUSIONS

Although UHTCs are considered to be poorly sinterable compounds, multiple processing options are available to achieve fully dense ceramics with controlled microstructures. All the possible MB$_2$/sintering additive combinations have not been tested for both ZrB$_2$ and HfB$_2$, but a tentative rationalization has been devised for the most important densification mechanisms. Densification can be accomplished by the following:

- Solid state sintering, in the presence of carbon, B$_4$C, and WC, at temperatures between 1850 and 1950°C; no amorphous phases are observed along grain boundaries. Sintered materials are very refractory, with 600 MPa of flexural strength up to 1500°C in air.
- Liquid-phase sintering in the presence of metals or Si$_3$N$_4$ at temperatures generally lower than 1850°C for both ZrB$_2$ and HfB$_2$. Low-melting metallic phases, residual Si–O-based amorphous phases, and low-melting silicides are found along grain boundaries upon cooling. At temperatures greater than 1000°C, significant strength degradation is observed.
- Transient liquid-phase sintering, in the presence of MeSi$_2$ (Me = Zr, Mo, Ta, W), at temperatures 1700–1900°C. A special case is ZrSi$_2$ enabling densification at 1550°C. In all cases, formation of solid solutions and cation interdiffusion are

both mechanisms that help densification. By using $ZrSi_2$, low-melting silicides are found at triple points, while amorphous grain boundaries are often observed by using $TaSi_2$. In contrast, residual amorphous phases are rarely found in MB_2–$MeSi_2$ composites (Me = Mo, W), which retained strength values close to 600 MPa at 1500°C in air, with appropriate tuning of processing and sintering conditions.

SiC is the most widely used additive for MB_2 to improve strength and oxidation resistance. In addition, SiC acts as a sintering aid that promotes ZrB_2 densification when its content is less than 20 vol%, mainly through activation of grain-boundary diffusion-controlled mechanisms. For increasing contents of SiC particles (up to 30 vol%), grain-boundary migration of the diboride matrix is progressively limited, refining the resulting microstructure.

Transition MC are affected by sintering issues similar to those of MB_2, but have not been studied as thoroughly. Metals, carbides, and $MeSi_2$ can be useful for the densification of MC, generally in combination with pressure-assisted techniques (HP and SPS). In particular, $MoSi_2$ is very effective for the densification of various MC even by PS. Densification mechanisms are supposed to be similar to those of MB_2, but for carbides, clean grain boundaries and crystalline triple points are generally found.

ACKNOWLEDGMENTS

The authors wish to thank Alida Bellosi for having sowed the seed of the UHTC challenge more than 15 years ago and encouraging the ISTEC research group since then with dedication and passion. Our colleagues C. Capiani (powder processing), D. Dalle Fabbriche (high-temperature sintering), C. Melandri and S. Guicciardi (mechanical testing) are also gratefully acknowledged for technical support, fruitful guidance, and helpful discussion. TEM analyses were performed thanks to the financial support of grant N. FA8655-12-1-3004, with Dr. Ali Sayir as contract monitor.

REFERENCES

1. Ushakov SV, Navrotsky A. Experimental approaches to the thermodynamics of ceramics above 1500°C. J Am Ceram Soc 2012;95 (5):1463–1482.
2. Sonber JK, Suri AK. Synthesis and consolidation of zirconium diboride: review. Adv Appl Ceram 2011;110:321–334.
3. Thompson M, Fahrenholtz WG, Hilmas GE. Effect of starting particle size and oxygen content on densification of ZrB_2. J Am Ceram Soc 2011;94 (2):429–435.
4. Jung S-H, Oh H-C. Pre-treatment of zirconium diboride powder to improve densification. J Alloys Compd 2013;548:173–179.
5. Zdaniewski WA, Brungard NL. X-ray photoelectron spectroscopy studies of metallic diborides. J Am Ceram Soc 1992;75 (10):2849–2856.
6. Kravchenko SE, V.I T, Shilkin SP. Nanosized zirconium diboride: synthesis and properties. Russ J Inorg Chem 2011;56 (4):506–509.

7. Jung E-Y, Kim J-H, Jung S-H, Choi S-C. Synthesis of ZrB_2 powders by carbothermal and borothermal reduction. J Alloys Compd 2012;538:164–168.

8. Ortiz AL, Zamora V, Rodriguez-Rojas F. A study of the oxidation of ZrB_2 powders during high-energy ball-milling in air. Ceram Int 2012;38:2857–2863.

9. Chamberlain A, Fahrenholtz WG, Hilmas GE. Reactive hot pressing of zirconium diboride. J Eur Ceram Soc 2009;29:3401–3408.

10. Guo Q. Densification of ZrB_2-based composites and their mechanical and physical properties: a review. J Eur Ceram Soc 2009;29:995–1011.

11. Fahrenholtz WG, Hilmas GE. Refractory diborides of zirconium and hafnium. J Am Ceram Soc 2007;90 (5):1347–1364.

12. Lee S-H, Sakka Y, Kagawa Y. Dispersion behavior of ZrB_2 powder in aqueous solution. J Am Ceram Soc 2007;90 (11):3455–3459.

13. Monteverde F, Bellosi A, Guicciardi S. Processing and properties of zirconium diboride-based composites. J Eur Ceram Soc 2002;22:279–288.

14. Monteverde F, Bellosi A. Effect of the addition of silicon nitride on sintering behaviour and microstructure of zirconium diboride. Scr Mater 2002;46:223–228.

15. Monteverde F, Bellosi A. Beneficial effect of AlN as sintering aid on microstructure and mechanical properties of hot-pressed ZrB_2. Adv Eng Mater 2003;5 (7):508–512.

16. Monteverde F. Hot pressing of hafnium diboride aided by different sinter additives. J Mater Sci 2008;43:1002–1007.

17. Sciti D, Monteverde F, Guicciardi S, Pezzotti G, Bellosi A. Microstructure and mechanical properties of ZrB_2-$MoSi_2$ composites produced by different sintering techniques. Mater Sci Eng A 2006;434:303–306.

18. Sciti D, Silvestroni L, Bellosi A. Fabrication and properties of HfB_2-$MoSi_2$ composites produced by hot pressing and spark plasma sintering. J Mater Res 2006;21 (6):1460–1466.

19. Sciti D, Silvestroni L, Celotti G, Melandri C, Guicciardi S. Sintering and mechanical properties of ZrB_2-$TaSi_2$ and HfB_2-$TaSi_2$ ceramic composites. J Am Ceram Soc 2008;91 (10):3285–3291.

20. Zhu S, Fahrenholtz WG, Hilmas GE, Zhang SC. Pressureless sintering of zirconium diboride using boron carbide and Carbon additions. J Am Ceram Soc 2007;90 (11):3660–3663.

21. Brown-Shaklee HJ, Fahrenholtz WG, Hilmas GE. Densification behavior and thermal properties of hafnium diboride with the addition of boron carbides. J Am Ceram Soc 2012;95 (6):2035–2043.

22. Guo W-M. High temperature deformation of ZrB_2 ceramics with WC additive in four-point bending. Int J Refract Metals Hard Mater 2011;29:705–709.

23. Monteverde F, Guicciardi S, Melandri C, Dalle Fabbriche D. In: Orlovskaya N, Lugovy M, editors. *Boron rich Solids—Sensors, Ultra-High Temperature Ceramics, Thermoelectrics, Armor*. Dordrecht: Springer; 2010. p 261–272.

24. Loehman R, Corral E, Dumm HP, Kotula P, Tandon R. Ultra high temperature ceramics for hypersonic vehicle applications. Sandia Report 2006–2925. Albuquerque, New Mexico and Livermore, CA: Sandia National Laboratories.; 2006.

25. Wang H. Preparation and characterization of high toughness ZrB_2/Mo composites by hot-pressing process. Int J Refract Metals Hard Mater 2009;27:1024–1026.

26. Sun X. Microstructure and mechanical properties of ZrB_2-Nb composite. Int J Refract Metals Hard Mater 2010;28:472–474.

27. Ran S. High strength ZrB_2-based ceramics prepared by reactive pulsed electric current sintering of ZrB_2-ZrH_2 powders. J Eur Ceram Soc 2012;32:2537–2543.

28. Melendez-Martınez JJ, Domınguez-Rodrıguez A, Monteverde F, Melandri C, de Portu G. Characterisation and high temperature mechanical properties of zirconium boride-based materials. J Eur Ceram Soc 2002;22:2543–2549.

29. Monteverde F, Bellosi A. Efficacy of HfN as sintering aid in the manufacture of ultrahigh-temperature metal diborides-matrix ceramics. J Mater Res 2004;19 (12):3576–3585.

30. Monteverde F, Bellosi A. Development and characterization of metal-diboride-based composites toughened with ultra-fine SiC particulates. Solid State Sci 2005;7:622–630.

31. Pastor H, Meyer R. An investigation of the effect of additions of metal silicides on titanium and zirconium borides from the point of view of their sintering behavior and their resistance to oxidation at high temperature. Rev Int Hautes Temp Refract 1974;2:41–54.

32. Sciti D, Guicciardi S, Bellosi A, Pezzotti G. Properties of a pressureless sintered ZrB_2-$MoSi_2$ ceramic composite. J Am Ceram Soc 2006;89 (7):2320–2322.

33. Silvestroni L, Sciti D. Effects of $MoSi_2$ additions on the properties of Hf- and Zr-B_2 composites produced by pressureless sintering. Scr Mater 2007;27:165–168.

34. Sciti D, Silvestroni L, Nygren M. Spark plasma sintering of Zr- and Hf-borides with decreasing amounts of $MoSi_2$ as sintering aid. J Eur Ceram Soc 2008;28:1287–1296.

35. Sciti D, Guicciardi S, Nygren M. Densification and mechanical behaviour of HfC and HfB_2 fabricated by spark plasma sintering. J Am Ceram Soc 2008;91 (5):1433–1440.

36. Sciti D, Bonnefont G, Fantozzi G, Silvestroni L. Spark plasma sintering of HfB_2 with low additions of silicides of molybdenum and tantalum. J Eur Ceram Soc 2010;30:3253–3258.

37. Guo S, Kagawa Y, Nishimura T, Tanaka H. Pressureless sintering and physical properties of ZrB_2-based composites with $ZrSi_2$ additive. Scr Mater 2008;58:579–582.

38. Guo S, Nishimura T, Kagawa Y. Low temperature Hot pressing of ZrB_2-based ceramics with $ZrSi_2$ additives. Int J Appl Ceram Technol 2011;8 (6):1425–1435.

39. Sonber JK, Murthy TSRC, Subramanian C, Kumar S, Fotedar RK, Suri AK. Investigations on synthesis of ZrB_2 and development of new composites with HfB_2 and $TiSi_2$. Int J Refract Metals Hard Mater 2011;29 (1):21–30.

40. Opila E, Levine S, Lorincz J. Oxidation of ZrB_2 and HfB_2-based Ultra-High temperature ceramics: effect of Ta. J Mater Sci 2004;39 (19):5969–5977.

41. Silvestroni L, Sciti D. Densification of ZrB_2–$TaSi_2$ and HfB_2–$TaSi_2$ Ultra-High-temperature ceramic composites. J Am Ceram Soc 2011;94 (6):1920–1930.

42. Toropov NA, Galakhov FY. Liquidation in the system ZrO_2-SiO_2. Uchenye Zapiski Kazanskogo Gosudarstvennogo Universiteta 1965;2:158.

43. Silvestroni L, Kleebe H-J, Lauterbach S, Müller M, Sciti D. Transmission electron microscopy on Zr- and Hf-borides with $MoSi_2$ addition: densification mechanisms. J Mater Res 2010;25 (5):828–834.

44. Katrych S, Grytsiv A, Bondar A, Rogl P, Velikanova T, Bohn M. Structural materials: metal–silicon–boron. On the melting behaviour of Mo–Si–B alloys. J Alloys Compd 2002;347:94–100.

45. Silvestroni L, Guicciardi S, Melandri C, Sciti D. TaB_2-based ceramics: microstructure, mechanical properties and oxidation resistance. J Eur Ceram Soc 2012;32:97–105.

46. Brown-Shaklee HJ, Fahrenholtz WG, Hilmas GE. Densification behavior and thermal properties of hafnium diboride with the addition of boron carbides. J Am Ceram Soc 2012;95 (6):2035–2043.

47. Pastor H. In: Matkovich VI, editor. *Boron and Refractory Borides*. New York: Springer Verlag; 1977. p 457–493.

48. Zhu S, Fahrenholtz WG, Hilmas GE, Zhang SC. Pressure-less sintering of carbon-coated zirconium diboride powders. Mater Sci Eng A 2007;459 (1):167–171.

49. Chamberlain AL, Fahrenholtz WG, Hilmas GE. Pressureless sintering of zirconium diboride. J Am Ceram Soc 2006;89 (2):450–456.

50. Zhang SC, Hilmas GE, Fahrenholtz WG. Pressureless densification of Zirconium diboride with boron carbon additions. J Am Ceram Soc 2006;89 (5):1544–1550.

51. Thompson MJ, Fahrenholtz WG, Hilmas GE. Elevated temperature thermal properties of ZrB_2 with carbon addition. J Am Ceram Soc 2012;95 (3):1077–1085.

52. Brown-Shaklee HJ, Fahrenholtz WG, Hilmas GE. Densification behavior and microstructure evolution of hot-pressed HfB_2. J Am Ceram Soc 2011;94 (1):49–58.

53. Monteverde F. Beneficial effects of an ultra-fine α-SiC incorporation on the sinterability and mechanical properties of ZrB_2. Appl Phys A Mater Sci Proc 2006;82:329–337.

54. Li W, Zhang Y, Zhang X, Hong C, Han W. Thermal shock behavior of ZrB_2-SiC ultra-high temperature ceramics with addition of zirconia. J Alloys Compd 2009;478:386–391.

55. Guo W-M, Vleugels J, Zhang G-J, Wang P-L, Van der Biest O. Effects of Re_2O_3 (Re = La, Nd, Y and Yb) addition in hot-pressed ZrB_2-SiC ceramics. J Eur Ceram Soc 2009;29: 3063–3068.

56. Zapata-Sovas E, Jayaseelan DD, Lin HT, Brown P, Lee WE. Mechanical properties of ZrB_2- and HfB_2-based ultra-high temperature ceramics fabricated by spark plasma sintering. J Eur Ceram Soc 2013;33:1373–1386.

57. Monteverde F, Guicciardi S, Bellosi A. Advances in microstructure and mechanical properties of zirconium diboride based ceramics. Mater Sci Eng A 2003;346:310–319.

58. Wang Y, Liang J, Han W, Zhang X. Mechanical properties and thermal shock behavior of hot-pressed ZrB_2-SiC-AlN composites. J Alloys Compd 2009;475 (1–2):762–765.

59. Monteverde F. The addition of SiC particles into a $MoSi_2$-doped ZrB_2 matrix: effects on densification, microstructure and thermo-physical properties. Mater Chem Phys 2009;113: 626–633.

60. Hu C, Sakka Y, Tanaka H, Nishimura T, Guo S, Grasso S. Microstructure and properties of ZrB_2-SiC composites prepared by spark plasma sintering using $TaSi_2$ as sintering additive. J Eur Ceram Soc 2010;30:2625–2631.

61. Grigoriev ON, Galanov BA, Kotenko VA, Ivanov SM, Korotnev A, Brodnikovsky NP. Mechanical properties of ZrB_2-SiC($ZrSi_2$) ceramics. J Eur Ceram Soc 2010;30 (11):2173–2181.

62. Zhang H, Yan Y, Huang Z, Liu X, Jiang D. Pressureless sintering of ZrB_2-SiC ceramics: the effects of B_4C. Scr Mater 2009;60:559–562.

63. Zou J, Sun S-K, Zhang G-J, Kan Y-M, Wang P-L, Ohij T. Chemical reactions, anisotropic grain growth and sintering mechanisms of self-reinforced ZrB_2-SiC doped with WC. J Am Ceram Soc 2012;94 (5):1575–1583.

64. Zou J, Zhang G-J, Hu C-F, Nishimura T, Sakka Y, Vleugels J, Van der Biest O. Strong ZrB_2-SiC ceramics at 1600°C. J Am Ceram Soc 2012;95 (3):874–878.

65. Zou J, Zhang G-J, Kan Y-M, Wang P-L. Hot-pressed ZrB_2-SiC ceramics with VC addition: chemical reactions, microstructure, and mechanical properties. J Am Ceram Soc 2009;92 (12):2838–2846.

66. Mallik M, Roy S, Ray KK, Mitra R. Effect of SiC content, additives and process parameters on densification and structure-property relations of pressureless sintered ZrB_2-SiC composites. Ceram Int 2013;39:2915–2932.

67. Bird MW, Aune RP, Yu F, Becher PF, White KW. Creep behavior of a zirconium diboride—silicon carbide composites. J Eur Ceram Soc 2013;33:2407–2420.

68. Watts J, Hilmas GE, Fahrenholtz WG. Mechanical characterization of annealed ZrB_2-SiC composites. J Am Ceram Soc 2013;96 (3):845–851.

69. Zou J, Zhang G-J, Zhang H, Huang Z-R, Vleugels J, Van der Biest O. Improving high temperature properties of ho pressed ZrB_2-20 vol% SiC ceramic using high purity powders. Ceram Int 2013;39:871–876.

70. Zhang X, Xu L, Han W, Weng L, Han J, Du S. Microstructure and properties of silicon carbide whisker reinforced zirconium diboride ultra-high temperature ceramics. Solid States Sci 2009;11:156–161.

71. Wang Y, Zhu M, Cheng L, Zhang L. Fabrication of SiC_w reinforced ZrB_2-based ceramics. Ceram Int 2010;36:1787–1790.

72. Du S, Xu L, Zhang X, Hu P, Han W. Effect of sintering temperature and holding time on the microstructure and mechanical properties of ZrB_2-SiC_w composites. Mater Chem Phys 2009;116:76–80.

73. Lin J, Zhang X, Han W, Jin H. The hybrid effect of SiC whisker coupled with ZrO_2 fiber on microstructure and mechanical properties of ZrB_2-based ceramics. Mater Sci Eng A 2012;551:187–191.

74. Sciti D, Silvestroni L, Saccone G, Alfano D. Effect of different sintering aids on thermomechanical properties and oxidation of SiC fibers—reinforced composites. Mater Chem Phys 2013;137:834–842.

75. Sciti D, Guicciardi S, Silvestroni L. SiC chopped fibers reinforced composites ZrB_2: effect of the sintering aid. Scr Mater 2011;64:769–772.

76. Sciti D, Silvestroni L. Processing, sintering and oxidation behavior of SiC fibers reinforced composites ZrB_2 composites. J Eur Ceram Soc 2012;32:1933–1940.

77. Walker LS, Pinc WR, Corral EL. Powder processing effects on the rapid low-temperature densification of ZrB_2–SiC Ultra-High temperature ceramic composites using spark plasma sintering. J Am Ceram Soc 2012;95 (1):194–203.

78. Patel M, Singh V, Reddy JJ, Bhanu Prased VV, Jayaram V. Densification mechanisms during hot pressing of ZrB_2-20 vol% SiC composite. Scr Mater 2013;69:370–373.

79. Zhang SC, Hilmas GE, Fahrenholtz WG. Improved oxidation resistance of zirconium diboride by tungsten carbide additions. J Am Ceram Soc 2008;91 (11):3530–3535.

80. Neuman EW, Hilmas GE, Fahrenholtz WG. Strength of zirconium diboride to 2300°C. J Am Ceram Soc 2013;96 (1):47–50.

81. Santoro G. Variation of some properties of Tantalum carbide with carbon content. Trans Metallur Soc AIME 1963;227:1361–1368.

82. Toth LE. Transition metal carbides and nitrides. In: Margrave JL, editor. *Refractory Materials*. New York: Academic Press; 1971. *A Series of Monographs*; p 6–10.

83. Sciti D, Guicciardi S, Nygren M. Spark plasma sintering and mechanical behavior of ZrC-based composites. Scr Mater 2008;59:638–641.

84. Sciti D, Guicciardi S, Nygren M. Densification and mechanical behavior of HfC and HfB_2 fabricated by spark plasma sintering. J Am Ceram Soc 2008;91 (5):1433–1440.

85. Silvestroni L, Liu JX, Sciti D, Dalle Fabbriche D, Melandri C, Zhang GJ, Bellosi A. Microstructure and properties of HfC and TaC-based ceramics using ultrafine powder. J Eur Ceram Soc 2011;31 (4):619–627.

86. Landwehr SE, Hilmas GE, Fahrenholtz WG, Talmy IG. Processing of ZrC–Mo cermets for high-temperature applications, part I: chemical interactions in the ZrC–Mo system. J Am Ceram Soc 2007;90:1998–2002.

87. Hamjian WG, Lidman HJ. Reactions during sintering of a zirconium carbide–niobium cermet. J Am Ceram Soc 1964;35:236–240.

88. Wang X-G, Liu J-X, Kan Y-M, Zhang G-J. Effect of solid solution formation on densification of hot-pressed ZrC ceramics with MC (M=V, Nb, and Ta) additions. J Eur Ceram Soc 2012;32 (8):1795–1802.

89. Silvestroni L, Sciti D. Sintering behavior, microstructure, and mechanical properties: a comparison among pressureless sintered ultra-refractory carbides. Adv Mater Sci Eng. doi:10.1155/2010/835018.

90. Silvestroni L, Sciti D. Effect of transition metal silicides on microstructure and mechanical properties of ultra-high temperature ceramics. In: Low J, Sakka Y, Hu C, editors. *MAX Phases and Ultra-High Temperature Ceramics for Extreme Environments*. Hershey, PA: IGI Global; 2013, pp.125–179.

91. Sciti D, Silvestroni L, Guicciardi S, Dalle Fabbriche D, Bellosi A. Processing, mechanical properties and oxidation behavior of TaC and HfC composites containing 15 vol% $TaSi_2$ or $MoSi_2$. J Mater Res 2009;24 (6):2056–2065.

92. Silvestroni L, Sciti D, Balat-Pichelin M, Charpentier L. Zirconium carbide doped with Tantalum Silicide: microstructure, mechanical properties and high temperature oxidation. Mater Chem Phys, 2013;143 (1):407–415.

93. Silvestroni L, Sciti D. Transmission electron microscopy on Hf- and Ta- carbides sintered with $TaSi_2$. J Eur Ceram Soc 2011;31:3033–3043.

94. Zhang X, Hilmas GE, Fahrenholtz WG. Densification, mechanical properties, and oxidation resistance of $TaC–TaB_2$ ceramics. J Am Ceram Soc 2008;91 (12):4129–4132.

95. Talmy IG, Zaykoski JA, Opeka MM. Synthesis, processing and properties of $TaC-TaB_2$-C ceramics. J Eur Ceram Soc 2010;30 (11):2253–2263.

96. Bakshi SR, Musaramthota V, Lahiri D, Singh V, Seal S, Agarwal A. Spark plasma sintered tantalum carbide: effect of pressure and nano-boron carbide addition on microstructure and mechanical properties. Mater Sci Eng A 2011;528:1287–1295.

97. Ma B, Zhang X, Han J, Han W. Fabrication of hot-pressed ZrC-based composites. J Aerosp Eng 2009;233:1153–1157.

98. Rasouliasiabi HR, Shahbahrami B. Pressureless sintering of TaC-HfC-VC. Int J Sci Emerg Technol 2012;3:31–36.

99. Silvestroni L, Sciti D, Kling J, Lauterbach S. H-J. Kleebe, Sintering mechanisms of zirconium and hafnium carbides doped with $MoSi_2$. J Am Ceram Soc 2009;92 (7):1574–1579.

100. Nöllmann I, Trigubo AB, Walsöe de Reca NE. Subgrain structure and dislocation density in annealed MCT. Jpn J Appl Phys 1991;30 (8R):1787–1791.

101. Silvestroni L, Sciti D, Bellosi A. Microstructure and properties of pressureless sintered HfB_2-based composites with additions of ZrB_2 or HfC. Adv Eng Mater 2007;9 (10):915–920.

102. Silvestroni L, Sciti D. Microstructure and properties of pressureless sintered ZrC-based materials. J Mater Res 2008;26 (7):1882–1889.

7

UHTC COMPOSITES FOR HYPERSONIC APPLICATIONS

Anish Paul[1], Jon Binner[1], and Bala Vaidhyanathan[2]

[1] *School of Metallurgy and Materials, University of Birmingham, Birmingham, UK*
[2] *Department of Materials, Loughborough University, Loughborough, UK*

7.1 INTRODUCTION

The development of structural materials for use in oxidizing environments at temperatures above 2000°C is of great interest to the defense and aerospace sectors. Materials for these environments are currently largely limited to silicon-based ceramics due to the beneficial formation of a protective SiO_2 surface film in a suitably oxygen-rich atmosphere [1, 2]. Although SiO_2 is an excellent oxidation barrier at temperatures below 1600°C, above this temperature it begins to soften dramatically, and in a low-oxygen atmosphere it develops a substantial vapor pressure [1, 2]. Relatively few refractory oxides are stable in an oxidizing environment at or above 2000°C, but among these, zirconia (ZrO_2) and hafnia (HfO_2) have the highest melting points, approximately 2700°C [3, 4] and approximately 2900°C [5], respectively. Although they are stable and chemically inert, they are susceptible to thermal shock and exhibit high creep rates and phase transitions at higher temperatures [2, 6].

In recent years, ultra-high temperature ceramics (UHTCs) have been extensively investigated as innovative thermal protection systems (TPS) [7–9] and sharp leading-edge

Ultra-High Temperature Ceramics: Materials for Extreme Environment Applications, First Edition.
Edited by William G. Fahrenholtz, Eric J. Wuchina, William E. Lee, and Yanchun Zhou.
© 2014 The American Ceramic Society. Published 2014 by John Wiley & Sons, Inc.

components [10–12] for aerospace vehicles as well as for other applications where oxidation and/or erosion resistance at temperatures up to 2000°C are required. These materials, which include HfB_2, ZrB_2, HfC, ZrC, TaC, HfN, ZrN, and TaN, have melting temperatures close to or above 3000°C [13–15] and retain strength and thermal shock resistance at moderate temperatures. Whilst a "sharp" configuration for the leading edges and nose of re-entry vehicles increases aerodynamic efficiency and vehicle maneuverability [16, 17], it causes higher thermal loads to be applied to the materials compared to a blunt configuration. This makes the thermal and chemical stability of UHTC compounds of great importance, causing them to be candidates for use in extreme environments including hypersonic flight (1400°C and above in air) and rocket propulsion (3000°C and above in reactive chemical vapors) [18].

Carbides and borides of transition metal elements such as Hf and Zr are widely studied due to their desirable combinations of mechanical and physical properties, including high melting points (>3000°C), high thermal and electrical conductivities, and chemical inertness against molten metals [2, 8]. Even though carbides have higher melting points than borides, the latter have much higher thermal conductivities making them more attractive for ultra-high temperature applications [4, 19]. The use of bulk single-phase materials for high-temperature structural applications is limited by their poor oxidation and ablation resistance, as well as poor damage tolerance. A multi-phase approach has been successfully adopted to improve densification [20–25], mechanical properties [26–33], physical properties [33–38], as well as oxidation and ablation resistance of ZrB_2 [39–54] and HfB_2 [11, 34, 46, 47, 49, 55–61] ceramics. The mechanical and physical properties of these are closely linked with the densification processes, compositions, starting powders, microstructures, and intergranular second phases. Although these multi-phase materials have many advantages over single-phase materials, their intrinsic characteristics such as low fracture toughness, poor thermal shock resistance, and poor sinterability limit their usage. One approach to overcome some of these disadvantages is to introduce reinforcing fibers, mainly as a toughening phase. It allows tailoring of the mechanical and thermal properties of the UHTC composites by careful selection of fibers, raw materials, and optimization of fiber architecture. Continuous-fiber-reinforced ceramic matrix composites have the potential to offer excellent toughness, good thermal shock resistance, and defect tolerance, and good mechanical properties at high temperatures. Three-dimensional (3D) woven composites can overcome the vulnerability of two-dimensional (2D) composites to undergo delamination at higher temperatures and loads. Carbon fibers (Cfs) and silicon carbide fibers (SiCfs) are two obvious reinforcement choices because of their relatively high temperature resistance and availability. Cfs exhibit many advantages compared to other reinforcements, including high specific modulus, high specific strength, high stiffness, ability to offer exceptional fracture toughness in composites, low density, outstanding fatigue properties, excellent weight-to-strength ratio, a negative coefficient of longitudinal thermal expansion, and low coefficient of thermal expansion. These fibers can be processed into complex-shaped preforms by braiding, filament winding, or needling, which can then be converted into UHTC composites. Although the poor oxidation resistance of Cfs above 500°C potentially limits their high temperature applications [62, 63] it has been demonstrated that it is possible to prepare Cf–UHTC powder composites, where the Cfs improve the toughness while the UHTC phase provides oxidation resistance [63–67].

Cf-reinforced silicon carbide matrix composites (Cf/SiCs) have been widely studied and used in a wide variety of demanding applications such as combustion chambers, thruster components and nozzles for rocket engines, and as thermal protection systems for space vehicles (including nose and leading edges) with service temperatures less than 1600°C, taking advantage of the formation of the protective silica layer referred to earlier [2, 4]. More demanding applications require the development of novel UHTC composites that can offer excellent high temperature resistance at greater than 2000°C. A number of processing routes are viable for the preparation of whisker or short-fiber-reinforced UHTC composites [53, 54, 68–72], but the discussion here is mainly focused on continuous-fiber-reinforced UHTC composites with a small section dedicated to whisker or short-fiber-reinforced UHTCs.

7.2 PREPARATION OF CONTINUOUS-FIBER-REINFORCED UHTC COMPOSITES

7.2.1 Precursor Infiltration and Pyrolysis

This process involves infiltration of a low-viscosity precursor into a fibrous preform, particulate bed or a porous solid followed by pyrolysis to yield the ceramic matrix [73] and can be employed for the preparation of large, complex-shaped parts. Sometimes the precursor is polymeric and then it is called polymer infiltration and pyrolysis (PIP). This is most commonly used for the preparation of SiC matrix composites from polycarbosilanes [74, 75] or Si_3N_4 or SiCN matrix composites from polysilazanes [76]. Thermoplastic resins or thermosetting resins can be used for the preparation of C matrix composites [77]. The number of infiltration and pyrolysis cycles to achieve a particular porosity level depends on the ceramic yield of the precursor and the infiltration efficiency decreases with the number of infiltrations [78, 79]. The precursor infiltration and pyrolysis approach can also be adopted for the preparation of UHTC composites. A mixture of zirconium butoxide ($Zr(OC_4H_9)_4$) and divinyl benzene ($C_{10}H_{10}$) can be used as the precursor for preparing ZrC matrix 3D Cf-reinforced composites [78, 80]. Repeated infiltration, curing, pyrolysis, and heat treatment of the preforms with the precursor is required to obtain a dense matrix. In one case, the final density reported for such a composite was approximately $1.98\,g\,cm^{-3}$, with a combined total porosity of 34.3% [78]. Complete densification was difficult to achieve due to the low ceramic yield, which also led to the formation of microcracks. The composites displayed an average room temperature flexural strength of approximately 108 MPa after 3-point bend testing [78]. A similar process has been used for the preparation of Cf-ZrC [81] and Cf/SiC-ZrC composites [82]; a ZrC powder/precursor–polycarbosilane–xylene slurry was used for the impregnation and pyrolysis. Again, a number of impregnation and pyrolysis cycles and high-temperature heat treatments were required to obtain the desired density. The use of a pyrolytic carbon (PyC) or PyC/SiC interface layer was beneficial for increasing the bending strength of the composites [81].

A precursor solution to obtain SiC–ZrB$_2$ matrix composites can be prepared by mixing colloidal silica, boric acid, zirconium oxychloride, and sucrose [83]. The ratio of the reactants can be adjusted to vary the amount of ZrB$_2$ in the final powder from 5 to 20 wt%. Strength measurements were carried out on minicomposites prepared by infiltrating this precursor solution into T300 Cf tows that were subsequently heat treated at 1600 or 1700°C. The formation of an additional ZrC phase was also observed and it was ascribed to the loss of boron at higher temperature due to the high volatility of boron oxide [83]. The tensile strength of the minicomposites increased with an increase in ZrB$_2$ content, but decreased with an increase in heat treatment temperature with the values ranging from 150 to 270 MPa. The preparation of a sol–gel-based precursor for ZrB$_2$ and its impregnation into C–C preforms has been reported. [84] Nanosized ZrB$_2$ particles were obtained at the relatively low-heat treatment temperature of 1150°C, but the ceramic yield was very low.

For preparing HfB$_2$ matrix composites using the PIP process, a polymer-based HfB$_2$–C precursor has been prepared using HfCl$_4$ (hafnium source), H$_3$BO$_3$ (boron source), and phenolic resin (carbon source) using ethanol as a medium [85]. This precursor was then employed to impregnate SiC preform panels, which were then heat treated at 1600°C for 5 h during which the precursor was carbothermally converted to HfB$_2$ with varying amounts of residual carbon. Microstructural analysis, however, indicated localized sparse coating of poorly adhered HfB$_2$ particles due to the very low ceramic yield of the polymer.

7.2.2 Chemical Vapor Deposition

Chemical vapor deposition (CVD) involves the deposition of a solid on a heated surface from a chemical reaction in the vapor phase [86]. The deposition species are atoms, molecules, or a combination of these. It is a relatively simple technique. CVD can be used for producing uniform coatings on complex-shaped material and is versatile in that the whole process can be controlled at an atomistic level. A variant of CVD, called CVI (chemical vapor infiltration), can be used to infiltrate fiber preforms to produce composites and is largely employed for the preparation of C or SiC matrix composites. This method is also viable for preparing UHTC composites by deposition of UHTC phases. Only a few reports have described the CVD/CVI of UHTC phases for ultra-high temperature applications. The most common system for deposition of UHTC carbides such as ZrC and HfC is the reaction of metal chlorides with a hydrocarbon that is usually methane, propane, propene, or toluene [87]. The chloride is generally produced by the *in situ* chlorination of elemental hafnium. Similarly, ZrB$_2$ and HfB$_2$ can be deposited by the hydrogen reduction of the metal halides in the temperature range of 800–1100°C [88]. Diborane can also be used as a source of boron instead of BCl$_3$.

Continuous-fiber UHTC composites with an HfC matrix have been prepared using CVI [65]. 2D woven carbon fabrics were wrapped on a suitable mandrel to produce the complex shape required. Using an industrial-scale facility, the fabric was coated with 5–20 μm pyrolytic graphite prior to deposition of HfC. More details on the deposition method are not available. The pyrolytic graphite layer rigidized the fabric and helped it retain the shape of the mandrel. This layer also protected the fibers from reaction with the HfC matrix and minimized fiber strength degradation. The composites displayed relatively high fracture strains of the order of 1–2%; however, the fracture strengths were very low with an average value of just 25 ± 8 MPa for 30 samples, but the samples retained this strength after annealing at 2200°C for 4 h under argon.

Cf–HfC/TaC refractory composites for rocket thruster applications have been prepared using CVD [89, 90]. Preliminary deposition was carried out on radio frequency (RF) heated graphite. The deposition temperature was kept constant at 1150°C with the Hf and Ta burn rates adjusted to obtain deposits with different HfC/TaC ratios [89, 90]. CVI was performed using triaxially braided Cf preforms. Hf and Ta metal sponges were directly chlorinated and fed into a reaction chamber where they reacted with C from CH_4 to form HfC and TaC. Small amounts of TaC can improve the oxidation performance of HfC, but an optimum composition was not reported as there were difficulties controlling the deposition rates of the individual carbides.

7.2.3 Reactive Melt Infiltration

Reactive melt infiltration (RMI) is a rapid and low-cost manufacturing process widely used for the manufacture of Cf/SiC structural components. The driving force behind infiltration is capillary force and the whole process is finished within minutes to a few hours. RMI can be used effectively when one of the ceramic matrix elements possesses a relatively low melting temperature and readily wets the fibers [91]. As mentioned already, UHTC materials are characterized by very high melting temperatures and, hence, it is extremely difficult to use them for liquid or melt infiltration because of the high temperature requirements and the reactions of the melt with the fibers at the infiltration temperatures. The process, in general, also leaves residual metal that could melt at higher temperatures and enhance creep or attack constituent phases. In a typical RMI process for the preparation of Cf–ZrC UHTC composite, the Cf is coated with a suitable interfacial material. The coated fibers are then woven into a 2D fabric preform, which is subsequently deposited with C to form a C/C skeleton. Molten Zr is then infiltrated into the porous skeleton using a wicking action. The molten metal reacts with the previously deposited C to form a ZrC matrix. Detailed microstructural analysis of the matrix phases indicated the presence of ZrC along with α-Zr+ZrC eutectic. Some of the ZrC grains contained nanosized α-Zr and/or α-Zr-ZrC eutectic inclusions (Fig. 7.1) [91].

A few low melting intermetallic compounds can also be used for the reactive melt infiltration of UHTC phases at relatively low temperatures. Two such compounds are Zr_2Cu, which has a melting temperature of 1025°C [92, 93], and 35 wt% Hf 75 wt% Cu, which has a melting temperature of approximately 970°C [94]. It can be infiltrated into a porous material or composite containing B or C, which can subsequently react together to form ZrB_2, ZrC, HfB_2, or HfC. The reaction of B with molten Zr_2Cu is favored at temperatures as low as 1100°C [92]. For preparing Cf–ZrC composites using Zr_2Cu, [95] C was introduced into the fiber preform by the infiltration and pyrolysis of phenolic resin. RMI was performed by melting Zr_2Cu at 1200°C under vacuum and then slowly moving the carbon-enriched preform into the melt. The preform was kept at this

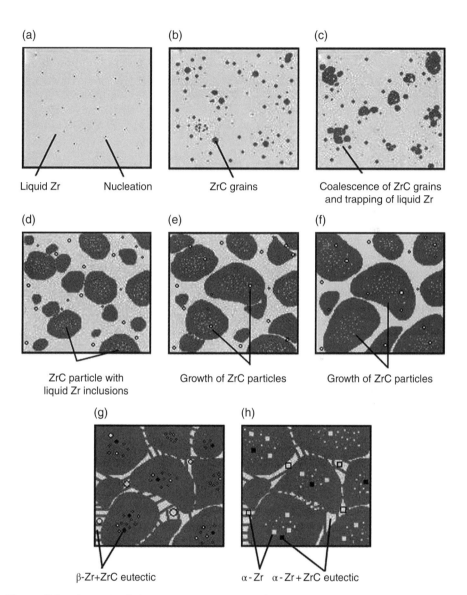

(a) Liquid Zr   Nucleation

(b) ZrC grains

(c) Coalescence of ZrC grains and trapping of liquid Zr

(d) ZrC particle with liquid Zr inclusions

(e) Growth of ZrC particles

(f) Growth of ZrC particles

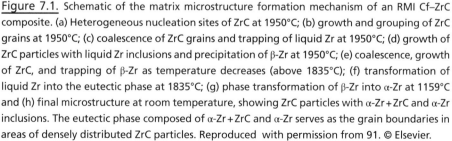

(g) β-Zr+ZrC eutectic

(h) α-Zr   α-Zr+ZrC eutectic

Figure 7.1. Schematic of the matrix microstructure formation mechanism of an RMI Cf–ZrC composite. (a) Heterogeneous nucleation sites of ZrC at 1950°C; (b) growth and grouping of ZrC grains at 1950°C; (c) coalescence of ZrC grains and trapping of liquid Zr at 1950°C; (d) growth of ZrC particles with liquid Zr inclusions and precipitation of β-Zr at 1950°C; (e) coalescence, growth of ZrC, and trapping of β-Zr as temperature decreases (above 1835°C); (f) transformation of liquid Zr into the eutectic phase at 1835°C; (g) phase transformation of β-Zr into α-Zr at 1159°C and (h) final microstructure at room temperature, showing ZrC particles with α-Zr+ZrC and α-Zr inclusions. The eutectic phase composed of α-Zr+ZrC and α-Zr serves as the grain boundaries in areas of densely distributed ZrC particles. Reproduced with permission from 91. © Elsevier.

temperature for 3 h to facilitate the infiltration and the reaction of Zr_2Cu with C to form ZrC [95]. No elemental Zr was detected after the RMI process and no damage was reported to the fibers. On high-temperature testing under an oxyacetylene flame, the samples survived without any surface erosion and this excellent performance was attributed to the evaporation of copper consuming a large amount of heat [95].

7.2.4 Slurry Infiltration and Pyrolysis

Slurry infiltration has been used by a number of researchers for preparing UHTC composites [63, 66, 85, 96]. Both aqueous and organic slurries can be used and the infiltration can be carried out using pressureless or pressure-assisted methods.

In the pressure-assisted aqueous slurry infiltration route reported by Tang *et al.* [63], 2D Cf (ex-PAN Toray T700) preforms were infiltrated separately with five different formulations of aqueous UHTC powder slurries based on ZrB_2, SiC, HfC, and TaC. Pyrolytic carbon deposition was used to hold the powders in place. The UHTC powder was mainly concentrated in an approximately 2 mm surface layer. These hybrid UHTC composites were tested using an oxyacetylene flame; different gas ratios were used to obtain different temperatures and heat fluxes. At 1800°C and 2380 kW m^{-2}, the compositions containing SiC demonstrated the lowest erosion depth. However, at the more aggressive conditions of 2700°C and 3920 kW m^{-2}, a $C/C-ZrB_2$ composite outperformed the other compositions (Fig. 7.2) and the superior performance was attributed to the formation of liquid ZrO_2 at the test temperature, which prevented the inward diffusion of oxygen.

Researchers at Loughborough University have studied the high-temperature oxidation performance of UHTC composites prepared using Cf preforms and a variety of UHTC powder combinations [66, 96]. The composites were fabricated using 30 mm diameter × 20 mm thick 2.5 D Cf preforms that contained 23 vol% fibers and the UHTC compositions included ZrB_2, ZrB_2–20 vol% SiC (ZS20), ZrB_2–20 vol% SiC–10 vol% LaB_6 (ZS20–1La), HfB_2 and HfC. The Cf preforms were separately impregnated with the slurries using either a squeeze or vacuum impregnation, dried, cured, and pyrolyzed at 900°C under flowing argon (99.998% pure) and then further densified using chemical vapor infiltration of carbon. Benchmark carbon/carbon composites were also prepared using chemical vapor infiltration of the preforms without UHTC powder impregnation. The powder distribution across the cross section of a Cf–HfB_2 UHTC composite

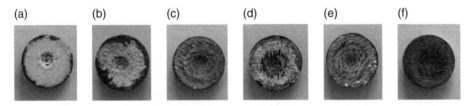

(a)   (b)   (c)   (d)   (e)   (f)

Figure 7.2. Comparison of C/C–UHTC composites ablated for a 30 s period under a 3920 kW m^{-2} heat flux: (a) C/C–ZrB_2, (b) C/C–$4ZrB_2$–1SiC, (c) C/C–$1ZrB_2$–2SiC, (d) C/C–2SiC–$1ZrB_2$–2HfC, (e) C/C–2SiC–$1ZrB_2$–2TaC, and (f) C/C. Reproduced with permission from Ref. 63. © Elsevier.

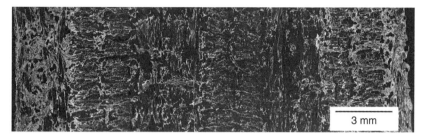

Figure 7.3. Cross section of a 30 mm diameter × 20 mm thick UHTC composite showing the distribution of UHTC powder. Reproduced from Ref. 66. © The American Ceramic Society.

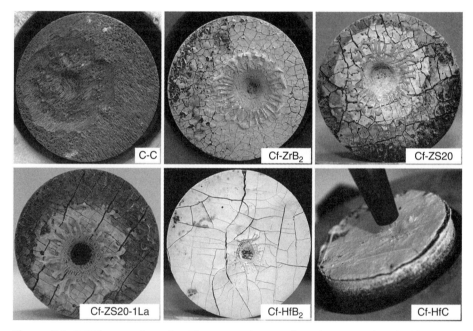

Figure 7.4. UHTC composites after 60 s oxyacetylene torch testing. Reproduced with permission from Ref. 96. © Elsevier.

(Fig. 7.3) revealed that the preforms were fully impregnated from top to bottom, although the composites were found to be slightly denser near the circumference because of the UHTC powder slurry impregnation from the sides of the preform. Powder distribution obtained using chemical analysis on the cross section of UHTC composites revealed uniform mixing and distribution of the powder. In general, the vacuum impregnation technique yielded superior results to the squeeze impregnation route.

The high-temperature performance of composites was determined using an oxyacetylene torch test facility for various durations ranging from 30 to 300 s. Images of composites after 60 s oxyacetylene torch testing are compared in Figure 7.4. The peak front-face temperatures reached by the samples ranged from 2300 to 2650°C even though the gas flow rates and ratios were similar. As expected, the CC composite was damaged over a much wider area, approximately 20 mm diameter, while the performances of the Cf–ZrB$_2$ and Cf–ZS20 composites were similar, with the damage being mainly focused over an approximately 5 mm diameter area. More erosion damage was observed for the Cf–ZS20–1La composite compared to the other ZrB$_2$-containing composites; the high-temperature flame penetrated through the impregnated layer and attacked the carbon-rich layer below it.

Cf–HfB$_2$ and Cf–HfC composites offered the best erosion protection, even though the UHTC powders were oxidized to HfO$_2$ as expected. The presence of molten phases on all of the impregnated composites independently confirmed the temperature of the flame. The amount of melting was much lower in HfB$_2$-based composites than in those based on ZrB$_2$ because of the higher melting temperature of HfO$_2$ (~2900°C vs. ~2700°C for ZrO$_2$) [5]. It was noteworthy that the oxide layer formed on the Cf–HfC composite was less adherent than those on the other composites and fell off after the test, possibly due to the absence of any glassy phases during the test and the buildup of CO/CO$_2$ gas below the oxide layer. Hence, subsequent research has been limited to HfB$_2$-based composites. With controlled fiber architecture and better impregnation, it is now possible to prepare Cf–HfB$_2$ UHTC composites that do not show any surface erosion after 60 s at ultra-high temperatures (Fig. 7.5).

Chemical mapping of the cross section of one of these composites after 60 s torch testing indicated that UHTC particles were oxidized only on the top layer, even directly below the flame tip (Fig. 7.6). Most of the carbon that was detected originated from epoxy resin used for mounting, not the tested sample.

Elevated temperature flexural strength was measured for Cf–HfB$_2$ composites using $140 \times 25 \times 10$ mm test bars at 1400°C under a flowing argon atmosphere in a 4-point bend configuration. The average strength was approximately 103 ± 25 MPa. The coefficient of thermal expansion was measured from room temperature up to 1700°C. The values were 1.63×10^{-6} °C^{-1} along the plane and 4.67×10^{-6} °C^{-1} perpendicular to it. This difference in CTE will require careful consideration when designing components using these materials [97].

For preparation of SiC–HfB$_2$ composites using slurry infiltration, 5 and 15 vol% HfB$_2$ slurries were prepared in ethanol using polyethyleneimine as a dispersant and

Figure 7.5. The 30 mm diameter × 20 mm thick, Cf–HfB$_2$ composites after 60 s oxyacetylene torch testing at >2500°C showing negligible surface erosion.

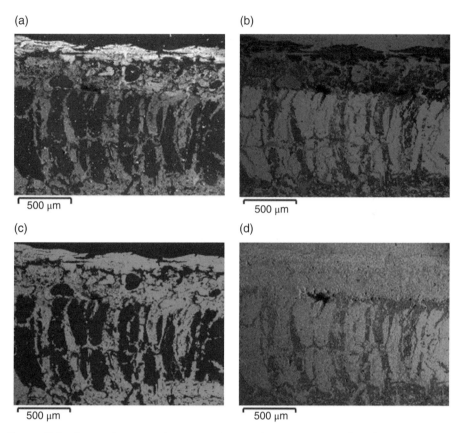

Figure 7.6. Electron image and EDS mapping on the cross section of a Cf–HfB$_2$ composite subjected to 60s oxyacetylene torch testing. (a) Back scattered electron image, (b) carbon, (c) hafnium, and (d) oxygen. The bright top layer in (a) indicates HfO$_2$.

binder [85]. A number of infiltration techniques were investigated, namely (i) vacuum infiltration, (ii) pressure infiltration, (iii) vibration infiltration, (iv) vibration-assisted vacuum infiltration, and (v) vibration-assisted pressure infiltration. After infiltration, SiC panels were heat treated at 1600°C under an inert atmosphere. Microstructural analysis revealed the presence of a significant amount of porosity within the fiber tows. The vacuum and vibration infiltrated samples showed partial infiltration, whereas pressure infiltrated panels (using 5 vol% slurry) showed excellent infiltration through the thickness, with most of the porosity present within tows. Increasing the slurry concentration to 15 vol% resulted in poorer pressure infiltration. Jayaseelan *et al.* [67] reported the infiltration of a porous C/C tube with three different types of ZrB$_2$ slurries, namely a nonaqueous ZrB$_2$ slurry, ZrB$_2$ gel and a "gel-composite" that contained a ZrB$_2$ gel and commercial ZrB$_2$ particles. They reported the formation of a chemical bond between the fibers and ZrB$_2$-gel-derived UHTC particles.

7.2.5 Combined Processes

A combination of one or more of the aforementioned processes can also be used for preparation of UHTC composites. SiC-fiber-reinforced ZrB_2–20 vol% SiC composites have been prepared using filament winding, slurry impregnation, and hot pressing [10]. The microstructure of such a composite highlighting the fiber distribution and matrix porosity is shown in Figure 7.7.

The high-temperature oxidation resistance of such a composite was compared against that of non-reinforced ZrB_2–20 vol% SiC at up to 1927°C for periods of up to 100 min [10]. Whilst the non-reinforced material showed the best oxidation protection at 1327 and 1627°C, at 1927°C both compositions underwent severe degradation and bloating. The authors expressed concerns about the thermal shock resistance of the nonreinforced materials in high-heat flux, aeroconvective environments. Attempts to sinter slurry impregnated $Cf–HfB_2$ composites using SPS at Queen Mary University, London were not completely successful, possibly due to the formation of a C coating on the surfaces of the HfB_2 particles during impregnation and pyrolysis. The C coating is also believed to have affected the conduction of current through the composite during SPS [97].

A slurry paste process combined with CVI has been used for the preparation of 2D Cf/ZrB_2–SiC [26] and 2D $Cf/SiC–ZrB_2$–TaC composites [98]. Cf cloths coated with a ZrB_2 or (ZrB_2+TaC)-liquid polycarbosilane slurry were stacked, cured under pressure, and pyrolyzed to form a 2D composite that was further densified by the CVI deposition of SiC using methyltrichlorosilane. The flexural strength of the composites was reported to be approximately 237 MPa with an interlaminar shear strength of approximately 16 MPa. The samples survived 20 s ablation under an oxyacetylene flame and showed very low linear ablation rates. The addition of TaC was reported to improve the oxidation resistance due to the formation of liquid phases above 1870°C.

A combination of slurry infiltration and precursor infiltration have been used for the preparation of $Cf/SiC–ZrB_2$ composite with ZrB_2 content ranging from 0 to approximately 25 vol% [99]. The porosity of the composites increased from 14 to 23 vol% with

(a) (b)

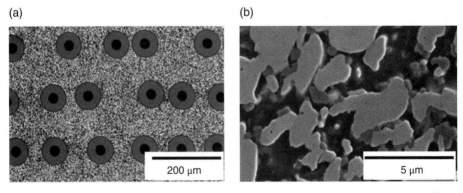

200 μm 5 μm

Figure 7.7. Microstructure of polished sections of ZrB_2 plus 20 vol% SiC plus SCS-9a fibers composite showing (a) representative fiber distribution and (b) matrix porosity. Reproduced with permission from Ref. 10. © Elsevier.

increasing ZrB_2 content in the slurry and the room temperature flexural strength decreased from approximately 370 to approximately 160 MPa. The addition of ZrB_2 was found to retain the strength at higher temperatures compared to composites containing only SiC. At 2000°C, Cf/SiC–ZrB_2 retained over 32% of its room temperature strength.

7.2.6 Functionally Graded UHTC Composites

Functionally graded composites can be prepared to offer different levels of high-temperature protection at different depths, to gradually adjust the CTE values between two different phases or to tailor the bulk density of a composite as per design requirements. Certain ultra-high temperature applications may only experience extreme temperatures for a few seconds and a relatively thin UHTC composite layer near the surface of the TPS may be sufficient to offer the required protection. Only limited studies have reported the preparation of functionally graded UHTC composites. In one process, Zoltek Panex® 30 carbon fabric, allylhydridopolycarbosilane (AHPCS) preceramic polymer, HfB_2 and SiC powders were used to prepare UHTC composites with a graded structure that had an HfB_2-rich surface through to a SiC-rich surface with Si-O-C preceramic polymer throughout [64]. This was achieved by stacking carbon fabric coated with either SiC/AHPCS or HfB_2/AHPCS and curing under pressure at 400°C and subsequent pyrolysis at 850°C. The HfB_2 side of the composite was then coated with HfB_2/AHPCS slurry and the whole composite was infiltrated and pyrolyzed a number of times with AHPCS assisted by vacuum. Microcracks were present in the final composites even after seven cycles of repeated infiltration and pyrolysis. Four-point flexural strength measurements on the composites showed strengths ranging from 100 to 162 MPa depending on whether the SiC side or the HfB_2 side was in tension, with the former configuration yielding higher values. Oxidation testing was carried out in a furnace at 1617°C and using an oxyacetylene flame at 1805–2015°C. Following cyclic heating in the furnace, a nonuniform $HfSiO_4$ and monoclinic HfO_2 surface was formed on the HfB_2-rich surface and a glassy SiO_2 layer was formed on the SiC-rich surface. Damage to the Cfs in the furnace testing was found to be lower at the HfB_2-rich surface compared to the SiC-rich surface. In contrast, during the oxyacetylene flame testing the HfB_2-rich surface suffered a greater degree of damage during a 4 min test compared to the SiC-rich surface. These results highlight significant differences arising from different test methods in which not only temperature but also gas flow rates differ and the effect this has upon the surface reactions and damage. As a result, direct comparisons between different high-temperature test methods are rarely meaningful. This technique demonstrated that the preparation of a functionally graded HfB_2–SiC UHTC composite can be adapted for preparation of other UHTC composites by varying the powder combinations.

7.3 UHTC COATINGS

Developing a UHTC coating on CC, Cf–SiC or even Cf–UHTC composites is another method of improving the high-temperature performance of materials for UHT applications. A good coating should offer very good adhesion and resist the formation of any cracks due to CTE mismatch at higher temperatures. Recent measurement of the CTE values of Cf–HfB$_2$ slurry impregnated composites at Loughborough University revealed that the CTE values of the composites are widely different from those of UHTC monoliths. They were also found to vary depending on the fiber orientation, which imposes additional complications for component design. Interface materials that can bridge the CTE mismatches while withstanding very high temperature need to be identified. SiC has a CTE between those of the composites and the monoliths, but the maximum service temperature will be limited by the performance of this interlayer. Coatings should also minimize diffusion of oxygen and prevent any spallation. The main disadvantage with this route is that once the coating layer is destroyed, the base material will have little protection unless UHTC composites are involved.

HfC coatings have been deposited on C/C composites using low-pressure chemical vapor deposition [100–102]; hafnium chloride (HfCl$_4$) powder, CH$_4$, H$_2$, and Ar were used as feed materials. A special powder feeder was needed to protect the HfCl$_4$ from moisture, while maintaining a continuous supply. The feed rate of HfCl$_4$ required was 1.15 g min^{-1} and the gas flow rates of CH$_4$, H$_2$, and Ar were 160, 800, and 800 ml min^{-1}. The deposition was performed at a reduced pressure of 5 kPa and a temperature of 1500°C for 3 h. A smooth coating free from cracks was obtained using this method, with grain sizes in the 100–200 nm range. On oxyacetylene torch ablation, the coatings developed a three-layer structure with a highly porous outer oxide layer, a dense transitional interlayer, and residual carbide layer (Fig. 7.8).

A ZrB$_2$ SiC coating was found to be useful for protecting CC composites at temperatures up to 1500°C under static conditions [103]. The coating with approximately 120 µm thickness was developed by vapor-phase infiltration of silicon in a ZrB$_2$/phenolic resin precoated CC composite. A similar ZrB$_2$–SiC coating was developed on Cf/SiC composites using a ZrB$_2$ powder–polycarbosilane– xylene slurry [104]. The latter was applied on top of the composite and then pyrolyzed at 1200°C to form the coating. A CVD layer of SiC was deposited on top of this layer, which penetrated the gaps between the ZrB$_2$/SiC powders forming a ZrB$_2$/SiC coating with a dense layer of SiC on top. A three-step process for the protection of Cf–SiC composites with a coating of ZrB$_2$/SiC/ZrC has been reported by Blum *et al.* [105] Most of the coating development work was carried out using monolithic SiC specimens. In a typical process, the preforms were dip coated with a slurry of ZrB$_2$ in phenolic resin. Both water- and alcohol-based phenolic slurries were used and sometimes a mixture of ZrB$_2$ and SiC powder was used. The coated specimens were pyrolyzed at 1000°C. The C formed as a result of pyrolysis was converted to SiC using a molten Si infiltration process. An Si top layer was observed after the liquid Si infiltration step, which was subsequently converted to SiC by high-temperature (1500°C) heat treatment in the presence of a carbon source. Preliminary oxidation studies at 1500°C indicated the formation of a 5–10 µm protective silica layer on the surface of the specimens.

Other reports described deposition of ZrC [106, 107] and HfC [108, 109], but the methods have not yet been applied for the preparation of coatings for UHTC applications and hence are not discussed here.

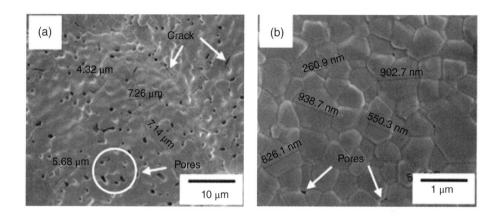

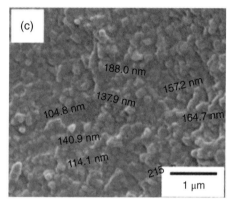

Figure 7.8. Microstructures of the ablated HfC coating on a C/C composite in different regions: (a) central; (b) transitional; and (c) outer ablation region. Reproduced with permission from Ref. 102. © Pergamon.

7.4 SHORT-FIBER-REINFORCED UHTC COMPOSITES

All the methods discussed so far used continuous-fiber reinforcements. It is also possible to prepare UHTC composites using chopped or discontinuous fibers as reinforcement. In a typical process, the UHTC powder is mixed with short fibers and ball milled in an organic solvent to obtain the required distribution. The solvent is removed, preferably by rotary evaporation, and then the dried powder is either hot pressed or spark plasma sintered under vacuum to achieve the final composite. Composites of ZrB_2–SiC containing 20 vol% chopped Cf were prepared using this approach [110]. The addition of Cf was found to improve the fracture toughness of the composites, but a reduction in fracture strength was observed. SiC-whisker-reinforced ZrB_2 composites have also been prepared by hot pressing [70, 71] or SPS [72]. In general, the composites displayed superior flexural strength and fracture toughness to those of monolithic ZrB_2 and SiC-particle-reinforced ZrB_2 ceramics. The SiC whiskers were not stable at higher temperatures (1900°C) and converted to SiC particles. SiC-whisker- or SiC-chopped-fiber-reinforced ZrB_2 composites were prepared at relatively lower temperatures of up to 1730°C by hot pressing using Si_3N_4 as a sintering aid [69]. The whiskers underwent undesirable reactions even at this temperature, but the fibers were more stable. The addition of reinforcing fibers increased the fracture toughness and high-temperature flexural strength at 1200°C, but reduced the room temperature strength, whereas the addition of whiskers increased the fracture toughness and high-temperature strength and the room temperature strength was largely unaffected. $ZrSi_2$ or $MoSi_2$ can also be used as sintering aids instead of Si_3N_4, and SPS can be employed for sintering instead of hot pressing [54]. The selection of sintering aid is crucial as it controls the matrix sintering temperature and, more importantly, influences the reactivity between matrix and fiber. This controls the formation of secondary phases and hence affects the high-temperature mechanical and thermal properties. Both furnace [54] and plasma wind tunnel [53, 54] performance of the composites were studied and reported to be similar to those of SiC-particle-reinforced ZrB_2 composites. HfB_2 powder synthesized using a special self-propagating high-temperature synthesis (SHS) technique was used for the SPS sintering of SiC-whisker-containing HfB_2 matrix composites [68]. At higher sintering temperatures of approximately 1830°C, the whiskers were converted to SiC platelets. Partial conversion was observed at1700°C sintering temperature.

7.5 HYBRID UHTC COMPOSITES

Hybrid UHTC composites are a novel concept conceived to produce UHTC Cf-based composites with a UHTC monolith layer on top. The idea is to utilize the Ultra-High temperature capabilities of the UHTC monolithic materials whilst the UHTC composite layer beneath it provides longer term protection and thermal shock resistance. Since the underlying layer is a UHTC composite and not C/C or Cf/SiC composite, this layer can also provide excellent high-temperature protection while providing a means of controlling the total mass of the thermal protection system. In the case of any failure of the monolithic layer, the composite layer is designed to provide the required protection to enable the function to be completed. However, joining of the UHTC monolith to the UHTC composite is very challenging. A number of approaches can be adopted and SPS joining and laser melting have been investigated by the team in collaboration with Imperial College and Queen Mary, both in London [97]. Preliminary experiments were carried out using ZrB_2–impregnated Cf composite and ZrB_2 powder. The impregnated and pyrolyzed composite was loaded into a graphite-foil-lined graphite die of the SPS press. Cf–ZrB_2 composites were used without any precompaction or presintering. The composite was surrounded by ZrB_2 powder and the temperature and pressure gradually increased to achieve sintering and joining simultaneously. The composite was held at the peak temperature for 10 min and then it was cooled naturally after releasing the pressure. The ZrB_2 monolithic layer was found to develop a good bond with the composite, but it developed many cracks on the UHTC monolith layer due to the difference in CTE values. On cross-sectional analysis (Fig. 7.9) it was found that the monolithic layer had many pores and the interface had a finer grain size than the outer layer. This is rather

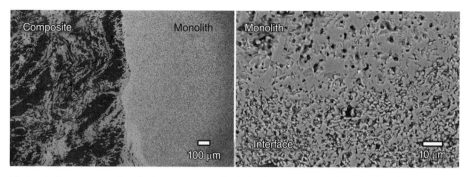

Figure 7.9. Cross-sectional microstructure of a hybrid UHTC composite. The bonding between the composite and monolith layers are fundamentally good as seen in the higher magnification image.

difficult to explain. No fiber degradation was observed and no sintering was observed for the ZrB_2 particles within the preforms. This could be due to the formation of a carbon coating on the surface of the ZrB_2 particles due to the use of phenolic resin during the slurry preparation or due to most of the current being conducted through the outer UHTC layer, resulting in lower heating for the composite part. The bulk density of the UHTC monolith layer was determined to be 76%. Preliminary laser melting and joining were not successful due to the oxidation of the molten layer formed.

7.6 SUMMARY AND FUTURE PROSPECTS

UHTC composites are promising materials for future hypersonic applications because of their advantages, including excellent defect tolerance, high fracture toughness, low density, excellent thermal shock resistance, high thermal conductivity, and low coefficient of thermal expansion. Various processing techniques used for their preparation have been discussed. Precursor infiltration and pyrolysis is a useful technique, but suffers from the lack of availability of good precursors for UHTC phases such as ZrB_2 or HfB_2. Chemical vapor infiltration is an excellent technique for preparing CC and Cf/SiC composites and has been used with some success for the preparation of ZrC- or HfC-based composites, but hardly any reports describe the deposition of ZrB_2 or HfB_2 for preparing UHTC composites. It is also a very lengthy and expensive process. Reactive melt infiltration is limited by the extremely high melting temperatures of the borides and carbides and the high temperatures required can destroy the fiber and/or any protective coatings. The use of low melting alloys such as Zr_2Cu is promising, but the presence of residual, low melting Cu is a potentially significant disadvantage. Development of UHTC coatings on composites is challenging due to the difference in CTE values.

So far the main focus has been on the development of UHTC composites using Cfs or, in rare cases, using SiC fibers. None of these can withstand the extreme temperatures that the composites have to withstand. So far, very limited efforts have attempted to prepare UHTC fibers, which may be required to realize the application of UHTC composites at very high temperatures. Recently, Matech GSM (Matech GSM, California, USA) reported the production of HfC fibers based on a patented process [111]. The use of short C or SiC fiber composites resulted in very little improvement in flexural strength and fracture toughness; there was also an additional complication due to the reaction of SiC fibers at higher temperature and their subsequent conversion to SiC particles.

The fiber–matrix interface is another critical area that needs attention. Only limited data are available on the room or high-temperature strength of UHTC composites as most of the studies so far have been focused on the high-temperature oxidation testing of UHTC composites, mainly using lab-scale techniques. Once the required oxidation performance is realized, it is likely that demand will increase for design data, including mechanical and thermal properties. Strength maximization will require the optimization of the matrix–fiber interface to maximize load transfer. There are also constraints with the high-temperature strength testing of UHTC composites. Unlike the UHTC monoliths, which require very small test bars, UHTC composite test specimens have to be sufficiently large to give a true representation of the complete composite. This is more crucial for gradient structures. Not many facilities can accommodate reasonably large UHTC composite bars and perform strength testing at very high temperature (>1800°C). An added complication with UHTC composites compared to UHTC monoliths is the fact that most of them have an anisotropic structure. So, for design purposes, the mechanical and especially the thermal properties need to be measured across all three axes.

The realization of hypersonic vehicles may also require flexible or shape-morphing UHTC parts. As listed by Miles *et al.* [112], high-temperature shape morphing materials could (i) enable continuous variation of the inlet area ratio to accommodate changing Mach number and air density associated with climb and descent; (ii) improve engine and isolator performance by accommodating area and shape changes associated with engine mode transition from subsonic to supersonic compression (RAM to SCRAM); (iii) optimize the compression ramp and inlet contour to minimize shock formation and reduce pressure losses; (iv) permit dynamic adaptation to compensate for flight maneuvers and correct for distortion due to thermal expansion as well as suppress upstream shock motion to reduce the possibility of engine unstart, and if unstart occurs, reduce the recovery time; and (v) improve nozzle performance by controlling the nozzle area ratio and contour for efficient thrust generation and minimum pressure loss. This can only be achieved through the development of continuously shape-morphing composite materials.

In summary, an ideal UHTC composite could be the one based on UHTC fiber reinforcement, with excellent high-temperature oxidation and mechanical properties, with preferably isotropic thermal properties. A more concerted effort from the UHTC research groups all over the world is required for the development of such a material.

REFERENCES

1. Berton B, Bacos MP, Demange D, Lahaye J. High-temperature oxidation of silicon carbide in simulated atmospheric re-entry conditions. J Mater Sci 1992;27 (12):3206–3210.
2. Upadhya K, Yang J, Hoffman WP. Materials for ultrahigh temperature structural applications. J Am Ceram Soc 1997;76 (12):51–56.

3. Hlavac J. Melting temperatures of refractory oxides: part I. Pure Appl Chem 1982;54: 681–688.

4. Upadhya K, Yang J, Hoffman W. Advanced materials for ultrahigh temperature structural applications above 2000°C. Air Force Research Laboratory (AFRL) Report No.: AFRL-PR-ED-TP-1998-007. Edwards AFB, CA: AFRL; 1997.

5. Ruh R, Garrett HJ, Domagala RF, Tallan NM. The system zirconia-hafnia. J Am Ceram Soc 1968;51 (1):23–28.

6. Scott HG. Phase relationships in the zirconia-yttria system. J Mater Sci 1975;10 (9): 1527–1535.

7. Opeka MM, Talmy IG, Zaykoski JA. Oxidation-based materials selection for 2000°C + hypersonic aerosurfaces: theoretical considerations and historical experience. J Mater Sci 2004;39 (19):5887–5904.

8. Chamberlain A, Fahrenholtz W, Hilmas G, Ellerby D. Characterization of zirconium diboride for thermal protection systems. Key Eng Mater 2004;264:493–496.

9. Savino R, De Stefano Fumo M, Paterna D, Serpico M. Aerothermodynamic study of UHTC-based thermal protection systems. Aerosp Sci Technol 2005;9 (2):151–160.

10. Levine SR, Opila EJ, Halbig MC, Kiser JD, Singh M, Salem JA. Evaluation of Ultra-High temperature ceramics for aeropropulsion use. J Eur Ceram Soc 2002;22 (14–15): 2757–2767.

11. Gasch M, Ellerby D, Irby E, Beckman S, Gusman M, Johnson S. Processing, properties and arc jet oxidation of hafnium diboride/silicon carbide ultra high temperature ceramics. J Mater Sci 2004;39 (19):5925–5937.

12. Zhang X, Hu P, Han J, Meng S. Ablation behavior of ZrB_2–SiC Ultra High temperature ceramics under simulated atmospheric re-entry conditions. Compos Sci Technol 2008;68 (7–8):1718–1726.

13. Wuchina E, Opila E, Opeka M, Fahrenholtz W, Talmy I. UHTCs: Ultra-High temperature ceramic materials for extreme environment applications. Electrochem Soc Interface 2007;16 (4):30.

14. Monteverde F, Bellosi A, Scatteia L. Processing and properties of Ultra-High temperature ceramics for space applications. Mater Sci Eng A 2008;485 (1):415–421.

15. Fahrenholtz WG, Hilmas GE, Talmy IG, Zaykoski JA. Refractory diborides of zirconium and hafnium. J Am Ceram Soc 2007;90 (5):1347–1364.

16. Bartuli C, Valente T, Tului M. Plasma spray deposition and high temperature characterization of ZrB_2–SiC protective coatings. Surf Coat Technol 2002;155 (2–3):260–273.

17. Tului M, Marino G, Valente T. High temperature characterization of a UHTC candidate materials for RLVs. In: Wilson A, editor. Hot Structures and Thermal Protection Systems for Space Vehicles, Proceedings of the 4th European Workshop; November 26–29, 2002; Palermo, Italy. European Space Agency; 2003, Vol. 521. p 161–165.

18. Fahrenholtz WG, Hilmas GE. Report on the NSF-AFOSR joint workshop on future Ultra-High temperature materials. Refractories Appl News 2005;10 (1):24–26.

19. Squire TH, Marschall J. Material property requirements for analysis and design of UHTC components in hypersonic applications. J Eur Ceram Soc 2010;30 (11):2239–2251.

20. Akin I, Hotta M, Sahin FC, Yucel O, Goller G, Goto T. Microstructure and densification of ZrB_2–SiC composites prepared by spark plasma sintering. J Eur Ceram Soc 2009;29 (11):2379–2385.

21. Medri V, Monteverde F, Balbo A, Bellosi A. Comparison of ZrB_2-ZrC-SiC composites fabricated by spark plasma sintering and hot-pressing. Adv Eng Mater 2005;7 (3):159–163.

22. Cao JL, Xu Q, Zhu SZ, Zhao JF, Wang FC. Microstructure of ZrB$_2$–SiC composite fabricated by spark plasma sintering. Key Eng Mater 2008;368:1743–1745.

23. Chamberlain AL, Fahrenholtz WG, Hilmas GE. Low-temperature densification of zirconium diboride ceramics by reactive hot pressing. J Am Ceram Soc 2006;89 (12):3638–3645.

24. Chen D, Xu L, Zhang X, Ma B, Hu P. Preparation of ZrB$_2$ based hybrid composites reinforced with SiC whiskers and SiC particles by hot-pressing. Int J Refract Metals Hard Mater 2009;27 (4):792–795.

25. Guo WM, Zhang GJ, Zou J, Kan YM, Wang PL. Effect of Yb$_2$O$_3$ addition on hot-pressed ZrB$_2$-SiC ceramics. Adv Eng Mater 2008;10 (8):759–762.

26. Wang Y, Liu W, Cheng L, Zhang L. Preparation and properties of 2D C/ZrB$_2$-SiC Ultra High temperature ceramic composites. Mater Sci Eng A 2009;524 (1–2):129–133.

27. Ramirez-Rico J, Bautista M, Martínez-Fernández J, Singh M. Compressive strength degradation in ZrB$_2$-SiC and ZrB$_2$-SiC-C ultra high temperature composites. In: Singh D, Kriven WM, editors. *Mechanical Properties and Performance of Engineering Ceramics and Composites IV*. Volume 30, Hoboken (NJ): John Wiley & Sons Inc; 2009. p 127.

28. Liu Q, Han W, Hu P. Microstructure and mechanical properties of ZrB$_2$-SiC nanocomposite ceramic. Scr Mater 2009;61 (7):690–692.

29. Zhang X, Wang Z, Sun X, Han W, Hong C. Effect of graphite flake on the mechanical properties of hot pressed ZrB$_2$-SiC ceramics. Mater Lett 2008;62 (28):4360–4362.

30. Liang J, Wang C, Wang Y, Jing L, Luan X. The influence of surface heat transfer conditions on thermal shock behavior of ZrB$_2$-SiC-AlN ceramic composites. Scr Mater 2009;61 (6):656–659.

31. Zhang H, Yan Y, Huang Z, Liu X, Jiang D. Pressureless sintering of ZrB$_2$-SiC ceramics: the effect of B$_4$C content. Scr Mater 2009;60 (7):559–562.

32. Monteverde F, Bellosi A. Microstructure and properties of an HfB$_2$-SiC composite for Ultra High temperature applications. Adv Eng Mater 2004;6 (5):331–336.

33. Guo SQ. Densification of ZrB$_2$ based composites and their mechanical and physical properties: a review. J Eur Ceram Soc 2009;29 (6):995–1011.

34. Zhang X, Weng L, Han J, Meng S, Han W. Preparation and thermal ablation behavior of HfB$_2$-SiC-based ultra-high-temperature ceramics under severe heat conditions. Int J Appl Ceram Technol 2009;6 (2):134–144.

35. Weng L, Han W, Li X, Hong C. High temperature thermo-physical properties and thermal shock behavior of metal–diborides-based composites. Int J Refract Metals Hard Mater 2010;28 (3):459–465.

36. Monteverde F. The addition of SiC particles into a MoSi$_2$ ZrB$_2$ matrix: effects on densification, microstructure and thermo-physical properties. Mater Chem Phys 2009;113 (2):626–633.

37. Guo SQ, Kagawa Y, Nishimura T, Chung D, Yang JM. Mechanical and physical behavior of spark plasma sintered ZrC–ZrB$_2$-SiC composites. J Eur Ceram Soc 2008;28 (6):1279–1285.

38. Zimmermann JW, Hilmas GE, Fahrenholtz WG, Dinwiddie RB, Porter WD, Wang H. Thermophysical properties of ZrB$_2$ and ZrB$_2$-SiC ceramics. J Am Ceram Soc 2008;91 (5):1405–1411.

39. Tripp WC, Graham HC. Thermogravimetric study of the oxidation of ZrB$_2$ in the temperature range of 800° to 1500°C. J Electrochem Soc 1971;118 (7):1195–1199.

40. Rezaie A, Fahrenholtz WG, Hilmas GE. Oxidation of zirconium diboride-silicon carbide at 1500°C at a low partial pressure of oxygen. J Am Ceram Soc 2006;89 (10):3240–3245.

41. Han W, Hu P, Zhang X, Han J, Meng S. High-temperature oxidation at 1900°C of ZrB_2-xSiC Ultrahigh-temperature ceramic composites. J Am Ceram Soc 2008;91 (10):3328–3334.

42. Talmy IG, Zaykoski JA, Martin CA. Flexural creep deformation of ZrB_2/SiC ceramics in oxidizing atmosphere. J Am Ceram Soc 2008;91 (5):1441–1447.

43. Talmy IG, Zaykoski JA, Opeka MM. High-temperature chemistry and oxidation of ZrB_2 ceramics containing SiC, Si_3N_4, Ta_5Si_3, and $TaSi_2$. J Am Ceram Soc 2008;91 (7):2250–2257.

44. Zhang X, Hu P, Han J. Structure evolution of ZrB_2–SiC during the oxidation in air. J Mater Res 2008;23:1961–1972.

45. Zhang X, Xu L, Du S, Han W, Han J. Preoxidation and crack-healing behavior of ZrB_2-SiC ceramic composite. J Am Ceram Soc 2008;91 (12):4068–4073.

46. Carney CM, Mogilvesky P, Parthasarathy TA. Oxidation behavior of zirconium diboride silicon carbide produced by the spark plasma sintering method. J Am Ceram Soc 2009;92 (9):2046–2052.

47. Licheri R, Oru R, Musa C, Locci A, Cao G. Spark plasma sintering of ZrB_2- and HfB_2-based ultra high temperature ceramics prepared by SHS. Int J Self-Propagating High-Temp Syn 2009;18 (1):15–24.

48. Hu P, Zhang X, Han J, Luo X, Du S. Effect of various additives on the oxidation behavior of ZrB_2-based Ultra-High-temperature ceramics at 1800°C. J Am Ceram Soc 2010;93 (2):345–349.

49. Mallik M, Ray KK, Mitra R. Oxidation behavior of hot pressed ZrB_2–SiC and HfB_2–SiC composites. J Eur Ceram Soc 2011;31 (1-2):199–215.

50. Wang M, Wang C, Yu L, Huang Y, Zhang X. Oxidation behavior of SiC platelet-reinforced ZrB_2 ceramic matrix composites. Int J Appl Ceram Technol 2011;9 (1):178–185.

51. Savino R, De Stefano Fumo M, Paterna D, Di Maso A, Monteverde F. Arc-Jet testing of Ultra-High-temperature-ceramics. Aerosp Sci Technol 2010;14 (3):178–187.

52. Monteverde F, Savino R. ZrB_2-SiC Sharp leading edges in high enthalpy supersonic flows. J Am Ceram Soc 2012;95 (7):2282–2289.

53. Sciti D, Savino R, Silvestroni L. Aerothermal behaviour of a SiC fibre-reinforced ZrB_2 sharp component in supersonic regime. J Eur Ceram Soc 2012;32 (8):1837–1845.

54. Sciti D, Silvestroni L. Processing, sintering and oxidation behavior of SiC fibers reinforced ZrB_2 composites. J Eur Ceram Soc 2012;32 (9):1933–1940.

55. Opila E, Levine S, Lorincz J. Oxidation of ZrB_2- and HfB_2-Based ultra-high temperature ceramics: effect of Ta additions. J Mater Sci 2004;39 (19):5969–5977.

56. Monteverde F, Bellosi A. The resistance to oxidation of an HfB_2–SiC composite. J Eur Ceram Soc 2005;25 (7):1025–1031.

57. Lespade P, Richet N, Goursat P. Oxidation resistance of HfB_2–SiC composites for protection of carbon-based materials. Acta Astronaut 2007;60 (10–11):858–864.

58. Carney C. Oxidation resistance of hafnium diboride-silicon carbide from 1400 to 2000°C. J Mater Sci 2009;44 (20):5673–5681.

59. Sciti D, Medri V, Silvestroni L. Oxidation behaviour of HfB_2–15 Vol.% $TaSi_2$ at low, intermediate and high temperatures. Scr Mater 2010;63 (6):601–604.

60. Carney CM, Parthasarathy TA, Cinibulk MK. Oxidation resistance of hafnium diboride ceramics with additions of silicon carbide and tungsten boride or tungsten carbide. J Am Ceram Soc 2011;94 (8):2600–2607.

61. Ni D, Zhang G, Xu F, Guo W. initial stage of oxidation process and microstructure analysis of HfB_2–20 Vol.% SiC composite at 1500°C. Scr Mater 2011;64 (7):617–620.

62. Tang S, Deng J, Wang S, Liu W. Fabrication and characterization of an Ultra-High-temperature carbon fiber-reinforced ZrB_2-SiC matrix composite. J Am Ceram Soc 2007;90 (10):3320–3322.

63. Tang S, Deng J, Wang S, Liu W, Yang K. Ablation behaviors of ultra-high temperature ceramic composites. Mater Sci Eng A 2007;465 (1–2):1–7.

64. Levine SR, Opila EJ, Robinson RC, Lorincz JA. Characterization of an ultra-high temperature ceramic composite. NASA Report No.: NASA TM-2004-213085. Cleveland, OH: NASA Glenn Research Centre; 2004.

65. Sayir A. Carbon fiber reinforced hafnium carbide composite. J Mater Sci 2004;39 (19): 5995–6003.

66. Paul A, Jayaseelan DD, Venugopal S, Zapata-Solvas E, Binner J, Vaidhyanathan B, Heaton A, Brown P, Lee WE. UHTC composites for hypersonic applications. J Am Ceram Soc 2012;91:22–29.

67. Jayaseelan DD, de Sa RG, Brown P, Lee WE. Reactive infiltration processing (RIP) of Ultra High temperature ceramics (UHTC) into porous C/C composite tubes. J Eur Ceram Soc 2011;31 (3):361–368.

68. Musa C, Orrù R, Sciti D, Silvestroni L, Cao G. Synthesis, consolidation and characterization of monolithic and SiC whiskers reinforced HfB_2 ceramics. J Eur Ceram Soc 2013;33 (3):603–614.

69. Silvestroni L, Sciti D, Melandri C, Guicciardi S. Toughened ZrB_2-based ceramics through SiC whisker Or SiC chopped fiber additions. J Eur Ceram Soc 2010;30 (11):2155–2164.

70. Zhang P, Hu P, Zhang X, Han J, Meng S. Processing and characterization of ZrB_2–SiC_W Ultra-High temperature ceramics. J Alloys Compd 2009;472 (1–2):358–362.

71. Zhang X, Xu L, Du S, Han J, Hu P, Han W. Fabrication and mechanical properties of ZrB_2–SiC_W ceramic matrix composite. Mater Lett 2008;62 (6–7):1058–1060.

72. Zhang X, Xu L, Du S, Liu C, Han J, Han W. Spark plasma sintering and hot pressing of ZrB_2–SiC_W Ultra-High temperature ceramics. J Alloys Compd 2008;466 (1–2):241–245.

73. Lee SG, Fourcade J, Latta R, Solomon AA. Polymer impregnation and pyrolysis process development for improving thermal conductivity of SiCp/SiC–PIP matrix fabrication. Fusion Eng Des 2008;83 (5–6):713–719.

74. Schilling CL, Wesson JP, Williams TC. Polycarbosilane precursors for silicon carbide. J Polym Sci 1983;70 (1):121–128.

75. Takeda M, Kagawa Y, Mitsuno S, Imai Y, Ichikawa H. Strength of a Hi-nicalon/silicon carbide matrix composite fabricated by the multiple polymer infiltration-pyrolysis process. J Am Ceram Soc 1999;82 (6):1579–1581.

76. Jones R, Szweda A, Petrak D. Polymer derived ceramic matrix composites. Compos Part A Appl Sci Manuf 1999;30 (4):569–575.

77. Manocha LM, Bahl OP, Singh YK. Mechanical behaviour of carbon-carbon composites made with surface treated carbon fibers. Carbon 1989;27 (3):381–387.

78. Zhao D, Zhang C, Hu H, Zhang Y. Ablation behavior and mechanism of 3D C/ZrC composite in oxyacetylene torch environment. Compos Sci Technol 2011;71 (11):1392–1396.

79. Ly HQ, Taylor R, Day RJ. Carbon fibre-reinforced CMCs by PCS infiltration. J Mater Sci 2001;36 (16):4027–4035.

80. Zhao D, Zhang C, Hu H, Zhang Y. Preparation and characterization of three-dimensional carbon fiber reinforced zirconium carbide composite by precursor infiltration and pyrolysis process. Ceram Int 2011;37 (7):2089–2093.

81. Wang Z, Dong S, Zhang X, Zhou H, Wu D, Zhou Q, Jiang D. Fabrication and properties of Cf/SiC-ZrC composites. J Am Ceram Soc 2008;91 (10):3434–3436.

82. Li Q, Dong S, Wang Z, He P, Zhou H, Yang J, Wu B, Hu J. Fabrication and properties of 3-D Cf/SiC-ZrC composites, using ZrC precursor and polycarbosilane. J Am Ceram Soc 2012;95 (4):1216–1219.

83. Padmavathi N, Ray K, Subrahmanyam J, Ghosal P, Kumari S. New route to process unidirectional carbon fiber reinforced (SiC+ZrB$_2$) matrix mini-composites. J Mater Sci 2009;44 (12):3255–3264.

84. Corral EL, Walker LS. Improved ablation resistance of C–C composites using zirconium diboride and boron carbide. J Eur Ceram Soc 2010;30 (11):2357–2364.

85. Leslie CJ, Boakye E, Keller KA, Cinibulk MK. Development of continuous SiC fiber reinforced HfB$_2$–SiC composites for aerospace applications. In: Bansal NP, Singh JP, Ko SW, et al., editors. *Processing and Properties of Advanced Ceramics and Composites V*. Volume 240, Hoboken (NJ): John Wiley & Sons; 2013. p 3–12.

86. Pierson HO. *Handbook of Chemical Vapor Deposition: Principles, Technology and Applications*. New York: William Andrew Publishing, LLC; 1999. p 1–11.

87. Pierson HO. *Handbook of Chemical Vapor Deposition: Principles, Technology and Applications*. New York: William Andrew Publishing, LLC; 1999. p 207–240.

88. Pierson HO. *Handbook of Chemical Vapor Deposition: Principles, Technology and Applications*. New York: William Andrew Publishing, LLC; 1999. p 299–320.

89. Patterson MCL, He S, Fehrenbacher LL, Hanigofsky J, Reed BD. Advanced HfC-TaC oxidation resistant composite rocket thruster. Mater Manuf Process 1996;11 (3):367–379.

90. Patterson MCL. Oxidation resistant HfC-TaC rocket thruster for high performance propellants. NASA Report No: NASA-27272. Cleveland, OH: NASA Glenn Research Centre; 1999.

91. Zou L, Wali N, Yang J, Bansal NP. Microstructural development of a Cf/ZrC composite manufactured by reactive melt infiltration. J Eur Ceram Soc 2010;30 (6):1527–1535.

92. Zhang S, Wang S, Li W, Zhu Y, Chen Z. Preparation of ZrB$_2$ based composites by reactive melt infiltration at relative low temperature. Mater Lett 2011;65 (19–20):2910–2912.

93. Lipke DW, Zhang Y, Liu Y, Church BC, Sandhage KH. Near net-shape/net-dimension ZrC/W-based composites with complex geometries via rapid prototyping and displacive compensation of porosity. J Eur Ceram Soc 2010;30 (11):2265–2277.

94. Liu Y, Lipke DW, Zhang Y, Sandhage KH. The kinetics of incongruent reduction of tungsten carbide via reaction with a hafnium–copper melt. Acta Mater 2009;57 (13):3924–3931.

95. Zhu Y, Wang S, Li W, Zhang S, Chen Z. Preparation of carbon fiber-reinforced zirconium carbide matrix composites by reactive melt infiltration at relative low temperature. Scr Mater 2012;67 (10):822–825.

96. Paul A, Venugopal S, Binner JGP, Vaidhyanathan B, Heaton ACJ, Brown PM. UHTC–carbon fibre composites: preparation, oxyacetylene torch testing and characterisation. J Eur Ceram Soc 2013;33 (2):423–432.

97. Paul A, Binner J, Vaidhyanathan B. UHTC composites for hypersonic applications, Internal project report to the Defence Science and Technology Laboratory (DSTL), Report: UHTC2. Porton Down, UK: Loughborough University; 2013.

98. Li L, Wang Y, Cheng L, Zhang L. Preparation and properties of 2D C/SiC–ZrB$_2$–TaC composites. Ceram Int 2011;37 (3):891–896.

99. Hu H, Wang Q, Chen Z, Zhang C, Zhang Y, Wang J. Preparation and characterization of C/SiC–ZrB$_2$ Composites by precursor infiltration and pyrolysis process. Ceram Int 2010;36 (3):1011–1016.

100. Wang Y, Xiong X, Li G, Zhang H, Chen Z, Sun W, Zhao X. Microstructure and ablation behavior of hafnium carbide coating for carbon/carbon composites. Surf Coat Technol 2012;206 (11–12):2825–2832.

101. Wang Y, Xiong X, Li G, Liu H, Chen Z, Sun W, Zhao X. Ablation behavior of HfC protective coatings for carbon/carbon composites in an oxyacetylene combustion flame. Corros Sci 2012;65:549–555.

102. Wang Y, Xiong X, Zhao X, Li G, Chen Z, Sun W. Structural evolution and ablation mechanism of a hafnium carbide coating on a C/C Composite in an oxyacetylene torch environment. Corros Sci 2012;61:156–161.

103. Zhou H, Gao L, Wang Z, Dong S. ZrB_2-SiC Oxidation protective coating on C/C composites prepared by vapor silicon infiltration process. J Am Ceram Soc 2010;93 (4):915–919.

104. Yang X, Wei L, Song W, Bi-feng Z, Zhao-hui C. ZrB_2/SiC as a protective coating for C/SiC composites: effect of high temperature oxidation on mechanical properties and anti-ablation property. Compos Part B Eng 2013;45 (1):1391–1396.

105. Blum YD, Marschall J, Hui D, Young S. Thick protective UHTC coatings for SiC-based structures: process establishment. J Am Ceram Soc 2008;91 (5):1453–1460.

106. Wang Y, Liu Q, Liu J, Zhang L, Cheng L. Deposition mechanism for chemical vapor deposition of zirconium carbide coatings. J Am Ceram Soc 2008;91 (4):1249–1252.

107. Liu Q, Zhang L, Cheng L, Wang Y. Morphologies and growth mechanisms of zirconium carbide films by chemical vapor deposition. J Coat Technol Res 2009;6 (2):269–273.

108. Emig G, Schoch G, Wormer O. Chemical vapor deposition of hafnium carbide and hafnium nitride. J Phys IV France 1993;03:C3-535–C3-540.

109. Ache HF, Goschnick J, Sommer M, Emig G, Schoch G, Wormer O. Chemical vapour deposition of hafnium carbide and characterization of the deposited layers by secondary-neutral mass spectrometry. Thin Solid Films 1994;241 (1–2):356–360.

110. Yang F, Zhang X, Han J, Du S. Characterization of hot-pressed short carbon fiber reinforced ZrB_2-SiC Ultra-High temperature ceramic composites. J Alloys Compd 2009;472 (1-2):395–399.

111. Pope EJ, Kratsch KM. US Patent 2002/0165332 A1. November 7, 2002.

112. Miles R, Howard P, Limbach C, Zaidi S, Lucato S, Cox B, Marshall D, Espinosa AM, Driemeyer D. A shape-morphing ceramic composite for variable geometry scramjet inlets. J Am Ceram Soc 2011;94 (s1):s35–s41.

8

MECHANICAL PROPERTIES OF ZIRCONIUM-DIBORIDE BASED UHTCs

Eric W. Neuman* and Greg E. Hilmas

Materials Science and Engineering Department, Missouri University of Science and Technology, Rolla, MO, USA

8.1 INTRODUCTION

This chapter focuses on the mechanical properties of zirconium diboride (ZrB_2)-based ultra-high temperature ceramics (UHTCs), including elastic modulus, strength, and fracture toughness at ambient and elevated temperatures. The materials covered include ZrB_2 ceramics with additions of various sintering aids, silicon carbide, and transition metal silicides. Reporting of carbide, nitride, and oxide additions is limited to instances where materials were added in small quantities or as sintering aids.

To produce structures using UHTCs, designers need to have a robust set of mechanical property data detailing their elastic properties, fracture toughness, and strength. These data will need to include microstructural relations, as well as specimen processing and testing procedures. Currently, much of the published mechanical property data are not accompanied by microstructural analysis. Further, a wide variety of testing methods are utilized for measuring the mechanical properties. Lack of microstructural characterization confounds the analysis, which is directly related to the microstructural features. Further, the use of various testing standards and specimen preparation

*Current affiliation: Intel Corporation, Beaverton, OR, USA.

Ultra-High Temperature Ceramics: Materials for Extreme Environment Applications, First Edition.
Edited by William G. Fahrenholtz, Eric J. Wuchina, William E. Lee, and Yanchun Zhou.
© 2014 The American Ceramic Society. Published 2014 by John Wiley & Sons, Inc.

methods make comparisons between studies difficult. Microstructural information is included in the proceeding discussion, when provided. Differences in testing methods will be noted, but only minimally discussed.

The chapter also discusses improvements in mechanical properties that have been obtained using modern powders and processing techniques compared to historic studies from the 1960s and 1970s. The main contributions to the improvement in properties of the modern materials are the use of powder with higher purity and finer particle sizes, along with the use of sintering aids and isothermal reaction holds to remove impurities, such as surface oxides, from the starting powders.

8.2 ROOM TEMPERATURE MECHANICAL PROPERTIES

Zirconium diboride (ZrB_2), a transition metal boride compound, is part of a class of materials known as UHTCs. This family of compounds is characterized by melting points in excess of 3000°C [1]. These ceramics are candidates for applications including molten metal crucibles [2, 3], furnace electrodes, cutting tools [4, 5], control rods for fission reactors, and wing leading edges on future hypersonic aerospace vehicles. Additionally, particulate-reinforced ZrB_2 matrix composites, especially those with silicon carbide (SiC) additives, have displayed enhanced properties. ZrB_2–SiC composites exhibit room temperature strengths in excess of 1000 MPa [6–8], fracture toughness values as high as 5.5 MPa·$m^{1/2}$ [6, 7, 9], and hardness values exceeding 22 GPa [6, 8, 10]. These properties are comparable to other commonly used structural ceramics (Al_2O_3, ZrO_2, Si_3N_4, etc.). In addition, ZrB_2 offers chemical stability and increased refractoriness. The following discussion highlights the room temperature elastic modulus, strength, and fracture toughness of ZrB_2-based UHTCs.

8.2.1 ZrB_2

Elastic modulus values reported for selected ZrB_2 ceramics with and without sintering additives are summarized in Table 8.1. Elastic modulus ranges from approximately 350 to approximately 530 GPa depending on porosity and additives (Fig. 8.1). Fitting the data to the Einstein (linear) [42], Spriggs [43], and Nielson [40, 41] models (Table 8.2) gives a value for fully dense ZrB_2 of 511 GPa compared to measured values of 490–500 GPa in previous studies [38, 44]. However, the materials in the historic studies were not fully dense, and fitting the historic data to similar models predicts an elastic modulus of 519 GPa for dense ZrB_2. Zhang et al. [45] calculated the elastic modulus of ZrB_2 from first principles to be 520 GPa, in good agreement with the fitted data. Recent work by Okamoto et al. calculated the polycrystalline elastic modulus of ZrB_2 from single crystal measurements to be 525 GPa [46]. The modulus of 99.8%-dense ZrB_2 was measured to be 489 GPa [26]. However, additions of AlN, Si_3N_4, B_4C, and C, as well as impurities picked up during powder milling (i.e., WC, ZrO_2), affect the elastic modulus of ZrB_2 ceramics. Small additions of B_4C and/or C increase the elastic modulus of ZrB_2 [20, 22, 29]. Additions of AlN and Si_3N_4 lower the elastic modulus [11, 32]. Additions affect the elastic modulus primarily through interactions with impurities on the surface of the starting powders. Additives such as C and B_4C remove low modulus surface

TABLE 8.1. Elastic modulus, Vickers hardness, fracture toughness by direct crack method, and four-point flexure strength of ZrB_2 ceramics with and without sintering additives

Composition (vol%)	Relative density (%)	Grain size (μm)	Elastic modulus (GPa)	Hardness (GPa)	Fracture toughness (MPa·m^{½})	Flexure strength (MPa)	References
ZrB_2	87	10	346±4	8.7±0.4	2.4±0.2[a]	351±31	[11–14]
ZrB_2	90	—	—	16.1±1.1	1.9±0.4	325±35[b]	[15]
ZrB_2	90.4	6.1	417	—	4.8±0.4	457±58	[16–18]
ZrB_2	95.8	10	—	16.5±0.9	3.6±0.3	450±40[b]	[19]
ZrB_2	97	8.1	479±8	16.7±0.6	2.8±0.1	452±27	[20]
ZrB_2	97.2	5.4±2.8	498	—	—	491±22	[21]
ZrB_2	98	9.1	454	14.5±2.6	—	444±30	[22, 23]
ZrB_2	>98	—	—	14.7±0.8	—	300±40	[24]
ZrB_2	~99	20	491±34	—	—	326±46	[25]
ZrB_2	99.8	~6	489	23±0.9	3.5±0.3[c]	565±53	[26–28]
ZrB_2 +0.5 wt% C	99.4	19±13	524±17	14.3±0.7	2.9±0.2[a]	381±41	[29, 30]
ZrB_2 +4 wt% B_4C	100	8	530	18	3.1	473	[22]
ZrB_2 +4 wt% Ni	98	5–15	496	14.4±0.8	3.4±0.4[a]	371±24	[12, 31]
ZrB_2 +4.6 AlN	~92	—	407±5	9.4±0.5	3.1±0.1[a]	580±80	[32]
ZrB_2 +5 Si_3N_4	98	3	419±5	13.4±0.6	3.7±0.1[a]	600±90	[11, 33–35]
ZrB_2 +3 YAG	95	7.5	—	—	5.4±0.2[d]	629±31[b]	[36]
ZrB_2 +2 wt% B_4C+1 wt% C	100	4.1	507	19.6±0.4	3.5	575±29	[22, 37]
ZrB_2 +2 wt% B_4C+1 wt% C	100	2.5	509±11	19.7±0.6	3.0±0.1	547±35	[20]

[a]Chevron notched beam.

[b]Three-point flexure.

[c]Indentation strength in bending.

[d]Single-edge notched beam.

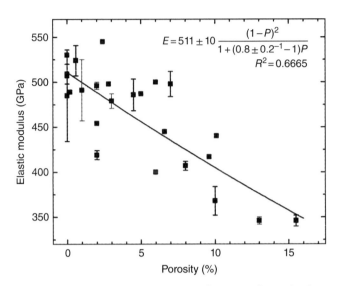

Figure 8.1. Room-temperature elastic modulus as a function of porosity for ZrB$_2$ (left) with and without sintering aids [11–14, 16, 17, 20–23, 25–29, 31–35, 37–39]. Line represents fitted relationship of elastic modulus to porosity according to Nielsen's relationship [40, 41].

TABLE 8.2. Summary of various fitted models of elastic modulus as a function of porosity for ZrB$_2$

Equation	E_0	b	R^2
$E = E_0(1 - bP)$	511 ± 10	2.0 ± 0.3	0.6416
$E = E_0 e^{-bP}$	512 ± 11	2.4 ± 0.4	0.6614
$E = E_0 \dfrac{(1-P)^2}{1 + (b^{-1} - 1)P}$	511 ± 11	0.8 ± 0.2	0.6665

oxides, which increases modulus, whereas additives such as AlN (308 GPa) [47] and Si$_3$N$_4$ (310 GPa) promote formation of low modulus grain boundary phases, which reduces modulus [44]. The elastic modulus of 99.4% dense ZrB$_2$ with 0.5 wt% C added as a sintering aid was 524 GPa [29]. This material was nearly phase pure (residual C not observed) and the modulus matched Okamoto's predicted polycrystalline value of 525 GPa. The reported modulus values for ZrB$_2$ are typically impacted porosity and/or impurities, but the intrinsic elastic modulus seems to be approximately 525 GPa.

Room temperature flexure strengths for ZrB$_2$ ceramics with and without sintering aids are also given in Table 8.1. The strength ranges from 250 to 630 MPa depending on grain size and additives (Fig. 8.2). In general, the strength of ZrB$_2$ follows an inverse square root relation with grain size (GS$^{-\frac{1}{2}}$), as expected for ceramics free of other larger flaws. The line in Figure 8.2 highlights the GS$^{-\frac{1}{2}}$ relationship, but is not intended as a fit

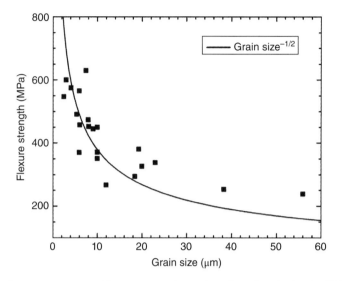

Figure 8.2. Room-temperature flexure strength as a function of grain size for ZrB$_2$ (left) with and without sintering additives [11–14, 16–29, 31, 33, 34, 36, 37, 48]. Line is not fitted to data, and is meant to guide the eye.

of the data. This relation indicates that flaw-free ZrB$_2$ with a fine grain size has improved strengths.

The fracture toughness of ZrB$_2$ is generally in the range of 3.0–4.5 MPa·m$^{1/2}$, with most values being reported near approximately 3.5 MPa·m$^{1/2}$. Unfortunately, the effects of porosity or grain size on fracture toughness have not been investigated. Also, the possibility of R-curve behavior has not been investigated, even though the CTE mismatch between the a-axis (6.9×10^{-6} K^{-1}) and c-axis (6.7×10^{-6} K^{-1}) [49] of the hexagonal unit cell leads to residual stresses that could induce R-curve behavior.

8.2.2 ZrB$_2$ with SiC Additions

8.2.2.1 *Size Effect of SiC Addition.* The four-point flexure strength of ZrB$_2$–SiC ceramics without additives is shown in Table 8.3. Although ZrB$_2$ grain size is a factor, the room temperature strength of ZrB$_2$–SiC is controlled by the size of the dispersed SiC particles or clusters [7–9]. Rezaie *et al.* [7] and Watts *et al.* [8] used linear elastic fracture mechanics to show that critical flaw sizes correlated to the SiC cluster size in ZrB$_2$–30 vol% SiC ceramics. This conclusion is consistent with the CTE mismatch between ZrB$_2$ and SiC (for α-SiC: $\alpha_{a\text{-axis}} = 4.3 \times 10^{-6}$ K^{-1}, $\alpha_{c\text{-axis}} = 4.7 \times 10^{-6}$ K^{-1}) [55]. Watts *et al.* [56] used neutron diffraction to measure the thermal residual stresses present in ZrB$_2$–30 vol% SiC. They found that the thermal residual stresses begin to accumulate at approximately 1400°C, resulting in residual tensile stresses in the ZrB$_2$ and compressive stresses in the SiC at room temperature. Assuming constant strain across the interface between SiC and ZrB$_2$, the maximum tensile stress in the ZrB$_2$ was determined to be approximately 1000 MPa [8]. Watts predicted a critical grain size range

TABLE 8.3. Elastic modulus, Vickers hardness, fracture toughness by direct crack method, and four-point flexure strength of ZrB_2–SiC ceramics

Composition	Relative density	ZrB_2 Grain size	SiC Grain size	Elastic modulus	Hardness	Fracture toughness	Flexure strength	References
(vol%)	(%)	(μm)	(μm)	(GPa)	(GPa)	(MPa·m$^{1/2}$)	(MPa)	
ZrB_2 + 10SiC	—	3	—	507±4	—	4.8±0.3[a]	835±35	[13]
ZrB_2 + 10SiC	93.2	~3	—	450	24±0.9	4.1±0.3[b]	713±48	[26, 28]
ZrB_2 + 10SiC	97.1	2.2	~0.2	—	—	5.7±0.2[c]	720±55[d]	[50]
ZrB_2 + 10SiC	97.4	4.5±1.6	0.8±0.4	—	—	—	524±63	[16, 18]
ZrB_2 + 10SiC	99.8	4.3±1.4	—	500±16	18±0.9	3.8±0.3	393±114[d]	[51]
ZrB_2 + 15SiC	96.5	4.4±1.7	0.9±0.5	—	—	—	714±59	[16, 18]
ZrB_2 + 15SiC	99	2	—	480±4	17.7±0.4	4.1±0.1[a]	795±105	[52]
ZrB_2 + 20SiC	—	1.8	~0.2	—	—	6.8±0.1[c]	1009±43[d]	[50]
ZrB_2 + 20SiC	97.3	4.2±1.9	0.9±0.5	—	—	—	608±93	[16, 18]
ZrB_2 + 20SiC	99.7	~3	—	466	24±2.8	4.4±0.2[b]	1003±94[d]	[26, 28]
ZrB_2 + 20SiC	99.7	4.0±1.1	—	506	22.1±0.1	4.2±0.8	487±68[d]	[51]
ZrB_2 + 20SiC	5.62 g/cm^3	4.1±0.9	—	505±3	21.3±0.7	3.9±0.3	937±84[d]	[53]
ZrB_2 + 30SiC	—	3	~0.2	—	—	5.9±0.2[c]	860±70[d]	[50]
ZrB_2 + 30SiC	97.2	2.2±1.2	1.2±0.6	516±3	20±2.0	5.5±0.3[b]	1063±91	[7]
ZrB_2 + 30SiC	97.5	3.9±0.9	—	487±12	24.4±0.6	4.4±0.5	425±35[d]	[51]
ZrB_2 + 30SiC	99.8	1.5±1.2	—	510	27.0±2.2	2.1±0.1	800±115	[54]
ZrB_2 + 30SiC	99.4	~3	—	—	24.0±0.7	5.3±0.5[b]	1089±152	[26, 28]
ZrB_2 + 30SiC	99.8	1.2±0.4	1.0±0.4	520±7	20.7±1.0	4.6±0.1[b]	909±136	[9]
ZrB_2 + 30SiC	>99	10.6	1.6	541±22	21.4±0.6	—	1150±115	[8]

[a]Chevron notched beam.
[b]Indentation strength in bending.
[c]Single-edge notched beam.
[d]Three-point flexure.

for microcracking between 6.5 and 13.8 μm. This matches well with the critical SiC particle size of approximately 11.5 μm observed by Watts experimentally (estimated by fitting the major axis of the clusters to ellipses, as discussed later).

Microstructure–mechanical property relationships for ZrB_2-based ceramics are complex. Because the size of the particulate-reinforcing phase controls the strength, a uniform dispersion of SiC is required to maximize strength, even for compositions and starting particle sizes that should be below the percolation threshold for the system. The method of estimating grain size from image analysis data also affects the interpretation of microstructure–property relations. Figure 8.3 shows the flexure strength of ZrB_2–30 vol% SiC as a function of cluster size estimated using equivalent area diameter, one of the most used methods for reporting grain size. This method assumes a spherical particle, and underestimates the critical grain size for microcracking as approximately 3 μm (Fig. 8.3). Figure 8.3 also shows flexure strength as a function of the maximum measured equivalent area diameter (EAD_{max}). Since failure occurs at the largest flaw, a measurement that reports the largest flaw size is more useful, but still may not accurately describe the behavior of the material. From Figure 8.3, below the critical grain size determined by EAD_{max}, the strength shows a linear trend with grain size, instead of the inverse square root relationship predicted by Griffith analysis.

Figure 8.4 shows flexure strength, elastic modulus, and hardness as a function of the SiC cluster size that was estimated using the maximum measured major axis from ellipses (ME_{max}) fitted to the SiC clusters. All three properties experience a sudden, discontinuous decrease at approximately 11.5 μm. Below the critical grain size, the strength follows an inverse square root relation with cluster size. Fitting ellipses to the clusters is preferred because it more closely resembles the morphology of the clusters. Taking this

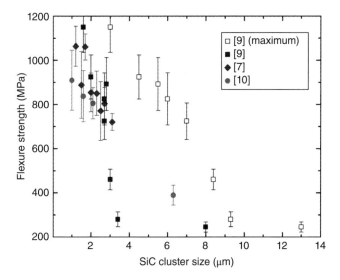

Figure 8.3. Room-temperature flexure strength as a function of SiC cluster size (equivalent area diameter) for ZrB_2–30 vol% SiC ceramics produced by hot pressing [7–9].

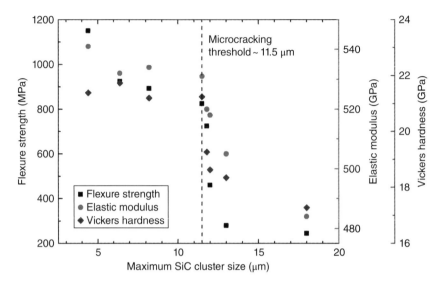

Figure 8.4. Room-temperature flexure strength, elastic modulus, and Vickers hardness as a function of maximum SiC cluster size (major axis of ellipse) for ZrB$_2$–30 vol% SiC ceramics prepared by hot pressing. The dashed line indicates the microcracking threshold that occurs at an SiC cluster size of approximately 11.5 μm [8].

analysis even further, measuring the maximum Feret's diameter of the cluster, should more accurately correlate to the flaw size by freeing the analysis from assumptions of circularity.

8.2.2.2 Effect of SiC Concentration. The effect of SiC concentration on the strength of ZrB$_2$–SiC ceramics is shown in Figure 8.5. In general, the strength (Table 8.3) increases with SiC content. From Chamberlain *et al.*, the strength of ZrB$_2$ increases from 560 to 1090 MPa with addition of up to 30 vol% SiC [26]. However, the strength is still controlled by the size of the SiC inclusions as discussed previously. Chamberlain was able to achieve high strengths through dispersion of the SiC particulate during processing. When percolation clusters begin to form, or the size of the SiC particles is large, strengths will decrease. Liu *et al.* produced ZrB$_2$–SiC ceramics using nanometer-sized SiC powder [50]. They produced strengths comparable to Chamberlain, except at the level of 30 vol% additions, where SiC began to form large percolation clusters, resulting in a material with lower strength than expected based solely on the submicron SiC particle size. Zhang *et al.* produced pressurelessly sintered ZrB$_2$–SiC that resulted in large asymmetric SiC grains from the long isothermal holding times (3 h) required for densification [57]. Even though Zhang achieved good dispersion of the SiC particles, the resulting SiC grain size (up to 13 μm long) was larger than the critical grain size for microcracking of approximately 11.5 μm, thus resulting in reduced strengths (490 MPa).

The fracture toughness of ZrB$_2$–SiC ceramics as a function of SiC additions is also shown in Figure 8.5. In general, fracture toughness (Table 8.3) increases with increasing SiC additions. Chamberlain reported a fracture toughness of 3.5 MPa·m$^{1/2}$ for pure ZrB$_2$,

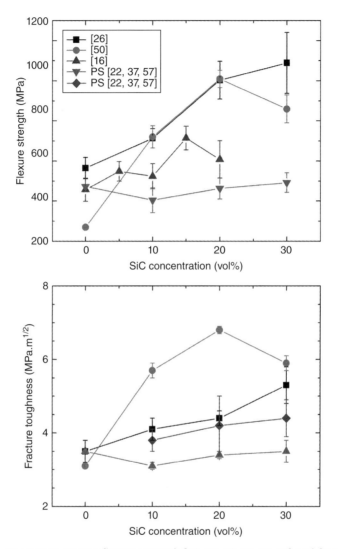

Figure 8.5. Room-temperature flexure strength [16, 22, 26, 37, 50, 57] and fracture toughness [16, 22, 26, 37, 50, 57] as a function of SiC concentration for ZrB_2–SiC ceramics produced by hot pressing and pressureless sintering.

increasing to $5.3 \, MPa \cdot m^{1/2}$ for ZrB2–30 vol% SiC. Liu's work [50] showed that toughness followed a similar trend to strength for nanosized SiC, increasing from $3.1 \, MPa \cdot m^{1/2}$ for ZrB_2 to $6.8 \, MPa \cdot m^{1/2}$ for 20 vol% SiC then decreasing to $5.9 \, MPa \cdot m^{1/2}$ for 30 vol% SiC (along with observed formation of percolation clusters). SiC additions increase the fracture toughness of ZrB_2 by increasing the tortuosity of the crack path. Figure 8.6 shows a polished and etched specimen of ZrB_2–SiC exhibiting crack bridging and crack deflection. The ZrB_2 grains typically fail by transgranular fracture, while cracks deflect at or

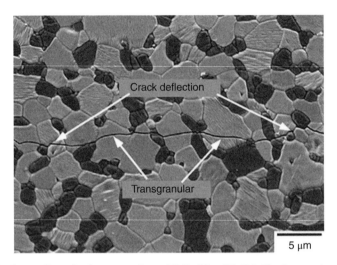

Figure 8.6. Thermally etched cross section of ZrB_2–30 vol% SiC. The image shows the crack path from a Vickers indent with arrows indicating predominantly transgranular fracture for the ZrB_2 grains and crack deflection near the ZrB_2–SiC interfaces. Reproduced from [58].

near ZrB_2–SiC interfaces. This behavior is consistent with residual stresses resulting from CTE mismatch between SiC and ZrB_2. Further, *R*-curve behavior has been observed in ZrB_2–SiC, as expected based on the CTE mismatch between the phases. Kurihara *et al.* [59] reported falling *R*-curve behavior for ZrB_2–10 vol% SiC and Bird *et al.* [60] reported rising *R*-curve behavior for ZrB_2–20 vol% SiC. Further increases in fracture toughness will likely require second-phase additions with higher aspect ratios, such as whiskers, rods, or platelets (discussed later), or the fabrication of laminate-type architectures [61].

The elastic modulus of ZrB_2 with SiC (475 GPa) [62] additions follows a simple linear volumetric rule of mixtures trend (Fig. 8.7). Small variations in elastic modulus from the expected values may be attributed to the impact of processing impurities on ZrB_2–SiC, such as WC from milling media or the effect of thermal residual stresses.

8.2.2.3 Effect of Additional Phases. The effect of various additions to ZrB_2–SiC is summarized in Table 8.4. Additions are typically introduced as sintering aids that may be designed to remove surface oxide impurities, as in the case of C and B_4C, or interacting with surface oxides to form softer grain boundary phases, as in the case of Si_3N_4. ZrB_2–SiC with other additions has similar room temperature mechanical properties to ZrB_2–SiC without additions. Nitride additions typically increase toughness by formation of a weaker grain boundary phase that increases crack bridging and deflection. Han [70] and Wang [71] observed an increased toughness for ZrB_2–20 vol% SiC (from ~4.2 to ~5.5 MPa·m$^{\frac{1}{2}}$) with 5–10 vol% additions of aluminum nitride, while retaining a room temperature flexure strength of 830 MPa. This increase was likely due to improved densification, which resulted in a finer grain size, and the presence of weaker grain boundary phases (BN and Al_2O_3) that enhanced crack deflection. Small additions of oxides (such as YAG [11] and La_2O_3 [19]) follow a similar trend to the

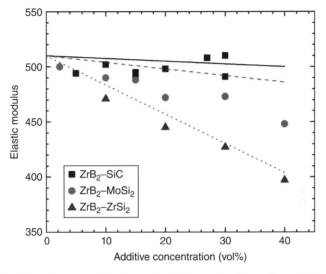

Figure 8.7. Elastic modulus as a function of additive content for selected hot-pressed ZrB$_2$-based composites with SiC [16, 18, 26, 28, 52, 54, 63] , MoSi$_2$ [14, 64–67] , and ZrSi$_2$ [17] additives. Values have been corrected for porosity using a linear relationship and b = 2.0.

nitride additions with minimal improvements in toughness. This shows that small additions of nitrides and oxides can be added to ZrB$_2$–SiC to provide additional means of controlling densification and microstructure without adversely affecting room temperature mechanical properties.

In contrast to oxide and nitride additions, the mechanical properties of ZrB$_2$–SiC ceramics with ZrC (Table 8.4) typically result in mechanical properties similar to ZrB$_2$–SiC without ZrC additions (Table 8.3). This is most likely the result of closer CTE match between ZrC (7.6×10^{-6} K^{-1}) [78] and ZrB$_2$ compared to SiC and ZrB$_2$, resulting in the SiC phase still dominating the mechanical behavior. With ZrC additions from 5 to 30 vol%, the flexure strengths (~500–750 MPa), fracture toughnesses (3.5–6.5 MPa · m^{½}) are nominally the same as ZrB$_2$–SiC. Similarly, it is reasonable to assume that the same microstructural controls used in ZrB$_2$–SiC to improve mechanical properties can be employed in the ZrB$_2$–SiC–ZrC system. Small additions of other carbides (C, B$_4$C, VC, and WC) have little effect on the mechanical properties of ZrB$_2$–SiC. However, these additions improve densification of ZrB$_2$–SiC in the same manner as when added to ZrB$_2$, through removal of surface oxides on the starting powders. Thus, small additions of carbides can also be used to improve the room temperature mechanical properties by allowing greater control of microstructure during sintering.

8.2.3 ZrB$_2$ with Disilicide Additions

8.2.3.1 Flexure Strength. The flexure strength of ZrB$_2$ with additions of ZrSi$_2$, MoSi$_2$, or TaSi$_2$ is summarized in Table 8.5. Figure 8.8 shows the effect of MoSi$_2$ and ZrSi$_2$ additives on the strength of ZrB$_2$. Guo [17] reported strengths for ZrB$_2$–ZrSi$_2$ in

TABLE 8.4. Elastic modulus, Vickers hardness, fracture toughness by direct crack method, and four-point flexure strength of ZrB_2–SiC ceramics with various additives

Composition (vol%)	Relative density (%)	ZrB_2 Grain size (µm)	SiC Grain size (µm)	Elastic modulus (GPa)	Hardness (GPa)	Fracture toughness (MPa·m$^{1/2}$)	Flexure strength (MPa)	References
ZrB_2+20SiC+1 wt% B	>99.0	7.3±0.8	3.3±0.5	—	16.0±0.4	—	519±31[a]	[68]
ZrB_2+27SiC+1B_4C	100	2	1	508±6	22.6±0.9	3.5	720±140	[63]
ZrB_2+30SiC+2 wt% B_4C[b]	100	1.9±0.9	1.2±0.5	513±24	20.2±0.5	4.9±0.4[c]	682±98	[69]
ZrB_2+20SiC+3wt%C+0.5wt%B_4C	99	~10	~8	374±25	14.7±0.2	5.5±0.5[d]	361±44[a]	[39]
ZrB_2+30SiC+4 wt% B_4C[b]+5 wt% C[e]	99	2.8±0.2	2±0.8	511±7	—	3.8±0.2[f]	604±69	[57]
ZrB_2+20SiC+2 wt% La_2O_3	99.6	3.5±0.4	—	—	19.3±0.6	5.2±0.5	600±70[a]	[19]
ZrB_2+20SiC+4Si_3N_4	98	2.4±0.1	—	—	14.6±0.3	—	730±100[a]	[11]
ZrB_2+20SiC+5AlN	100	3	—	—	19.4±0.6	5.4±0.3[d]	835±26[a]	[70]
ZrB_2+20SiC+10AlN	100	2.5	—	—	—	5.6±0.5[d]	831±12[a]	[71]
ZrB_2+15SiC+4.5ZrN	99	—	—	467±4	15.6±0.3	5.0±0.1[c]	635±60	[72]
ZrB_2+20SiC+6ZrC	99.1	~4	—	—	19.0±0.5	6.5±0.4[d]	622±64[a]	[73]
ZrB_2+21SiC+5ZrC	97.3	<2	—	—	17.2±0.8	5.2±0.4	747±101	[74]
ZrB_2+13SiC+15ZrC	>99	—	—	—	18.2	4.3	512±50[a]	[75]
ZrB_2+20SiC+10ZrC	99.8	2.7±0.5	1.6±0.2	—	18.1±0.6	5.3	662±64[a]	[76]
ZrB_2+10SiC+30ZrC	99	2	—	474	18.8±0.5	3.5±0.2	723±136	[67]
ZrB_2+20SiC+5VC	>99	1.9±0.3	1.4±0.3	—	15.8±0.3	5.5±0.5	804±90[a]	[77]
ZrB_2+10HfB_2+15SiC	98.2	3	—	508±4	18.2±0.5	4.1±0.8	765±75	[52]
ZrB_2+35HfB_2+15SiC+4.5ZrN	99.5	—	—	494±4	16.7±0.7	4.8±0.2[c]	590±25	[72]
ZrB_2+9.6SiC+28.9ZrC+3.7Si_3N_4	99.5	2	—	450	21.1±0.8	3.8±0.1	510±160	[67]
ZrB_2+18.5SiC+3.7Si_3N_4+1Al_2O_3+ 0.5Y_2O_3	98	2.5±0.1	—	421±5	14.2±0.6	4.6±0.1	710±110	[11]

[a]Three-point flexure.
[b]ZrB_2 basis.
[c]Chevron notched beam.
[d]Single-edge notched beam.
[e]SiC basis.
[f]Indentation strength in bending.

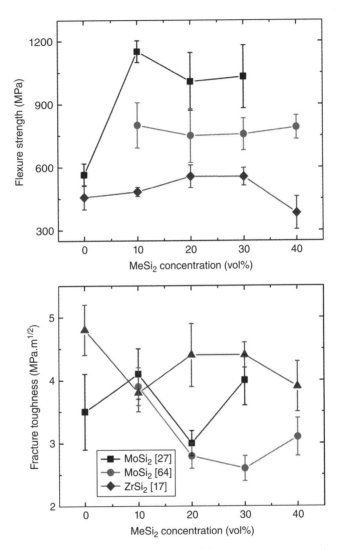

Figure 8.8. Room-temperature flexure strength and fracture toughness as a function of disilicide concentration for ZrB_2–$MeSi_2$ ceramics produced by hot pressing [17, 27, 64].

the range of 380 to 555 MPa depending on $ZrSi_2$ content. Chamberlain [27] reported the effect of $MoSi_2$ on strength of ZrB_2, increasing from 565 MPa for ZrB_2 to 1150 MPa for 10 vol% $MoSi_2$, decreasing to approximately 1020 MPa for 20 and 30 vol% additions of $MoSi_2$. Similarly, Guo [64] also reported the strength of ZrB_2 to increase from 460 MPa for ZrB_2 [17] to between 750 and 800 MPa for 10–40 vol% $MoSi_2$ additions.

The strength of ZrB_2 with $MoSi_2$ additions has been reported by numerous researchers. This has resulted in a wide range of reported strengths: ZrB_2–10 vol% $MoSi_2$, 560–1150 MPa; ZrB_2–20 vol% $MoSi_2$, 460–1010 MPa; ZrB_2–30 vol% $MoSi_2$, 550–1030 MPa. These values highlight the importance of microstructure and processing

TABLE 8.5. Elastic modulus, Vickers hardness, fracture toughness by direct crack method, and four-point flexure strength of $ZrB_2–(Zr, Mo, Ta)$ Si_2 ceramics

Composition	Relative density	ZrB_2 Grain size	$MeSi_2$ Grain size	Elastic modulus	Hardness	Fracture toughness	Flexure strength	References
(vol%)	(%)	(μm)	(μm)	(GPa)	(GPa)	(MPa·m$^{1/2}$)	(MPa)	
$ZrB_2 + 10ZrSi_2$	96.6	2.3±0.9	0.6±0.4	432	—	3.8±0.3	483±22	[17]
$ZrB_2 + 20ZrSi_2$	99.1	2.5±0.8	0.7±0.6	445	—	4.4±0.5	556±54	[17]
$ZrB_2 + 30ZrSi_2$	99.8	2.6±1.1	0.7±0.6	427	—	4.4±0.2	555±42	[17]
$ZrB_2 + 40ZrSi_2$	99.2	2.7±1.0	0.9±0.7	397	—	3.9±0.4	382±78	[17]
$ZrB_2 + 2.3MoSi_2$	6.04 g/cm³	5	—	500±2	18.1±0.4	3.4±0.3[a]	750±160	[14]
$ZrB_2 + 5MoSi_2$	96	2.6	—	516±4	15.2±1.0	2.9±0.1[a]	569±54	[79]
$ZrB_2 + 10MoSi_2$	99.7	1.9±0.6	1.8±0.5	490±7	15.8±0.7	3.7±0.3	800±108	[64]
$ZrB_2 + 10MoSi_2$	6.19 g/cm³	~3	—	516	20.4±2.2	4.1±0.4[b]	1151±52	[27, 28]
$ZrB_2 + 10MoSi_2$	99.9	—	—	—	17.5±0.4	3.8±0.4	560±115[c]	[80]
$ZrB_2 + 15MoSi_2$	5.99 g/cm³	1.9±0.6	—	452±4	14.7±0.6	3.5±0.6[a]	780±87	[66]
$ZrB_2 + 15MoSi_2$	98.1	1.4	—	479±4	16.2±0.5	2.6±0.3	643±97	[65–67]
$ZrB_2 + 20MoSi_2$	99.9	—	—	—	14.9±0.1	3.2±0.2	680±40[c]	[80]
$ZrB_2 + 20MoSi_2$	6.17 g/cm³	~3	—	523	18.5±2.7	3.0±0.2[b]	1008±137	[27, 28]
$ZrB_2 + 20MoSi_2$	99.8	1.6±0.6	2.7±0.9	472±6	16.3±0.9	2.8±0.2	750±128	[64]
$ZrB_2 + 20MoSi_2$	99.1	~2.5	—	489±4	16.0±0.4	2.3±0.2	531±46	[65, 79, 81, 82]
$ZrB_2 + 30MoSi_2$	99.7	—	—	—	14.3±0.2	3.2±0.3	550±65[c]	[80]
$ZrB_2 + 30MoSi_2$	99.8	2.1±0.7	2.4±0.6	473±3	15.4±0.7	2.6±0.2	757±76	[64]
$ZrB_2 + 30MoSi_2$	6.29 g/cm³	~3	—	494	17.7±1.6	4.0±0.4[b]	1031±150	[27, 28]
$ZrB_2 + 40MoSi_2$	99.7	1.9±0.7	2.6±0.8	448±4	13.2±0.7	3.1±0.3	790±57	[64]
$ZrB_2 + 15TaSi_2$	99	2	—	444±24	17.8±0.5	3.8±0.1[a]	840±33	[83]

[a]Chevron notched beam.
[b]Indentation strength in bending.
[c]Three-point flexure.

control similar to the ZrB_2–SiC system. Unfortunately, microstructural effects of disilicide additions on the strength of ZrB_2 have not been thoroughly investigated. This is further complicated by the tendency of the ZrB_2–disilicide systems to form complex grain boundary phases and solid solution phases with the ZrB_2. Silvestroni *et al.* [84] reported the formation of a ZrB_2 core–$(Zr,Mo)B_2$ rim structures during sintering. In addition, ZrC, ZrO_2, MoB, and SiO_2 were detected in the final microstructures. Further, they observed various Mo–Zr–B–Si phases that may have also contained C and O, as well as the core–rim structures. These complex chemistries and architectures make analysis of microstructure–mechanical properties' effects difficult. Though detailed microstructural analysis has not been performed on the other transition metal disilicide systems, similar processes presumably occur in these systems as well.

8.2.3.2 Fracture Toughness. Table 8.5 shows the fracture toughness of ZrB_2 with the addition of $ZrSi_2$, $MoSi_2$, or $TaSi_2$. The effect of $ZrSi_2$ and $MoSi_2$ additions on the fracture toughness of ZrB_2 is also shown in Figure 8.8. In general, $MoSi_2$ additions result in fracture toughness values in the range of 3.2–4.0 MPa·m$^{1/2}$. Chamberlain [27] reported indentation strength in bending (ISB) fracture toughness values of 3–4 MPa·m$^{1/2}$ for ZrB_2–$MoSi_2$, but without a trend with $MoSi_2$ content. Guo [64] measured fracture toughness of ZrB_2–$MoSi_2$ using the direct crack method (DCM) and reported values that decreased from 4.8 MPa·m$^{1/2}$ for ZrB_2 [17] to approximately 2.8 MPa·m$^{1/2}$ for additions of 20–40 vol% $MoSi_2$. Guo [17] also reported DCM toughness for ZrB_2–$ZrSi_2$ as 3.8 MPa·m$^{1/2}$ for 10 vol% additions, increasing to 4.4 MPa·m$^{1/2}$ for 20 and 30 vol% additions, and then decreasing to 3.9 MPa·m$^{1/2}$ for 40 vol% $ZrSi_2$ additions. Unlike SiC additions to ZrB_2, no trend was observed for fracture toughness with disilicide additions.

Using the Griffith criteria and a Y-parameter of 1.99 (long, semi-elliptical surface flaw), the ZrB_2–$MoSi_2$ composites produced by Chamberlain and Guo both exhibit critical flaw sizes that correlated to the observed grain sizes. For example, Chamberlain's ZrB_2–10 vol% $MoSi_2$ has a calculated flaw size of 3.2 μm and a reported grain size of approximately 3 μm. However, Guo observed that the calculated critical flaw size was much larger than the observed grain size, suggesting that something other than grain size was controlling the strength of the ZrB_2–$ZrSi_2$ composites. From the microstructural analysis of polished cross sections and fracture surfaces, Guo observed that the $ZrSi_2$ segregated to the grain boundaries during sintering, exhibiting a "string of pearls" type morphology and clustering of $ZrSi_2$ grains. Further, small amounts of porosity were observed in the $ZrSi_2$ clusters between the grains. For ZrB_2–$ZrSi_2$, the size and morphology of the $ZrSi_2$ phase controls the failure behavior of the composite.

8.2.3.3 Elastic Modulus. Table 8.5 also summarizes the hardness and elastic modulus of ZrB_2 with the addition of $ZrSi_2$, $MoSi_2$, or $TaSi_2$. The effect of additive concentration on the elastic modulus of ZrB_2–$MoSi_2$ and ZrB_2–$ZrSi_2$ is shown in Figure 8.7. The dashed lines represent the expected modulus based on linear volumetric rule of mixtures calculations assuming moduli of 510 GPa for ZrB_2, 440 GPa for $MoSi_2$ [85], and 235 GPa for $ZrSi_2$ [86]. From Figure 8.7, it can be seen that the ZrB_2–$MoSi_2$ composites follow the rule of mixtures trend, but the values are consistently 10–20 GPa lower than expected. This is likely due to the formation of complex phases and solid solutions during densification ZrB_2–$MoSi_2$ composites. Thermal residual stresses resulting from the CTE mismatch between ZrB_2 ($\alpha_{a\text{-axis}} = 6.9 \times 10^{-6}\,K^{-1}$, $\alpha_{c\text{-axis}} = 6.7 \times 10^{-6}\,K^{-1}$) [49] and $MoSi_2$ ($\alpha_{a\text{-axis}} = 8.2 \times 10^{-6}\,K^{-1}$, $\alpha_{c\text{-axis}} = 9.4 \times 10^{-6}\,K^{-1}$) [87] may also play a role in the observed difference between the predicted and observed modulus. In the case of ZrB_2–$ZrSi_2$, the observed modulus is in good agreement with the predicted modulus. Since a tendency to form complex phases has not been observed in the ZrB_2–$ZrSi_2$ system, this likely explains why the elastic modulus of ZrB_2–$ZrSi_2$ follows predictions while ZrB_2–$MoSi_2$ has a more complex behavior.

8.2.4 ZrB_2–$MeSi_2$–SiC

The flexure strength and fracture toughness of several ZrB_2–disilicide ceramics with SiC additions is shown in Figure 8.9. Guo [64] investigated the effect of SiC additions to ZrB_2–20 vol% $MoSi_2$ and ZrB_2–40 vol% $MoSi_2$ ceramics. The strength of both ZrB_2–20 vol% $MoSi_2$ and ZrB_2–40 vol% $MoSi_2$ increased with the addition of 5 vol% SiC, but decreased with SiC contents of more than 5 vol%. Further, toughness increased with the addition of SiC, but was relatively insensitive to the amount. The improvement in properties was the result of a finer grain size for the ZrB_2 and $MoSi_2$, with the addition of 5 vol% SiC. Further additions of SiC increased the observed average and maximum grain sizes for ZrB_2, $MoSi_2$, and SiC. A Griffith-type failure analysis revealed that the critical flaw size increased with the SiC content, and is consistent with the maximum observed $MoSi_2$ grain size. Elastic modulus increased with a 5 vol% SiC addition, then decreased with further additions. Additions of SiC to ZrB_2–$TaSi_2$ [88, 89] resulted in similar trends in mechanical properties as additions to ZrB_2–$MoSi_2$. Additions of SiC to ZrB_2–5 vol% $TaSi_2$ and ZrB_2–10 vol% $TaSi_2$ increased the toughness. In general, additions of SiC to ZrB_2 with disilicides results in improved toughness.

8.3 ELEVATED-TEMPERATURE MECHANICAL PROPERTIES

While several studies have reported elevated-temperature mechanical properties of ZrB_2 ceramics, most have only reported a single property (i.e., strength) for a limited number of temperatures and have not provided a systematic evaluation of the effects of temperature on mechanical behavior. The following discussion highlights studies of the elevated-temperature elastic modulus, strength, and fracture toughness of various ZrB_2-based UHTCs.

8.3.1 Elastic Modulus of ZrB_2-Based Ceramics

The elevated-temperature elastic modulus of hot-pressed ZrB_2 with and without additives is shown in Figure 8.10. Only limited data on elastic behavior of ZrB_2 have been published for elevated temperatures. Okamoto *et al.* [49] reported elastic modulus from single crystal values as being 525 GPa at room temperature, decreasing linearly to 490 GPa at 1100°C. Experimental results of bulk ZrB_2 have resulted in lower values of elastic modulus than those reported by Okamoto. The historical work of Rhodes *et al.* [25] provided the only reported values for the elastic modulus of bulk ZrB_2 tested in inert atmosphere. The modern studies by Zhu [20] and Neuman *et al.* [29] were conducted in

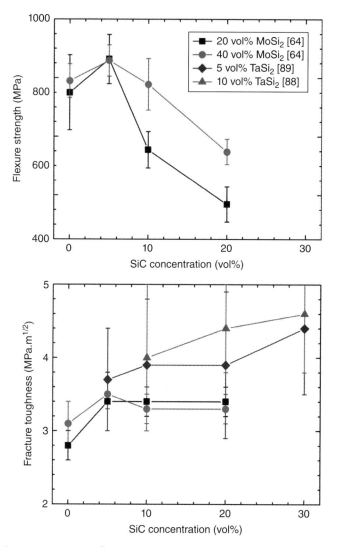

Figure 8.9. Room-temperature flexure strength and fracture toughness as a function of SiC content for ZrB_2–$MoSi_2$–SiC and ZrB_2–$TaSi_2$–SiC ceramics [64, 88, 89].

air, and thus are affected by the formation of an oxide scale on the test articles. Unfortunately, the effect of the oxide scale on the measured elastic modulus was not investigated. Regardless, the elastic modulus of bulk ZrB_2 decreased more rapidly than expected based on single crystal measurements presumably due to a grain boundary effect. The modulus decreased linearly from room temperature to approximately 1200°C (450 GPa). Above this temperature, the elastic modulus decreased more rapidly with temperature. The change in slope was thought to be a result of softening of grain boundary phases combined with the activation of grain boundary sliding and diffusional creep mechanisms [20, 25, 29, 69, 90]. Reported modulus values were similar up to 1600°C

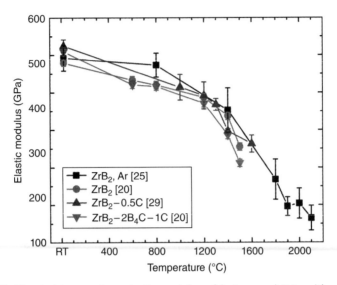

Figure 8.10. Elevated-temperature elastic modulus of hot-pressed ZrB_2 with and without additives [20, 25,29].

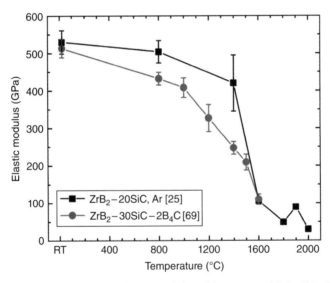

Figure 8.11. Elevated-temperature elastic modulus of hot- pressed ZrB_2–SiC with and without additives [25, 69].

(250 MPa), despite the differences in testing methods. To date, Rhodes [25] is the only study to report the modulus above 1600°C, where it decreases to 100 GPa at 2000°C.

The elevated-temperature elastic modulus of ZrB_2–SiC ceramics is shown in Figure 8.11. Rhodes *et al.* [25] observed a steady decrease in modulus from 530 GPa at room temperature to 420 GPa at 1400°C. Above 1400°C, the modulus decreased dramatically to approximately 100 GPa at 1600°C. Neuman *et al.* [69] observed a steady decrease from 510 GPa at room temperature to 410 GPa at 1000°C. The slope of the modulus then changed, decreasing to 210 GPa at 1500°C, and then more rapidly to 110 GPa at 1600°C. As with ZrB_2, the changes in slope of the elastic modulus are likely the result of softening of grain boundary phases, and grain boundary sliding and diffusional creep. In ZrB_2–SiC, the softening was expected to occur at a lower temperatures to the presence of SiO_2 (surface oxide impurity from SiC) and its interaction with the B_2O_3 and ZrO_2 present from the ZrB_2. Zou *et al.* [90] measured the internal damping in ZrB_2–SiC prepared by milling with Si_3N_4 balls, and found that the damping peaks just above 800°C and again around 1400°C. The reduction in modulus with temperature for ZrB_2–SiC ceramics is enhanced by the presence of oxides at the grain boundaries and triple junctions as well as the polycrystalline nature of the material.

8.3.2 Strength and Fracture Toughness

8.3.2.1 *Strength of ZrB_2*.
The elevated-temperature four-point flexure strength of ZrB_2 with and without additives is shown in Figure 8.12. In the historical study of Rhodes [25], a room-temperature strength of 325 MPa was measured for fully dense ZrB_2 with an approximately 20 μm grain size. Strength increased between room temperature and 800°C (420 MPa), as a result of relief of thermal residual stresses. Strength decreased to 145 MPa at 1400°C, increased to 200 MPa at 1900°C, and then decreased to 50 MPa at 2200°C. The observed increase in strength from 1400 to 1900°C was thought to be caused by stress relief during testing through plastic flow, arising from diffusional creep. A general trend was observed that finer grain size, and lower porosity resulted in improved elevated-temperature strength, which was offset by enhanced creep

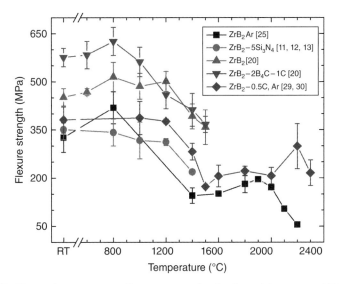

Figure 8.12. Elevated-temperature flexure strength of selected hot-pressed ZrB_2 ceramics with and without additives in air and argon [11–13, 20, 25, 29, 30].

in the finer-grained material. Similar results have been observed in other structural ceramic systems such as SiC, ZrO_2, Al_2O_3, and Si_3N_4.

Since the historical work of Rhodes [25], several modern studies have investigated the strength of ZrB_2 at elevated temperatures. Melendez-Martinez et al. [12] measured the flexure strength of an 87%-dense ZrB_2 (350 MPa at room temperature) up to 1400°C in air (220 MPa). Zhu [20] produced dense ZrB_2 and ZrB_2–2 wt% B_4C–1 wt% C with grain sizes of 8.1 and 2.5 μm, respectively. At room temperature, the ZrB_2 had a strength of 450 MPa compared to 575 MPa for ZrB_2–B_4C–SiC. The ZrB_2 strength increased up to 1200°C (500 MPa) before decreasing to 360 MPa at 1500°C. The strength of ZrB_2–B_4C–C increased to 630 MPa at 800°C, then steadily decreased to 370 MPa at 1500°C. Zhu performed his testing in air, and even though he used a protective SiO_2 coating, the strength above 1200°C was likely controlled by flaws induced by oxidation damage (discussed in more detail later). For dense ZrB_2 ceramic with 0.5 wt% carbon added as sintering aid and a grain size of approximately 19 μm [29] the room temperature strength of the modern ZrB_2 was 380 MPa, compared to 325 MPa for the Rhodes material. The strength was maintained up to 1200°C before decreasing to a minimum of 170 MPa at 1500°C. Above 1600°C, the strength was approximately 220 MPa up to 2300°C. Chemical compatibility between ZrB_2 and the graphite test fixture ($T_e = 2390$°C) prevented testing at higher temperatures. The improvement in strength, especially above 2000°C, was most likely due to improved purity of modern ZrB_2 powders (the ZrB_2 used by Rhodes contained several percent ZrC and ZrO_2). Additionally, the carbon sintering aid helped remove the surface oxides present on the ZrB_2 starting powders [22], presumably leaving less oxide phase at the grain boundaries, which would improve high-temperature strength and stiffness.

8.3.2.2 Strength of ZrB₂–SiC.
The elevated-temperature flexure strength of several hot-pressed ZrB_2–SiC ceramics with and without additives is shown in Figure 8.13. The ZrB_2–20 vol% SiC studied by Rhodes [25] exhibited an increase in strength from 390 MPa at room temperature to 420 MPa at 800°C. The strength steadily decreased to 245 MPa at 1800°C and decreased more rapidly to 115 MPa at 2000°C. Modern ZrB_2–SiC exhibited higher strengths at room and elevated temperature compared to Rhodes' material and similarly for all temperatures measured thus far. Zou et al. [90, 91] have reported the three-point bend strengths at elevated temperature for several compositions up to 1600°C in argon. Zou reported a strength of 550 MPa for ZrB_2–20 vol% SiC at room temperature, increasing to 680 MPa at 1000°C, then steadily decreasing to 460 MPa at 1600°C. Zou added 5 vol% WC, which resulted in a room-temperature strength of 605 MPa that steadily increased to 675 MPa at 1600°C. Interestingly, the composition ZrB_2–20SiC–5ZrC (not shown) exhibited a room-temperature strength of 1100 MPa, only slightly decreasing to 1020 MPa at 1000°C. Increasing temperatures resulted in a steep decrease in strength to 320 MPa at 1600°C. Grigoriev [15] measured the strength of a ZrB_2–19 vol% SiC composite in air (not shown), reporting a room-temperature strength of 500 MPa that was maintained up to 1200°C, decreasing to 430 MPa at 1400°C. Not only do the room-temperature strengths vary widely for ZrB_2–SiC, but the change in strength with temperature also varies.

Zou showed that the addition of WC to ZrB_2–20SiC resulted in finer ZrB_2 and SiC grain sizes plus enhanced removal of surface oxides from the starting powders. Reducing

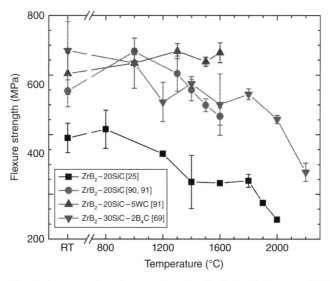

Figure 8.13. Elevated-temperature flexure strength of selected hot-pressed ZrB_2–SiC ceramics with and without additives in argon [25, 69, 90, 91].

the amount of low-melting-point oxides present at the grain boundaries should result in a higher stiffness at elevated temperatures, thus also improving strength and resistance to creep. Similarly, the increase in strength for ZrB_2–20SiC–5ZrC at room temperature was attributed to grain refinement. The steep drop in strength at elevated temperatures was attributed to the softening of residual oxides present at grain boundaries, since no favorable reactions with the surface oxides or ZrC were identified up to the sintering temperature. Further, creep was observed in the material during testing at 1600°C. Hence, the presence of oxide impurities reduces strength and increases creep at elevated temperatures.

Neuman *et al.* reported the elevated-temperature four-point flexure strengths of ZrB_2–30SiC with 2 wt% B_4C in both an inert argon atmosphere and air (not shown) [30, 69]. The strength of ZrB_2–30SiC–2B_4C in argon was 680 MPa at room temperature, steadily decreased to 540 MPa at 1800°C, then more rapidly decreased to 450 MPa at 2000°C, and finally reached 275 MPa at 2200°C. In air, the strength increased from 680 MPa at room temperature to approximately 740 MPa at 800 and 1000°C and decreased to approximately 370 MPa at 1500 and 1600°C. The increase in strength between room temperature and 1000°C was attributed to the healing of surface flaws from the formation of the ZrO_2–B_2O_3 oxide scale. The size of the SiC clusters controlled the strength up to 1800°C in the inert atmosphere. In air, the size of the SiC clusters controlled strength up to 1000°C, while oxidation damage controlled strength above 1200°C, as discussed in the following. The elevated temperature testing performed by Neuman was different than other studies in that increasing testing rates were used at increasing temperatures to stay in the fast fracture regime and avoid creep effects, which is consistent with the methods described in ASTM C1211. The increased testing rate typically results in higher measured strengths compared to tests using a slower rate where nonlinear-elastic behavior is observed.

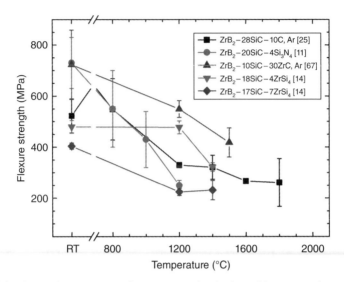

Figure 8.14. Elevated-temperature flexure strength of selected hot-pressed ZrB$_2$–SiC ceramics with various additives in argon [11, 15, 25, 67].

Figure 8.14 also shows the elevated-temperature flexure strengths of some ZrB$_2$–SiC ceramics with various additives. Rhodes added graphitic carbon to ZrB$_2$–SiC to improve the thermal shock resistance. The strength increased from 520 MPa at room temperature to 640 MPa at 600°C in argon and then steadily decreased to 260 MPa at 1800°C. Monteverde [11] added 4 vol% Si$_3$N$_4$ as a sintering aid to ZrB$_2$–20 vol% SiC. The strength decreased from 730 MPa at room temperature to 250 MPa at 1200°C due to formation of various phases containing Zr, Si, and/or B with O and/or N at the grain boundaries and triple junctions. These phases begin to soften at temperatures as low as 800°C, leading to strength degradation with increasing temperatures. Bellosi [67] measured the strength of a ZrB$_2$–10SiC–30ZrC composite in argon, showing a room-temperature strength of 720 MPa, which steadily decreased to 420 MPa at 1500°C. The decline in strength between 1200 and 1500°C was similar to the decline observed by Zou over a similar temperature range. Grigoriev made additions of 4 and 7 vol% ZrSi$_2$ to ZrB$_2$–18SiC and ZrB$_2$–17SiC. In the case of the 4% addition, a strength of approximately 480 MPa was maintained up to 1200°C, before decreasing to 230 MPa. Adding 7 vol% ZrSi$_2$ resulted in a decrease in strength from 400 MPa at room temperature to approximately 230 MPa at 1200 and 1400°C. Grigoriev did not speculate as to the effect of ZrSi$_2$ on the response of strength at elevated temperatures, but it is likely due to the relatively low melting point of ZrSi$_2$ of 1620°C and the formation of complex phases at the grain boundaries and triple junctions as seen with in ZrB$_2$–MoSi$_2$ as discussed in the following.

8.3.2.3 Strength of ZrB$_2$ with MoSi$_2$ and TaSi$_2$. Figure 8.15 shows the four-point flexure strengths at elevated temperatures in air for several ZrB$_2$ ceramics containing MoSi$_2$ or TaSi$_2$. ZrB$_2$–20 vol% SiC prepared by Rhodes is included as a comparison. Silvestroni [79] measured the strength of a pressurelessly sintered ZrB$_2$–5 vol% MoSi$_2$

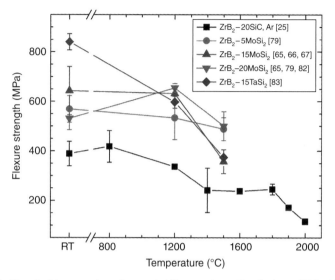

Figure 8.15. Elevated-temperature four-point flexure strength of selected ZrB_2–$MeSi_2$ ceramics in air [25, 65–67, 79, 82, 83].

ceramics, finding that the strength decreased only slightly from 570 MPa at room temperature to 490 MPa at 1500°C. Bellosi *et al.* [67] reported properties for a spark-plasma-sintered ZrB_2–$15MoSi_2$ ceramic, finding that the strength remained approximately 635 MPa from room temperature up to 1200°C, before decreasing to 360 MPa at 1600°C. In a follow-up study, Balbo and Sciti [66] reported similar results for a hot-pressed material of the same composition, speculating that the decrease in strength at 1500°C was the result of softening of silicate phases present in the material. Sciti *et al.* [81] also measured the strength of pressurelessly sintered ZrB_2–$20MoSi_2$ ceramics, in this case observing an increase in strength from 530 MPa at room temperature to 655 MPa at 1200°C, declining to 500 MPa at 1500°C. They attributed the increase in strength at 1200°C to healing of surface flaws due to formation of an oxide scale. Sciti *et al.* [83] also investigated the properties of hot-pressed ZrB_2–$TaSi_2$, reporting a strength of 840 MPa at room temperature that decreased to 375 MPa at 1500°C. This strength was slightly higher than they observed for a comparably processed ZrB_2–$15MoSi_2$ (705 MPa at room temperature, 335 MPa at 1500°C), but the strength followed a similar trend. The increased strength of ZrB_2–$TaSi_2$ was likely a result of the observed increase in fracture toughness between the $MoSi_2$- and $TaSi_2$-containing composites, but no further speculation was offered by the authors.

The ZrB_2–transition metal disilicide systems show promise for their retention of strength at elevated temperatures. Unfortunately, few data are available for these systems above 1500°C. In addition, no elastic properties or fracture toughness behavior have been reported at elevated temperatures. Chemical stability issues may also restrict the use of these composites in the ultra-high temperature regimes. These are discussed in more detail later, but should not be meant to diminish the promising properties these materials have demonstrated thus far.

8.3.2.4 Fracture Toughness of ZrB₂ and ZrB₂–SiC. Measuring the fracture toughness allows for the critical flaw size for failure to be calculated using the measured strength. Figure 8.16 shows the fracture toughness of ZrB_2 [30] and ZrB_2–30 vol% SiC [69] determined by chevron notched beam in flexure at elevated temperatures. The fracture toughness of ZrB_2 with 0.5 wt% C increased from 2.9 MPa·m^{½} at room temperature to 5.2 MPa·m^{½} at 1200°C, above which toughness steadily decreased to 3.7 MPa·m^{½} at 2300°C. Using the Griffith criterion and a Y-parameter of 1.29 (semicircular surface crack), the critical flaw size was similar to the maximum grain size for nearly all temperatures. The critical flaw size calculated for the material at 1400°C was slightly higher than any features observed in the microstructure. The cause of this discrepancy is currently unknown, but could be due to a change in the stress state, since Watts [56] found that stresses begin to accumulate in ZrB_2 below 1400°C and Rhodes [25] observed plasticity at 1400°C and higher.

The fracture toughness of ZrB_2–30SiC with 2 wt% B_4C, was measured up to 1600°C in air [69]. The fracture toughness was approximately 4.8 MPa·m^{½} from room temperature to 800°C, steadily decreasing to 3.3 MPa·m^{½} at 1600°C. Failure analysis revealed that SiC clusters acted as critical flaws below 1000°C, while oxidation damage was the critical flaw above 1200°C. While the oxide scale is a logical choice for failure origin in these materials, the study by Neuman showed that regions of enhanced ingress of oxide scale were the critical flaws. Thus, control of the microstructure is only part of the solution to improving the performance of these ceramics at elevated temperatures in oxidizing atmospheres. Control of the oxidation behavior and the morphology of the oxide scale are also critical for improving their performance in structural applications in extreme environments.

Kalish *et al.* [92] reported an increase in the amount of transgranular fracture in ZrB_2 from approximately 20% at room temperature up to approximately 60% at 1000°C. The

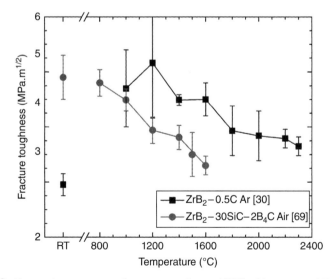

Figure 8.16. Elevated-temperature fracture toughness (CNB) of hot-pressed ZrB_2 and ZrB_2–SiC ceramics [30, 69].

amount of transgranular fracture then decreased to approximately 10% at 1200 and 1400°C. Similarly, Bird *et al.* [60] observed an increase in the amount of intergranular fracture in ZrB_2–20 vol% SiC from approximately 5% at room temperature to approximately 95% at 1400°C. The amount of intergranular fracture increased between 800 and 1000°C. The increase in the amount of intergranular fracture with temperature corresponded to an increase in the observed plateau in toughness, which increased from approximately 2.8 MPa·m$^{1/2}$ at room temperature to approximately 5.8 MPa·m$^{1/2}$ at 1400°C.

8.4 CONCLUDING REMARKS

This chapter presented a review of the current state-of-the-art of the mechanical properties of zirconium diboride-based ceramics. These ceramics offer a combination of properties that include high refractoriness, modulus, hardness, strength, and moderate fracture toughness. Their mechanical properties are controlled by their microstructures. The control of grain size, the dispersion and cluster size of second phases, and impurity phases are critical for the production of materials for use in extreme environment applications. Lack of sufficient control over the microstructure and impurities reduces the properties of these materials—for example, exceeding the microcracking threshold. The mechanical properties of these materials at elevated temperatures were also reviewed. To date, there has been a lack of fundamental research studies of the mechanical properties of these materials at elevated temperatures. Since the main driver for research into the ZrB_2 system is for use in ultra-high temperature (>2000°C) applications, further effort is needed to evaluate the mechanical properties of these materials in relevant environments. It was shown that there is currently a lack of robust microstructure—property relations for the ZrB_2-based systems. Microstructure and impurity characterization needs to be performed, and controlled, in future studies.

This chapter has shown that ceramics based on ZrB_2 offer promise for use as structural materials in extreme environments. It was shown that ZrB_2 ceramics offer mechanical properties at ultra-high temperatures that should allow designers to incorporate these materials for use in future applications for hypersonic vehicles and in other industrial applications where ceramics that can maintain strength (hundreds of MPa) at temperatures exceeding 1500°C are needed.

REFERENCES

1. Fahrenholtz WG, Hilmas GE, Talmy IG, Zaykoski JA. Refractory diborides of zirconium and hafnium. J Am Ceram Soc 2007;90 (5):1347–1364.
2. Jin ZJ, Zhang M, Guo DM, Kang RK. Electroforming of copper/ZrB_2 composite coating and its performance as electro-discharge machining electrodes. Key Eng Mater 2005;291–292: 537–542.
3. Sung J, Goedde DM, Girolami GS, Abelson JR. Remote-plasma chemical vapor deposition of conformal ZrB_2 films at low temperature: a promising diffusion barrier for ultralarge scale integrated electronics. J Appl Phys 2002;91 (6):3904–3911.

4. Mishra SK, Das S, Das SK, Ramachandrarao P. Sintering studies on ultrafine ZrB_2 powder produced by a self-propagating high-temperature synthesis process. J Mater Res 2000;15 (11):2499–2504.

5. Murata Y. *Cutting Tool Tips and Ceramics Containing Hafnium Nitride and Zirconium Diboride.* U.S. patent 3,487,594. 1970.

6. Chamberlain AL, Fahrenholtz WB, Hilmas GE. High strength ZrB_2-based ceramics. J Am Ceram Soc 2004;87 (6):1170–1172.

7. Rezaie A, Fahrenholtz WG, Hilmas GE. Effect of hot pressing time and temperature on the microstructure and mechanical properties of ZrB_2–SiC. J Mater Sci 2007;42 (8):2735–2744.

8. Watts J, Hilmas GE, Fahrenholtz WG. Mechanical characterization of ZrB_2–SiC composites with varying SiC particle sizes. J Am Ceram Soc 2011;94 (12):4410–4418.

9. Zhu S, Fahrenholtz WG, Hilmas GE. Influence of silicon carbide particle size on the microstructure and mechanical properties of zirconium diboride-silicon carbide ceramics. J Eur Ceram Soc 2007;27:2077–2083.

10. Chamberlain AL, Fahrenholtz WG, Hilmas GE. Low-temperature densification of zirconium diboride ceramics by reactive hot pressing. J Am Ceram Soc 2006;89 (12): 3638–3645.

11. Monteverde F, Guicciardi S, Bellosi A. Advances in microstructure and mechanical properties of zirconium diboride based ceramics. Mater Sci Eng A 2003;346:310–319.

12. Melendez-Martinez JJ, Dominguez-Rodriguez A, Monteverde F, Melandri C, Portu GD. Characterization and high temperature mechanical properties of zirconium boride-based materials. J Eur Ceram Soc 2002;22:2543–2549.

13. Monteverde F. Beneficial effects of an ulta-fine α-SiC incorporation on the sinterability and mechanical properties of ZrB_2. Appl Phys A 2006;82:329–337.

14. Monteverde F. The addition of SiC Particles into a $MoSi_2$-doped ZrB_2 matrix: effects on densification, microstructure and thermo-physical properties. Mater Chem Phys 2009;113:626–633.

15. Grigoriev ON, Galanov BA, Kotenko VA, Ivanov SM, Koroteev AV, Brodnikovsky NP. Mechanical properties of ZrB_2–SiC($ZrSi_2$) ceramics. J Eur Ceram Soc 2010;30:2173–2181.

16. Guo SQ, Yang JM, Tanaka H, Kagawa Y. Effect of thermal exposure on strength of ZrB_2-based composites with nano-sized SiC particles. Compos Sci Tech 2008;68:3033–3040.

17. Guo SQ, Kagawa Y, Nishimura T. Mechanical behavior of two-step hot-pressed ZrB_2-based composites with $ZrSi_2$. J Eur Ceram Soc 2009;29:787–794.

18. Guo SQ. Densification of ZrB_2-based composites and their mechanical and physical properties: a review. J Eur Ceram Soc 2009;29:995–1011.

19. Zapata-Solvas E, Jayaseelan DD, Lin PBHT, Lee WE. Mechanical properties of ZrB_2- and HfB_2-based ultra-high temperature ceramics fabricated by spark plasma sintering. J Eur Ceram Soc 2013;33 (7):1373–1386.

20. Zhu S. Densification, microstructure, and mechanical properties of zirconium diboride based ultra-high temperature ceramics [PhD thesis]. Rolla (MO): Missouri University of Science and Technology; 2008.

21. Guo S, Nishimura T, Kagawa Y. Preparation of zirconium diboride ceramics by reactive spark plasma sintering of zirconium hydride-boron powders. Scripta Mater 2011;65: 1018–1021.

22. Fahrenholtz WG, Hilmas GE, Zhang SC, Zhu S. Pressureless sintering of zirconium diboride: particle size and additive effects. J Am Ceram Soc 2008;91 (5):1398–1404.

23. Chamberlain AL, Fahrenholtz WG, Hilmas GE. Pressureless sintering of zirconium diboride. J Eur Ceram Soc 2006;89 (2):450–456.

24. Talmy IG, Zaykoski JA, Opeka MM, Smith AH. Properties of ceramics in the system ZrB_2-Ta_5Si_3. J Mater Res 2006;21 (10):2593–2599.

25. Rhodes WH, Clougherty EV, Kalish D. Research and development of refractory oxidation-resistant diborides part II. volume IV: mechanical properties. Technical Report AFML-TR-68-190. Wright-Patterson Air Forse Base (OH): ManLabs Incorporated and Avco Corporation; 1970.

26. Chamberlain AL, Fahrenholtz WG, Hilmas GE. High-strength zirconium diboride-based ceramics. J Am Ceram Soc 2004;87 (6):1170–1172.

27. Chamberlain AL, Fahrenholtz WG, Hilmas GE, Ellerby DT. Characterization of zirconium diboride-molybdenum disilicide ceramics. In: Bansal NP, Singh JP, Kriven WM, Schneider H, editors. *Advances in Ceramic Matrix Composites IX*. Westerville (OH): The American Ceramic Society; 2003. p 299–308.

28. Chamberlain AL, Fahrenholtz WG, Hilmas GE, Ellerby DT. Characterization of zirconium diboride for thermal protection systems. Key Eng Mater 2004;264–268:493–496.

29. Neuman EW, Hilmas GE, Fahrenholtz WG. Strength of zirconium diboride to 2300 °C. J Am Ceram Soc 2013;96 (1):7–50.

30. Neuman EW. Elevated temperature mechanical properties of zirconium diboride based ceramics [PhD thesis]. Rolla (MO): Missouri University of Science and Technology; 2014.

31. Monteverde F, Bellosi A, Guicciardi S. Processing and properties of zirconium diboride–based composites. J Eur Ceram Soc 2002;22:279–288.

32. Monteverde F, Bellosi A. Beneficial effects of AlN as sintering aid on microstructure and mechanical properties of hot-pressed ZrB_2. Adv Eng Mater 2003;5 (7):508–512.

33. Guicciardi S, Silvestroni L, Nygren M, Sciti D. Microstructure and toughening mechanisms in spark plasma-sintered ZrB_2 ceramics reinforced by SiC whiskers or SiC-chopped fiber. J Am Ceram Soc 2010;93 (8):2384–2391.

34. Silvestroni L, Sciti D, Melandri C, Guicciardi S. Toughened ZrB_2-based ceramics through SiC whisker or SiC chopped fiber additions. J Eur Ceram Soc 2010;30 (11):2155–2164.

35. Sciti D, Silvestroni L. Processing, sintering and oxidation behavior of SiC fibers reinforced ZrB_2 composites. J Eur Ceram Soc 2012;32:1933–1940.

36. Zhang X, Xu L, Han W, Weng L, Han J, Du S. Microstructure and properties of silicon carbide whisker reinforced zirconium diboride ultra-high temperature ceramics. Solid State Sci 2009;11 (1):156–161.

37. Zhu S, Fahrenholtz WG, Hilmas GE, Zhang SC. Pressureless sintering of zirconium diboride using boron carbide and carbon additions. J Am Ceram Soc 2007;90 (11):3660–3663.

38. Wiley DE, Manning WR, Hunter O. Elastic properties of polycrystalline TiB_2, ZrB_2 and HfB_2 from room temperature to 1300°K. J Less-Common Metals 1969;18 (2):149–157.

39. Zhang H, Yan Y, Huang Z, Liu X, Jiang D. Properties of ZrB_2-SiC ceramics by pressureless sintering. J Am Ceram Soc 2009;92 (7):1599–1602.

40. Nielsen LF. Elastic properties of two-phase materials. Mater Sci Eng 1982;52 (1):39.

41. Nielsen LF. Elasticity and damping of porous materials. J Am Ceram Soc 1984;67 (2):93–98.

42. Dean EA, Lopez JA. Empirical dependence of elastic moduli on porosity for ceramic materials. J Am Ceram Soc 1983;66 (5):366–370.

43. Spriggs RM. Expression for effect of porosity on elastic modulus of polycrystalline refractory materials, particularly aluminum oxide. J Am Ceram Soc 1961;44 (12):628–629.

44. Cutler RA. Engineering properties of borides. In: Schneider SJS Jr, editor. *Ceramics and Glasses: Engineered Materials Handbook.* Volume 4, Materials Park, OH: ASM International Materials; 1991. p 787–803.

45. Zhang X, Luo X, Han J, Li J, Han W. Electronic structure, elasticity and hardness of diborides of zirconium and hafnium: first principles calculations. Comput Mater Sci 2008;44 (2):411–421.

46. Okamoto NL, Kusakari M, Tanaka K, Inui H, Otani S. Anisotropic elastic constants and thermal expansivities in monocrystal CrB_2, TiB_2, and ZrB_2. Acta Mater 2010;58 (1):76–84.

47. Wachtman JB. *Mechanical Properties of Ceramics.* New York, NY: John Wiley & Sons, Inc.; 1996.

48. Li CW, Lin YM, Wang MF, Wang CA. Preparation and mechanical properties of ZrB_2-based ceramics using $MoSi_2$ as sintering aids. Front Mater Sci China 2010;4 (3):271–275.

49. Okamoto NL, Kusakari M, Tanaka K, Inui H, Yamaguchi M, Otani S. Temperature dependence of of thermal expansion and elastic constants of single crystals of ZrB_2 and the suitability of ZrB_2 as a substrate for GaN Film. J Appl Phys 2003;93 (1):88–93.

50. Liu Q, Han W, Han J. Influence of SiC_{np} Content on the microstructure and mechanical properties of ZrB_2-SiC nanocomposite. Scripta Mater 2010;63:581–584.

51. Patel M, Reddy JJ, Prasad VVB, Jayaram V. Strength of hot pressed ZrB_2-SiC composite after exposure to high temperatures (1000–1700°C). J Eur Ceram Soc 2012;32 (16):4455–4467.

52. Monteverde F, Scatteia L. Resistance to thermal shock and to oxidation of metal diborides-SiC ceramics for aerospace applications. J Am Ceram Soc 2007;94 (4):1130–1138.

53. Ran S, Biest OVD, Vleugels J. ZrB_2-SiC composites prepared by reactive pulsed electric current sintering. J Eur Ceram Soc 2010;30 (12):2633–2642.

54. Chamberlain AL, Fahrenholtz WG, Hilmas GE. Low-temperature densification of zirconium diboride ceramics by reactive hot pressing. J Am Ceram Soc 2006;89 (12):6368–6375.

55. Kern EL, Hamil DW, Deam HW, Sheets HD. Thermal properties of silicon carbide from 20 to 2000°C. Mater Res Bull 1969;4:S25–S32.

56. Watts J, Hilmas G, Fahrenholtz WG, Brown D, Clausen B. Measurement of thermal residual stresses in ZrB_2–SiC composites. J Eur Ceram Soc 2011;31 (9):1811–1820.

57. Zhang SC, Hilmas GE, Fahrenholtz WG. Mechanical properties of sintered ZrB_2-SiC ceramics. J Eur Ceram Soc 2011;31 (5):893–901.

58. Fahrenholtz WG, Hilmas GE, Talmy IG, Zaykoski JA. Refractory diborides of zirconium and hafnium. J Am Ceram Soc 2007;95 (5):1347–1364.

59. Kurihara JK, Tomimatsu T, Liu YF, Guo SQ, Kagawa Y. Mode I fracture toughness of SiC particel-dispersed ZrB_2 matrix composied measured using DCDC specimen. Ceram Int 2010;36:381–384.

60. Bird MW, Aune RP, Thomas AF, Becher PF, White KW. Temperature-dependent mechanical and long crack behavior of zirconium diboride–silicon carbide composite. J Eur Ceram Soc 2012;32 (12):3453–3462.

61. Kovar D, King BH, Trice RW, Halloran JW. Fibrous monolithic ceramics. J Am Ceram Soc 1997;80 (10):2471–2487.

62. Shaffer PTB, Jun CK. The elastic modulus of dense polycrystalline silicon carbide. Mater Res Bull 1972;7 (1):63–70.

63. Zimmerman JW, Hilmas GE, Fahrenholtz WG, Monteverde F, Bellosi A. Fabrication and properties of reactively hot pressed ZrB_2-SiC ceramics. J Eur Ceram Soc 2007;27 (7):2729–2736.

64. Guo SQ, Nishimura T, Mizuguchi T, Kagawa Y. Mechanical properties of hot-pressed ZrB_2-$MoSi_2$-SiC composites. J Eur Ceram Soc 2008;28:1891–1898.

65. Sciti D, Monteverde F, Guicciardi S, Pezzotti G, Bellosi A. Microstructure and mechanical properties of ZrB_2-$MoSi_2$ ceramic composites produced by different sintering techniques. Mater Sci Eng A 2006;434:303–309.

66. Balbo A, Sciti D. Spark plasma sintering and hot pressing of ZrB_2-$MoSi_2$ ultra-high-temperature ceramics. Mater Sci Eng A 2008;475:108–112.

67. Bellosi A, Monteverde F, Sciti D. Fast densification of ultra-high-temperature ceramics by spark plasma sintering. Int J Appl Ceram Technol 2006;3 (1):32–40.

68. Wang XG, Guo WM, Zhang GJ. Pressureless sintering mechanism and microstructure of ZrB_2-SiC ceramics doped with boron. Scr Mater 2009;61 (2):177–180.

69. Neuman EW, Hilmas GE, Fahrenholtz WG. Mechanical behavior of zirconium diboride-silicon carbide ceramics at elevated temperature in air. J Eur Ceram Soc 2013;33 (15):2889–2899.

70. Han W, Li G, Zhang X, Han J. Effect of AlN as sintering aid on hot-pressed ZrB_2-SiC ceramic composite. J Alloys Compd 2009;471 (1):488–491.

71. Wang Y, Liang J, Han W, Zhang X. Mechanical properties and thermal shock behavior of hot-pressed ZrB_2-SiC-AlN composites. J Alloys Compd 2009;475 (1):762–765.

72. Monteverde F, Bellosi A. Development and characterization of metal-diboride-based composites toughened with ultra-fine SiC particulates. Solid State Sci 2005;7 (5):622–630.

73. Qiang Q, Xinghong Z, Songhe M, Wenbo H, Changqing H, Jiecai H. Reactive hot pressing and sintering characterization of ZrB_2–SiC–ZrC composites. Mater Sci Eng A 2008;491 (1):117–123.

74. Wu WW, Zhang GJ, Kan YM, Wang PL. Reactive hot pressing of ZrB_2–SiC–ZrC composites at 1600°C. J Am CeramSoc 2008;91 (8):2501–2508.

75. Wu WW, Zhang GJ, Kan YM, Wang PL. Reactive synthesis and mechanical properties of ZrB_2-SiC-ZrC composites. Key Eng Mater 2008;368:1758–1760.

76. Guo WM, Zhang GJ. Microstructures and mechanical properties of hot-pressed ZrB_2-based ceramics from synthesized ZrB_2 and ZrB_2-ZrC powders. Adv Eng Mater 2009;11 (3):206–210.

77. Zou J, Zhang GJ, Kan YM, Wang PL. Hot-pressed ZrB_2–SiC ceramics with VC addition: chemical reactions, microstructures, and mechanical properties. J Am Ceram Soc 2009;92 (12):2838–2846.

78. Richardson JH. Thermal expansion of three group IVA carbides to 2700°C. J Am Ceram Soc 1965;48 (10):497–499.

79. Silvestroni L, Sciti D. Effects of $MoSi_2$ additions on the properties of Hf–and Zr–B_2 composites produced by pressureless sintering. Scr Mater 2007;57 (2):165–168.

80. Liu HT, Wu WW, Zou J, Ni DW, Kan YM, Zhang GJ. In situ synthesis of ZrB_2–$MoSi_2$ platelet composites: reactive hot pressing process, microstructure and mechanical properties. Ceram Int 2012;38 (6):4751–4760.

81. Sciti D, Guicciardi S, Bellosi A, Pezzotti G. Properties of a pressureless-sintered ZrB_2–$MoSi_2$ ceramic composite. J Am Ceram Soc 2006;89 (7):2320–2322.

82. Silvestroni L, Sciti D, Bellosi A. Microstructure and properties of pressureless sintered HfB_2-based composites with additions of ZrB_2 or HfC. Adv Eng Mater 2007;9 (10): 915–920.

83. Sciti D, Silvestroni L, Celotti G, Melandri C, Guicciardi S. Sintering and mechanical properties of ZrB_2–$TaSi_2$ and HfB_2–$TaSi_2$ ceramic composites. J Am Ceram Soc 2008;91 (10): 3285–3291.

84. Silvestroni L, Kleebe HJ, Lauterbach S, Müller M, Sciti D. Transmission electron microscopy on Zr- and Hf-borides with $MoSi_2$ addition: densification mechanisms. J Mater Res 2010;25 (5):828–834.

85. Nakamura M, Matsumoto S, Hirano T. Elastic constants of $MoSi_2$ and WSi_2 single crystals. J Mater Sci 1990;25:3309–3313.

86. Rosenkranz R, Frommeyer G. Microstructures and properties of the refractory compounds $TiSi_2$ and $ZrSi_2$. Z Metl 1992;83 (9):685–689.

87. Thomas O, Senateur JP, Madar R, Laborde O, Rosencher E. Molybdenum disilicide: crystal growth, thermal expansion and resistivity. Solid State Commun 1985;55 (7):629–632.

88. Hu C, Sakka Y, Tanaka H, Nishimura T, Guo S, Grasso S. Microstructure and properties of ZrB_2–SiC composites prepared by spark plasma sintering using $TaSi_2$ as sintering additive. J Eur Ceram Soc 2010;30 (12):2625–2631.

89. Hu C, Sakka Y, Jang B, Tanaka H, Nishimura T, Guo S, Grasso S. Microstructure and properties of ZrB_2–SiC and HfB_2–SiC composites fabricated by spark plasma sintering (SPS) using $TaSi_2$ as sintering aid. J Ceram Soc Jpn 2010;118 (1383):997–1001.

90. Zou J, Zhang GJ, Hu CF, Nishimura T, Sakka Y, Tanaka H, Vleugels J, Van der Biest O. High-temperature bending strength, internal friction and stiffness of ZrB_2–20 vol% SiC ceramics. J Eur Ceram Soc 2012;32 (10):2519–2527.

91. Zou J, Zhang GJ, Hu CF, Nishimura T, Sakka Y, Vleugels J, Biest O. ZrB_2–SiC–WC ceramics at 1600°C. J Am Ceram Soc 2012;95 (3):874–878.

92. Kalish D, Clougherty EV, Kreder K. Strength, fracture mode, and thermal stress resistance of HfB_2 and ZrB_2. J Am Ceram Soc 1969;52 (1):30–36.

9

THERMAL CONDUCTIVITY OF ZrB$_2$ AND HfB$_2$

Gregory J. K. Harrington and Greg E. Hilmas

Department of Materials Science and Engineering, Missouri University of Science and Technology, Rolla, MO, USA

9.1 INTRODUCTION

ZrB$_2$ and HfB$_2$ are proposed for use in hypersonic vehicles and reentry spacecraft where they will be used as thermal management materials for extreme thermal environments. These applications typically include sharp control surfaces (leading/trailing edges) as well as scramjet engine components [1–5]. Hypersonic control surfaces need thermal conductivities (k) that are as high as possible to conduct heat generated from atmospheric friction through the material so that it can be dissipated elsewhere [1, 3]. In addition, higher k values increase thermal shock resistance in high heat flux situations. Because engine components cannot radiate to the atmosphere, lower k refractory materials could be beneficial, depending on the cooling capabilities of the vehicle [3]. From a materials engineering perspective, the effects of processing, composition (intentional additions or from impurities), and the resulting microstructure need to be characterized to design materials with appropriate properties. The goal of this chapter is to review the state-of-the-art with respect to thermal conductivity for single-phase and composite ZrB$_2$ and HfB$_2$ ceramics.

Ultra-High Temperature Ceramics: Materials for Extreme Environment Applications, First Edition.
Edited by William G. Fahrenholtz, Eric J. Wuchina, William E. Lee, and Yanchun Zhou.
© 2014 The American Ceramic Society. Published 2014 by John Wiley & Sons, Inc.

9.2 CONDUCTIVITY OF ZrB$_2$ AND HfB$_2$

9.2.1 Pure ZrB$_2$

Few nominally phase-pure ZrB$_2$ materials (referred to hereafter as "pure ZrB$_2$") have been produced for measurement of k. As summarized in Table 9.1, k has been reported for fewer than 25 materials starting with the work by Sindeband et al. [6] in 1950, up through the publication of this chapter. This is likely due to the difficulty of densifying ZrB$_2$ without sintering aids and the frequency with which second phases are added to ZrB$_2$ to improve strength, fracture toughness, resistance to oxidation, and thermal shock performance. From this point on, reports before 1980 are referred to as "historic" while anything after is considered "current." Most studies have evaluated k indirectly using the bulk density (ρ), thermal diffusivity (D), and the constant pressure heat capacity (C_p), along with Equation 9.1 [8, 10–18, 20–23] (Jason Lonergan, Missouri University of Science and Technology, personal communication). However, a few studies have measured k directly using steady-state methods [7–9] or a direct dynamic method [19]. These exceptions have been noted in Table 9.1.

$$k = \rho * D * C_p \qquad (9.1)$$

For studies utilizing Equation 9.1, D was evaluated by dynamic methods including xenon flash [16, 17], laser flash [8, 10, 13–15, 20–23] (Jason Lonergan, Missouri University of Science and Technology, personal communication), or a plane temperature wave approach [12, 18]. Overall, D is the most important parameter for calculating k. Compared to C_p and ρ, D is the property that is most affected by changes in grain size and purity. In some cases, C_p values have either been measured by drop calorimetry [8], DSC [14, 15], or one of the flash techniques [13, 16, 17]. More likely, C_p values have been taken from thermochemical tables [10, 12, 18, 20–23] (Jason Lonergan, Missouri University of Science and Technology, personal communication), historic literature by either Schick [24] or Bolgar et al. [25], or sources such as the NIST-JANAF [26] tables or HSC chemistry [28] (HSC primarily derives its ZrB$_2$ equations from the NIST-JANAF). Figure 9.1 shows C_p for ZrB$_2$ as a function of temperature from several of studies [8, 14–17] as well as the thermochemical tables [24, 26–28].

Figure 9.1 shows obvious discrepancies among reported values. Despite divergence at higher temperatures, data from Schick et al., the NIST-JANAF tables, HSC, and a more recent DSC data by Zimmermann et al. [15] appear to agree between room temperature (RT) and approximately 200°C (430–550 J/kg-K). In contrast, room-temperature values reported by Guo et al. [17] and Ikegami et al. [16], which were both obtained by a comparison method using xenon flash, are approximately 20 and approximately 29% higher, respectively. The cause of the discrepancies in the C_p values is unclear, but it goes without saying that reliable C_p values are of utmost importance when calculating k. Finally, density is almost always evaluated using the Archimedes method. For cases in which density values are corrected for the effects of thermal expansion, tabulated data [20–23] (Jason Lonergan, Missouri University of Science and Technology, personal communication) or experimental dilatometry results [7, 8, 10, 13–15, 22] were used.

TABLE 9.1. Thermal conductivity of pure ZrB_2 with information on starting/final materials, processing, and density.

Powder processing	Purity (%)	Densification technique	Special notes (gs in μm)	$\rho_{relative}$ (%)	Thermal conductivity (k, W/m-K)		Ref.
					25°C	at max temp	
Fused-salt electrolysis	—	PS	—	85	24	—	[6]
—	96.3	HP	—	70	—	29 at 2200°C	[7]
"Fluid energy milled"	99.2	HP	gs = 28.5 μm	100	—	82 at 1000°C	[8]
"Fluid energy milled"	99.2	HP	gs = 28.5 μm	100	—	82 at 2000°C	[8]
—	—	HP	—	100^a	—	133 at 2020°C	[9]
—	99.1 (final)	HP	gs = 11 μm	97.4	—	92 at 1143°C	[10]
—	—	HP	gs = 13 μm	100^a	—	84 at 885°C	[11]
"Vibrogrinding"	—	CVD	Pyrolytic	100^a	—	84 at 2150°C	[12]
"Vibrogrinding"	—	PS 2100°C	Ni sintering aid	100^a	—	55 at 2200°C	[12]
—	—	RF FZR	SX—a-direction	100	132-145	—	[13]
—	—	RF FZR	SX—c-direction	100	95-102	78-82 at 200°C	[13]
BM w/ZrO$_2$	99.5	HP at 2000°C	gs = 5 μm	75	—	43 at 2000°C	[14]
AM w/WC	—	HP at 1900°C	gs = 6 μm^a	100	56	67 at 1325°C	[15]
BM w/SiC	98	SPS at 1900°C	gs = 3.0 μm	95.6	113	—	[16]

(Continued)

TABLE 9.1. (*Continued*)

Powder processing	Purity (%)	Densification technique	Special notes (gs in μm)	$\rho_{relative}$ (%)	Thermal conductivity (k, W/m-K)		Ref.
					25°C	at max temp	
BM ZrH$_2$+B w/SiC	99 ZrH$_2$, 95.9 B	rxn SPS at 1800°C	gs = 5.35 μm	97	133	—	[17]
HM Zr+B	99.7 Zr, 99 B	rxn SPS	gs = 5.6 μm	92.5	—	92 at 435°C	[18]
BM w/ZrO$_2$	—	SPS at 2000°C	gs = 6–11 μm	95	50	—	[19]
AM w/WC	—	HP at 1900°C	gs = 3.3 μm	100	—	59 at 2000°C	[21]
BM w/ZrB$_2$	—	HP 2100°C	gs = 22 μm	97	93	97 at 2000°C	[20]
BM w/ZrB$_2$	98.7	HP at 2150°C	gs = 25, 0.2 wt%C sintering aid	100[a]	99	75 at 2100°C	[22]
BM ZrH$_2$+B w/ZrB$_2$	99.7 ZrH$_2$, >99 B	HP at 1900°C	gs = 5.5 μm	100[a]	107	66 at 2000°C	(Jason Lonergan, Missouri University of Science and Technology, personal communication)
BM w/ZrB$_2$	98.7	HP at 2150°C	gs = 21, 0.5 wt%C sintering aid	100[a]	89	87 at 200°C	[23]

AM, attrition milling; BM, ball milling; HM, hand milling; HP, hot pressing; PS, pressureless sintering; RF FZR, radio frequency float zone refinement; SPS, spark plasma sintering; gs, grain size.

[a]density has been corrected to 100% by author.

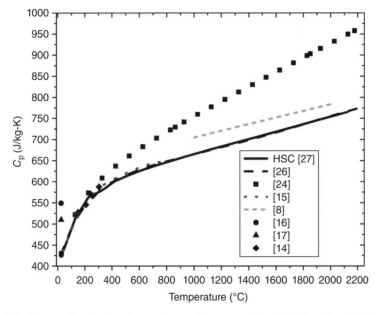

Figure 9.1. Thermochemical and experimental heat capacity (C_p) values for ZrB$_2$ [8, 14–17, 24, 26, 27].

Most studies have used some type of milling procedure for particle size reduction, to mix the starting components, or to blend-in sintering aids. Contamination from milling media is common and may affect the properties of the densified ceramics. Only the modern reports provide much detail on the types of media used, with WC [15, 20], SiC [16, 17], ZrO$_2$ [14, 19], and ZrB$_2$ [21–23] (Jason Lonergan, Missouri University of Science and Technology, personal communication) being the most common media types. Pressure-assisted densification techniques such as hot pressing (HP) or spark plasma sintering (SPS; also known as field-assisted sintering (FAST) or pulsed electric current sintering (PECS)) were most often employed for the densification of ZrB$_2$ [7–11, 14–23] (Jason Lonergan, Missouri University of Science and Technology, personal communication), but pressureless sintering (PS) [6, 12], chemical vapor deposition [12], and float zone refinement [13] have also been utilized. From these densification techniques, researchers were often successful at producing greater than or equal to 95% dense ceramics with several achieving full density. Figures 9.2 and 9.3 summarize thermal conductivity data from historic and current literature. All data have been presented "as published" unless otherwise noted in the figures.

9.2.1.1 Historic ZrB$_2$. Historically, ceramics with the lowest thermal conductivities were produced by Sindeband *et al.* (~25 W/m-K) [6] and Neel *et al.* (20–46 W/m-K) [7], likely due to low relative densities, which were 85 and 70%, respectively. The effect of density is especially evident at elevated temperatures where *k* begins to rise as temperature increases, presumably due to photon heat transport across pores. Despite the low density of these ceramics, research by Neel *et al.* was significant in that ZrB$_2$ was

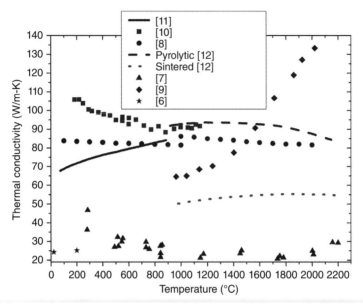

Figure 9.2. Historic thermal conductivity as a function of temperature for ZrB$_2$. Data for Clougherty changed testing method at 1000°C [6–12].

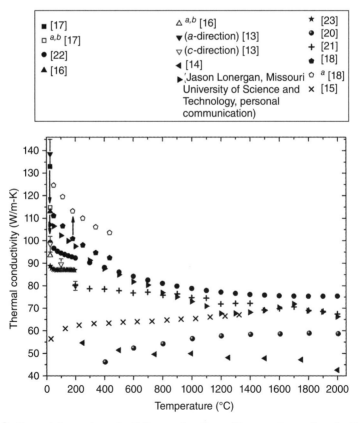

Figure 9.3. Current thermal conductivity as a function of temperature values for ZrB$_2$. [a]Data corrected for ρ. [b]Data corrected for C_p [13–18, 20–23] (Jason Lonergan, Missouri University of Science and Technology, personal communication).

tested up to 2200°C, which has since only been equaled by Fridlender *et al.* [12]. In addition to high testing temperatures, Neel *et al.*, along with Samsonov *et al.* [9] and Clougherty *et al.* [8], are the only researchers to perform direct steady-state measurements of thermal conductivity for any ZrB_2-based materials.

Other historic *k* values for ZrB_2 were obtained for ceramics that were >97% dense [10], fully dense [8], or had thermal conductivities corrected for porosity [9, 11, 12]. No clear trend in thermal conductivity as a function of temperature emerges as *k* increases for some and decreases for others as temperature increases. The *k* values also vary widely, from as low as approximately 35 W/m-K at 200°C to approximately 80 W/m-K at 2000°C. Both ZrB_2 and HfB_2 should exhibit *k* values that decrease with increasing temperature because they behave similar to metals whereby heat is transferred by phonons and electrons, both of which scatter more at elevated temperatures due to increased thermal vibrations. Based on this, and in comparison to the other historic literature values in Figure 9.2, values reported by Samsonov *et al.* are likely not representative of ZrB_2 as *k* steadily increases from approximately 60 W/m-K at 1000°C to approximately 130 W/m-K at 2000°C.

Andrievskii *et al.* [11] also reported *k* values that increased with temperature, which was not explained. In contrast, Fridlender reported the addition of 1 wt% carbonyl nickel for densification of "sintered" ZrB_2. The presence of Ni most likely caused the relatively low *k* values (50–54 W/m-K) and increasing *k* with temperature. Fridlender *et al.* also investigated pyrolytic ZrB_2 produced by chemical vapor deposition (CVD) that led to a "columnar microstructure with semicoherent crystallite boundaries." No overall temperature trend was apparent for *k*, but the higher conductivity (84–93 W/m-K) compared to sintered material was attributed to decreased phonon scattering from more coherent grain boundaries [12].

Finally, ZrB_2 produced by Branscomb and Hunter [10] and Clougherty *et al.*[1] [8] had the highest *k* values of the historic ceramics and had *k* values that decreased with temperature. This behavior is consistent with high relative density and purity. Further, Branscomb and Hunter reported the highest *k* at low temperature (105 W/m-K at 200°C) and the highest conductivity above 1000°C (~91 W/m-K at 1143°C, excluding Fridlender's pyrolytic ZrB_2). Branscomb and Hunter analyzed the metallic impurity content and found their ceramic to be approximately 99.1 at% pure with the largest impurities being Fe (0.7 at%), Si (0.07 at%), and Nb (0.05 at%) with other impurities <0.02 at% [10]. However, the analysis did not include Hf (common impurity in Zr), C (typically introduced during processing), or O (from native oxide layers), which were likely also present. Despite this, their ZrB_2 was one of the purest historic ceramics resulting in its high *k*.

9.2.1.2 Current ZrB₂

Current studies summarized in Figure 9.3 are similar to historic studies (Fig. 9.2) as a large range of values have been reported. However, values are consistent with the main factors (i.e., density, purity, and grain size) that control *k*.

Thermal conductivity at the highest temperatures measured for current ZrB_2 ceramics [15, 20–22] (Jason Lonergan, Missouri University of Science and Technology, personal communication) ranged from 43 W/m-K [14] to 75 W/m-K [22]. Room-temperature *k* values are all >85 W/m-K [13, 16–18, 21–23] (Jason Lonergan, Missouri University of Science and Technology, personal communication) with one exception at 56 W/m-K [15]. Despite the range of *k* values for ZrB_2, consensus has not been reached on the intrinsic *k* of ZrB_2.

---

[1]The discontinuity at 1000°C for data by Clougherty *et al.* [6] is likely due to a change in testing method.

To investigate ZrB$_2$ as a lattice-matched substrate for GaN, Kinoshita *et al.* produced single-crystal (SX) ZrB$_2$ by float zone refinement (Table 9.1). ZrB$_2$ is hexagonal and has an anisotropic *k*. Measurements ranged from 132–145 W/m-K parallel to the *a*-axis to 95–102 W/m-K parallel to the *c*-axis (Fig. 9.3) [13]. Differences can be explained by different phonon and electron scattering behavior for different directions. Phonon scattering increases as the ratio of atomic masses increases between constituent elements [29]. Therefore, phonon conduction will be highest for the direction with B–B and Zr–Zr bonds, which is the basal plane. In addition, the electron contribution to *k* is higher along the basal plane due to electrical conduction in the close-packed planes of Zr atoms [30].

The SX values reported by Kinoshita *et al.* provided insight into the theoretical limits for *k* of ZrB$_2$ ceramics. Equations 9.2 and 9.3 have been used to estimate the arithmetic-mean (k_{AM}, upper bounds) and harmonic-mean (k_{HM}, lower bounds) polycrystalline thermal conductivities from SX values [31, 32]:

$$k_{AM} = \frac{1}{3}\left(k_1 + k_2 + k_3\right) \tag{9.2}$$

$$\frac{1}{k_{HM}} = \frac{1}{3}\left(\frac{1}{k_1} + \frac{1}{k_2} + \frac{1}{k_3}\right) \tag{9.3}$$

Using values from Kinoshita, the theoretical *k* for ZrB$_2$ is between 117 and 131 W/m-K. These values do not account for grain boundary resistances, which are influenced by both grain size and impurities, and would likely decrease thermal conductivity. Therefore, the predicted *k* values should be upper limits for polycrystalline ZrB$_2$ ceramics.

The *k* of one material from Figure 9.3 has a value within the SX bounds [17], while three others are within 10 W/m-K of the lower limit [16, 18] (Jason Lonergan, Missouri University of Science and Technology, personal communication). To more thoroughly compare these materials, corrections for heat capacity and density have been applied in the present analysis. The C_p values were corrected because of the uncharacteristically high values reported by Guo *et al.* [17] and Ikegami *et al.* [16], and for the low relative densities (97.2% [17], 95.6% [16], and 92.5% [18]), which were corrected using the Maxwell–Eucken relation [33, 34]. The modified C_p values decreased *k* considerably, while accounting for porosity in the ceramic produced by Zhang *et al.* [18] increased *k* to 125 W/m-K at 50°C. These corrections left Zhang's as the only material with a *k* value that fell in the theoretical range, which makes it the best representation of the intrinsic *k* of polycrystalline ZrB$_2$.

The three highest *k* materials from the corrected values (after C_p and ρ corrections) were all reaction processed [17] (Jason Lonergan, Missouri University of Science and Technology, personal communication). All other current ZrB$_2$ was produced from commercial ZrB$_2$ powder. An obvious division in room-temperature *k* values occurs at approximately 67 W/m-K. Above this value, the reported *k* values decrease as temperature increases [18, 21–23] (Jason Lonergan, Missouri University of Science and Technology, personal communication). Besides the material by Loehman *et al.*, which may be anomalous due to its low density (75%) [14], the two other materials below 67 W/m-K exhibited an increase in *k* with temperature [15, 20]. These ceramics contained WC impurities from milling and, in both cases the authors indicated that W from the media was a likely cause for the slope and low k values. Despite the differences in room-temperature k and the change in k with temperature [21, 22] (Jason Lonergan, Missouri University of Science and Technology, personal communication), the k values converge to a narrow range (within ~10 W/m-K) above 1300°C. This indicates that factors resulting in large changes in the k at lower temperatures may give way to a common mechanism that controls k at elevated temperatures.

Purity may be the most significant factor affecting k for current ZrB₂ ceramics. Conductivities for fully dense ZrB₂ [15, 21], or values for porous ceramics corrected to full density [22, 23] (Jason Lonergan, Missouri University of Science and Technology, personal communication), span a significant range of k values (more than 50 W/m-K) and appear to be independent of grain size. Grain sizes can be divided into two separate ranges, fine with average grain sizes below 10 μm [14–18, 20] (Jason Lonergan, Missouri University of Science and Technology, personal communication) or coarse with average grain sizes over 20 μm [21–23]. Fine grains can result in either high k (>107 W/m-K at 25°C) for reaction processed ZrB₂ [17, 18] (Jason Lonergan, Missouri University of Science and Technology, personal communication) or low k (<57 W/m-K at 25°C) for conventionally processed ZrB₂ ceramics [15, 20]. Thermal conductivities for coarse-grained materials all fall between those values [21–23]. Therefore, the difference apparent in Figure 9.3 is not significantly affected by grain size. Differences in purity, which are reported much less often, are, therefore, the most likely explanation for the observed variations. Unfortunately, none of the current studies report chemical analysis of the final ceramics, although assays of starting powders are typically provided [14, 16–18, 22, 23] (Jason Lonergan, Missouri University of Science and Technology, personal communication). As a result, analysis of the effect of purity is difficult. However, some studies have systematically evaluated the effects of solid solution (SS) additions of various species as discussed in the next section.

9.2.2 ZrB₂ with Solid Solution Additions

Thermal conductivities from three studies evaluating the effects of SS additions to ZrB₂ [21–23] are summarized in Table 9.2 and plotted in Figure 9.4. Values for reaction hot-pressed ZrB₂ ceramics reported by Zhang et al. [18] are included for comparison. These studies were self-consistent since similar processing was used for pure, baseline ZrB₂ ceramics and those with subsequent SS additions. However, if SS additions [21, 23] were compared to the corrected data for reaction-processed ZrB₂ [18] instead of the baseline material produced in each study, the effects could be misleading.

SS additions decrease the k values of ZrB₂ ceramics from the respective baselines, especially at room temperature. The trend in k as a function of temperature depends on both the type and amount of additive. For example, Thompson et al. showed that an SS addition of 10 vol% TiB₂ reduced k resulting in an almost linear trend with temperature. In contrast, the addition of 50 vol% TiB₂ resulted in a decrease in k of approximately 23 W/m-K at 25°C compared to the addition of 10 vol% TiB₂ [21]. Both Thompson et al. [20] and McClane et al. [23] found that smaller additions of TiB₂ (data by McClane not shown in Figure 9.4) had a minimal effect on k. In contrast, other transition metal additions such as W or Nb evaluated by McClane had a much larger effect at the same concentration [23]. McClane confirmed that SS additions of tungsten in ZrB₂ decreased

TABLE 9.2. Summary of solid solution effects on thermal conductivity of ZrB_2

Material (vol% or otherwise stated)	Powder processing	Densification technique	Special notes	Relative density (%)	Thermal conductivity (k, W/m-K)		Ref.
					25°C	at max temp	
ZrB_2 + 2.8 mol%NbB$_2$	BM w/ZrB$_2$	HP at 2150°C	gs = 7 μm, 0.5 wt%C sintering aid	100	74	76 at 200°C	[23]
ZrB_2 + 3.1 mol%W$_2$B$_5$	BM w/ZrB$_2$	HP at 2150°C	gs = 18 μm, 0.5 wt%C sintering aid	100[a]	35	40 at 200°C	[23]
ZrB_2 + 10%TiB$_2$	BM w/ZrB$_2$	HP at 2100°C	gs = 10 μm	100[a]	81	72 at 2000°C	[21]
ZrB_2 + 50%TiB$_2$	BM w/ZrB$_2$	HP at 2100°C	gs = 7 μm	100[a]	58	65 at 1900°C	[21]

[a]Corrected to 100% in this work.

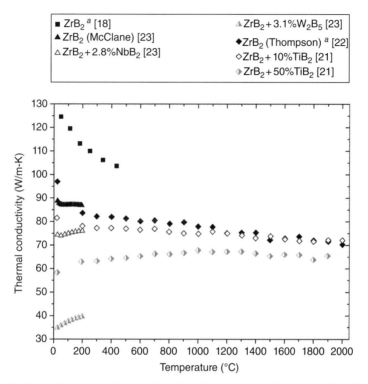

Figure 9.4. Thermal conductivity as a function of temperature for ZrB$_2$ with solid solution additions. [a]Data corrected for ρ [18, 21–23].

k significantly (~53 W/m-K drop at 25°C for 3.1 mol% W). Evaluation of k for different SS additions was also shown to correlate to changes in lattice parameters [23].

9.2.3 Pure HfB$_2$

Research on k of HfB$_2$ is summarized in Table 9.3. Reports of k for HfB$_2$ are more scarce than those of ZrB$_2$, likely due to the higher cost and density (11.21 g/cm^3 [38] vs. 6.12 g/cm^3 [39]) of Hf, which may make it less attractive for aerospace applications. The k values of HfB$_2$, like ZrB$_2$, were typically calculated from D, ρ, and C_p using Equation 9.1 [10, 14, 18, 35, 37], but some direct measurements have been reported [9, 36]. Also similar to ZrB$_2$, C_p was either evaluated experimentally [14, 35], taken from thermochemical tables (Schick [24]) [10], or an electronic database (HSC [27]) [18, 37] and these values are plotted as a function of temperature in Figure 9.5. The tabulated data by Schick and values obtained from HSC are nearly identical, but C_p values used by Loehman et al. [14] and Gosset et al. [35] were lower and may have led to lower k values compared to studies using typical values.

Of the studies reporting powder-processing methods, hand mixing (HM) was most common [18, 37], while one study used ZrO$_2$ media and ball milling [14]. The two studies utilizing HM also performed reaction processing using SPS to produce HfB$_2$

TABLE 9.3. Thermal conductivity of pure HfB_2 with information on starting/final materials, processing, and density

Powder processing	Purity (%)	Densification technique	Special notes	Relative density (%)	Thermal conductivity (k, W/m-K)		Ref.
					25°C	at max temp	
—	—	HP	—	100[a]	—	144 at 2000°C	[9]
—	98.2	HP	gs = 15 μm	95.2	—	74 at 1200°C	[10]
—	99.5	HP at 2000°C	—	95	72	—	[35]
—	—	HP at 2160°C	—	95	104	72 at 820°C	[36]
BM w/ZrO$_2$	99.5 ZrB$_2$	HP at 2000°C	—	69	—	31 at 2000°C	[14]
HM Hf+B	99.8 Hf, 99 B	rxn SPS	gs = 4.0 μm	95.4	107	91 at 600°C	[37]
HM Hf+B	99.8 Hf, 99 B	rxn SPS	gs = 10.7 μm	98.1	—	101 at 420°C	[18]
HM Hf+1.9B	99.8 Hf, 99 B	rxn SPS	Zr rich, gs = 10 μm	98.1	—	97 at 430°C	[18]
HM Hf+2.1B	99.8 Hf, 99 B	rxn SPS	B rich, gs = 10.9 μm	101	—	110 at 420°C	[18]

[a]Corrected to 100% in this work.

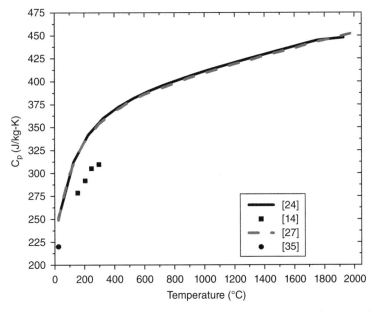

Figure 9.5. Heat capacity as a function of temperature for HfB$_2$ [24, 27, 35].

from Hf and B. All other HfB$_2$ studies used commercial HfB$_2$ powder and densification by hot pressing [9, 14, 35, 36] or evaluated commercially sintered material [10]. Aside from HfB$_2$ studied by Loehman *et al.* [14], which had a relative density of 69%, all other reports used HfB$_2$ that was greater than or equal to 95% relative density.

The *k* values presented in Table 9.3 are plotted in Figure 9.6. Similarities are apparent between trends discussed for ZrB$_2$ and HfB$_2$. First, both display a wide variation in *k* values across the temperature range. Second, *k* should decrease with increasing temperature, porosity, and impurity content.

The *k* values for HfB$_2$ reported by Samsonov *et al.* [9] increase with increasing temperature from 1000 to 2000°C, as did ZrB$_2$ reported in the same study. Again, this trend does not fit with other reports. As mentioned previously, Loehman *et al.* reported *k* for highly porous HfB$_2$, resulting in low values (37–31 W/m-K from 250–2000°C) [14]. Branscomb and Hunter [10] reported the only other historic HfB$_2$, which had a slightly negative slope of *k* with temperature (78 W/m-K at 200°C to 74 W/m-K at 1200°C), but *k* was lower compared to HfB$_2$ produced by Zhang *et al.* [18] and Gasch *et al.* [37] due to lower purity (98.2%) and density (95.2%).

Like the highest *k* for polycrystalline ZrB$_2$, the highest *k* for HfB$_2$ was also a result of high purity starting materials (i.e., 99.8% Hf and 99% B) [18, 37]. Zhang *et al.* also produced nonstoichiometric compositions of HfB$_{1.9}$ and HfB$_{2.1}$, but only HfB$_{2.1}$, was fully dense. To better evaluate these *k* values, and estimate a theoretical maximum *k* for HfB$_2$, reported values were adjusted for ρ (hollow shapes in Fig. 9.6). The two stoichiometric compositions had similar *k* values of approximately 114 W/m-K at the lowest temperatures (30–50°C). The values steadily decreased to approximately 103 W/m-K at 400°C. The similarities were likely due to the nearly identical processing procedures

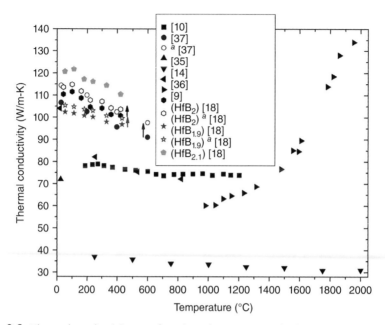

Figure 9.6. Thermal conductivity as a function of temperature for historic and current HfB$_2$.
[a]Data corrected for ρ [9, 10, 14, 18, 37, 39, 40].

(i.e., HM of the constituents and densification via SPS). Unless higher purity starting materials were used, the compositions by Gasch et al. and by Zhang et al. are likely the best representation of the intrinsic k for polycrystalline HfB$_2$.

Interestingly, the change in stoichiometry studied by Zhang et al. had a significant effect on k. The k for the boron-rich composition was approximately 6% higher than the stoichiometric composition. In contrast, the boron deficient material had k values that were approximately 6% lower. No explanation was provided for the reported differences, but both the electron and phonon contributions to k were higher for HfB$_{2.1}$ [18].

9.2.4 Conclusions Regarding Phase-Pure ZrB$_2$ and HfB$_2$

The k values for nominally pure ZrB$_2$ and HfB$_2$ were affected more strongly by purity than grain size or porosity. Porosity decreased k; however, as these materials are mainly intended for use as structural ceramics, densities are typically maximized. Grain size had little effect on k for ZrB$_2$ as either higher or lower k materials were produced with fine grain sizes, with larger-grained material bridging the gap in conductivity. As these materials are all single phase, changes in purity were due to elements in SS. Reaction processing from pure elemental constituents (>99.7% metals, 99% B) has been used to produce the highest conductivity polycrystalline ZrB$_2$ and HfB$_2$. Comparisons with single-crystal ZrB$_2$ values (117–131 W/m-K after averaging a- and c-axis values) revealed that the k for ZrB$_2$ produced by reactive SPS were in good agreement with values predicted from single-crystal values, which may indicate that grain boundary resistances for ZrB$_2$ are low.

Due to their proposed use temperatures, many studies report k values up to 2000°C. In contrast, SX ZrB$_2$ was only evaluated to 200°C. Evaluation up to 2000°C would have been very valuable. Research on transition metal additions indicated the effect of SSs on k. Depending on the element (i.e., W has a much larger effect than Ti) and concentration, the k values of ZrB$_2$ (and HfB$_2$ by corollary) can change significantly. With SS additions, k drops more dramatically at room temperature compared to higher temperatures. This results in a change in slope in k versus T from negative to positive. The observed changes in k are most likely tied to changes in bonding and distortions in the lattice as they have been shown to correlate to shifts in measured lattice parameters.

9.3 ZrB$_2$ AND HfB$_2$ COMPOSITES

9.3.1 Thermal Conductivity of ZrB$_2$ Composites

Researchers have used additions of secondary, tertiary, and quaternary phases to improve densification and important properties (i.e., strength, thermal shock resistance, and oxidation performance) of ZrB$_2$ ceramics. Along with these possible improvements, the transport properties also influence the thermal management performance and further contribute to thermal shock behavior. Therefore, the k of the final ceramic is important in its own right and has been studied to understand the effects of additives. Table 9.4 (ZrB$_2$–SiC), Table 9.5 (other second-phase additions), and Table 9.6 (tertiary and quaternary phases) summarize the ZrB$_2$ composites that have been evaluated.

9.3.1.1 ZrB$_2$-X Composites. ZrB$_2$–SiC ceramics have been studied by a host of researchers (Table 9.4) [8, 14–16, 18, 46–46], along with other ZrB$_2$ based composites with additions of C [8, 11, 20], MoSi$_2$ [48, 49], ZrC [11, 12], or ZrSi$_2$ [50] (Table 9.5). The compositions span a large range of additive contents. However, the present review was limited to additions less than or equal to 50 vol% so the materials may be regarded as ZrB$_2$-based. Values of k were plotted as a function of temperature for ZrB$_2$–SiC (Fig. 9.7)[2] and ZrB$_2$–X (Fig. 9.8), where X is C, MoSi$_2$, or ZrC. All of the studies but Clougherty *et al.* [8] measured D and calculated k using Equation 9.1.

ZrB$_2$–SiC As seen in Table 9.4 and Figure 9.7, SiC additions to ZrB$_2$ range from 5 to 50 vol%. In general, k decreases as SiC content increases. Therefore, SiC appears to act as a lower k phase (with respect to ZrB$_2$). However, whether this is an expected outcome depends on the inherent conductivity of SiC in the composite.

The room-temperature k of high-purity single-crystal SiC (ppm impurity levels) can be as high as 490 W/m-K, with a theoretical upper limit predicted to be as high as 700 W/m-K. This high k is attributed to phonon transport since SiC is not electrically conductive [57]. Despite high k values at room temperature, k of SiC drops off quickly with increasing impurity content due to phonon scattering [57–62]. When N is present as an impurity, SiC can have k as low as 60 W/m-K at room temperature [57, 63]. Further, SSs with elements like Be, B, and Al, which are introduced from common sintering aids, can result in SiC ceramics

---

[2] Values for ZrB$_2$ prepared by reactive SPS by Zhang *et al.* [18] were corrected for porosity.

TABLE 9.4. Thermal conductivity of ZrB$_2$-SiC with information on starting/final materials, processing, and density

SiC vol%	Powder processing	Purity (%)	Densification technique	Special notes (gs in μm)	$\rho_{relative}$	Thermal conductivity (k, W/m-K)		Ref.
						25°C	at max temp	
20%	VM w/plastic	—	HP	—	100%	—	78 at 1000°C	[8]
20%	VM w/plastic	—	HP	—	100%	—	69 at 2000°C	[8]
10%	BM w/SiC	98.0 ZrB$_2$, 99.0 SiC	SPS at 1900°C	gsZrB$_2$ = 3.5, gsSiC = 8.9	99.4%	130	—	[16]
20%	BM w/SiC	98.0 ZrB$_2$, 99.0 SiC	SPS at 1900°C	gsZrB$_2$ = 3.4, gsSiC = 8.9	99.6%	131	—	[16]
30%	BM w/SiC	98.0 ZrB$_2$, 99.0 SiC	SPS at 1900°C	gsZrB$_2$ = 3.2, gsSiC = 8.9	99.6%	138	—	[16]
5%	AM w/SiC	99.5 ZrB$_2$, 99.9 SiC	HP at 2000°C	gsZrB$_2$ = 2, gsSiC = 8	101%	—	82 at 2000°C	[14]
10%	AM w/SiC	99.5 ZrB$_2$, 99.9 SiC	HP at 2000°C	gsZrB$_2$ = 1, gsSiC = 4	101%	—	80 at 2000°C	[14]
20%	AM w/SiC	99.5 ZrB$_2$, 99.9 SiC	HP at 2000°C	gsZrB$_2$ = 3, gsSiC = 4	100%	—	67 at 2000°C	[14]
20%	BM w/WC	>99.5	HP at 2000°C	—	>98%	90	72 at 1200°C	[41]
20%	PBM w/Si$_3$N$_4$	—	HP at 1900°C	—	?	95	76 at 400°C	[42]
14%	Coarse mixing	99.5 ZrB$_2$, 99 SiC	Arc melting	Eutectic structure	100%	95	83 at 750°C	[43]
31%	Coarse mixing	99.5 ZrB$_2$, 99 SiC	Arc melting	Eutectic structure	100%	77	64 at 770°C	[43]
50%	Coarse mixing	99.5 ZrB$_2$, 99 SiC	Arc melting	Eutectic structure	100%	72	44 at 760°C	[43]
20%	BM w/SiC	—	HP at 1900°C	—	98.8%	68	56 at 1000°C	[40]
20%	PBM w/WC	99.5 ZrB$_2$, 99.8 SiC	SPS	gs = 4.5, SiC whiskers	99.3%	—	79 at 420°C	[18]
10%	BM w/ZrO$_2$	99 ZrB$_2$, 99 SiCw	HP at 1800°C	gs = 4.5, SiC whiskers	98%	—	61 at 1800°C	[44]
20%	BM w/ZrO$_2$	99 ZrB$_2$, 99 SiCw	HP at 1800°C	gs = 4, SiC whiskers	98%	—	61 at 1800°C	[44]
30%	BM w/ZrO$_2$	99 ZrB$_2$, 99 SiCw	HP at 1800°C	gs = 4, SiC whiskers	96%	—	58 at 1800°C	[44]
20%	BM w/SiC	>96 ZrB$_2$, >99 SiC	PS at 2100°C	gsZrB$_2$ = 10, gsSiC = 8	99%	94	62 at 1200°C	[45]
30%	AM w/WC	—	HP at 1900°C	—	100%	62	51 at 1330°C	[15]
20%	BM w/Si$_3$N$_4$	98 ZrB$_2$, 98.5 SiC	HP at 1900°C	gsZrB$_2$ = 2.7	>99%	94	77 at 1200°C	[46]
20%	PBM w/SiC	—	HP at 2000°C	gsZrB$_2$ = 2.6, gsSiC = 1.3	99.8%	107	—	[47]

TABLE 9.5. Thermal conductivity of two-phase ZrB_2-based composites with information on starting/final materials, processing, and density

Material (vol% or otherwise stated)	Powder processing	Purity (%)	Densification technique	Special notes (gs in µm)	$\rho_{relative}$ (%)	Thermal conductivity (k, W/m-K)		Ref.
						25°C	at max temp	
26%C	—	—	HP	gsZrB$_2$ = 20	100^a	—	75 at 900°C	[11]
50%C	VM w/plastic	—	HP	—	100	—	64.0 at 1000°C	[8]
50%C	VM w/plastic	—	HP	—	100	—	64.0 at 2000°C	[8]
1 wt%C	AM w/WC	—	HP at 1900°C	gsZrB$_2$ = 2.4	100	—	53.9 at 2000°C	[20]
3 wt%C	AM w/WC	—	HP at 1900°C	gsZrB$_2$ = 1.8	100	—	63.7 at 2000°C	[8]
10%MoSi$_2$	BM w/SiC	—	HP at 1800°C	—	100^a	88	—	[48]
20%MoSi$_2$	BM w/SiC	—	HP at 1800°C	—	100^a	83	—	[48]
30%MoSi$_2$	BM w/SiC	—	HP at 1800°C	—	100^a	81	—	[48]
40%MoSi$_2$	BM w/SiC	—	HP at 1800°C	—	100^a	76	—	[48]
2.3%MoSi$_2$	BM w/ZrO$_2$	—	HP at 1900°C	gsZrB$_2$ = 1–10	?	66	74.5 at 1200°C	[49]
50%ZrC	—	—	HP	gsZrB$_2$ = 10	100^a	—	54.3 at 905°C	[11]
5%ZrC	—	—	HP	gsZrB$_2$ = 27	100^a	—	79.3 at 910°C	[11]
20%ZrC	"Vibrogrinding"	—	PS 2100°C	gsZrB$_2$ = 5, gsZrC = 6, 1% Ni sintering aid	100^a	—	76.1 at 2200°C	[12]
40%ZrC	"Vibrogrinding"	—	PS 2100°C	gsZrB$_2$ = 3, gsZrC = 4, 1% Ni sintering aid	100^a	—	55.5 at 2200°C	[12]
45%ZrC	"Vibrogrinding"	—	PS 2100°C	gsZrB$_2$ = 1, gsZrC = 3, 1% Ni sintering aid	100^a	—	53.2 at 2150°C	[12]
10%ZrSi$_2$	BM w/SiC	—	PS at 1650°C	—	100^a	107	—	[50]
20%ZrSi$_2$	BM w/SiC	—	PS at 1650°C	—	100^a	97	—	[50]
30%ZrSi$_2$	BM w/SiC	—	PS at 1650°C	—	100^a	90	—	[50]
40%ZrSi$_2$	BM w/SiC	—	PS at 1650°C	—	100^a	74	—	[50]

VM, vibratory mixing. ^{a}Corrected to 100% in this work.

TABLE 9.6. Thermal conductivity of three- and 4-phase ZrB_2-based composites with information on starting/final materials, processing, and density

Material (vol% or otherwise stated)	Powder processing	Purity (%)	Densification technique	Special notes (gs in μm)	$\rho_{relative}$ (%)	Thermal conductivity k (W/m-K)		Ref.
						25°C	at max temp	
5%SiC+20%MoSi$_2$	BM w/SiC	—	HP at 1800°C	—	100^a	82	—	[48]
10%SiC+20%MoSi$_2$	BM w/SiC	—	HP at 1800°C	—	100^a	94	—	[48]
20%SiC+20%MoSi$_2$	BM w/SiC	—	HP at 1800°C	—	100^a	90	—	[48]
30%SiC+20%MoSi$_2$	BM w/SiC	—	HP at 1800°C	—	100^a	98	—	[48]
5%SiC+40%MoSi$_2$	BM w/SiC	—	HP at 1800°C	—	100^a	81	—	[48]
10%SiC+40%MoSi$_2$	BM w/SiC	—	HP at 1800°C	—	100^a	85	—	[48]
10.7%SiC+8.9%B$_4$C	BM w/B$_4$C	—	PS, HIP at 2000°C, 1800°C	—	100	94	57 at 1940°C	[48]
21.9%SiC+7.8%B$_4$C	BM w/B$_4$C	—	PS, HIP at 2000°C, 1800°C	—	100	90	47.3 at 2000°C	[51]
48.7%SiC+5.1%B$_4$C	BM w/B$_4$C	—	PS, HIP at 2000°C, 1800°C	—	100	101	33.4 at 1950°C	[51]
18%SiC+10%C	VM w/plastic	—	HP	—	100	—	73.6 at 1000°C	[8]
18%SiC+10%C	VM w/plastic	—	HP	—	100	—	61.1 at 2000°C	[8]
14%SiC+30%C	VM w/plastic	—	HP	—	100	—	57.7 at 1000°C	[8]
14%SiC+30%C	VM w/plastic	—	HP	—	100	—	59.2 at 2000°C	[8]
20%SiC+2wt%CNT	PBM w/Si$_3$N$_4$	—	HP at 1900°C	—	?	91	72.4 at 400°C	[42]
20%SiC+15% C_g	BM w/SiC	—	HP at 1900°C	—	100	81	58.7 at 1000°C	[40]
30%SiC(10%SiC$_w$)+10% C_g	BM w/ZrO$_2$	99.5 ZrB2/SiCp, 99 SiCw/Cg	HP at 1800°C	—	99.6	100	65.5 at 1200°C	[52]

20%SiC+10% C_f	PBM w/SiC	—	HP at 2000°C	gsZrB2 = 2.55, gsSiC = 1.47	99.6	95	—	[47]
20%SiC+20% C_f	PBM w/SiC	—	HP at 2000°C	gsZrB2 = 2.79, gsSiC = 1.65	99.4	83	—	[47]
20%SiC+30% C_f	PBM w/SiC	—	HP at 2000°C	gsZrB2 = 2.82, gsSiC = 1.35	99.5	68	—	[47]
2%MoSi$_2$+15%SiC	BM w/ZrO$_2$	—	HP at 1820°C	gsZrB$_2$ = 1–5	?	62	64.5 at 1200°C	[49]
20%SiC+5%Si$_3$N$_4$+20%ZrC	BM w/WC	All powder >99.5	HP at 2000°C		>98	82	50.0 at 1200°C	[41]
20%SiC+5%Si$_3$N$_4$	BM w/WC	All powder >99.5	HP at 2000°C	—	>98	103	61.2 at 1200°C	[41]
15%SiC+2%Si$_3$N$_4$	BM w/SiC	—	HP at 1900°C	—	98.1	80	58 at 1500°C	[53]
5%SiC+5%TaSi$_2$	BM w/SiC	—	SPS at 1800°C	—	>98.2	42	—	[54]
10%SiC+5%TaSi$_2$	BM w/SiC	—	SPS at 1800°C	—	>98.2	40	—	[54]
20%SiC+5%TaSi$_2$	BM w/SiC	—	SPS at 1800°C	—	>98.2	43	—	[54]
30%SiC+5%TaSi$_2$	BM w/SiC	—	SPS at 1800°C	—	>98.2	50	—	[54]
10%SiC+10%TaSi$_2$	BM w/SiC	99 (all powders)	SPS at 1600°C	—	?	43	—	[55]
20%SiC+10%TaSi$_2$	BM w/SiC	99 (all powders)	SPS at 1600°C	—	?	45	—	[55]
20%SiC+20%TaSi$_2$	BM w/SiC	99 (all powders)	SPS at 1600°C	—	?	45	—	[55]
30%SiC+10%TaSi$_2$	BM w/SiC	99 (all powders)	SPS at 1600°C	—	?	40	—	[55]
30%SiC+20%TaSi$_2$	BM w/SiC	99 (all powders)	SPS at 1600°C	—	?	46	—	[55]
10%SiC+10%TaSi$_2$	BM w/SiC	99 (all powders)	SPS at 1800°C	—	?	46	—	[55]
20%SiC+10%TaSi$_2$	BM w/SiC	99 (all powders)	SPS at 1800°C	—	?	47	—	[55]

(Continued)

TABLE 9.6. (*Continued*)

Material (vol% or otherwise stated)	Powder processing	Purity (%)	Densification technique	Special notes (gs in μm)	$\rho_{relative}$ (%)	Thermal conductivity k (W/m-K)		Ref.
						25°C	at max temp	
20%SiC+20%TaSi$_2$	BM w/SiC	99 (all powders)	SPS at 1800°C	—	?	53	—	[55]
30%SiC+10%TaSi$_2$	BM w/SiC	99 (all powders)	SPS at 1800°C	—	?	47	—	[55]
30%SiC+20%TaSi$_2$	BM w/SiC	99 (all powders)	SPS at 1800°C	—	?	51	—	[55]
33.3%SiC+33.3%ZrC	BM w/SiC	—	SPS at1950°C	—	98.7	73	—	[56]
15%SiC+15%ZrC	BM w/SiC	—	SPS at1950°C	—	98.5	86	—	[56]
30%SiC+15%ZrC	BM w/SiC	—	SPS at1950°C	—	98.7	89	—	[56]
15%SiC+30%ZrC	BM w/SiC	—	SPS at1950°C	—	98.8	74	—	[56]
18%SiC+8%ZrC	BM w/ZrO$_2$	—	SPS at 2000°C	—	85.7	71	—	[19]
25%SiC+12%ZrC	BM w/ZrO$_2$	—	SPS at 2000°C	—	81–90	71	—	[19]
33%SiC+16%ZrC	BM w/ZrO$_2$	—	SPS at 2000°C	—	94–99	60	—	[19]
4%SiC+35%ZrC	BM w/ZrO$_2$	—	SPS at 2000°C	—	97–99	62	—	[19]

PBM, planetary ball milling. ^{a}Corrected to 100% in this work.

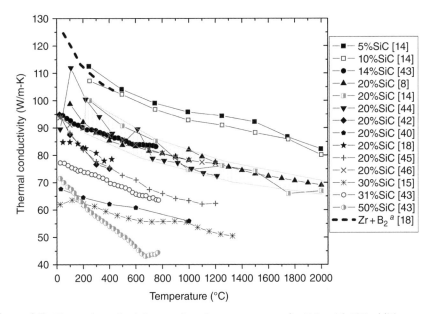

Figure 9.7. Thermal conductivity as a function temperature for ZrB$_2$ with SiC additions ranging from 5 to 50 vol%. (Note: Clougherty changed testing methods at 1000°C.) ^{a}Data corrected for ρ [8, 14, 15, 18, 36, 41–45].

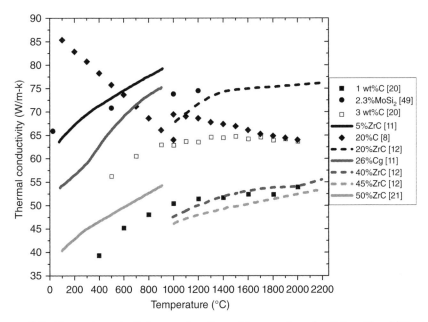

Figure 9.8. Thermal conductivity as a function of temperature for ZrB$_2$ with additions of carbon, MoSi$_2$, or ZrC. Clougherty changed testing methods at 1000°C. [8, 11, 12, 20, 21, 48].

with k values in the range of 250–270 W/m-K [59, 64], 100–150 W/m-K, and 80–120 W/m-K [62], respectively. In addition, k decreases with increasing amounts of each constituent [61]. Given the effect of impurities, SiC in ZrB$_2$–SiC composites could be below the k of ZrB$_2$ due to starting impurities or by picking up B from the surrounding diboride.

Other explanations are possible for observed reductions in k values for ZrB$_2$–SiC. For example, microcracks can form due to residual stresses from SiC particle or cluster sizes greater than approximately 11.5 μm [65]. Microcracking is similar to porosity, reducing k by decreasing the effective conduction path, thus disrupting phonon and electron transport. Depending on SiC distribution, size, morphology, and volume fraction, microcracks can be present in the final ceramic, but have rarely been discussed. Another explanation for a drop in k due to SiC additions is the presence of a second phase at the grain boundaries. An SiO$_2$-based phase can form at the grain boundaries in ZrB$_2$–SiC, but appropriate additions of B$_4$C and C can remove it prior to densification [66].

Loehman *et al.* reported one of the best examples of decreasing k with increasing SiC content in diborides. As SiC content increase from 5 vol% SiC to 20 vol% SiC, k at 250°C dropped from 112 to 100 W/m-K [14]. For their ceramic containing 5 vol% SiC, k was 112 W/m-K, which is quite high and similar to the value of 110 W/m-K reported by Zhang *et al.* [18] for reaction-processed ZrB$_2$. As the ZrB$_2$ comprises the majority of this material, the high conductivity may be an indication of the diboride purity.

For SiC contents of 20 vol%, a majority of k values [8, 14, 41, 46] lie in a range (highlighted in Fig. 9.7) between 88–102 W/m-K at 200°C and 66–70 W/m-K at 2000°C. Because reported k values vary for both ZrB$_2$ and SiC, it is difficult to define one k value for ZrB$_2$–20% SiC. This region is likely adequate based on the convergence of multiple research efforts into a narrow set of k values. Four ceramics containing 20 vol% SiC fell below this region [18, 40, 42, 45] and deserve further discussion. Ceramics produced by both Wang *et al.* [40] and Zhang *et al.* [45] likely had microcracking. Image analysis by both researchers showed that SiC cluster sizes were greater than or equal to 11.5 μm, indicating that both materials had SiC inclusions large enough to promote microcracking. In addition, modulus values reported were lower than expected at 480 GPa (~492 GPa if corrected for porosity) for Wang and 374 GPa (~382 GPa if corrected for ~1% porosity) for Zhang, further supporting the case for microcracking.

Microcracking does not appear to be an issue with the other two lower k compositions. Hence, purity may be the cause. Zhang *et al.* milled the ZrB$_2$ and SiC using WC media [18]. As discussed earlier, W decreases the k of ZrB$_2$. Finally, the material evaluated by Tian *et al.* used lower purity SiC (98.5%) and Si$_3$N$_4$ media for milling. Nitrogen [57], and specifically Si$_3$N$_4$, [67] contamination have been shown to decrease the conductivity of SiC.

Arc melted ZrB$_2$–SiC evaluated by Tu *et al.* may be the least practical material from a production standpoint, but it is a unique material from a scientific perspective. The material containing 50 vol% SiC had a very fine eutectic microstructure with ZrB$_2$ and SiC features approximately 100 nm wide [43]. The fine microstructure likely led to a higher phonon scattering (compared to other traditionally processed ZrB$_2$–SiC compositions), which may have resulted in a steeper slope in k with temperature.

ZrB$_2$-(C, MoSi$_2$, ZrC, AND ZrSi$_2$). In addition to diboride–SiC composites, k has been evaluated for two-phase ZrB$_2$-based compositions containing C [8, 11, 20],

MoSi$_2$ [48, 49], ZrC [11, 12], and ZrSi$_2$ [50] (Table 9.5). Figure 9.8 shows k as a function of temperature for these materials. Note that the disilicides by Guo [48, 50] were only evaluated at room temperature. A broad range of k values were reported depending on the specific additive and its amount. For ZrB$_2$ containing 20 vol% C, the material from Clougherty *et al.* is the only one with k values that decreased with temperature. Like other materials evaluated by Clougherty, this one had a discontinuity at 1000°C where the test method switched from a steady-state cut bar method to calculation from D, but overall k dropped from approximately 85 W/m-K at 100°C to approximately 64 W/m-K at 2000°C [8]. The form of the carbon in the material, either before or after processing, is not discussed. Given the small decease in k at room temperature between this material and the ZrB$_2$ studied by Clougherty, the C is likely to be graphitic in nature. Amorphous carbon has a lower k than ZrB$_2$ and would have decreased k considerably more than for graphitic carbon, which has a k value about the same as ZrB$_2$ [68, 69].

Thompson *et al.* [20] and Andrievskii *et al.* [11] also investigated ZrB$_2$ with carbon additions. The k values increased with increasing temperature, indicating that the diboride was the primary influence on k. For Andrievskii *et al.*, 26 vol% carbon was added as graphite, which reduced the conductivity approximately 10 W/m-K at all temperatures [11]. Thompson *et al.* added carbon in the form of phenolic resin, resulting in 1 and 3 wt% amorphous carbon after pyrolysis. After hot pressing, graphitic carbon was identified at the grain boundaries. ZrB$_2$ containing 1 wt% C had k values lower than pure ZrB$_2$. They hypothesized that graphite was aligned, with its basal plane parallel to the grain boundaries, thus disrupting phonon transport between ZrB$_2$ grains. However, k did not continue to decrease when C content was increased to 3 wt% C due to the segregation of W into ZrC after testing up to 2000°C. The ZrC phase, produced due to carbon additions, served as a sink for W, reducing the W content of the ZrB$_2$ grains and boosting the overall k of the composite [20].

The addition of ZrC to ZrB$_2$ typically reduces k. Andrievskii *et al.* evaluated both 5 and 50 vol% ZrC from 100 to 900°C, while Fridlender *et al.* added 20, 40, and 45 vol% ZrC and tested up to 2150 or 2200°C. The higher ZrC additions resulted in k values between 40 and 55 W/m-K, while the lower additions resulted in k values between 64 and 80 W/m-K, despite the different testing temperatures and compositions. The differences, and the decrease in k with increasing ZrC content, were probably due to the variability in k values for ZrC (20–45 W/m-K at room temperature and increasing to 30–50 W/m-K at 2000°C), which are well below the values for pure ZrB$_2$ [70].

Monteverde *et al.* evaluated ZrB$_2$ containing 2.3 vol% MoSi$_2$ as a sintering aid. The k for this material was approximately 66 W/m-K at 25°C and increased to approximately 75 W/m-K at 1200°C [49]. If it were only a second phase, this small addition would not likely decrease k of the composite much, even though the k of MoSi$_2$ is significantly lower than ZrB$_2$ [70–72]. However, Mo also dissolves into ZrB$_2$, which decreases k and results in k values that increase as temperature increases [23]. Therefore, even small additions of MoSi$_2$ have a significant effect on k.

9.3.1.2 ZrB₂ Ternary- and Quaternary-Phase Composites.

Several three-phase, and even a few four-phase, ZrB$_2$-based composites have been produced and evaluated (Table 9.6). One common feature of these multiphase composites is the addition of SiC, which was present in varying amounts. Once past this common addition, the tertiary

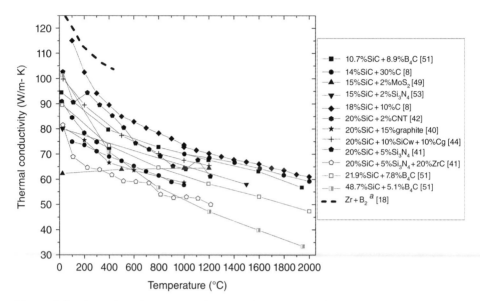

Figure 9.9. Thermal conductivity as a function of temperature for ZrB$_2$-SiC-based composites with additions of, B$_4$C, C (elemental, nanotubes (CNT) and graphite (Cg)), MoSi$_2$, Si$_3$N$_4$, SiCw (whiskers), or ZrC. Clougherty changed testing methods at 1000°C. [a]Data corrected for ρ [8, 18, 36, 41, 43, 48, 60, 72].

and quaternary phases were MoSi$_2$ [48], B$_4$C [51], C_g [8, 40, 52], CNTs [42], carbon fiber (C_f) [47], SiC whiskers (SiC$_w$) [52], Si$_3$N$_4$ [41, 53], TaSi$_2$ [55], or ZrC [41, 56, 73]. Only two studies, [8] and [73], evaluated k other than by using Equation 9.1, and all but one study ([51] performed PS) employed HP or SPS for densification. From Table 9.6, k was evaluated beyond room temperature for 14 of these materials (Fig. 9.9) [8, 40–42, 49, 51–53].[3]

Because of the complex effects of the addition of two, or even three, phases to ZrB$_2$, it is difficult to understand how each species affects k. Therefore, instead of evaluating k for each composition, more generalities will be surmised with some specific examples brought out when clear. Overall, k values presented in Figure 9.9 lie between 80 and 103 W/m-K at room temperature. Further, all of k values decrease with increasing temperature, except for ZrB$_2$ with 15 vol% SiC and 2 vol% MoSi$_2$ [49]. This is likely due to the dissolution of Mo from the MoSi$_2$ into ZrB$_2$. At the highest test temperatures (1940–2000°C), the range of reported k values increased. This is due to the low k (31 W/m-K) of ZrB$_2$ containing 48.7 vol% SiC and 5.1 vol% B$_4$C produced by Speyer et al. [51] The k of this material was likely diminished at high temperatures compared to the other materials due to a higher SiC content and the addition of B$_4$C, which has a k of approximately 7 W/m-K above 1700°C [74]. At lower temperatures (<1000°C), the four-phase composite containing 20 vol% SiC, 5 vol% Si$_3$N$_4$, and 20 vol% ZrC had the lowest k and a negative slope [41]. The presence of ZrC, which has an increase in k with temperature [75], may be the reason that this material does not have as negative a slope as Speyer's composition.

The highest k was reported for a composite containing 18 vol% SiC and 10 vol% C (presumably graphitic C because of the higher conductivity). At 100°C, k was approximately

---

[3]Corrected values for rxn-SPS ZrB$_2$ by Zhang et al. [18] were included for comparison.

115 W/m-K, which approached the value for the reaction-processed ZrB$_2$ by Zhang *et al.* [18]. However, compared to the pure ZrB$_2$, *k* of the 18% SiC–10% C composition dropped off much faster with temperature, eventually reaching 74 W/m-K at 2000°C.

9.3.2 Thermal Conductivity of HfB$_2$ Composites

Tables 9.7 and 9.8 summarize *k* values reported for HfB$_2$ composites. Like ZrB$_2$ composites (Tables 9.4 and 9.5), HfB$_2$ composites were primarily HfB$_2$-SiC (Table 9.7). [8, 14, 18, 28, 37, 41, 54, 76] The other HfB$_2$ composites are two-phase materials with additions of boron carbides (B$_3$C or B$_{12}$C) [78] or C (graphite) [8], or three-phase materials containing SiC and C [8], SiC and B$_4$C [76], or SiC and TaSi$_2$ [54] (Table 9.8). Densification was carried out using either HP or SPS for materials in Table 9.7 while only HP was used for those in Table 9.8. Besides the steady-state evaluation of *k* performed by Clougherty *et al.* [8] (below 1000°C), all of the other researchers evaluated *D* to calculate *k*. Plots of *k* as a function of temperature for HfB$_2$–SiC composites, and the other two- and three-phase materials, are included in Figures 9.10 and 9.11.[4]

9.3.2.1 HfB$_2$–SiC. Unlike ZrB$_2$–SiC (Fig. 9.7), where increasing SiC additions decreased *k*, HfB$_2$–SiC does not show the same trends (Fig. 9.10). Loehman *et al.* [14] and Clougherty *et al.* [8] evaluated SiC additions of 2, 5, and 20 vol%, and 5, 20, and 30 vol%, respectively. These systematic studies should be useful for evaluating trends. However, despite the variation in SiC content, *k* values did not vary considerably (≤12 W/m-K for Loehman *et al.* and ≤16 for Clougherty *et al.*). Further, no distinct trends were apparent with respect to SiC content, with both data sets exhibiting several crossover points.

Conductivities were reported for three HfB$_2$–20 vol% SiC composites in two separate studies ([28] and [37]) by Gasch *et al.* The highest *k* material was produced for the SHARP-B2 flight experiment. At room temperature, *k* was approximately 107 W/m-K and testing was performed up to 1800°C, where *k* dropped to 69 W/m-K. The other material in this study had significantly lower *k* values at room temperature and showed an increase in *k* with temperature (45 W/m-K at 25°C and 51 W/m-K at 1800°C) [28]. From the other Gasch study, *k* decreased slightly from 79 to 76 W/m-K between room temperature and 600°C [37]. Little information was given about the processing conditions for the SHARP-B2 material, so it is hard to say why its *k* was higher. However, powders for the lower *k* compositions were milled with WC media, which led to marked decreases in *k* for other diborides, as discussed earlier.

Similar to ZrB$_2$–SiC materials, microcracking appears to also be a consideration for HfB$_2$–SiC. The HfB$_2$–20 vol% SiC evaluated by Weng *et al.* has the lowest *k* of all compositions in Figure 9.10 (44–33 W/m-K from 300 to 1800°C) and the reported modulus is only 346 GPa (352 GPa, corrected for ρ), which is much lower than expected based on the composition [77]. Other researchers report modulus values of approximately 520 GPa for HfB$_2$ [28, 80] and 437–455 GPa for SiC. Conditions for microcracking in HfB$_2$–SiC composites have not been reported, but based on the modulus, and what appear to be cracks, microcracking is a likely contributing factor.

---

[4]Corrected values for rxn-SPS HfB$_2$ by Zhang *et al.* [18] were included for comparison.

TABLE 9.7. Thermal conductivity of HfB$_2$–SiC with information on starting/final materials, processing, and density

SiC Vol%	Powder processing	Purity (%)	Densification Technique	Special Nnotes	$\rho_{relative}$ (%)	Thermal Conductivity k (W/m-K) 25°C	Thermal Conductivity k (W/m-K) at max temp	Ref.
20%	VM w/plastic	—	HP	—	100	—	62 at 1000°C	[8]
30%	VM w/plastic	—	HP	—	100	—	58 at 1000°C	[8]
30%	VM w/plastic	—	HP	—	100	—	60 at 2000°C	[8]
5%	VM w/plastic	—	HP	—	100	—	59 at 1000°C	[8]
5%	VM w/plastic	—	HP	—	100	—	68 at 2000°C	[8]
20%	PBM or AM w/WC	98.8	HP	25 mm billet	100	42	49 at 2000°C	[28]
20%	PBM or AM w/WC	98.8	HP	50 mm billet	100	45	51 at 2000°C	[28]
2%	BM w/ZrO$_2$	99.5 ZrB$_2$, 99 SiC	HP at 2000°C	—	101	—	77 at 2000°C	[14]
20%	BM w/ZrO$_2$	99.5 ZrB$_2$, 99.9 SiC	HP at 2000°C	—	103	—	65 at 2000°C	[14]
5%	BM w/ZrO$_2$	99.5 ZrB$_2$, 99 SiC	HP at 2000°C	—	101	—	76 at 2000°C	[14]
20%	PBM w/WC	98.8	HP	—	99.7	79	76 at 600°C	[37]
20%	PBM w/WC	—	SPS	—	99.1	78	73 at 600°C	[37]
20%	BM w/ZrO$_2$	97.6 HfB$_2$, 99.5 SiC, 98.1 B$_4$C	HP at 2000°C	gsHfB$_2$ = 6 μm	98.4	—	55 at 1800°C	[76]
20%	—	—	HP at 2150°C	—	99.1	—	33 at 1800°C	[77]
5%	PBM w/WC	99.5 HfB$_2$, 99.8 SiC	SPS	—	99	—	104 at 425°C	[18]
5%	HM Hf+B+SiC	99.8 Hf, 99 B, 99.8 SiC	rxn SPS	—	101	—	110 at 425°C	[18]
20%	BM w/WC	All powder >99.5	HP at 2000°C	—	>96	141	86 at 1200°C	[41]
20%	—	—	—	SHARP-B2 material	100	107	69 at 1800°C	[28]

TABLE 9.8. Thermal conductivity of HfB_2-based composites with information on starting/final materials, processing, and density

Material (vol%)	Powder processing	Purity (%)	Densification technique	Special notes (gs in µm)	$\rho_{relative}$ (%)	Thermal conductivity k (W/m-K)		Ref.
						25°C	at max temp	
20%C	VM w/plastic	—	HP	—	100	—	70 at 1000°C	[8]
20%C	VM w/plastic	—	HP	—	100	—	69 at 2000°C	[8]
12%B_{12}C	BM w/ZrB$_2$	99.5 HfB$_2$, 99% B	HP at 2000°C	—	99.1	106	89 at 1000°C	[78]
12%B_3C	BM w/ZrB$_2$	99.5 HfB$_2$, 99% B	HP at 2000°C	—	99.1	112	84 at 2000°C	[78]
20%SiC+10%B_4C	BM w/ZrO$_2$	—	HP at 1850°C	gsZrB$_2$=4	98.7	—	35 at 1800°C	[76]
18%SiC+10%C	VM w/plastic	—	HP	—	100	—	66 at 1000°C	[8]
18%SiC+10%C	VM w/plastic	—	HP	—	100	—	61 at 2000°C	[8]
20%SiC+20%C_f	PBM w/SiC	—	HP at 2100°C	gsHfB$_2$ = 2.6, gsSiC = 1.3	99.5	94	—	[79]
20%SiC+30%C_f	PBM w/SiC	—	HP at 2100°C	gsHfB$_2$ = 2.6, gsSiC = 1.5	99.3	74	—	[79]
30%SiC+5%TaSi$_2$	BM w/SiC	—	SPS at 1800°C	—	>98.2	118	—	[54]
5% SiC+5%TaSi$_2$	BM w/SiC	—	SPS at 1800°C	—	>98.2	78	—	[54]
10% SiC+5%TaSi$_2$	BM w/SiC	—	SPS at 1800°C	—	>98.2	90	—	[54]
20% SiC+5%TaSi$_2$	BM w/SiC	—	SPS at 1800°C	—	>98.2	87	—	[54]

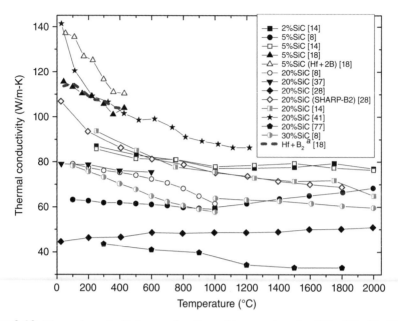

Figure 9.10. Thermal conductivity as a function of temperature for HfB$_2$–SiC with SiC contents ranging from 2 vol% to 30 vol%. Clougherty changed testing methods at 1000°C. [a]Data corrected for ρ [8, 14, 18, 28, 36, 39, 78].

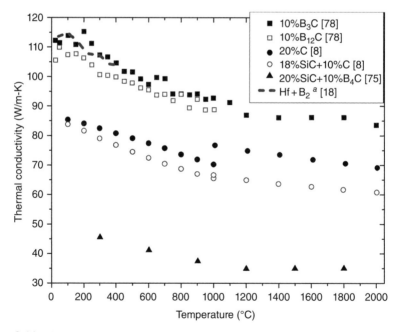

Figure 9.11. Thermal conductivity as a function of temperature for HfB$_2$ with additions of B$_x$C (x=3 or 12), C, or SiC with C or B$_4$C. Clougherty changed testing methods at 1000°C. [a]Data corrected for ρ [8, 18, 54, 75].

The three HfB$_2$–SiC composites with the highest k values are all at least as high as the pure reaction-processed HfB$_2$ evaluated by Zhang *et al.* [18] Both 5 vol% SiC compositions were characterized in the same study as the pure HfB$_2$, and the more conductive of the two was also produced using reaction processing. The purity of the reaction-processed diboride, along with the possibility of SiC sequestering impurities, resulted in HfB$_2$–5 vol% SiC with a k of 137 W/m-K at 50°C [18]. Overall, the HfB$_2$–20 vol% SiC presented by Mallik *et al.* was similar to the high k HfB$_2$ containing 5 vol% SiC at 25°C (141 W/m-K). However, k decreased to that of the reaction-processed HfB$_2$ by 200°C (~110 W/m-K) [41]. Of all the HfB$_2$–SiC materials, the material by Mallik is surprising based on what was observed for increasing SiC additions to ZrB$_2$ (Fig. 9.7) (especially given that the material was milled with WC) but no explanation is provided for the remarkably high k of this ceramic.

9.3.2.2 Other Secondary and Tertiary Additions to HfB$_2$.

The k values for HfB$_2$ with other second and third phases are plotted in Figure 9.11. The k values vary from as low as approximately 35 W/m-K to as high as approximately 115 W/m-K. The highest k values were reported by Brown-Shaklee *et al.* [78]. A slight variation in the thermal conductivity of these HfB$_2$–B$_x$C composites was observed, as can be seen by the decrease in conductivity of the 12 vol% B$_{12}$C material between 3 and 7 W/m-K across the temperature range. The researchers mentioned that the differences in k were less than the uncertainty of the test. It was also concluded that the purity of the materials in this study led to the high k values, nearly matching the reaction-processed material by Zhang *et al.* [18].

Microcracking was identified in HfB$_2$ containing 20 vol% SiC and 10 vol% B$_4$C by Weng *et al.* [76]. Therefore, k of this composite should be higher without such flaws. The final two compositions characterized by Clougherty *et al.* (HfB$_2$–20 vol% C (presumably graphite) and HfB$_2$–18 vol% SiC–10 vol% C) had similar k values at 100°C (84–85 W/m-K). Their k values deviated more at 2000°C (~10 W/m-K), but based on the trend for HfB$_2$–20% C, and the apparent discontinuity in values from switching testing methods at 1000°C, this difference was probably not significant. These materials were HfB$_2$ analogues to ZrB$_2$-based materials evaluated within the same study [8]. A comparison with the ZrB$_2$ compositions (Figs. 9.7 and 9.8) shows little difference between the 20% SiC compositions. At 2000°C the 18% SiC–10% C composites were identical (~61 W/m-K) but at 100°C the k values differed by 32 W/m-K with the HfB$_2$ composition having the lower k. This is unexpected since ZrB$_2$ and HfB$_2$ have similar behavior and was likely due to composition (purity) and microstructure, which could be evaluated to elucidate the cause for unexpected changes.

9.3.3 Conclusions Regarding Composites

Second-phase additions decrease k for diborides. The same decrease in k with temperature, as expected for the pure diborides, is observed, but the addition of second and third phases with lower k reduced k for the composites. One exception was the addition of MoSi$_2$, which changed the slope of k as a function of temperature, as observed for dissolution of other transition metals in diborides.

While the addition of SiC could increase the k of composites due to its high intrinsic k, all reported SiC additions lower conductivity, likely due to the purity of SiC used in the diboride systems. Additions of SiC can also be detrimental to k due to microcracking. The threshold for SiC particle or cluster size in ZrB₂ is approximately 11.5 μm. Larger sizes lead to microcracking in ZrB₂–SiC, but no research has been examined microcracking in HfB₂. Overall, the conductivities of the composites are still primarily controlled by the diboride. Whereas k may decrease with the addition of other phases, densification behavior, strength, thermal shock resistance, or oxidation resistance are often improved by these additions. Hence, understanding the factors that control k may enable design of compositions to balance the trade-offs for specific applications.

9.4 ELECTRON AND PHONON CONTRIBUTIONS TO THERMAL CONDUCTIVITY

The total thermal conductivity of the diborides of zirconium and hafnium, (k_t, i.e., the experimental values discussed to this point) can be separated into the electron (k_e) and phonon (k_p) contributions. Several researchers have used Equations 9.4 and 9.5 to evaluate k_t ZrB₂ [15, 18, 20, 21, 81], ZrB₂–SiC [15, 18, 46, 81], HfB₂ [18], and HfB₂–SiC [18, 81].

$$k_t = k_e + k_p \tag{9.4}$$

$$k_e = \frac{L_o T}{\rho} \tag{9.5}$$

Equation 9.5, the Wiedemann-Franz law, was used to estimate k_e from measured electrical resistivity (ρ) at temperature (T) using the theoretical Lorenz number ($L_o = 2.45e^{-8} \, W/\Omega\text{-}K^{-2}$), which was derived for metals.

9.4.1 ZrB₂ and HfB₂

Values of k_t, k_e, and k_p for several ZrB₂ and HfB₂ ceramics are plotted as a function of temperature in Figure 9.12. Comparing k_t and k_e, the positive or negative trends with temperature for k_e are identical to those of the k_t, except for data from Tye and Clougherty[5] [81]. On the other hand, k_p, in general, shows little variation with temperature.

The relatively small differences between k_e and k_t (excluding Tye and Clougherty and Zimmermann et al. [15]) indicate that electrons are the dominant thermal carrier for diborides. No indication is given as to why k_p is so much higher and has a positive trend (and conversely k_e so much lower) for the material studied by Tye and Clougherty. When compared with other reports, this does not seem to be characteristic of ZrB₂. Zimmermann et al. found k_p values as high as approximately 24 W/m-K below 200°C. However, most other research [18, 21], and the ZrB₂ produced by Thompson et al. [20], indicated that k_p was less than approximately 12 W/m-K for either ZrB₂ or HfB₂.

---

[5]This reference was used for separation of the contributions to thermal conductivity in place of Clougherty et al. [8] which is the original reference for the k_t data.

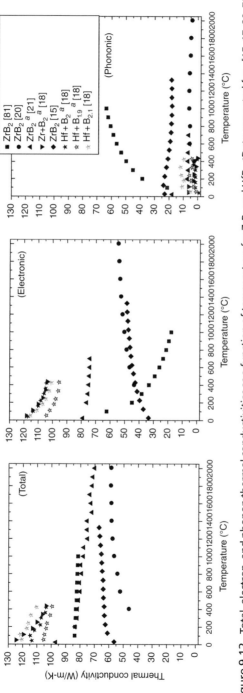

Figure 9.12. Total, electron, and phonon thermal conductivities as a function of temperature for ZrB$_2$ and HfB$_2$. [a]Data corrected for ρ [15–17, 20, 79].

Because k_e and k_p are estimated using Equations 9.4 and 9.5, this method may be prone to error. One assumption that permeates the studies to date is that the theoretical Lorenz number is appropriate for diborides. Calculation of negative k_p values for materials produced by Zhang *et al.* in Figure 9.12, indicate that a lower value for the Lorenz number may be more representative of the behavior of diborides. However, research still indicates that k_e is the dominant thermal conduction mechanism for ZrB₂ and HfB₂.

9.4.2 ZrB₂ and HfB₂ Composites with SiC

Figure 9.13 highlights the contributions of k_e and k_p to k_t for ZrB₂–SiC and HfB₂–SiC. Overall, some of the differences observed for materials with SiC additions are not explained and some trends appear to contradict prevailing theories. For example, k_e increases with increasing temperature for ZrB₂ + 20% SiC reported by Zhang *et al.* despite electrical resistivity values that increase with temperature for metallic conductors. Likewise, Tye and Clougherty reported significantly higher k_p values than other researchers and their k_p values increased with increasing temperature. The most important observation from Figure 9.13 is the decreasing contribution of k_e with increasing SiC content. This is expected due to the higher electrical resistivity of SiC (>1000 μΩ-cm) [82] compared to ZrB₂ (<25 μΩ-cm at 25°C) [15]. Based on this difference, electron conduction, which dominated k_t for diboride–SiC composites, is likely entirely supported by ZrB₂, with SiC acting as an electrically insulating phase.

In general, k_p increases as SiC content increases since phonon conduction is the primary thermal transport pathway for SiC. One clear exception is ZrB₂–30 vol% SiC, which should have a higher k_p due to its higher SiC content. However, as mentioned earlier, factors other than SiC content likely control this behavior since that composition contained WC contamination from milling. Hence, a direct comparison to other ZrB₂–SiC compositions is not possible without considering purity. Like some pure diboride material, such as those evaluated by Zhang *et al.* [18], the k_p of HfB₂–5 vol% SiC dropped below zero, which is unrealistic. This is another indication that the theoretical Lorenz number is not appropriate for diborides.

9.4.3 Conclusions Regarding k_e and k_p Research

The total k of diborides is dominated by electron transport. Therefore, to understand and control thermal conductivity of the diborides, factors that influence k_e are most important. The electron portion of k_t is clearly altered by transition metal impurities as indicated by decreased k_e values for materials containing W. SiC, which is electrically insulating, reduces k_e and increases k_p. The key to evaluating k_e and its contribution to k_t is the Wiedemann–Franz law. In some cases, the Wiedemann–Franz law predicts k_e values that are higher than k_t, resulting in negative k_p. Not all specimens exhibit this behavior, but it is an indication that the theoretical Lorenz number may not be valid for the diborides.

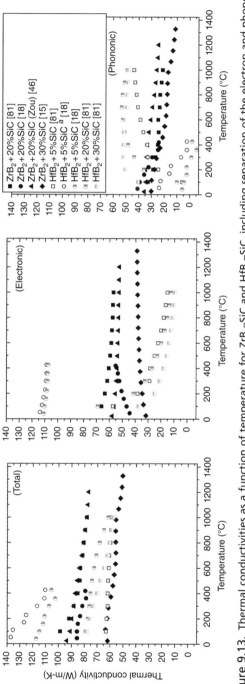

Figure 9.13. Thermal conductivities as a function of temperature for ZrB$_2$–SiC and HfB$_2$–SiC, including separation of the electron and phonon contributions to total conductivity.[a]Data corrected for ρ [16, 20, 50, 84].

9.5 CONCLUDING REMARKS

The thermal conductivity of ZrB_2- and HfB_2-based ceramics is affected by a number of factors, perhaps none more important than processing. In particular, contaminants from milling, impurities inherent to starting powders, or intentional additives almost always decrease k for ZrB_2 and HfB_2. The final microstructure, relative density, and phase distribution are critical and cannot be neglected in the evaluation of k. Additional phases or SSs that form during densification also strongly affect k. In particular, SSs appear to have the largest impact. Careful reporting of initial powder purity, contaminants from processing, and additives along with the resulting microstructure and phase distribution is necessary for direct comparison of k values among different materials. Further, future research on k of diborides would benefit from systematic studies designed to isolate specific compositional or processing effects rather than drawing conclusions solely based on comparison with reported values. In addition to performing more systematic research, separating electron and phonon contributions to k_t will be critical for gaining insight into what controls conductivity. The field would also benefit from experimental or computational investigation of Lorenz numbers to improve the accuracy of k_e and k_p. The electron contribution to k_t is dominant in ZrB_2 and HfB_2. Therefore, more detailed studies of factors that affect k_e would also be beneficial.

Research to date on additions of alloying elements or second phases has shown that k_t is maximized for phase pure diborides. SiC additions, despite their high k_t in single crystal form, do not increase k for either ZrB_2 or HfB_2, especially at higher temperatures. Combined with research suggesting that ZrB_2–SiC has poor oxidation resistance above approximately 1600°C [28, 83], SiC additions (as well as other silica formers) will not result in improved performance of UHTCs under the conditions they are being proposed to operate (i.e., hypersonic flight). Most other second-phase additions also decrease k_t of the diborides, so their use requires consideration of trade-offs such as improved strength or oxidation resistance for lower k_t. More research may also be useful in the area of decreasing k_t to produce ultra-high temperature insulators. SS additions such as W have been shown to impact k_p, especially at low temperatures. However, k_e dominates k_t and has the largest potential to be diminished, but is the least understood.

REFERENCES

1. Squire TH, Marschall J. Material property requirements for analysis and design of UHTC components in hypersonic applications. J Eur Ceram Soc 2010;30 (11):2239–2251.
2. Wuchina E, Opila E, Opeka M, Fahrenholtz W, Talmy I. UHTCs: ultra-high temperature ceramic materials for extreme environment applications. Electrochem Soc Interface 2007;16 (4):30.
3. Van Wie DM, Drewry DG, King DE, Hudson CM. The hypersonic environment: required operating conditions and design challenges. J Mater Sci 2004;39 (19):5915–5924.
4. Walker SP, Sullivan BJ. Sharp refractory composite leading edges on hypersonic vehicles. AIAA Paper, 6915; 2003.

5. Malone JE. Materials may allow spacecraft design change. Aerosp Technol Innov 2000;8 (6):1.

6. Sindeband SJ, Schwarzkopf P. The metallic nature of metal diborides. Powder Metall Bull 1950;5 (3):42–43.

7. Neel DS, Pears CD, Oglesby S, Jr. The thermal properties of thirteen solid materials to 5000°F for thier destruction temperatures. WADD TR 60-924. Southern Research Institute; February 1962.

8. Clougherty EV, Wilkes KE, Tye RP. Research and development of refractory oxidation-resistant diborides, part II, volume V: thermal, physical, electrical, and optical properties, Technical report AFML-TR-68-190. Wright-Patterson Air Force Base (OH): Air Force Materials Laboratory, Air Force Systems Command; 1969.

9. Samsonov GV, Kovenskaya BA, Serebryakova TI, Tel'nikov EY. Thermal conductivity of diborides of group IV-VI transition metals. High Temp 1972;10 (6):1193–1195.

10. Branscomb TM, Hunter JO. Improved thermal diffusivity method applied to TiB_2, ZrB_2, and HfB_2 from 200° to 1300°C. J Appl Phys 1971;42 (6):2309–2315.

11. Andrievskii RA, Korolev LA, Klimenko VV, Lanin AG, Spivak II, Taubin IL. Effect of zirconium carbide and carbon additions on some physicomechanical properties of zirconium diboride. Powder Metall Metal Ceram 1980;19 (2):93–94.

12. Fridlender BA, Neshpor VS, Ordan'yan SS, Unrod VI. Thermal conductivity and diffusivity of binary alloys of the ZrC-ZrB_2 system at high temperatures. High Temp 1980;17 (6):1001–1005.

13. Kinoshita H, Otani S, Kamiyama S, Amano H, Akasaki I, Suda J, Matsunami H. Zirconium diboride (0001) as an electrically conductive lattice-matched substrate for gallium nitride. Jpn J Appl Phys Part 2 Lett 2001;40 (12 A):L1280–L1282.

14. Loehman R, Corral E, Dumm HP, Kotula P, Tandon R. Ultra high temperature ceramics for hypersonic vehicle applications. Sandia Report SAND 206-2925. Albuquerquie, New Mexico, and Livermore, California: Sandia National Laboratories; 2006.

15. Zimmermann JW, Hilmas GE, Fahrenholtz WG, Dinwiddie RB, Porter WD, Wang H. Thermophysical properties of ZrB_2 and ZrB_2–SiC ceramics. J Am Ceram Soc 2008;91 (5):1405–1411.

16. Ikegami M, Matsumura K, Guo SQ, Kagawa Y, Yang JM. Effect of SiC particle dispersion on thermal properties of SiC particle-dispersed ZrB_2 matrix composites. J Mater Sci 2010;45 (19):5420–5423.

17. Guo S, Nishimura T, Kagawa Y. Preparation of zirconium diboride ceramics by reactive spark plasma sintering of zirconium hydride-boron powders. Scr Mater 2011;65 (11):1018–1021.

18. Zhang L, Pejaković DA, Marschall J, Gasch M. Thermal and electrical transport properties of spark plasma-sintered HfB_2 and ZrB_2 ceramics. J Am Ceram Soc 2011;94 (8):2562–2570.

19. Snyder A, Bo Z, Hodson S, Fisher T, Stanciu L. The effect of heating rate and composition on the properties of spark plasma sintered zirconium diboride based composites. Mater Sci Eng A 2012;538 (0):98–102.

20. Thompson MJ, Fahrenholtz WG, Hilmas GE. Elevated temperature thermal properties of ZrB_2 with carbon additions. J Am Ceram Soc 2012;95 (3):1077–1085.

21. Thompson MJ. Densification and thermal properties of zirconium diboride based ceramics [Ph.D. thesis]. Rolla (MO) : Missouri University of Science and Technology; 2012.

22. Harrington GJK, Hilmas GE, Fahrenholtz WG. Effect of carbon on the thermal and electrical transport properties of zirconium diboride, Submitted to J Eur Ceram Soc 2014.

23. McClane D, Fahrenholtz WG, Hilmas GE. Thermal properties of zirconium diboride with transition metal diboride additions. J Am Ceram Soc 2013;97 (5):1552–1558.

24. Schick HL, Avco Corporation Research and Advanced Development Division, Air Force Materials Laboratory. Materials Physics Division. *Thermodynamics of Certain Refractory Compounds*. New York: Academic Press; 1966.

25. Bolgar AS, Guseva EA, Turchanin AG, Fesenko VV. Thermodynamic properties of zirconium diboride. In: *Thermophysical Propeties of Solid Substances [in Russian]*. Moscow: Nauka; 1976. p 130–132.

26. Chase MW. *NIST-JANAF Thermochemical Tables*. 4th ed. Woodbury (NY): American Chemical Society and the American Institute of Physics; 1998.

27. Roine A. HSC chemistry for Window, Version 5.11 [Computer Program]. Pori: Outokumpu Research, Oy; 2006.

28. Gasch M, Ellerby D, Irby E, Beckman S, Gusman M, Johnson S. *Processing, Properties and Arc Jet Oxidation of Hafnium Diboride/Silicon Carbide Ultra High Temperature Ceramics*. Volume 39, Heidelberg: Springer; 2004. p 55–60.

29. Kingery WD, Mcquarrie MC. Thermal conductivity: I, concepts of measurement and factors affecting thermal conductivity of ceramic materials. J Am Ceram Soc 1954;37 (2):67–72.

30. Vajeeston P, Ravindran P, Ravi C, Asokamani R. Electronic structure, bonding, and ground-state properties of AlB_2-type transition-metal diborides. Phys Rev B 2001;63 (4):045115.

31. Voigt W. *Lehrbuch der kristallphysik*. Volume 34, Leipzig: BG Teubner; 1910.

32. Nichols JL. Orientation and temperature effects on the electrical resistivity of high-purity magnesium. J Appl Phys 1955;26 (4):470–472.

33. Maxwell JC. *A Treatise on Electricity and Magnetism*. Volume 1, Oxford: Clarendon Press; 1873. p 361–73.

34. Eucken A. Wärmeleitfähigkeit keramischer feuerfester Stoffe (Thermal conductivity of ceramic refractory materials). Forschung auf dem Gebiete des Ingenieurwesens 1932; Ausgabe B3 (353):6–21.

35. Gosset D, Decroix G-M, Kryger B. Improvement of thermo-mechanical properties of boron-rich compounds. Japanese Journal of Applied Physics, Proceedings of the 11th International Symposium on Boron, Borides, and Related Compounds; Tsukuba, Japan; 1993. p 216–219.

36. Opeka MM, Talmy IG, Wuchina EJ, Zaykoski JA, Causey SJ. Mechanical, thermal, and oxidation properties of refractory hafnium and zirconium compounds. J Eur Ceram Soc 1999;19 (13–14):2405–2414.

37. Gasch M, Johnson S, Marschall J. Thermal conductivity characterization of hafnium diboride-based ultra-high-temperature ceramics. J Am Ceram Soc 2008;91 (5):1423–1432.

38. Card Number 89-3651, Powder Diffraction File for HfB_2, International Centre for Diffraction Data.

39. Card Number 75-1050, Powder Diffraction File for ZrB_2, International Centre for Diffraction Data.

40. Wang Z, Hong C, Zhang X, Sun X, Han J. Microstructure and thermal shock behavior of ZrB_2–SiC–graphite composite. Mater Chem Phys 2009;113 (1):338–341.

41. Mallik M, Kailath AJ, Ray KK, Mitra R. Electrical and thermophysical properties of ZrB_2 and HfB_2 based composites. J Eur Ceram Soc 2012;32 (10):2545–2555.

42. Tian W-B, Kan Y-M, Zhang G-J, Wang P-L. Effect of carbon nanotubes on the properties of ZrB_2–SiC ceramics. Mater Sci Eng A 2008;487 (1–2):568–573.

43. Tu R, Hirayama H, Goto T. Preparation of ZrB_2-SiC composites by arc melting and their properties. J Ceram Soc Jpn 2008;116 (1351):431–435.

44. Zhang X, Xu L, Han W, Weng L, Han J, Du S. Microstructure and properties of silicon carbide whisker reinforced zirconium diboride ultra-high temperature ceramics. Solid State Sci 2009;11 (1):156–161.

45. Zhang H, Yan Y, Huang Z, Liu X, Jiang D. Properties of ZrB_2–SiC ceramics by pressureless sintering. J Am Ceram Soc 2009;92 (7):1599–1602.

46. Zou J, Zhang G-J, Zhang H, Huang Z-R, Vleugels J, Van Der Biest O. Improving high temperature properties of hot pressed ZrB_2–20vol% SiC ceramic using high purity powders. Ceram Int 2013;39 (1):871–876.

47. Guo S. Thermal and electrical properties of hot-pressed short pitch-based carbon fiber-reinforced ZrB_2–SiC matrix composites. Ceram Int 2013;39 (5):5733–5740.

48. Guo S, Kagawa Y, Nishimura T, Tanaka H. Thermal and electric properties in hot-pressed ZrB_2–$MoSi_2$–SiC composites. J Am Ceram Soc 2007;90 (7):2255–2258.

49. Monteverde F. The addition of SiC particles into a $MoSi_2$-doped ZrB_2 matrix: effects on densification, microstructure and thermo-physical properties. Mater Chem Phys 2009;113 (2–3):626–633.

50. Guo S-Q, Kagawa Y, Nishimura T, Tanaka H. Pressureless sintering and physical properties of ZrB_2-based composites with $ZrSi_2$ additive. Scr Mater 2008;58 (7):579–582.

51. Speyer RF. Oxidation resistance, electrical and thermal conductivity, and spectral emittance of fully dense HfB_2 and ZrB_2 with SiC, $TaSi_2$, and LaB_6 additives. Air Force Office of Scientific Research Report. Arlington, VA: Air Force Office of Scientific Research;2012.

52. Zhang XH, Wang Z, Hu P, Han WB, Hong CQ. Mechanical properties and thermal shock resistance of ZrB_2–SiC ceramic toughened with graphite flake and SiC whiskers. Scr Mater 2009;61 (8):809–812.

53. Monteverde F, Savino R. ZrB_2–SiC sharp leading edges in high enthalpy supersonic flows. J Am Ceram Soc 2012;95 (7):2282–2289.

54. Hu C, Sakka Y, Jang B, Tanaka H, Nishimura T, Guo S, Grasso S. Microstructure and properties of ZrB_2-SiC and HfB_2-SiC composites fabricated by spark plasma sintering (SPS) using $TaSi_2$ as sintering aid. J Ceram Soc Jpn 2010;118 (1383):997–1001.

55. Hu C, Sakka Y, Tanaka H, Nishimura T, Guo S, Grasso S. Microstructure and properties of ZrB_2–SiC composites prepared by spark plasma sintering using TaSi2 as sintering additive. J Eur Ceram Soc 2010;30 (12):2625–2631.

56. Guo S-Q, Kagawa Y, Nishimura T, Chung D, Yang J-M. Mechanical and physical behavior of spark plasma sintered ZrC–ZrB_2–SiC composites. J Eur Ceram Soc 2008;28 (6): 1279–1285.

57. Slack GA. Thermal conductivity of pure and impure silicon, silicon carbide, and diamond. J Appl Phys 1964;35 (12):3460–3466.

58. Slack GA. Nonmetallic crystals with high thermal conductivity. J Phys Chem Solids 1973;34 (2):321–335.

59. Watari K, Nakano H, Sato K, Urabe K, Ishizaki K, Cao S, Mori K. Effect of grain boundaries on thermal conductivity of silicon carbide ceramic at 5 to 1300 K. J Am Ceram Soc 2003;86 (10):1812–1814.

60. Burgemeister EA, Von Muench W, Pettenpaul E. Thermal conductivity and electrical properties of 6H silicon carbide. J Appl Phys 1979;50 (9):5790–5794.

61. Nakamura K, Maeda K. Hot-pressed SiC ceramics. In: Shigeyuki Somiya and Yoshizo Inomata, editors. *Silicon Carbide Ceramics: Gas Phase Reactions, Fibers and Whisker, Joining.* Volume 2, New York: Elsevier Applied Science; 1991. p 139–162.

62. Sigl LS. Thermal conductivity of liquid phase sintered silicon carbide. J Eur Ceram Soc 2003;23 (7):1115–1122.

63. Vasilos T, Kingery WD. Thermal conductivity: XI, conductivity of some refractory carbides and nitrides. J Am Ceram Soc 1954;37 (9):409–414.

64. Takeda Y. Development of high-thermal-conductive SiC ceramics. Ceram Bull 1988; 67:1961–1963.

65. Watts J, Hilmas G, Fahrenholtz WG. Mechanical characterization of ZrB$_2$–SiC composites with varying SiC particle sizes. J Am Ceram Soc 2011;94 (12):4410–4418.

66. Zhang SC, Hilmas GE, Fahrenholtz WG. Pressureless sintering of ZrB$_2$–SiC ceramics. J Am Ceram Soc 2008;91 (1):26–32.

67. Ogihara S, Maeda K, Takeda Y, Nakamura K. Effect of impurity and carrier concentrations on electrical resistivity and thermal conductivity of sic ceramics containing BeO. J Am Ceram Soc 1985;68 (1):C-16–C-18.

68. Silva SRP. *Properties of Amorphous Carbon.* London: INSPEC; 2003. p 158–162.

69. Lide D. *CRC Handbook of Chemistry and Physics.* 88th ed. Boca Raton (FL): CRC; 2007.

70. Dasgupta T, Umarji AM. Thermal properties of MoSi$_2$ with minor aluminum substitutions. Intermetallics 2007;15 (2):128–132.

71. Bai G, Jiang W, Chen L. Effect of interfacial thermal resistance on effective thermal conductivity of MoSi$_2$/SiC composites. Mater Trans 2006;47 (4):1247–1249.

72. Neshpor VS. The thermal conductivity of the silicides of transition metals. J Eng Phys 1968;15 (2):750–752.

73. Snyder A, Quach D, Groza JR, Fisher T, Hodson S, Stanciu LA. Spark plasma sintering of ZrB$_2$–SiC–ZrC ultra-high temperature ceramics at 1800°C. Mater Sci Eng A 2011;528 (18):6079–6082.

74. Wood C, Emin D, Gray PE. Thermal conductivity behavior of boron carbides. Phys Rev B 1985;31 (10):6811–6814.

75. Jackson HF, Lee WE. 2.13—properties and characteristics of ZrC. In: Konings JM, editor. *Comprehensive Nuclear Materials.* Oxford: Elsevier; 2012. p 339–372.

76. Weng L, Zhang X, Han J, Han W, Hong C. The effect of B$_4$C on the microstructure and thermo-mechanical properties of HfB$_2$ based ceramics. J Alloys Compd 2009;473 (1–2): 314–318.

77. Weng L, Han W, Li X, Hong C. High temperature thermo-physical properties and thermal shock behavior of metal-diborides-based composites. Int J Refract Metals Hard Mater 2010;28 (3):459–465.

78. Brown-Shaklee HJ, Fahrenholtz WG, Hilmas GE. Densification behavior and microstructure evolution of hot-pressed HfB$_2$. J Am Ceram Soc 2011;94 (1):49–58.

79. Guo S, Naito K, Kagawa Y. Mechanical and physical behaviors of short pitch-based carbon fiber-reinforced HfB$_2$-SiC matrix composites. Ceram Int 2013;39 (2):1567–1574.

80. Monteverde F. Progress in the fabrication of ultra-high-temperature ceramics: "in situ" synthesis, microstructure and properties of a reactive hot-pressed HfB$_2$–SiC composite. Compos Sci Technol 2005;65 (11–12):1869–1879.

81. Tye RP, Clougherty EV. The thermal and electical conductivities of some electrically con-ducting compounds. Proceedings of the Fifth Symposium on Thermophysical Properties; September 30–October 32, 1970; New York. Boston (MA): American Society of Mechanical Engineers; 1970. p 396–401.

82. Schneider SJ Jr, Committee AH. *Enginered Materials Handbook, Volume 4, Ceramics and Glasses*. Volume 4, Materials Park (OH): ASM International; 1991.

83. Zhang SC, Fahrenholtz WG, Hilmas GE. Oxidation of ZrB_2 and ZrB_2-SiC ceramics with tungsten additions. ECS Trans 2009;16 (44):137–145.

10

DEFORMATION AND HARDNESS OF UHtcs AS A FUNCTION OF TEMPERATURE

J. Wang and L. J. Vandeperre

Centre for Advanced Structural Ceramics, Department of Materials,
Imperial College London, London, UK

10.1 INTRODUCTION

According to Fahrenholtz and coworkers, ultra-high-temperature ceramics (UHTCs) are the materials with a melting point above 3000°C [1]. The transition metal carbides, nitrides, and borides belong to the group of UHTCs and will be the main focus of the discussion.

When crystalline materials are loaded with small forces, resistance to deformation stems from bond stretching and a reversible return to the original state when the load is released—the elastic response. As the load increases, irreversible deformation mechanisms are eventually activated in which matter is displaced permanently by dislocations or diffusion to relieve the applied stress by deformation. Alternatively, if the required stresses for such mechanisms are very high, crack propagation can lead to failure of the material. All these aspects of the mechanical behavior of crystalline materials are, to some extent, temperature dependent, and this dependence will be discussed. Since the elastic response is an important precursor to all other mechanisms and often the stresses or strains needed scale with the elastic properties, the chapter will start by exploring the elastic properties. Following this, the discussion will turn to hardness measurements

Ultra-High Temperature Ceramics: Materials for Extreme Environment Applications, First Edition.
Edited by William G. Fahrenholtz, Eric J. Wuchina, William E. Lee, and Yanchun Zhou.
© 2014 The American Ceramic Society. Published 2014 by John Wiley & Sons, Inc.

and, specifically, the use of hardness measurements as a convenient experiment to investigate the resistance to plastic deformation. As the temperature increases, deformation becomes increasingly easy as a multitude of creep mechanisms aid in the progression of deformation. These will be discussed as well. Finally, failure by fracture will not be discussed much in this chapter as it has already been discussed in Chapter 9. However, the reader will be reminded that fracture is a competing mechanism to deformation, that is, deformation will only occur to a large extent if fracture can be avoided.

10.2 ELASTIC PROPERTIES

Bulk ceramics contain many grains in random orientations and can, therefore, be considered isotropic in their response to elastic loading. However, since this isotropic response is a weighted average of the anisotropic response of the single-crystal grains from which the body is made, it remains instructive to consider the elastic response of a single crystal. The carbides of the early transition metals with melting point above 3000°C (ZrC, HfC, TaC, and NbC) are all cubic with the rock salt structure (space group $Fm\bar{3}m$). This means that the contracted relationship between stresses and strains

$$
\begin{bmatrix} \sigma_{11} \\ \sigma_{22} \\ \sigma_{33} \\ \tau_{23} \\ \tau_{31} \\ \tau_{12} \end{bmatrix} = \begin{bmatrix} C_{11} & C_{12} & C_{13} & C_{14} & C_{15} & C_{16} \\ C_{21} & C_{22} & C_{23} & C_{24} & C_{25} & C_{26} \\ C_{31} & C_{32} & C_{33} & C_{34} & C_{35} & C_{36} \\ C_{41} & C_{42} & C_{43} & C_{44} & C_{45} & C_{46} \\ C_{51} & C_{52} & C_{53} & C_{54} & C_{55} & C_{56} \\ C_{61} & C_{62} & C_{63} & C_{64} & C_{65} & C_{66} \end{bmatrix} \cdot \begin{bmatrix} \varepsilon_{11} \\ \varepsilon_{22} \\ \varepsilon_{33} \\ \gamma_{23} \\ \gamma_{31} \\ \gamma_{12} \end{bmatrix}
$$

which describes the elastic response of single crystals, only contains 3 independent elastic constants—$C_{11}(=C_{22}=C_{33})$, $C_{12}(=C_{13}=C_{21}=C_{23}=C_{31}=C_{32})$, and $C_{44}(=C_{55}=C_{66})$—and all other constants are zero. Zeener defined the anisotropy number, A, of cubic crystals as [2]

$$
A = \frac{2C_{44}}{C_{11} - C_{12}} \tag{10.1}
$$

Calculation of the shear modulus in all possible orientations using the transformations given in Ref. [3] shows that for cubic crystals this is the ratio of the extremes of the shear moduli and therefore a good measure for the anisotropy in the elastic response.

Experimental values for the elastic constants are reported in bold in Table 10.1 alongside values reported from first-principles and other theoretical calculations. The experimental values show that A is in between 0.71 and 0.95. This means that shear across the {100} planes in ⟨010⟩ directions is easier than shear across the {111} planes in any direction and that pulling along <111> directions has the lowest Young's modulus for any direction in the crystal, while pulling in <100> directions gives rise to the highest

TABLE 10.1. Elastic constants (C_{11}, C_{12}, and C_{44}) and elastic properties of UHTC carbides with the cubic rock salt structure (NaCl, Fm-3 m), with K as the bulk modulus and E_{avg} the average Young's modulus over all orientations

Substance	C_{11}	C_{12}	$\dfrac{C_{44}}{G}$	$\dfrac{(C_{11}-C_{12})/2}{G}$	K	A	E_{avg}	Method	References
ZrC	548	87	87	231	241	0.377	346	(-)	[4]
ZrC	643	74	121	285	264	0.425	442	LDA	[4]
ZrC	483	109	110	187	234	0.558	355	GGA	[5]
ZrC	512	95	128	209	234	0.614	396	GGA	[4]
ZrC	452	92	112	180	212	0.622	347	MIPMC$_v$	[4]
ZrC	**441**	**60**	**151**	**191**	**187**	**0.793**	**391**	**EXP**	**[6]**
ZrC	**472**	**99**	**159**	**187**	**223**	**0.853**	**411**	**EXP**	**[5]**
ZrC	468	100	159	184	233	0.864	409	3BFP	[5]
ZrC	**470**	**100**	**160**	**185**	**223**	**0.865**	**411**	**EXP**	**[7]**
HfC	454	87	87	184	209	0.471	311	(-)	[4]
HfC	474	88	98	193	217	0.508	367	MIPMC$_v$	[4]
HfC	556	97	143	230	250	0.623	436	LDA	[4]
HfC	548	99	142	225	249	0.633	430	GGA	[5]
HfC	556	105	152	226	255	0.674	446	GGA	[4]
HfC	498	113	179	193	241	0.939	443	3BFP	[5]
HfC	**500**	**114**	**180**	**193**	**243**	**0.933**	**449**	**EXP**	**[5]**
HfC	**500**	**114**	**180**	**193**	**243**	**0.933**	**449**	**EXP**	**[7]**

TaC	505	91	79	207	229	0.382	316	(-)	[8]
TaC	694	127	127	284	316	0.448	467	MIPMC$_v$	[4]
TaC	486	115	130	186	239	0.701	383	GGA	[4]
TaC	621	155	167	233	311	0.716	487	3BFP	[8]
TaC	547	119	159	214	262	0.743	448	EXP	[5]
TaC	**550**	**150**	**190**	**200**	**283**	**0.950**	**479**	**EXP**	[7]
TaC	370	110	153	130	197	1.177	341	GGA	[5]
TaC	**595**		**153**					**EXP**	[8]
NbC	648	109	109	270	289	0.404	420	(-)	[4]
NbC	593	126	132	234	282	0.565	433	MIPMC$_v$	[4]
NbC	460	159	103	151	259	0.684	321	GGA	[5]
NbC	**620**	**200**	**150**	**210**	**340**	**0.714**	**453**	**EXP**	[7]
NbC	617	199	159	209	338	0.761	463	3BFP	[5]
NbC	587	121	220	233	276	0.944	534	LDA	[4]
NbC	627	159	224	234	315	0.957	553	GGA	[4]

All values are in GPa except A, which is dimensionless; experimental results are shown in bold.
EXP, experimental; GGA, generalized gradient approximation; LDA, local density approximation; MIPMC, modified interaction potential model with covalency; 3BFP, 3-body force potential.

Young's modulus. However, the variations are limited to 40% between maximum and minimum values.

Theoretical estimates for elastic constants are variable. Clearly, two common simplifying assumptions made to enable the calculation (local density approximation, generalized gradient approximation) tend to overestimate the anisotropy of crystals. Bhardwaj and Singh [4] attribute this to many first-principles calculations ignoring the dipole–dipole and dipole–quadruple interactions. However, **modified interaction potential model with covalency** overall does not seem to give much better agreement with the experimental values, whereas the three-body force potential does capture the anisotropy of the crystals and seems to give estimates closest to experimental values.

The nitrides, which also have a rock salt structure, appear to be more anisotropic (see Table 10.2), and here, theoretical calculations appear to capture the anisotropy better. Hence, although no experimental information for TaN was found, the theoretically calculated high anisotropy ($A = 0.17$), which means the maximum shear modulus (384 GPa) is six times larger than the lowest shear modulus (64 GPa), **could** be true. Such high anisotropy is not very common, but it is of the same order of magnitude as in the alkali metal group (Li, 9.1; Na, 6.8; K, 6.7; Rb, 6.2; and Cs 7.2) [6] and, therefore, well within what has been observed for crystals.

The borides are hexagonal crystals (space group **P6/mmm**). The lower symmetry of hexagonal crystals means that there are five independent elastic constants: C_{11} ($=C_{22}$), C_{12} ($=C_{21}$), C_{13} ($=C_{31}=C_{32}=C_{23}$), C_{33}, C_{44} ($=C_{55}$). In addition to these independent constants, one other value is different from zero but dependent on the value of others ($C_{66} = (C_{11}-C_{12})/2$). The anisotropy number, A, can be defined in analogy with the one for cubic crystals by taking the ratio of the maximum to minimum value of the shear modulus, but is not defined directly by the elastic constants. It must therefore be evaluated by calculating G for all possible orientations of shear deformation. As shown in Table 10.3, calculations and experiments for diborides appear to agree quite well that anisotropy is limited (maximum shear modulus 40% higher than the minimum shear modulus), which is similar to what was observed experimentally for carbides. Young's modulus is lowest perpendicular to the basal plane, that is, along the c-axis, and the highest shear modulus is for shear in the c-direction in a plane perpendicular to any of the a-directions.

In summary, near room temperature, the elastic properties of the UHTC materials tend to be fairly isotropic. As a rule, UHTCs have a high stiffness with Young's moduli of the diborides (~500–600 GPa) exceeding those of the carbides and nitrides (~300–500 GPa). Theoretical predictions of the elastic properties are currently reasonable and capture the order of magnitude of the stiffness quite well. However, comparison with experiments shows that the predictions tend to overestimate elastic anisotropy in these materials.

While theoretical predictions for the effect of increased pressure on elastic constants and phase stability of UHTCs are available (see, e.g., [4, 14]), no predictions of the variation of the elastic constants with temperature were found. Experimentally, the most complete set of measurements are for ZrB$_2$ for which Okamoto and coworkers measured the variation of all five elastic constants with temperature up to 1400 K [12]. As shown in Figure 10.1a, C_{11}, C_{33}, and C_{44} decrease noticeably with temperature, while C_{13} and

TABLE 10.2. Elastic constants (C_{11}, C_{12}, and C_{44}) and elastic properties of UHTC nitrides with the cubic rock salt structure (NaCl, Fm-3 m), with K as the bulk modulus and E_{avg} the average Young's modulus over all orientations

Substance	C_{11}	C_{12}	$\dfrac{C_{44}}{G}$	$\dfrac{(C_{11}-C_{12})/2}{G}$	K	A	E_{avg}	Method	References
ZrN	523	111	116	206	248	0.563	381	**EXP**	[9]
ZrN	**471**	**88**	**138**	**192**	**216**	**0.721**	**389**	**EXP**	[10]
ZrN	547	150	147	199	282	0.741	426	LDA	[10]
ZrN	462	141	143	161	248	0.891	378	GGA	[10]
HfN	659	141	102	259	314	0.394	407	LDA	[10]
HfN	561	147	100	207	285	0.483	365	GGA	[10]
HfN	588	113	120	238	271	0.505	415	GGA	[9]
HfN	**679**	**119**	**150**	**280**	**306**	**0.536**	**498**	**EXP**	[10]
TaN	848	133	62	358	371	0.173	367	GGA	[10]
TaN	898	131	64	384	387	0.166	384	GGA	[11]
TaN	906	140	64	383	395	0.167	385	LDA	[10]

All values are in GPa except A, which is dimensionless; experimental results are shown in bold.
EXP, experimental; GGA, generalized gradient approximation; LDA, local density approximation; MIPMC, modified interaction potential model with covalency; 3BFP, 3-body force potential.

TABLE 10.3. Elastic constants and elastic properties of UHTC borides with the hexagonal structure (P6/mmm), with K as the bulk modulus and E_{avg} the average Young's modulus over all orientations

Substance	C_{11}	C_{12}	C_{13}	C_{33}	C_{44}	K	A	E_{avg}	Method	References
ZrB$_2$	**568**	**57**	**121**	**436**	**248**	**240**	**1.38**	**495**	**EXP**	[12]
ZrB$_2$	553	64	113	419	234	232	1.35	476	GGA	[13]
ZrB$_2$	563	64	133	446	253	248	1.40	496	GGA	[13]
HfB$_2$	578	71	118	430	228	242	1.35	485	GGA	[13]
HfB$_2$	609	73	135	470	266	263	1.36	530	GGA	[13]
HfB$_2$	583	98	132	456	258	259	1.36	508	GGA	[14]
HfB$_2$	593	100	141	481	262	269	1.35	523	GGA	[14]
TiB$_2$	**660**	**67**	**99**	**462**	**260**	**252**	**1.33**	**549**	**EXP**	[15]
TiB$_2$	**671**	**64**	**101**	**473**	**267**	**256**	**1.33**	**561**	**EXP**	[15]
TiB$_2$	**660**	**48**	**93**	**432**	**260**	**240**	**1.42**	**538**	**EXP**	[15]

All values are in GPa except A, which is dimensionless; experimental results are shown in bold.
EXP, experimental; GGA, generalized gradient approximation.

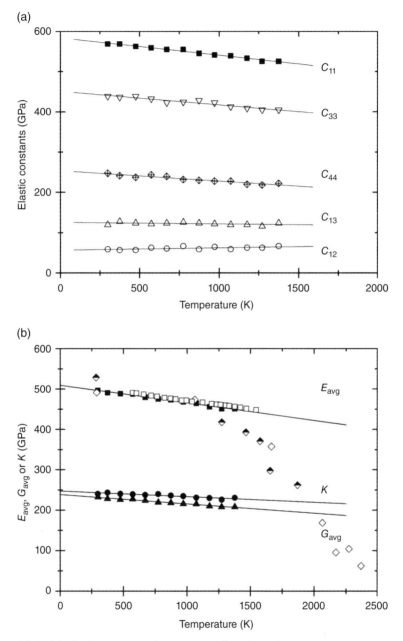

Figure 10.1. (a) Elastic constants for ZrB$_2$ as a function of temperature as measured by Okamoto *et al.* [12] and (b) the average Young's modulus, E_{avg} (filled squares); shear modulus, G_{avg} (filled triangles); and bulk modulus, K (filled circles), as a function of temperature using the same data. Also shown are the variation of Young's modulus as measured in flexure by Neuman *et al.* (half-filled diamonds) [16] and Rhodes *et al.* (open diamonds) [17] and the variation of Young's modulus of ZrB$_2$ as measured from the natural resonance frequency (open squares) [18].

C_{12} depend less on temperature. Evaluation of the ratio of maximum to minimum shear modulus as a function of temperature shows that the elastic anisotropy remains constant with temperature.

Figure 10.1b illustrates that the average shear modulus, obtained by averaging G for all possible orientations in 5° steps, decreases more strongly with temperature than the bulk modulus, K. Also, the decrease in average Young's modulus calculated from elastic constants agrees with the measured variation of Young's modulus by means of measuring the natural vibration frequency of a bar [18]. Below 1250 K, the data obtained in flexure [16, 17] also agrees with the single-crystal data. However, the latter measurements show a strong decrease for temperatures in excess of 1250 K. This suggests that the lower values are a result of creep influencing the measured slope or other problems. Indeed, such a dramatic decrease in elastic properties far away from any phase transition would be surprising, and the elastic modulus of ZrB_2 should really only decrease significantly on approaching its melting point.

This interpretation is confirmed when measurements for other borides and carbides obtained by measuring the natural frequency of vibration [19] are compared with flexural measurements (see Fig. 10.2). Again, a much more dramatic decrease with temperature is found for Young's modulus derived from flexural data than for the data obtained by impulse excitation of a natural vibration frequency. Especially the data for ZrC, which extends beyond 2000 K, indicates only a modest decrease in stiffness up to 2000 K. The slope by which the elastic modulus decreases with temperature divided by the value of Young's modulus at 298 K is reported in Table 10.4 and in all cases is of the order of -0.01% K^{-1}, that is, the elastic properties decay by about 1% every 100 K.

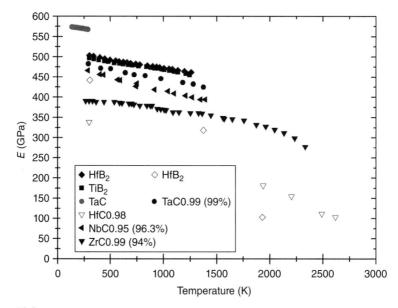

Figure 10.2. Young's modulus versus temperature for a range of UHTCs. Closed symbols are measurements based on vibration, whereas open symbols were obtained from flexural tests. Data from Refs. [18, 20–23].

TABLE 10.4. Rate of decay of the elastic modulus
with temperature

Substance	$1/E_{298\,K} \times dE/dT$ (K^{-1})
ZrB$_2$	−0.009%
TiB$_2$	−0.004%
HfB$_2$	−0.008%
ZrC	−0.012%
TaC	−0.015%
NbC	−0.014%

10.3 HARDNESS

Hardness measurements are often used to characterize mechanical behavior because the experiment is relatively easy and can be conducted on small specimens. Typically, a shape manufactured out of a very deformation-resistant material is pressed into the surface of the sample, and the ratio of the applied load to the size or depth of the (residual) imprint is taken as a measure of the hardness. In this chapter, hardness will be defined as the ratio of applied force, F, to projected area, A_{pr}, of the residual imprint after unloading:

$$H = \frac{F}{A_{pr}} \tag{10.2}$$

Vickers hardness is the ratio of applied force to pyramidal contact area. Therefore, to convert Vickers hardness values to projected hardness, the Vickers hardness needs to be divided by 0.9272 to account for the fact that the projected area is smaller than the true contact area [24].

While the experiment is easy to conduct, interpretation of hardness values is more difficult. Not only do hardness values depend on the method chosen to measure it, but it is often observed that the hardness is lower if the applied load is increased [25]. Numerous empirical equations and models have been put forward to quantify or explain this size effect [25–31]. Many of these explanations were first put forward in the time when microindenters became popular and enabled the analysis of small indents. At the time, it was assumed that deformation became more difficult when the indentation became smaller and, hence, the hardness obtained from larger indentations should be the true hardness reflecting the plastic properties. However, progress toward instrumentation, allowing even smaller indents (nanoindenters), suggests that these explanations might be incorrect for ceramics—at least at this scale. For example, data for slightly porous ZrB$_2$ and a dense ZrB$_2$ with 20 vol% SiC collected across a wide range of loads in Figure 10.3 shows that hardness at low loads is more or less constant and that the hardness decreases from this constant level as the load is increased. Interestingly, the transition between a constant hardness and a decreasing hardness tends to coincide with the onset of cracking around the indentation. Therefore, the data in Figure 10.3 supports Quinn's [31] suggestion that cracking has at least some role to play in explaining size effects in indentation of ceramics.

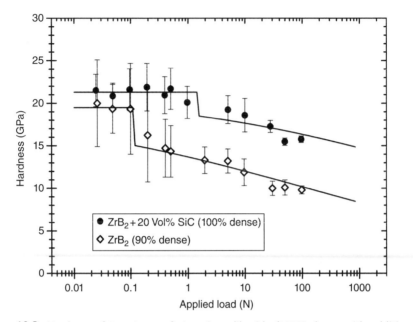

Figure 10.3. Hardness of two types of zirconium diboride (100% dense with addition of 20 vol% SiC and 90% dense with no additions) as a function of the size of the applied load. Data from Ref. [32].

Hence, hardness obtained from small indentations (100–200 mN) is a measure for the resistance to plastic deformation, whereas the hardness obtained from larger indents is a complex composite of the resistance to plastic deformation and fracture. The idea that fracture plays a role is not new and has been exploited in methodologies where the toughness of the material is derived from the size of the cracks relative to the size of the indents [33–35].

The data in Figure 10.3 also illustrate that at larger loads, the hardness of porous materials tends to be lower than that of dense materials in keeping with empiric relationships, which predict an exponential decrease of hardness with porosity [36].

Other microstructural features can play a role in hardness measurements too. Especially relevant is measurement of hardness in the case of ceramics containing several phases such as in ZrB_2 with SiC and B_4C [32]. For small indents, which are of the same size as the phases in the microstructure, three distinct hardness groupings with hardness values close to the hardness of the different phases can be seen in Figure 10.4, whereas for larger indents, a composite hardness is obtained, which is slightly higher than that of the ZrB_2. However, even when an indent is apparently made in a single phase, it is difficult to judge what might be present underneath the indentation. The TEM cross section through an indent in Figure 10.5 shows how the SiC grain appears to have resisted deformation better than the surrounding ZrB_2 and looks as if it was being pushed into the ZrB_2 grain, which suggests that the hardness measured will have been influenced by the SiC grain present underneath the indent.

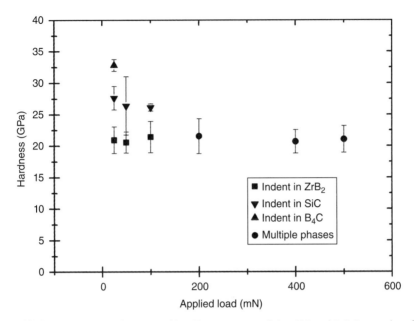

Figure 10.4. Hardness as a function of load in a ZrB_2 containing SiC and B_4C. Data taken from Ref. [32]. Where indents were apparently made in a single phase, they have been grouped accordingly.

Figure 10.5 also shows, quite clearly, that tremendous dislocation activity occurs underneath the indentation, even at room temperature. Other evidence of dislocation activity is the appearance of slip lines on the surface (see Fig. 10.6). Similar slip lines in ZrB_2 have also been reported by Ghosh et al. [38, 39]. This confirms that indentation can be used for studies of dislocation plasticity in ceramics, and the methodology will be explained further in Section 10.4. To avoid complications due to the microstructure, ideally, large-grained, single-phase materials or single crystals should be used to also eliminate any influence of grain size (see [40]).

Before discussing plasticity, the variation of the hardness of UHTCs with temperature will be described. Figure 10.7a–c shows hardness data collected from the literature for borides, carbides, and nitrides as a function of temperature. Many more measurements of room-temperature hardness were reported, but only measurements as a function of temperature have been included here for clarity.

Near room temperature, the hardness of all UHTCs is quite high (20–35 GPa). As the temperature increases, the decrease in hardness is rapid; by 1000 K, the remaining hardness is of the order of 5 GPa. Data for diborides suggests that the hardness decreases at a reduced rate between 1000 and 1750 K and reduces more rapidly again thereafter. The same trend is not as clear for the carbides. Not enough data for the nitrides is available to come to any conclusion. The initial rapid decay is related to the resistance of the lattice to dislocation motion. Therefore, first, the relation between hardness and yield stress will need to be explored before continuing the discussion in terms of the resistance to dislocation flow.

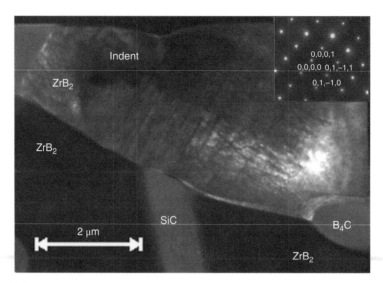

Figure 10.5. TEM bright-field micrograph of a cross section through a Berkovich indent in ZrB$_2$ containing SiC. It appears that the SiC has resisted deformation more and has been pushed into the ZrB$_2$ grain above it. Reproduced with permission from Ref. 37. © The American Ceramic Society.

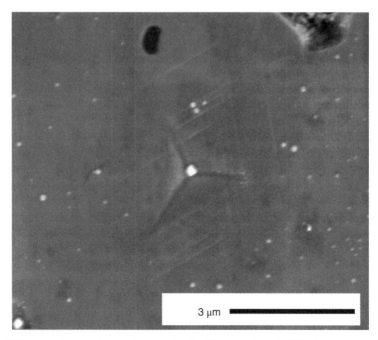

Figure 10.6. Scanning electron micrograph of a 50 mN indent in ZrB$_2$ showing multiple slip lines in three orientations formed by dislocations surfacing close to the indent.

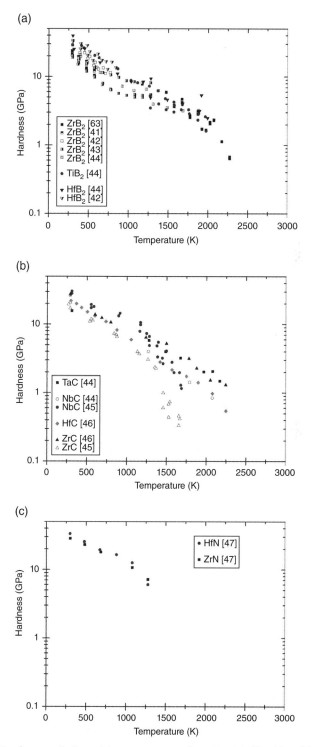

Figure 10.7. Hardness variation with temperature of UHTCs: (a) diborides, (b) carbides, and (c) nitrides. Data from Refs. [37, 41–47].

10.4 HARDNESS AND YIELD STRENGTH

Tabor showed using slip line theory that the hardness of metals is approximately three times the yield strength [24]. Since yield strength is a kinetic property of the material that is a function of deformation rate and history, this definition required more care in further work. Tabor and Atkins [48] showed that the amount of strain hardening varies with the geometry of the indenter. For a Vickers or Berkovich indenter, hardness is a measure for the yield strength after 8–10% additional plastic strain. This correlation has later been confirmed by extensive finite element calculations [49].

When Marsh [50] tried to apply the same proportionality to the hardness of glass fibers, yield strength was predicted to be below the observed brittle fracture strength, which means that glass should deform rather than fracture. Fortunately, he realized that the theory derived for metals, which have a relatively modest yield strength, Y, compared to their stiffness, E, might not apply to ceramics where the yield strength is a more substantial fraction of the stiffness. As a result in metals, elastic deformation before the onset of plasticity is negligible, and the material displaced by the indenter is forced out of the indented surface forming a pileup of material next to the indentation. When elastic strains preceding plasticity are substantial, the material surrounding the indentation will recede by elastic strain and create space into which deformed material can flow. This led him to propose that indentation in materials where Y/E is high can be thought of as the expansion of a cavity under gas pressure in an infinite body for which Hill had derived a theoretical solution [51]. However, Hill's solution can't be used directly because indentation only produces half a cavity and, therefore, a free surface that can move in the direction of indentation. Johnson [52] was the first to propose a theoretical way to modify the spherical cavity to account for the free surface but overestimated the effect. The treatment by Vandeperre et al. [53] agrees much better with available relationships derived by finite element calculations (Fig. 10.8). Their analytical relationship between hardness and yield strength is

$$\frac{H}{Y} = \frac{2}{3}\left\{1 + \frac{3}{3-6\lambda}\ln\left(\frac{(3+2\mu)\cdot(2\lambda\cdot(1-\zeta)-1)}{(2\lambda\mu-3\mu-6\lambda)\cdot\zeta}\right)\right\} \tag{10.3}$$

with

$$\zeta = \frac{E_r}{E_r - 2H\tan\alpha} \tag{10.4}$$

$$\lambda = \frac{(1-2v)Y}{E} \tag{10.5}$$

$$\mu = \frac{(1+v)Y}{E} \tag{10.6}$$

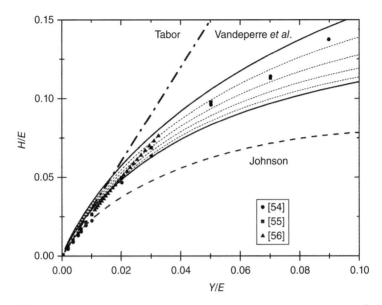

Figure 10.8. Hardness, H, over elastic modulus, E, as a function of the ratio of the yield strength, Y, to the elastic modulus according to the models of Tabor [24], Johnson [52], and Vandeperre, Giuliani, and Clegg [53]. For the latter, the effect of Poisson's ratio is shown with the lower curve for $\nu = 0$ and the upper curve for $\nu = 0.5$. Also shown is the finite element calculation of the relationship taken from Ref. [54–56].

where E is Young's modulus, ν is Poisson's ratio, α is the included equivalent semiangle of the indenter, and E_r is the reduced modulus of the contact defined itself as

$$\frac{1}{E_r} = \frac{1-\nu_i^2}{E_i} + \frac{1-\nu^2}{E} \qquad (10.7)$$

with E_i and ν_i as Young's modulus and Poisson's ratio of the indenter.

The key to the success of the latter treatment is that it recognizes that elastic deformation of the surface reduces the amount of material that needs to be displaced by plastic flow for a fixed penetration of the indenter. So, for low Y/E (<0.01), no elastic surface deformation occurs, and the entire penetration of the indenter must be accommodated by plastic flow of material. In contrast, when Y/E becomes high (>0.1), then the entire penetration of the indenter into the material is accommodated by deflection of the surface. The latter is observed in rubbers in which the relation between hardness and yield strength is lost and hardness becomes a measure of E and indenter shape alone. For very low values of Y/E, the expanding cavity-based model predicts that hardness would be more than three times the yield strength, which shows that the hemispherical flow of material would require a higher hardness than the formation of a pileup, thereby explaining the change in flow patterns from hemispherical flow to flow toward the surface when Y/E decreases. Once the change in displaced volume is accounted for, the hardness becomes a smaller multiple of the yield strength for Y/E values typical for ceramics, which explains why glass need not be very hard for its yield strength to be well above its fracture strength.

10.5 DEFORMATION MECHANISM MAPS

Plastic flow in ceramics, like in any other crystalline material, is normally carried by the movement of dislocations. Although other deformation mechanisms such as twinning [57, 58], phase transformations [59, 60], and densification [61] can play a role, these have not been reported for UHTCs and are therefore not included in this discussion. Deformation mechanisms maps are concise way of representing the dominant deformation mechanisms as a function of temperature, strain rate, and shear stress. These maps, introduced by Ashby and Frost [62], are plots containing lines of constant value of one of the three parameters as a function of the two other parameters. The treatment given here follows mostly the approach of Ashby and Frost, and their book contains much more information. The treatment here differs in that the lattice resistance is described slightly differently and more creep mechanisms are considered. To aid the understanding of the maps, the equations that govern the different mechanisms controlling deformation will be presented and related to hardness data for ZrB_2. Following this, maps for ZrB_2 and ZrC as assessed in the literature [62, 63] will be compared as examples for the behavior of diborides and carbides. Unfortunately, for nitrides, experimental data are too limited to make a sensible stab at a deformation mechanism map.

10.6 LATTICE RESISTANCE TO DISLOCATION GLIDE

Dislocations can only move if they can at least overcome the resistance of the lattice. The process can be described by considering the rate at which dislocations will be able to flow. A simple theory can therefore be derived starting from Orowan's relation between the velocity of the dislocations, v, and the strain rate, $d\gamma/dt$:

$$\frac{d\gamma}{dt} = \rho_m \cdot \mathbf{b} \cdot v \tag{10.8}$$

where ρ_m is the mobile dislocation density and $\mathbf{b}$ the Burgers vector. The velocity of the dislocations will be determined by the rate at which they can overcome the rate-determining obstacle. Treating the movement of the dislocation as any other kinetic process involving an activation energy barrier and activation by stress and by counting both the jumps over a distance b in the direction of the applied stress and the jumps over a distance b against the applied stress, the following expression for the velocity is obtained:

$$v = \upsilon_a \cdot \mathbf{b} \cdot \left[\exp\left(-\frac{[Q - \tau \cdot V]}{kT} \right) - \exp\left(-\frac{[Q + \tau \cdot V]}{kT} \right) \right] \tag{10.9}$$

where v_a is the attempt frequency by which the dislocation tries the move, Q is the activation energy barrier, τ is the applied shear stress, V is the activation volume, k is Boltzmann's constant, and T is temperature.

The resistance of the lattice, also termed Peierls stress, can be substantial, but the activation energy associated with it, Q, is normally not very high so that thermal energy alone, kT, can have a substantial influence on the stress required to make the dislocations move [62]. Indeed, fcc metals remain ductile even when tested in liquid nitrogen (77 K) because their lattice resistance is limited and reduced to effectively zero very rapidly. In contrast, bcc metals, for which the activation energy of the lattice resistance is in the range 0.5–1.13 eV [62], show a brittle-to-ductile transition temperature (BDTT) caused, in part, by the rapid rise in lattice resistance around the BDTT. Other obstacles to plastic flow such as forest dislocations and second-phase particles tend to have fairly high activation energy barriers (~6 to 40 eV [62]), and their resistance is, therefore, not very temperature dependent [62].

Hence, the high hardness of ceramics near room temperature and the rapid decrease in hardness are entirely consistent with it being controlled by lattice resistance.

Since the lattice resistance or Peierls stress is defined as the stress needed to make the dislocation move at absolute zero, the activation energy barrier can be equated to

$$Q = \tau_p \cdot V \tag{10.10}$$

Substitution of (Eq. 10.4) into (Eq. 10.3) and (Eq. 10.3) into (Eq. 10.2) and solving for the flow stress yields

$$\tau = \frac{kT}{V} \sinh^{-1} \left[\frac{\dfrac{d\gamma}{dt}}{2 \cdot \rho_m \cdot v \cdot \mathbf{b}^2} \exp\left(\frac{\tau_p V}{kT} \right) \right] \tag{10.11}$$

This expression can be simplified if it is assumed that the movement of the dislocations against the applied stress is negligible compared to movement along with the stress and that the activation volume can be considered to be independent of temperature. These assumptions lead to the following expression:

$$\tau = \tau_p - \frac{kT}{V} \ln\left(\frac{\rho_m \cdot \mathbf{b}^2 \cdot v_a}{d\gamma / dt} \right) \tag{10.12}$$

which shows that lattice resistance decreases linearly with temperature, in agreement with the steep decrease in hardness. The activation volume can be determined from measurements of the flow stress for different strain rates at fixed temperature and the product $\rho_m \cdot v_a \cdot b^2$ from the slope of the decay in flow stress with temperature for a fixed strain rate, and the Peierls stress can be found from the intersection at 0 K from such experiments.

Hence, full characterization of the kinetics of plastic flow controlled by the lattice resistance requires measurements as a function of strain rate. Bhakhri *et al.* [41] discuss how this is possible using modern indenters, which record load and displacement during an indentation experiment so that the strain rate dependence of hardness can be determined. In deriving the strain rate for ceramic materials, again, the elastic nature of the indentation process needs to be accounted for as only the strain rate due to plastic deformation should be used. Their results for ZrB_2 are shown in Figure 10.9 and illustrate that strain rate does influence the hardness at moderate temperatures (<1000 K) as expected for lattice resistance-controlled hardness. They estimate the Peierls stress of ZrB_2 as 6.6 ± 0.7 GPa and the activation energy as 1.6 ± 1 eV.

The hardness data available for UHTCs, in general, does not include information on the strain rate dependence. However, estimates for the Peierls stress and activation energy can still be made by fitting Equation 10.11 to flow stress data derived from the hardness. Since standard hardness measurement procedures tend to use larger loads and short times (10 s), it was assumed in the analysis that the strain rate was fairly high (10^{-1} s^{-1}), which can affect estimates for the activation energy, but not the Peierls stress. Estimates for the Peierls stress, on the other hand, might have been affected by the presence of rather limited data, so the values given in Table 10.5 are really only estimates.

What is clear is that the Peierls stress is of the same order of magnitude for most UHTCs (6–8 GPa). Within one group of materials, this could have been expected. The Peierls model for the lattice resistance predicts that it should only depend on the ratio of the Burgers vector, **b**, and the slip plane spacing, **d**, and the shear modulus [68, 69].

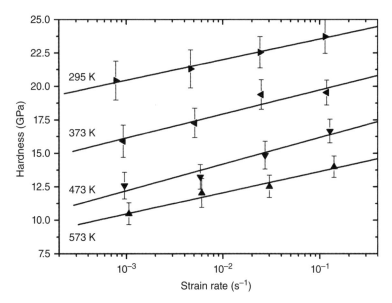

Figure 10.9. Hardness of ZrB_2 as a function of plastic strain rate. Data taken from Bhakhri *et al.* [41].

TABLE 10.5. Average shear modulus (G), Peierls stress (τ_p), activation energy (Q), slip planes, and Burgers vector (**b**) for UHTCs

Material	G (GPa)	τ_p (GPa)	Q (eV)	τ_p/G	Slip plane	Burgers vector (nm)
Carbides						
TaC	195	4.0	1.7	0.021	$\{111\}, \{110\}$	$\frac{1}{2}a\langle 1\bar{1}0\rangle$	0.321
NbC	167	6.3	1.5	0.037	$\{111\}, \{110\}$	$\frac{1}{2}a\langle 1\bar{1}0\rangle$	0.316
ZrC	168	6.0	1.6	0.036	$\{111\}, \{110\}$	$\frac{1}{2}a\langle 1\bar{1}0\rangle$	0.332
HfC	186	6.0	1.2	0.032	$\{111\}, \{110\}$	$\frac{1}{2}a\langle 1\bar{1}0\rangle$	0.328
Nitrides						
ZrN	153	8.2	1.6	0.054	$\{110\}$	$\frac{1}{2}a\langle 1\bar{1}0\rangle$	0.325
TaN	102	8.2	1.6	0.080	$\{110\}$	$\frac{1}{2}a\langle 1\bar{1}0\rangle$	0.325
HfN	153	8.5	1.6	0.056	$\{110\}$	$\frac{1}{2}a\langle 1\bar{1}0\rangle$	0.320
Diborides						
ZrB$_2$	257	6.6	1.6	0.025	$(0001), \{10\bar{1}0\}$	$\frac{1}{3}a\langle 1\bar{2}10\rangle$	0.317
TiB$_2$	259	6.9	1.7	0.026	$(0001), \{10\bar{1}0\}$	$\frac{1}{3}a\langle 1\bar{2}10\rangle$	0.303
HfB$_2$	236	7.5	1.6	0.032	$(0001), \{10\bar{1}0\}$	$\frac{1}{3}a\langle 1\bar{2}10\rangle$	0.314

Main slip system information from Refs. [37, 39, 64–67].

Assuming that slip in all the carbides occurs by partial dislocations ½a <1–10> on (111) or (110) planes [2, 62], the ratio **b/d** will be constant, and hence, the Peierls model predicts that the Peierls stress should only scale with the shear modulus. Since the latter is quite similar for all, similar Peierls stresses are to be expected. The exceptionally low value for the Peierls stress of TaC is probably lower than it should be as a result of very limited hardness data so that the fit is unreliable. Although the nitrides also have rock salt structure and, hence, can be expected to behave like the carbides, their bonding is more ionic, and therefore, slip along {110} planes is favored over slip along {111} planes [70], that is, over planes with large spacing **d**. This is expected to give a lower Peierls stress, but experimentally, their Peierls stress appears to be a larger fraction of the shear modulus than the carbides. However, some care is needed as all nitride data were obtained from thin films rather than from bulk ceramics and these can be affected by residual stresses.

The crystal structure and slip system of the borides are quite different. The Burgers vector has been identified as $\frac{1}{3}a\langle 1\bar{2}10\rangle$ [65], that is, one of the **a** vectors of the hexagonal unit cell of the crystal, and slip occurs on both the basal plane and the (10-10) prism planes [37, 38, 65]. The fact that the ratio of the Peierls stress to the shear modulus is different for these materials is, therefore, not surprising, and the boride values are reasonably close to each other as are their shear moduli.

The activation energies are all around 1.6 eV for the simple reason that, without strain rate information, it is difficult to estimate this parameter more accurately. However, given that the decay in hardness with temperature is similar across all systems, this estimate is certainly of the right order of magnitude.

The high values of the shear stress needed to move dislocations even only one Burgers vector in the crystal explain why at room temperature plasticity is only observed under exceptional circumstances such as during indentation where a compressive stress field for small indents suppresses cracking. This is not without significance as loading by small particles is quite common in wear and the good resistance to dislocation flow can contribute to a high wear resistance [71]. Nevertheless, the stresses needed to grow a crack—maximum fracture strengths are of the order of 1 GPa [72]—are much lower than those needed to make dislocations move, so in many cases, the UHTCs will be brittle at room temperature.

10.7 DISLOCATION GLIDE CONTROLLED BY OTHER OBSTACLES

The dashed lines marked **lattice resistance** in Figure 10.10 were calculated using (Eq. 10.11) and illustrate that the lattice resistance drops so strongly with temperature that even at fairly low temperatures (0.1–0.3 T_m), it becomes insignificant. It would be tempting to predict that at relatively low temperature, a transition from brittle to ductile should be observed.

However, the lattice resistance is not the only resistance that dislocations must overcome to move through a material as otherwise fcc metals would always be extremely soft and would not be useful structural materials. Other obstacles to the movement of

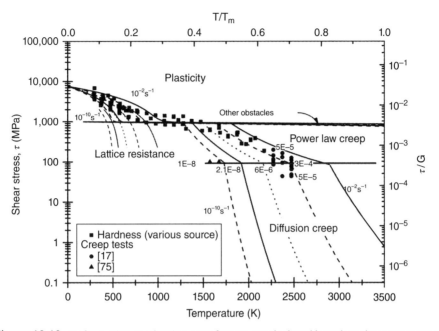

Figure 10.10. Deformation mechanism map for ZrB$_2$ recalculated based on the assessment of Wang [63]. Experimental data taken from the following sources: hardness [41, 44, 73, 74] and creep and flow stress measurements [65, 75].

dislocations include other dislocations forming locks, solute atoms, secondary-phase precipitates, and grain boundaries [76]. Hence, when the lattice resistance decays, the resistance to dislocation motion becomes dominated by the resistance due to these other obstacles. A transition to ductile behavior will, therefore, only be observed if the stresses needed to overcome the latter are lower than the stresses needed to activate crack growth.

Since dislocation densities underneath indentations tend to be high (see Fig. 10.5), it is reasonable to expect that in hardness tests the main strengthening mechanism would be strain hardening with an associated shear flow stress given by

$$\tau = G \cdot \mathbf{b} \cdot \sqrt{\rho} \tag{10.13}$$

Given that dislocation densities in heavily deformed material easily reach $10^{14}\,\text{m}^2$ [77, 78], τ/G is estimated to be of the order 3×10^{-3}, which agrees reasonably with the value for the section where the shear flow stress appears to show limited temperature dependence in Figure 10.10. Since this corresponds to a uniaxial applied stress of 1.54 GPa, which is higher than the highest strengths measured, reaching gross plasticity is still not expected.

An alternative approach to describing the resistance of other obstacles, taken by Ashby and Frost [62], is to use a single descriptive equation similar in nature to the one derived for glide controlled by the lattice resistance:

$$\gamma^{\bullet} = \gamma_0^{\bullet} \cdot \exp\left(-\frac{Q_{\text{obstacle}}}{kT}\left(1 - \frac{\tau}{\tau_{\text{obstacle}}} \right) \right) \qquad (10.14)$$

where τ_{obstacle} as before is the shear stress needed to overcome the obstacle resistance without thermal energy. To account for the more a thermal nature of the resistance by these other obstacles, the value of the activation energy, Q_{obstacle}, is set to a much higher value than that of the lattice resistance. For dislocation-strengthened materials, Ashby and Frost use $0.2–1 \times G\mathbf{b}^3$. The solid lines for the plasticity region in Figure 10.10 were calculated by summing the shear stress needed to overcome the lattice resistance and the shear stress needed to overcome the obstacles as flow is only possible if both resistances are overcome at the same rate.

10.8 DEFORMATION BY CREEP

At even higher temperatures, dislocations can escape from obstacles by climb, and, therefore, deformation can occur at lower stresses at a rate normally determined by diffusion of vacancies to aid the climb. This regime, termed power law creep, can be described effectively by

$$\gamma^{\bullet} = \frac{A \cdot D_{\text{eff}} \cdot G \cdot \mathbf{b}}{k \cdot T} \cdot \left(\frac{\tau}{G} \right)^n \qquad (10.15)$$

where A is a constant and the effective coefficient of diffusion, D_{eff}, is given by

$$D_{\text{eff}} = D_{\text{v}} + \frac{10a_{\text{c}}}{\mathbf{b}^2}\left(\frac{\tau}{G} \right)^2 D_{\text{c}} \qquad (10.16)$$

The latter accounts for both bulk diffusion through the bulk diffusion coefficient D_{v} and diffusion along the dislocation core through the second term, which takes into account the core area of the dislocation a_{c} and the diffusion coefficient through the core D_{c}. Hence, at low temperatures, diffusion through the core of the dislocation dominates and the power law has an exponent equal to $n+2$, whereas at higher temperatures, bulk diffusion dominates and the exponent is only n. As can be seen in Figure 10.10, the second region where a strong decrease in flow stress occurs in diborides is believed to coincide with the onset of power law creep [63].

Power law creep is not the only creep mechanism. An obvious alternative to mass transport by dislocations is mass transport by diffusion under influence of a stress, which can be described by [62]

$$\gamma^{\bullet} = \frac{42 \cdot \tau \cdot \Omega}{k \cdot T \cdot d^2} \cdot D'_{\text{eff}} \qquad (10.17)$$

where d is the grain size, Ω is the atomic volume, and the effective diffusion coefficient, D'_{eff}, is given by the weighted average of bulk and grain boundary diffusion:

$$D'_{eff} = D_v + \frac{\pi\delta}{d} D_b \qquad (10.18)$$

with δ as the grain boundary thickness. Again, two mechanisms of creep are captured in a single equation: when grain boundary diffusion dominates (Coble creep), the creep rate is inversely proportional to the third power of the grain size, whereas if bulk diffusion dominates (Nabarro–Herring creep), the creep rate is only inversely proportional to the second power of the grain size. In both cases, the creep rate is directly proportional to the stress, that is, the stress exponent can be said to be equal to 1.

Another creep mechanism that is very common in ceramic materials is creep due to the presence of viscous material on the grain boundaries [79, 80]. In such instances, creep can be caused by dissolution in a region of compressive stress, transport of the material through the viscous grain boundary, and precipitation in the tensile regions or by viscous flow of the grain boundary glass [81]. Indeed, a stress exponent of 1 is also consistent with creep accommodated entirely by viscous flow of the glassy material from the compressive to the tensile surfaces of the grains with creep rates given by [80–83].

$$\gamma^{\cdot} = \sqrt{3} \cdot \frac{\tau}{\eta} \cdot \left(\frac{\delta}{d}\right)^3 \qquad (10.19)$$

where τ is the applied shear stress, η is the viscosity, δ is the thickness of the grain boundary film, and d is the grain size. Note that in terms of stress and grain size dependence, the behavior is similar to what is observed for Coble creep. This mechanism is not recognized as a true deformation mechanism by all because it is not self-sustaining: once all viscous liquid has been squeezed out of the compressive regions, deformation by this mechanism comes to a halt. However, strains in excess of the typical failure strains can be accumulated for quite normal contents of viscous material [81]. These viscous mechanisms have not been added to the maps since TEM evidence suggests that such viscous phases are limited in **pure** UHTCs, provided surface oxides are removed during processing [63].

Since each of these creep rates occurs independently, the creep rates from all mechanisms should in principle be added, but in the maps shown in this chapter, only the dominant mechanism is shown. This only causes small differences to the map as the contributions of minor mechanism are only important in the transition regions.

10.9 DEFORMATION OF CARBIDES VERSUS BORIDES

The deformation mechanism map for ZrB_2 shown in Figure 10.10 should be contrasted with the deformation mechanism map for ZrC as assessed by Ashby and Frost [62] in Figure 10.11. The parameters used for constructing both maps are summarized in Table 10.6.

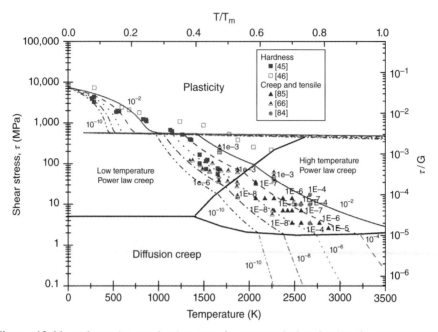

Figure 10.11. Deformation mechanism map for ZrC recalculated using the assessment of Ashby and Frost [62]. Experimental data taken from the following sources: hardness [45] and creep and flow stress measurements [66, 84, 85].

Two major differences exist. The first is that in the map for ZrB₂, deformation by diffusion is estimated to dominate for much higher stresses than for ZrC. The second is that ZrC shows creep deformation at a much reduced temperature with two very distinct regions of low- and high-temperature creep. For ZrB₂, power law creep requires higher temperatures, and it is not clear whether much distinction must be made between high- and low-temperature creep—in the assessment, only high-temperature creep was considered.

The fact that the transition metal carbides have a poor creep resistance is well established [62]. The map for ZrB₂ does not include creep at low temperature because the hardness variation with temperature is so different as commented on before: for all of the diborides, the hardness is found to almost not decrease between 500 and 1750 K. The fact that several authors observed this plateau-like region suggests strongly that the creep resistance of the diborides is indeed much better than that of the carbides, whose hardness drops almost at a fixed rate until below 1 GPa.

The reasons given for the extended role for diffusion in the ZrB₂ assessment were [63] (i) the fact that ZrB₂ sinters at 2273 K is evidence that flow by diffusion is possible in that region; (ii) that TEM observation of samples from hardness measurements showed limited dislocations when deformed at 2273 K, whereas dislocations were seen at lower temperatures; and (iii) the fact that the stress exponent in creep measurements by Rhodes *et al.* [17] (~2) was consistent with diffusion creep transiting into power law creep and, hence, a transition into diffusional creep at relatively high stress. Of these, the

TABLE 10.6. Parameters used to construct the deformation mechanism maps. Parameters for ZrB_2 from Ref. [63] and for ZrC from Ref. [62]. Where the treatment deviates from Ashby and Frost, parameters were adapted to get coincidence of the maps.

Parameter	ZrB_2	ZrC
General information		
Melting point, T_m	3505 K	3530 K
Lattice parameters		
a	0.317 nm	0.4698 nm
c	0.353 nm	(n/a)
Burgers vector, **b**	0.317 nm	0.334 nm
Atomic volume, Ω	0.0154 nm	0.0264 nm³
Shear modulus at RT, G	257 GPa	173 GPa
dG/dT	$-0.025\,GPa\,K^{-1}$	$-0.0139\,GPa\,K^{-1}$
Grain size	20 μm	10 μm
Bulk diffusion		
Pre-exponential factor $D_{v,0}$	$5\,m^2\,s^{-1}$	$1.03 \times 10^{-1}\,m^2\,s^{-1}$
Activation energy, Q_v	$678\,kJ\,mol^{-1}$	$720\,kJ\,mol^{-1}$
Grain boundary diffusion		
Pre-exponential factor $\delta \times Dg_{b,0}$	*	$5.0 \times 10^{-12}\,m^3 s^{-1}$
Activation energy, Q_{gb}		$468\,kJ\,mol^{-1}$
Dislocation core diffusion		
Pre-exponential factor $a_c \times D_{c,0}$	*	$1.0 \times 10^{-21}\,m^4 s^{-1}$
Activation energy, Q_c		$468\,kJ\,mol^{-1}$
Power law creep		
Dorn constant	10^{13}	10^{12}
Exponent	7.38	5
Lattice resistance-controlled glide		
Peierls stress, τ_p	6.6 GPa	6.92 GPa
Activation energy, Q	1.61 eV	1.40 eV
Attempt frequency, ν_a	$10^{11}\,s^{-1}$	$10^{11}\,s^{-1}$
Mobile dislocation density, ρ_m	$10^{14}\,m^{-2}$	$10^{14}\,m^{-2}$
Obstacle-controlled glide		
Critical shear stress at 0 K, $\tau^\wedge$	1 GPa	578 MPa
Pre-exponential γ_0	$10^8\,s^{-1}$	$10^8\,s^{-1}$
Activation energy	51 eV	40 eV

*Not included in assessment.

latter was the more important as it strongly confirms that diffusion does play a role to much higher stresses. The stress exponent of the creep data from Talmy [75] shown in Figure 10.10 was again indicative of creep controlled by diffusion, and therefore, this additional data point largely confirms the position of the creep controlled by diffusion in the map.

It is worth pointing out that α-Al_2O_3 is a trigonal crystal (space group $R\bar{3}c$) and, therefore, has a certain level of structural similarity to the hexagonal diborides. This compound shows deformation on similar slip systems to the transition metal diborides (on the basal and prism planes) [86]. It also shows a much larger region in which diffusion controls the deformation in the map by Ashby and Frost [62], indicating that diffusion does play a role to much higher stresses in similar materials too.

Finally, Miloserdin *et al.* [84] claimed diffusional creep as the mechanism for their results, while on the ZrC map, their data is lodged firmly in the power law creep region. Ashby and Frost did not include this data in their assessment, and hence, perhaps the diffusional region of carbides should extend to higher stresses, which would reduce the difference somewhat.

10.10 CONCLUSIONS

The carbides, diborides, and nitrides of the transition metals are all reasonably isotropic in their response to elastic loading and quite stiff and show only a limited reduction of about 1% per 100 K in elastic properties.

Small-scale measurements of hardness (100–200 mN) give a good indication of the deformation behavior at moderately elevated temperatures (~600 K) and in combination with strain rate-dependent measurements can give good estimates for the lattice resistance to dislocation motion. Measurements of the hardness at higher temperatures can also aid in the identification of transitions toward other deformation mechanisms such as power law creep and creep by diffusion and therefore are a powerful tool for determination of deformation mechanism maps. A comparison of maps for carbides and for diborides revealed that the latter are much more creep resistant and that diffusion appears to play a more important role in their deformation behavior.

REFERENCES

1. Fahrenholtz WG, Hilmas GE, Talmy IG, Zaykoski JA. Refractory diborides of zirconium and hafnium. J Am Ceram Soc 2007;90 (5):1347–1364.

2. Kelly A, Groves GW, Kidd P. *Crystallography and Crystal Defects*. 2nd ed. Chichester: John Wiley & Sons, Ltd; 2000. p 470.

3. Hearmon RFS. Equations for transforming elastic and piezoelectric constants of crystals. Acta Cryst 1957;10 (2):121–125.

4. Bhardwaj P, Singh S. Structural phase stability and elastic properties of refractory carbides. Int J Refract Metals Hard Mater 2012;35:115–121.

5. Srivastava A, Diwan BD. Elastic and thermodynamic properties of divalent transition metal carbides MC (M = Ti, Zr, Hf, V, Nb, Ta). Can J Phys 2012;90:331–338.

6. Landolt H, Elektrische B. Piezoelektrische, pyroelektrische, piezooptische, elektrooptische konstanten und nichtlineare dielektrische suszeptibilitaten. Zahlenwerte Funktionen Naturwissenschaft Technik 1979;Gruppe(III):11.

7. Weber W. Lattice dynamics of transition metal carbides. Phys Rev B 1973;8 (11):5082–5092.

8. Lopez-de-la-Torre L, Winkler B, Schreuer J, Knorr K, Avalos-Borja M. Elastic properties of tantalum carbide (TaC). Solid State Commun 2005;134:245–250.

9. Holec D, Friak M, Neugebauer J, Mayrhofer PH. Trends in the elastic response of binary early transition metal nitrides. Phys Rev B 2012;85:064101.

10. Chen W, Jiang JZ. Elastic properties and electronic structures of 4d- and 5d-transition metal mononitrides. J Alloys Compd 2010;499:243–254.

11. Chang J, Zhao G-P, Zhou X-L, Ke L, Lu L-Y. Structure and mechanical properties of tantalum mononitride under high pressure: a first principles study. J Appl Phys 2012;112: 083519.

12. Okamoto NL, Kusakari M, Tanaka K, Inui H, Yamaguchi M. Temperature dependence of thermal expansion and elastic constants of single crystals of ZrB_2 and the suitability of ZrB_2 as a substrate for GaN films. J Appl Phys 2003;93 (1):88–93.

13. Lawson JW, Bauschlicher CH Jr, Daw MS. Ab initio computations of electronic, mechanical and thermal properties of ZrB_2 and HfB_2. J Am Ceram Soc 2011;94 (10):3494–3499.

14. Zhang J-D, Cheng X-L, Li D-H. First-principles study of the elastic and thermodynamic properties of HfB_2 with AlB_2 structure under high pressure. J Alloys Compd 2011;509:9577–9582.

15. Waskowska A, Gerward L, Olsen JS, Babu KR, Vaitheeswaran G, Kanchana V, Svane A, Filipov VB, Levchenko G, Lyaschenko A. Thermoelastic properties of ScB_2, TiB_2, YB_4 and HoB_4: experimental and theoretical studies. Acta Mater 2011;59:4886–4894.

16. Neuman EW, Hilmas GE, Fahrenholtz WG. Strength of zirconium diboride to 2300°C. J Am Ceram Soc 2013;96 (1):47–50.

17. Rhodes WH, Clougherty EV, Kalish D. Research and development of refractory oxidation-resistant diborides. Part II. Volume IV. Mechanical properties. Technical Report. September 15, 1967–May 15, 1969. *Other Information: UNCL.* Original Receipt Date: June 30, 1973. 1970. Medium: X; Size: p. 153.

18. Wiley DE, Manning WR, Hunter O. Elastic properties of polycrystalline TiB_2, ZrB_2 and HfB_2 from room temperature to 1300 K. J Less Common Metals 1969;18:149–157.

19. Roebben G, Bollen B, Brebels A, Van Humbeeck J, Van der Biest O. Impulse Excitation apparatus to measure resonant frequencies, elastic moduli, and internal friction. Rev Sci Instrum 1997;68 (12):4511–4515.

20. Wuchina E, Opeka M, Causey S, Buesking K, Spain J, Cull A, Routbort J. Guitierrez-m-Morea. Designing for ultrahigh-temperature applications: the mechanical and thermal properties of HfB_2, HfNx. J Mater Sci 2004;39:5939–5949.

21. Dodd SP, Cankurtaran M. Ultrasonic determination of the elastic and nonlinear acoustic properties of transition-metal carbide ceramics: TiC and TaC. J Mater Sci 2003;8: 1007–1115.

22. Jun CK, Shafffer PTB. Elastic moduli of niobium carbide and tantalum carbide. J Less Common Metals 1971;23:367–373.

23. Baranov VM, Knyazev VI. The temperature dependence of the elastic constants of nonstoichiometric zirconium carbides. Probl Prochn 1973;9:45–47.

24. Tabor D. The physical meaning of indentation and scratch tests. Br J Appl Phys 1956;7:159–166.

25. Sangwal K. Review: indentation size effect, indentation cracks and microhardness measurement of brittle crystalline solids—some basic concepts and trends. Cryst Res Technol 2009;44 (10):1019–1037.

26. Krell A. A new look at grain size and load effects in the hardness of ceramics. Mater Sci Eng 1998;A245 (2):277–284.

27. Krell A. A new look at the influences of load, grain size, and grain boundaries on the room temperature hardness of ceramics. Int J Refract Metals Hard Mater 1998;16 (4–6):331–336.

28. Li H, Bradt RC. The microhardness indentation load/size effect in rutile and cassiterite single crystals. J Mater Sci 1993;28:917–926.

29. Sangwal K. On the reverse indentation size effect and microhardness measurement of solids. Mater Chem Phys 2000;63:145–152.

30. Chen J, Bull SJ. A critical examination of the relationship between plastic deformation zone size and Young's modulus to hardness ratio in indentation testing. J Mater Res 2006;21 (10):2617–2627.

31. Quinn GD, Green P, Xu K. Cracking and the indentation size effect for knoop hardness of glasses. J Am Ceram Soc 2003;86 (3):441–448.

32. Wang J, Giuliani F, Vandeperre LJ. The effect of load and temperature on hardness of ZrB_2 composites. Ceram Eng Sci Proc 2010;31 (2):59–68.

33. Anstis GR, Chantikul P, Lawn B, Marshall DB. A critical evaluation of indentation techniques for measuring fracture toughness: I. Direct crack measurements. J Am Ceram Soc 1981;64 (9):533–538.

34. Niihara K, Morena R, Hasselman DPH. Evaluation of K_{IC} of brittle solids by the indentation method with low crack-to-indent ratios. J Mater Sci Lett 1982;1:13–16.

35. Morris DJ, Myers SB, Cook RF. Sharp probes of varying acuity: instrumented indentation and fracture behavior. J Mater Res 2004;19 (1):165–175.

36. Rice RW. *Porosity of ceramics. Materials Engineering.* New York: Marcel Dekker Inc.; 1998. p 539.

37. Wang J, Giuliani F, Vandeperre L. Temperature and strain-rate dependent plasticity of ZrB_2 composites from hardness measurements. Ceram Eng Sci Proc 2011;32:137–149.

38. Ghosh D, Subhash G, Orlovskaya N. Slip-line spacing in ZrB_2-based ultrahigh-temperature ceramics. Scr Mater 2010;62:839–842.

39. Ghosh D, Ghatu S, Bourne GR. Room-temperature dislocation activity during mechanical deformation of polycrystalline ultra-high-temperature ceramics. Scr Mater 2009;61: 1075–1078.

40. Rice RW, Wu C, Borchelt F. Hardness—grain size relations in ceramics. J Am Ceram Soc 1994;77 (10):2539–2553.

41. Bhakhri V, Wang J, Ur-rehman N, Ciurea C, Giuliani F, Vandeperre L. Instrumented nanoindentation investigation into the mechanical behaviour of ceramics at moderately elevated temperatures. J Mater Res 2012;27 (1):65–75.

42. Bsenko L, Lundström T. The high-temperature hardness of ZrB_2 and HfB_2. J Less Common Metals 1974;34 (2):273–278.

43. Chen CH, Xuan Y, Otani S. Temperature and loading time dependence of hardness of LaB6, YB6 and TiC single crystals. J Alloys Compd 2003;350 (1–2):L4–L6.

44. Koester RD, Moak DP. Hot hardness of selected borides, oxides, and carbides to 1900°C. J Am Ceram Soc 1967;50 (6):290–296.

45. Kumashiro Y, Nagai Y, Kato H. The Vickers micro-hardness of NbC, ZrC and TaC single crystals up to 1500°C. J Mater Sci 1982;1:49–52.

46. Koval'chenko MS, Daemelinski VV, Borisenko V. Temperature dependence of the hardness of Titanium, Zirconium, and Hafnium carbides. Probl Prochnosti 1969;5:63–66.

47. Quinto DT, Wolfe GJ, Jindal PC. High temperature microhardness of hard coatings produced by physical and chemical vapor deposition. Thin Solid Films 1987;153:19–36.

48. Atkins AG, Tabor D. Plastic indentation in metals with cones. J Mech Phys Solids 1965;13:49–164.

49. Cheng YT, Cheng CM. Scaling approach to conical indentation in elastic-plastic solids with work hardening. J Appl Phys 1998;84 (3):1284–1291.

50. Marsh DM. Plastic flow in glass. Proc R Soc A 1963;279:420–435.

51. Hill R. *The Mathematical Theory of Plasticity*. Oxford: Clarendon Press; 1950.

52. Johnson KL. The correlation of indentation experiments. J Mech Phys Solids 1970;18:115–126.

53. Vandeperre LJ, Giuliani F, Clegg WJ. Effect of elastic surface deformation on the relation between hardness and yield strength. J Mater Res 2004;19 (12):3704–3714.

54. Cheng YT, Cheng CM. What is indentation hardness? Surf Coat Technol 2000;133–134:417–424.

55. Dao M, Chollacoop N, Vliet KJV, Venkatesh TA, Suresh S. Computational modeling of the forward and reverse problems in instrumented sharp indentation. Acta Mater 2001;49:3899–3918.

56. Mata M, Anglada M, Alcala J. A hardness equation for sharp indentation of elastic-power law strain-hardening materials. Philos Magaz 2002;A82 (10):1831–1839.

57. Giuliani F, Lloyd SJ, Vandeperre LJ, Clegg WJ. Deformation of GaAs under nanoindentation. In: McVitie S, McComb D, editors. *Electron Microscopy and Analysis*. Oxford: Institute of Physics; 2003. p 123–126.

58. Lloyd SJ, Castellero A, Giuliani F, Long Y, McLaughlin KK, Milina-Aldareguia JM, Stelmashenko NA, Vandeperre LJ, Clegg WJ. Observations of nanoindents via cross-sectional transmission electron microscopy: a survey of deformation mechanisms. Proc R Soc Lond A 2005;61 (2060):2521–2543.

59. Lloyd SJ, Molina-Aldareguia JM, Clegg WJ. Under nanoindents in Si, Ge, and GaAs examined through transmission electron microscopy. J Mater Res 2001;16 (12):3347–3350.

60. Vandeperre LJ, Giuliani F, Lloyd SJ, Clegg WJ. The hardness of silicon and germanium. Acta Mater 2007;55 (18):6307–6315.

61. Suzuki K, Benino Y, Fujiwara T, Komatsu T. Densification energy during nanoindentation of silica glass. J Am Ceram Soc 2002;85 (12):3102–3104.

62. Frost HJ, Ashby MF. *Deformation Mechanism Maps: The plasticity and Creep of Metals and Ceramics*. Oxford: Pergamon Press; 1982.

63. Wang J. Processing and deformation of ZrB_2. London: Department of Materials, Imperial College London; 2012. p 129.

64. Vahldiek FW, Mersol SA. Slip and microhardness of IVa to VIa refractory materials. J Less Common Metals 1977;55 (2):265–278.

65. Haggerty JS, Lee DW. Plastic deformation of ZrB_2 single crystals. J Am Ceram Soc 1971;54 (11):572–576.

66. Lee DW, Haggerty JS. Plasticity and creep in single crystals of zirconium carbide. J Am Ceram Soc 1969;52 (12):641–647.

67. Molina-Aldareguia JM, Lloyd SJ, Barber ZH, Clegg WJ. Lack of hardening effects in TiN/NbN multilayers. Materials Research Society Fall Meeting November 27-December 1; Boston, MA; Materials Research Society; 2001.

68. Cottrell AH. Plastic flow in crystals. In: Fowler RH *et al.*, editors. *International Monographs on Physics*. Oxford: Clarendon Press; 1961. p 223.

69. Peierls P. The size of a dislocation. Proc Phys Soc 1940;52:34–37.

70. Oden M, Ljungcrantz H, Hultman L. Characterisation of the induced plastic zone in a single crystal TiN(001) film by nanoindentation and transmission electron microscopy. J Mater Res 1997;12 (8):2134–2142.

71. Hutchings IM. *Tribology: Friction and Wear of Engineering Materials*. London: Edward Arnorld; 1992.

72. Chamberlain AL, Fahrenholtz WG, Hilmas GE. High-strength zirconium diboride-based ceramics. J Am Ceram Soc 2004;87 (6):1170–1172.

73. Xuan Y, Chen C-H, Otani S. High temperature microhardness of ZrB_2 single crystals. J Phys D Appl Phys 2002;35:L98–L100.

74. Wang J, Feilden-Irving E, Giuliani F, Vandeperre LJ. The Hardness of zirconium diboride between 1323 K and 2273 K. Ceram Eng Sci Proc 2013;33 (3):187–196.

75. Talmy IG, Zaykoski JA, Martin CA. Flexural creep deformation of ZrB_2/SiC ceramics in oxidizing atmosphere. J Am Ceram Soc 2008;91 (5):1441–1447.

76. Dieter GE. *Mechanical Metallurgy*. London: McGraw-Hill; 1998. SI Metric edition; p 751.

77. Barabash OM, Santella M, Barabash RI, Ice GE, Tischler J. Measuring depth-dependent dislocation densities and elastic strains in an indented Ni-based superalloy. J Miner Metals Mater Soc 2010;62 (12):29–34.

78. Dragomir-Cernatescu I, Gheorghe M, Thadhani N, Snyder RL. Dislocation densities and character evolution in copper deformed by rolling under liquid nitrogen from x-ray peak profile analysis. J Int Centre Diffract Data 2005;48:67–72.

79. Biswas K, Rixecker G, Aldinger F. Creep and visco-elastic behaviour of LPS-SiC sintered with Lu2O3-AlN additive. Mater Chem Phys 2007;104 (1):10–17.

80. de Arellano-Lopez AR, Melendez-Martinez JJ, Cruse TA, Koritala RE, Routbort JL, Goretta KC. Compressive creep of mullite containing Y_2O_3. Acta Mater 2002;50:4325–4338.

81. Dryden JR, Kucerovsky D, Wilkinson DS, Watt DF. Creep deformation due to a viscous grain boundary phase. Acta Metall 1989;7 (7):2007–2015.

82. Lofaj F, Wiederhorn SM, Dorcakova F, Hoffmann KJ. The effect of glass composition on creep damage development in silicon nitride ceramics. 11th International Conference on Fracture; Turin; 2005.

83. Yoon KJ, Wiederhorn SM, Luecke WE. Comparison of tensile and compressive creep behavior in silicon nitride. J Am Ceram Soc 2000;83 (8):2017–2022.

84. Miloserdin YV, Naboichenko KV, Laveikin LI, Bortsov AG. The high-temperature creep of zirconium carbide. Probl Prochn 1972;3:50–53.

85. Leipold MH, Nielsen TH. Mechanical properties of hot-pressed zirconium carbide tested to 2600°C. J Am Ceram Soc 1964;47 (9):419–424.

86. Lagerlöf KPD, Heuer AH, Castaing J, Rivière JR, Mitchell TE. Slip and twinning in sapphire (α-alumina). J Am Ceram Soc 1994;77 (2):385–397.

11

MODELING AND EVALUATING THE ENVIRONMENTAL DEGRADATION OF UhTCs UNDER HYPERSONIC FLOW

Triplicane A. Parthasarathy[1], Michael K. Cinibulk[2], and Mark Opeka[3]

[1] *UES, Inc., Dayton, OH, USA*
[2] *Materials and Manufacturing Directorate, Air Force Research Laboratory, Wright-Patterson AFB, OH, USA*
[3] *Naval Surface Warfare Center, West Bethesda, MD, USA*

11.1 INTRODUCTION

Over the past decade, a worldwide interest has grown in developing aircraft that can travel at or greater than Mach 6 [1]. Several short-duration flight tests by NASA and the U.S. Air Force have shown that a scramjet engine can be used to achieve aircraft speeds of Mach 5 to Mach 10 [2–4]. Extending these concepts toward the development of a vehicle for reusable or sustained flight now awaits the development of novel materials that can handle the aerothermal conditions imposed on some of the critical parts [5]. The sharp leading edges of the vehicle and cowl at the engine inlet as well as the fuel injection struts in the combustion section of the scramjet engine are considered to be critical components with anticipated material temperatures that could reach as high as 2000°C through aerothermal heating at Mach 8 [6]. These components are required by current design to have a geometrically sharp leading edge (762 μm or 30 mil radius of curvature) that must be retained while bearing the extreme environment resulting from hypersonic flight. Both material development and modeling activities toward such an application were carried out in the 1950s and 1960s.

Ultra-High Temperature Ceramics: Materials for Extreme Environment Applications, First Edition.
Edited by William G. Fahrenholtz, Eric J. Wuchina, William E. Lee, and Yanchun Zhou.
© 2014 The American Ceramic Society. Published 2014 by John Wiley & Sons, Inc.

The conditions experienced by a sharp body under hypersonic flow were initially modeled by scientists at, and/or funded by, NASA [7–12]. Material development and modeling efforts were pursued under Air Force funding by Kauffman *et al.* [13–17]. Based on the promising results from prior research and the recent worldwide interest for hypersonic aircraft development, a renewed effort has commenced in research and development of ultra-high temperature ceramics (UHTCs), which are typically defined as nonoxide ceramics with chemical and structural stability above 2000°C [18]. Environmental degradation and especially oxidation under hypersonic flow conditions are widely recognized as key life-limiting factors for leading-edge applications [18–21]. Thus, studying the oxidation kinetics and tailoring compositions to improve oxidation resistance have been the focus of the most recent work.

The most studied UHTC compositions are diborides of hafnium and zirconium with additives to control grain growth and/or improve oxidation resistance, the most common being SiC. The environmental response, especially oxidation, has been evaluated using several different methods. The most common has been the measurement of oxidation kinetics in a furnace atmosphere under isothermal conditions [18, 22–31]. Methods that also impose realistically high heating rates include arcjet testing [32, 33], laser-based heating [34], electric heating [35, 36] or oxyacetylene torch testing [37], and exposure inside a direct-connect scramjet engine [38–40]. Among these, arcjet testing is the most widely used historically by the Air Force and NASA for simulating reentry conditions. However, none of the tests used to date, including the arcjet test, reproduce the actual conditions that a leading-edge material will experience during hypersonic flight. Key parameters that represent these conditions with respect to material survivability are heat flux, total or stagnation temperature, total or stagnation pressure, dynamic pressure, fluid velocity at the material surface, fluid composition, degree of dissociation of gaseous elements, and catalytic recombination at the surface of the material. In addition, under realistic conditions, resistance to acoustic and mechanical vibrations and thermal shock will also be important. Most of the oxidation experiments described in the literature are conducted in laboratory furnaces, thus assigning total temperature as the key parameter of interest by default with all other factors ignored. Tests that use a laser as the heating source focus on reproducing the appropriate heat flux and, in some cases, include fluid flow. Arcjet tests generally focus on reproducing the heat flux, a significant fraction of which depends on catalytic recombination of gases, which itself varies with material surface chemistry and morphology. The scramjet rig reproduces several of the aerothermal conditions predicted to be experienced during free flight, but it uses gases that differ in chemistry from air, and the actual gas flow velocities are not sufficient to cause dissociation of gases behind the bow shock.

The lack of an accepted test to evaluate the suitability of a candidate material for hypersonic applications has necessitated the need to develop models to interpret and compare experimental data from the various testing techniques. Such models would then enable the prediction of performance in actual hypersonic environments. Advances toward this objective are presented in this chapter. Figure 11.1 is a schematic that summarizes the approach that has been taken to address this objective. Progress in environmental modeling efforts, their limitations, and possible future directions are discussed.

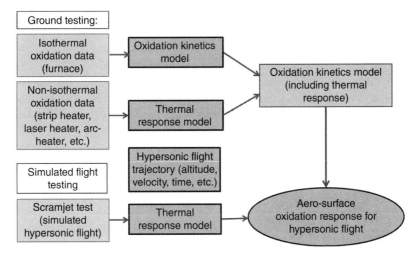

Figure 11.1. Overview of the approach used to model and evaluate UHTC leading-edge materials under simulated hypersonic flow conditions by Parthasarathy *et al.* [39, 40, 51–53, 57].

11.2 OXIDATION MODELING

The ideal model for oxidation kinetics of UHTCs will be able to include all of the control parameters of the experiments as input and predict any of the various parameters measured as outcomes of the experiments. Further, the model will be able to translate data across different testing techniques and use them to predict performance in an actual application. Thermodynamic modeling of the oxidation process is the natural starting point in the development of such a model, since it will involve the least assumptions and could be considered as modeling from first principles. It also provides information on key chemical species that will clearly guide kinetic modeling. In the literature, relatively simple yet elegant oxidation kinetic models have been identified, such as the classic parabolic scale growth model formulated by Wagner with or without accounting for evaporation of scale at the surface [41, 42]. However, UHTCs have been found to form a complex scale that is heterogeneous and defies the use of simple models. In particular, the complex morphology, chemical composition, and nature of the scale formation require modifications to the Wagner model. Further, parameters that affect the kinetics require material properties that are not readily available or known with certainty. Thus, models for oxidation kinetics of UHTCs can be expected to include approximations. Given these difficulties, it is best to think of the resulting model as a methodology to interpret data rather than one derived from first principles. For this approach, a large database of experimental results is useful. Fortunately, a large body of oxidation results is available in the literature from the 1960s, which is described briefly before modeling concepts are presented.

The oxidation behavior of transition metal diboride-based UHTCs has been characterized using a variety of methods and approaches. The earliest research identified the diborides of zirconium and hafnium as the best performers with respect to oxidation in flowing air at temperatures above 1600°C [13, 15, 16]. Later studies included composition modifications, which resulted in the conclusion that HfB_2–20 vol% SiC was the most oxidation-resistant UHTC at the highest temperatures. These studies utilized furnaces, forced flow air heaters, and arc heaters [17, 43]. This early research also included tests on leading-edge geometries of candidate materials using arc heaters [44]. These were followed by more careful studies in air or argon–oxygen mixtures in furnaces as well as thermogravimetric analysis (TGA) [22]. Typically, these evaluations reported scale thicknesses and sometimes recession rates, while TGA studies reported weight gain and weight of oxygen consumed as a function of time and temperature.

Oxidation studies on the most promising diborides were reinitiated in the 1990s [26, 31]. Much of the data reported on oxidation kinetics since then has been on the zirconium and hafnium diborides with SiC and other additions [18, 26, 30, 45–47]. The first modeling efforts were undertaken by Opeka et al. [18] and Fahrenholtz [48, 49]. Opeka et al. derived condensed/vapor-phase equilibrium diagrams for the Zr–O, Si–O, and B–O systems at 2227°C and assembled them to gain a thermodynamics-based description of the oxidation of the ZrB_2 + SiC composition, which provided insight into the synergistic benefit of each oxide in the multicomponent scale. While pure SiO_2-scale-forming materials (e.g., SiC) are limited to an approximately 1800°C usage due to the onset of active oxidation, the invariant vapor pressure of B_2O_3 with pO_2 was proposed to play a significant role by mitigating the active oxidation of the SiO_2-forming component above this temperature. This study was followed by the work of Fahrenholtz who derived the volatility diagrams for ZrB_2 at 1000, 1800, and 2500 K [49] with the P_{O2} calculated for the equilibrium coexistence of ZrB_2, ZrO_2, and B_2O_3, in contrast with the unit activities assumed by Opeka et al. His work provided an expanded thermodynamics-based description of the Zr–O and B–O condensed/vapor-phase reactions and associated vapor pressures as a function of the three selected temperatures (1000, 1800, and 2500 K) and showed that $B_2O_3(g)$ was the favored gaseous species at temperatures and pressures of engineering interest (Fig. 11.2a). He further provided a temperature-dependent, vapor pressure-based description of the B–O system to gain insight into the weight fraction of liquid B_2O_3 retained in the condensed but porous ZrO_2 scale. The retained B_2O_3 was also experimentally determined by Talmy et al. [50], and their results are shown in Figure 11.2b (graphically) and 11.2c (microstructurally). Fahrenholtz expanded the thermodynamics-based analysis to the ZrB_2–SiC UHTC composition and proposed a reaction sequence for the oxidation process, associated scale composition and growth, and formation of a depleted zone of the SiC phase in the base ceramic [20]. These largely thermodynamics-based efforts were followed by the development of an oxidation kinetic model of these refractory diboride UHTCs by Parthasarathy et al. [51–53].

The initial modeling effort on the pure diborides of Zr and Hf by Parthasarathy et al. was based on a schematic representation of the experimentally observed microstructural observations, as shown in Figure 11.3 [52]. The model assumed that the scale is made of two regions, an external glassy layer of liquid boria and an internal region consisting of two phases, a solid, but porous oxide of zirconia (or hafnia) with the pores filled with liquid boria. The model assumed that the solid oxide phase was impermeable to oxygen based on prior experimental observations of very low electronic conductivity, which would be required to ensure ambipolar diffusion where the transport of positively charged oxygen vacancies are accompanied by an appropriate (neutralizing) current

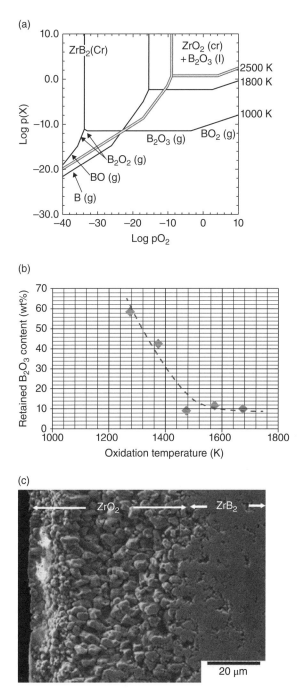

Figure 11.2. (a) Thermodynamic model for the oxidation of ZrB$_2$ by Fahrenholtz [49] plotted in terms of phase stability regions for various stoichiometric compounds. (b) Observation by Talmy *et al.* of the decrease in retained boria in the scale with temperature shown replotted using data [50] and (c) microstructural cross section of ZrB$_2$ after oxidation at 1500°C in air for 30 min, presented by Fahrenholtz [49].

of holes/electrons. The porous channels within the solid oxide were taken to be continuous and open such that oxygen can permeate through these channels to the substrate and oxidizes it. The oxygen permeation will be through a liquid of boria if the pores are filled, which happens at low temperatures. The molar ratio of boria to oxide formed from oxidation was taken to be proportional to the stoichiometry of the substrate diboride. Since the solid oxide was occupying most of the volume of the inner region, excess boria must be expelled as an external glassy layer (Fig. 11.3b). At the surface of this external boria layer, boria could evaporate by diffusion across a gaseous boundary layer whose width is determined by the flow conditions imposed. At intermediate temperatures, the evaporation of boria is sufficiently fast that no external layer forms. In fact, some boria from within the oxide pores evaporates leaving behind a partially porous oxide scale. Thus, in the intermediate temperature regime, oxygen permeates partially through these pores by gaseous diffusion or Knudsen diffusion and then through the

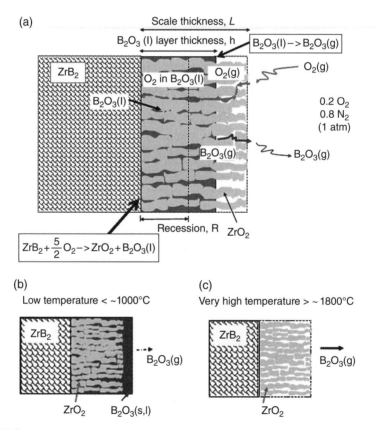

Figure 11.3. (a) A schematic of the microstructure interpreted from experimental data and used to build the oxidation model for ZrB_2. The loss of boria was taken to be limited by diffusion through porous channels in the intermediate temperature regime. (b) At lower temperatures, an external liquid boria forms and remains stable. (c) At very high temperatures, the boria evaporates as fast as it can form leading to rapid oxidation.

(d)

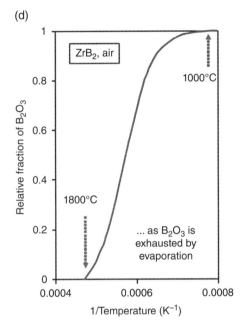

Figure 11.3. (*Continued*) (d) The transition between the three regimes as a function of temperature as predicted by the model of Parthasarathy *et al.* [52].

remaining depth by permeation in liquid boria, as illustrated in Figure 11.3a. At very high temperatures, the boria is essentially lost by evaporation as rapidly as it forms (Fig. 11.3c). Using literature values for oxygen permeability and vapor pressures for boria, the model was able to capture oxidation data reported in the literature. In particular, the model predicted the temperatures of the transitions as shown in Figure 11.3d, in close agreement with experimental observations.

A key complication in both measuring and predicting oxidation kinetics of the refractory diborides is the fact that the oxidation product, boria, is a glass with rapidly decreasing viscosity with increasing temperature. At temperatures of engineering interest, which are 1200–1600°C, the boria viscosity is sufficiently low to cause it to flow under gravity [54]. Thus, oxidized samples show significant variation in glass thickness depending on the orientation of the surface [23]. These effects were included in a modified version of the model for oxidation of diborides [51]. In this model, the effect of fluid flow was taken into account using a model adapted from models of water flooding from rainfall. As liquid boria forms by oxidation, it is added to the surface while gravitational forces tend to drain the glass in inverse proportion to its viscosity. The differential rate determines the actual thickness that is measured at the end of the experiment. This effect dominates in the low temperature regime where boria forms an external layer, but is not a factor once boria recedes into the porous oxide scale due to evaporation.

In the final iteration of the model for diboride oxidation kinetics, two more assumptions of the prior model were relaxed. First, inclusion of the effect of volume change associated with monoclinic to tetragonal phase change of the MeO_2 phases is found to rationalize the observations by several investigators of abrupt changes in weight gain, recession, and oxygen consumed, as the temperature is raised through the phase transformation temperatures for ZrO_2 and HfO_2. Second, the inclusion of oxygen permeability in ZrO_2 is found to rationalize the enhancement in oxidation behavior at very high temperatures (>1800°C) of ZrB_2, while the effect of oxygen permeability in HfO_2 is negligible. An important outcome of this model was that the pore fraction of the HfO_2 and ZrO_2 selected (0.03 and 0.04, respectively) to rationalize the data was very close to the volumetric shrinkage associated with the phase transformations of HfO_2 and ZrO_2 from monoclinic to tetragonal. Based on these considerations, the significant advantage of HfB_2 over ZrB_2 was credited to the higher transformation temperature and lower oxygen permeability of HfO_2 compared with ZrO_2. As shown in Figure 11.4, the model can rationalize the observed behavior both with respect to recession and with respect to

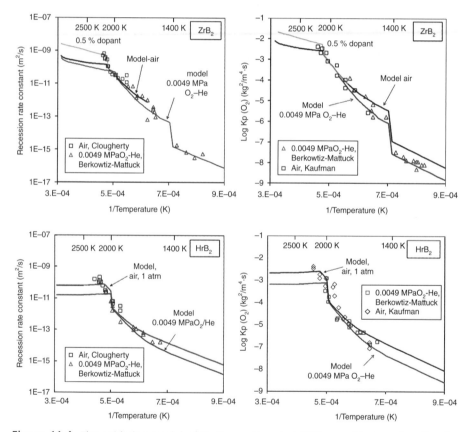

Figure 11.4. The oxidation model of Parthasarathy et al. [53] was able to rationalize the sudden jumps in oxidation kinetics with temperature in the ZrB_2 and HfB_2 systems, by proposing that the volume change associated with phase transformation from monoclinic to tetragonal oxide (ZrO_2 or HfO_2) opens the porous channels in the scale. At the highest temperatures, the possible effect of unintentional dopant on the electronic conductivity of the oxide and thus the oxidation kinetics was estimated and was found to correlate with experimental data.

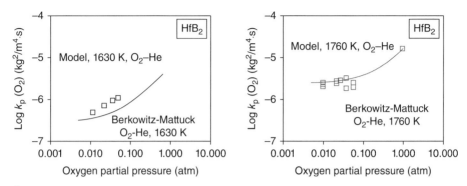

Figure 11.5. The proposed oxidation model [52] appears to capture the effect of oxygen partial pressure, but data in the literature are rather limited.

oxygen consumed simultaneously. The inclusion of the effect of dopants (impurities) in ZrO_2 or HfO_2 can rationalize the enhanced oxidation rates seen at the highest temperatures. This arises from the enhanced electronic conductivity of the oxides by permitting ambipolar oxygen transport [52, 53, 55]. At the highest temperatures of the tests, it is possible that the oxide scales become contaminated by impurities from the furnace itself. Such observations have been made by Carney *et al.* and reported in recent work, where significant amount of calcium was found in samples oxidized at the highest temperatures [56]. Finally, the model included the effect of oxygen partial pressure through the dependence of permeability in glass and pores on the oxygen partial pressure and was able to rationalize the limited data available reasonably well, as shown in Figure 11.5.

The model for pure diboride oxidation was later extended to include SiC-containing compositions [57]. In this work, once again, modeling started with recognition of the oxidation product morphology and chemistry as shown in Figure 11.6. Opila *et al.* first reported on the observation of a SiC-depleted zone beneath the substrate/oxide interface, during oxidation of ZrB_2–SiC [20, 31]. This phenomenon was confirmed and rationalized using thermodynamic modeling by Fahrenholtz [48]. As shown in Figure 11.7, thermodynamic calculations provided justification for the formation of a depleted zone. The kinetic modeling by Parthasarathy *et al.* [57] started with a reduction of the reported microstructures to a schematic representation (Fig. 11.8) superimposed with reactions known from prior thermodynamic modeling by Fahrenholtz [48]. The key conclusion of Fahrenholtz was that the internal depletion zone was due to internal active oxidation of SiC to SiO. The extension of this required a transport mechanism that is able to transfer oxygen fast enough to cause SiC oxidation faster than the oxidation rate of ZrB_2. This was achieved in the work of Parthasarathy *et al.* [57] by proposing a CO/CO_2 countercurrent mechanism, as had been derived earlier by Holcomb and St. Pierre [58]. The rest of the modeling involved including the effect of silica on the viscosity and oxygen permeability of the glassy region. These values are available in the literature for pure boria and silica and some but not all intermediate compositions. The actual composition of borosilicate glass in the oxidation product is difficult to determine due to difficulties in measuring B content. The model assumed that the composition is the same as what might be predicted from stoichiometry of

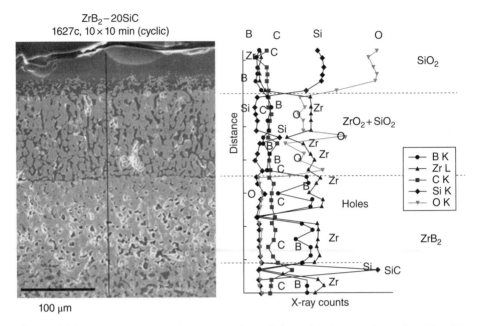

Figure 11.6. Microstructure and EDS mapping of the oxidation product of a 20 vol% SiC-containing ZrB$_2$ sample showing the phases present along with morphology and a depleted zone lacking SiC in a matrix of ZrB$_2$ after exposure in air at 1627°C for 100 min [20, 31].

reactants and products and the volume fraction of the phases in the substrate. The viscosity and permeability for this composition was, in turn, estimated by assuming a logarithmic mean as the composition varied from pure boria to pure silica. With these assumptions, the model was able to capture the effect of SiC addition on the oxidation kinetics of ZrB$_2$–SiC as shown in Figure 11.9. The model was also able to capture kinetics of oxidation of HfB$_2$–SiC. The weight gain and scale thicknesses were captured well, except at the highest temperatures. The depletion zone thickness is predicted to be significant (>1 μm in 2.5 h) at temperatures above 1550°C. The experimentally observed depletion zone thicknesses are sometimes significantly larger than that predicted, implying an effect not yet captured by the model.

11.3 UHTC BEHAVIOR UNDER SIMULATED HYPERSONIC CONDITIONS

The application of UHTC compositions is at present focused on the cowl leading edge and fuel injection struts of scramjet engine for hypersonic flight. These two applications involve a unique geometry, thermal loading with large temperature gradients, very high fluid velocities with its associated mechanical vibrations, and aerodynamic and aerothermal loads. While a number of tests have evolved over the past several decades as mentioned in the introduction, this section focuses on a recent study that examined the

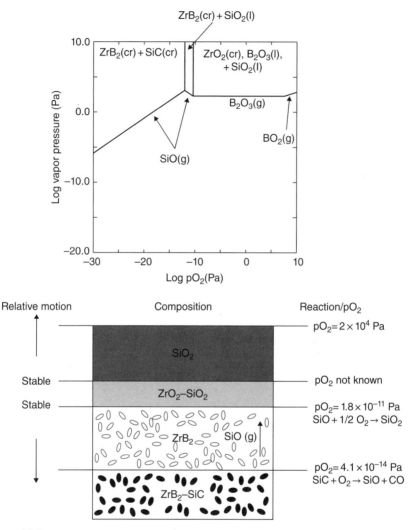

<u>Figure 11.7.</u> Thermodynamic model for the oxidation of ZrB_2–SiC calculated by Fahrenholtz [48] along with a schematic sketch showing a rationale for the formation of a depleted zone as observed by Opila *et al.* (Fig. 11.6).

use of a direct-connect scramjet engine as a wind tunnel to simulate conditions of free flight of leading-edge samples of a UHTC (HfB_2–20 vol% SiC) composition.

The details of the purpose, design, construction, and evaluation of the direct-connect scramjet engine have been described by Gruber *et al.* [38, 59, 60]. A brief description is given here. Ambient air is compressed before being expanded through an isolator nozzle resulting in a supersonic flow at Mach 1.8–2. The cooling that results from the expansion is augmented by adding heat using a combustion heater. Oxygen lost during combustion heating is compensated by adding oxygen to ensure the correct

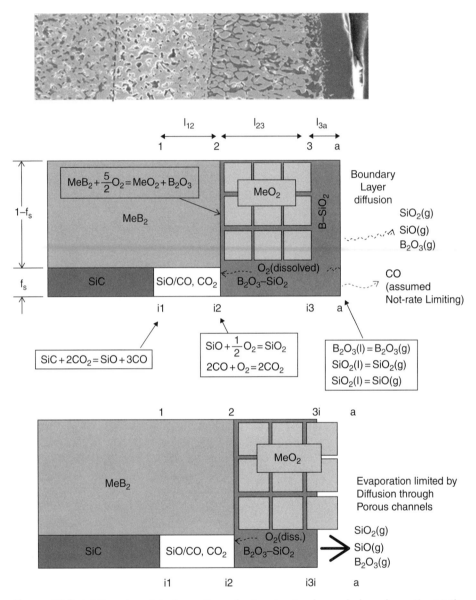

Figure 11.8. (a) Experimental observation of microstructural morphology (e.g., Fig. 11.6), (b, c) Schematics of the microstructural morphology and phase content of the oxidation product used in the model of Parthasarathy *et al.* [57].

partial pressure of oxygen in the resulting fluid that enters the supersonic combustor. The postcombustion section employs an exit nozzle before reaching a probe housing. It is within the probe housing that leading-edge samples were introduced. The gas composition at this location varies with combustor section but typically includes 5–10% moisture with a total static pressure near 1 atm. Samples were mounted on a sample

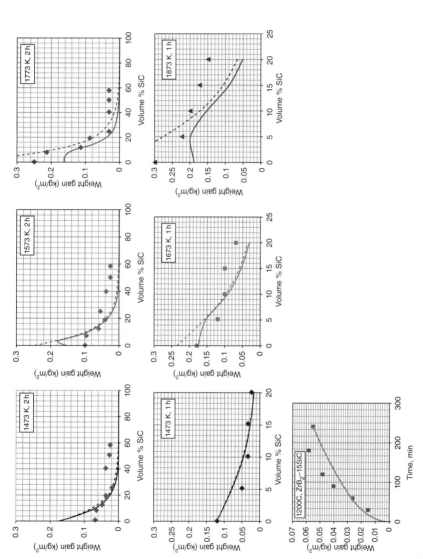

Figure 11.9. Predicted variation in oxidation kinetics of SiC-containing ZrB$_2$ as a function of SiC content compared with (a) experimental data of Talmy [66] and (b, c) Wang et al. [67]. Figure reproduced from the work of Parthasarathy et al. [57].

holder, constructed of water-cooled thermal barrier-coated Inconel [39, 40]. The samples were visible during the run through quartz windows. The scramjet rig was fully calibrated, and important parameters including enthalpy, total temperatures, gas flow velocities, gas composition, and pressures were available as a function of time during the run. The fluid flow parameters such as velocity (U) and static pressure (p_1), along with static temperature (T_1) and specific heat ratio of the gas composition (γ), were used to calculate the heat flux (Q_w) experienced by the leading-edge samples [7, 10]:

$$M = \frac{U}{\sqrt{\gamma R T_1}} \tag{11.1}$$

$$T_t = T_1\left(1 + \frac{\gamma - 1}{2} M^2\right) \tag{11.2}$$

$$p_{t2} = p_1 \left(\frac{\gamma + 1}{2}\right)^{\frac{\gamma}{\gamma-1}} \left(\frac{(\gamma+1)M^2}{2\gamma M^2 - (\gamma - 1)}\right)^{\frac{1}{\gamma-1}} M^2 \tag{11.3}$$

$$\left(h_{aw} - h_w\right), \quad kJ/kg = C_p\left(T_t - T_{wall}\right) \tag{11.4}$$

$$\dot{Q}_w, \quad W/m^2 = 3.88 \times 10^{-4} \sqrt{\frac{p_{t2}(Pa)}{r(m)}} \left(h_{aw} - h_w\right) \tag{11.5}$$

In the above set of equations, M is the Mach number, T_t is the total temperature, T_{wall} is the material surface temperature, p_{t2} is the total pressure behind the bow shock, r is the radius of curvature of the leading edge, C_p is the specific heat of fluid, h_{aw} is the total enthalpy of the fluid ($= C_p T_t$), the h_w is the enthalpy of the fluid at the material wall temperature ($= C_p T_{wall}$). The heat flux varies with orientation of the material wall as well as distance from the bow shock. These variations are given by the following:

$$\dot{Q}(\theta) = \dot{Q}_w \cos(\theta) \tag{11.6}$$

$$\dot{Q}(x) = \frac{\dot{Q}_w \cos(\theta_c)}{\sqrt{x}} \tag{11.7}$$

where θ gives the angle between the direction of fluid flow and the normal to the wall surface (=0 at the tip of the leading edge) and x is the distance from the tip of the leading edge. θ_c and Q_w refer to the orientation and the heat flux at the tangential point where the cylindrical surface of the tip meets the slant face of the leading-edge wedge. It must be noted that while the Equations 11.6 and 11.7 capture the dependences accurately, they are approximations to the actual variation of the heat flux. Further note that the net heat flux at any location is a function of the wall temperature and, thus, will vary with time. The steady-state material temperature will depend on radiation heat flux back to the atmosphere and thermal conduction within the solid. The wall temperature will depend on the static temperature of the fluid and material thermal properties. Thus, a thermal model that includes the aerothermal heating, radiation, and conduction for a given

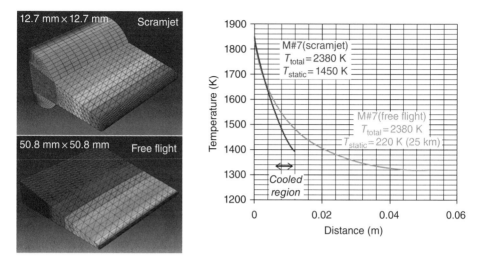

Figure 11.10. Thermal model of a leading-edge sample of UHTC was compared for two cases, one for the sample used in the scramjet tests and another simulating free flight. The results showed that the thermal profiles are nearly the same for the two cases near the leading-edge tip, which were of experimental interest [40].

geometry is required. Using such a model, the thermal profiles were calculated for a large ($50.8 \times 50.8\,mm$) leading-edge sample under free flight and compared to the thermal response of a smaller ($12.7 \times 12.7\,mm$) leading-edge sample used in the scramjet rig [40]. The results from these calculations are shown in Figure 11.10. An important outcome of the thermal model was that the thermal properties of the material significantly affect the temperature gradient. Materials based on diboride UHTCs have higher thermal conductivities at high temperatures than other ceramics including SiC, potentially making them better material candidates for these applications.

Figure 11.11 shows a typical oxidation scale formed on a UHTC (HfB_2–20 vol% SiC) leading-edge sample after exposure in the scramjet. The microstructural characterization of the UHTC showed that the scale was different than what is normally observed in samples after furnace oxidation. First, the external glassy layer is absent. Second, no depletion layer was observed. Third, silica was absent in the outer regions of the hafnium-rich oxide scale. Fourth, the formation of hafnon ($HfSiO_4$) was observed in the outer regions. While the reasons for these differences are not clear, it is suspected that the high flow rate of the environment along with the presence of moisture in the atmosphere may be responsible for these differences.

11.4 COMPARING MODEL PREDICTIONS TO LEADING-EDGE BEHAVIOR

The modeling work by Parthasarathy *et al.* [57] included the effect of fluid flow on the evaporation rates of boria and silica. Thus, the model was considered sufficient for predicting behavior of leading edges exposed to simulated hypersonic conditions. The

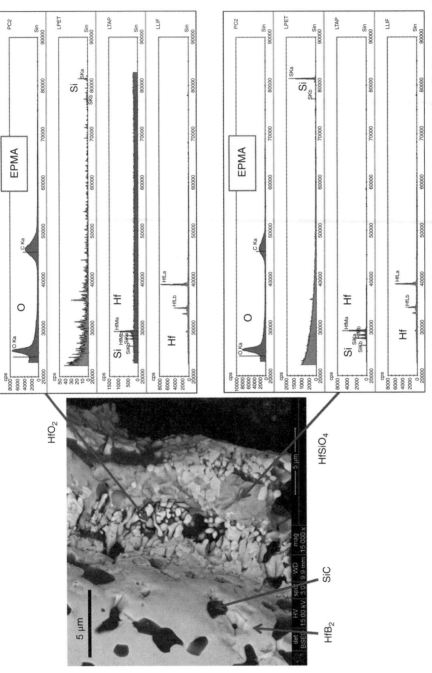

Figure 11.11. The oxidation scale formed on the UHTC sample tested in the scramjet under conditions representing free flight Mach numbers of 6.2–7. No external glassy layer or depleted zone was observed; however, hafnon (HfSiO$_4$) was present in the oxide scale as evidenced by both X-ray diffraction and EPMA analysis shown here [40].

model, however, did not include the physical shear mechanism by which the external glassy layer could be thinned down. Under most conditions of tests, evaporation rates were high enough to result in little or no external glassy layer; thus, shear forces were irrelevant. The thermal model described in the earlier section was used to obtain the thermal profile (time vs. temperature) at the tip of the leading edge and fed into the oxidation model to predict the oxidation scale thickness, depletion thickness, and recession. The aerothermal calculations were used to obtain the total pressure and fluid velocity behind the bow shock and used for the model predictions. The results obtained showed that predicted oxide thicknesses were close to the experimental results, although they were consistently lower. Impurities and moisture in the environment could have altered the oxygen permeation across the scales. The formation of hafnon could have increased the volume fraction of the porous channels. Finally, the possibility of glassy material being drawn out of the porous channels by the high velocity fluid flow was suggested.

Based on the reasonable predictions presented above, the model was used to estimate the behavior of a UHTC leading-edge behavior under hypersonic free flight conditions. Figure 11.12 summarizes the predictions. The results are shown as a function of free flight Mach numbers. The leading-edge tip temperatures, steady-state heat flux, thermal gradients in the sample, and oxidation thickness with time are shown. It must be noted that the results are sensitive to the thermal conductivity and specific heat of the composition used. A significant variation in thermal properties of UHTCs with processing method has been noted by several groups [32]. For example, the use of SiC results in higher tip temperatures due to the lower thermal conductivity at high temperatures, when compared with nominally pure UHTCs.

In closing, the following limitations are worth pointing out with a view to direct future work. Oxidation modeling of UHTCs includes some assumptions that are discussed in the following. First, the refractory oxide, ZrO_2 or HfO_2, is assumed to be impermeable to oxygen due to low electronic conductivity. This is a good approximation if the oxide is pure, but impurities can enhance the conductivity significantly especially at high temperatures (>2000 K). Impurities may be present in the raw materials or may be acquired during testing or use. Second, the variation of boria content with distance in the scale has been approximated, and boria content can significantly affect viscosity and oxygen permeation in borosilicate glasses. Third, the effect of mechanical shear on the external glassy layer has been ignored.

The scramjet rig offers a unique approach to evaluate leading-edge candidate materials using simulated hypersonic conditions but has some limitations. First, the gas chemistry is not that of ambient air. The gas contains a significant amount of water and carbon dioxide (about 10% each). In addition, impurities from the components of the rig cannot be easily avoided. The use of water as coolant for parts of the rig interfered with observation as it collected in the recess of the window; however, future design changes could alleviate this. The glass windows currently in use prevented the use of optical pyrometry. The use of a suitable material such as ZnSe for the window will alleviate this difficulty provided the material is stable at the temperatures of the rig walls. The actual velocity of the gases is not hypersonic even though the heat content of the gases simulates hypersonic flight conditions. The rig used in the reported work is limited to a maximum of Mach 7. Finally, the scramjet rig is a unique facility available only to a very limited group of researchers and only at periodic intervals. However, if the community recognizes it to be a useful method, a dedicated rig might be constructed that is smaller in scale, but it will be expensive.

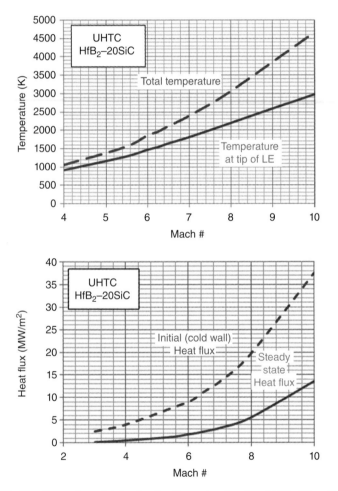

Figure 11.12. The thermal model for the leading edge (made of HfB$_2$–20 vol% SiC) under hypersonic flight conditions when combined with oxidation models can predict parameters of interest for design. The temperature at the tip of the leading edge is lower than the total temperature often assumed. Similarly, the steady-state heat flux is lower than the cold wall heat flux. The thermal profiles show a steep drop in temperature away from the tip, and the oxidation rates appear to be tolerable for UHTCs at Mach 6.5 [40].

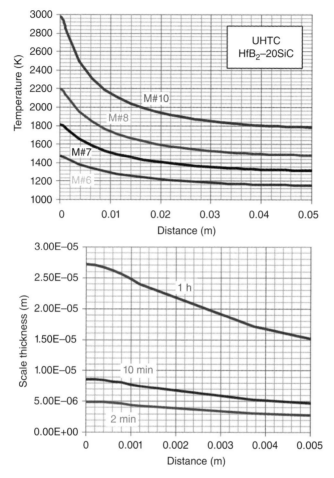

Figure 11.12. (*Continued*)

11.5 BEHAVIOR OF UHTCS UNDER OTHER TEST METHODS

Various compositions of UHTCs are being studied by different investigators to explore the possibility of enhancing the environmental degradation resistance at high temperatures. The evaluations have largely been based on furnace oxidation tests due to the ease and low cost of the test method. It is appropriate for initial screening of compositions and for understanding mechanisms of oxidation in static laboratory air; however, using these data to predict performance under hypersonic flight conditions is not trivial. Using a mechanistic model that captures the furnace data in combination with scramjet exposure tests shows promise. In the literature, one can find several other approaches that have been pursued to evaluate performance of candidate materials for thermal protection under hypersonic flight conditions. The merits and demerits of these approaches are discussed briefly in the following.

11.5.1 Arcjet Test

The arcjet test is a widely accepted method for evaluating performance of materials under reentry conditions, developed and used widely by NASA to design space vehicles. It uses a plasma-generated gas mixture, which contains a high fraction of ionized atoms of oxygen and nitrogen. Recombination of the charged species is catalyzed by the sample surface to release heat. The plasma is accelerated to high velocities (supersonic but not hypersonic) using an electric field before it is directed onto the sample. The enthalpy of the gases per unit area of the sample is reported as well as the surface temperature of the sample, which is measured using an optical pyrometer. The key advantage of this method is that it provides a simulation of dissociation of air that occurs behind the bow shock at hypersonic speeds. The catalytic coefficient for recombination varies with sample composition and can also change during the test since the surface composition changes as the sample oxidizes complicating interpretation and modeling of test results. Until recently, the test has used mostly flat samples or large-radius conical samples, and thus, the effect of a sharp leading-edge geometry had not yet been captured. Recent works are beginning to study samples with sharp geometries, but the aerothermal heat flux distribution during free flight is not reproduced in these tests [61–63]. The test is also not best for evaluating materials for hypersonic speeds in the range of Mach 5–7, where very little dissociation of gases is expected. Finally, the very high cost of the test makes it prohibitive for most studies.

11.5.2 Laser Test

The availability of laser facilities at reasonable cost compared to arcjets has resulted in tests that evaluate UHTC compositions using lasers as the heat source. These tests use a prescribed heat flux, calibrated using optical standards, on a flat geometry of the sample for a predetermined length of time. In some cases, an external tangential flow of air typically at subsonic speeds is included. The advantage of this method is that it requires a short setup and run time and several samples can be tested within a short period of time. The ambient fluid composition is nearly exact as its air and material surface temperature can be monitored accurately during the test using optical pyrometry. The key disadvantage of the method is that the actual heat flux absorbed by the sample varies with composition, wavelength of the laser, and temperature of the sample. Thus, laser absorptance and emittance of each sample must be measured as a function of temperature and accounted for when interpreting the results. Very few studies report these properties of UHTCs, with most studies assuming the absorptance to be unity. Emissivity has been reported in some recent work, but factors such as machining method can influence emissivity [64, 65]. The laser method also uses a constant heat flux, whereas in real flight heat flux varies from nearly the cold wall heat flux to the steady-state heat flux gradually as the sample heats. The steady-state heat flux will be affected by the emissivity of the material and ambient temperature during flight; this is not captured in the laser test.

11.5.3 Oxyacetylene Torch Test

The oxyacetylene torch has long been used as a heat source and has recently been revived as a low-cost alternative to arcjet and laser tests. In this test, a flat sample is exposed to hot combustion gases of an oxyacetylene torch by holding the sample at a calibrated distance from the torch tip. The calibration is usually made based on the measured optical pyrometer reading on the hottest section of the exposed sample. The heat flux is usually not measured or reported. The advantages of this method are that it provides high heat fluxes, is very quick and inexpensive, and, as such, can be used as a good oxidation screening tool for the relative comparison of material compositional variations. The disadvantage is the lack of standardization of the test procedure and lack of calibration with respect to heat flux. The exact composition of the combustion product is also unknown and depends on the initial oxygen–acetylene ratio, as well as on location within the flame. Water is present as a major product of combustion, which can complicate interpretation of the results.

11.6 SUMMARY

The status of mechanistic understanding of the degradation of UHTC leading-edge components under hypersonic flight conditions was reviewed. The oxidation behavior studied under conventional methods and reported over the past several decades could be interpreted reasonably well using analytical models that assume the glassy regions as the only pathway for oxygen permeation. Extension of this model to hypersonic flow conditions shows promise, but more experimental and modeling work is required to develop a comprehensive understanding. UHTC-based leading-edge samples appear to withstand the simulated hypersonic conditions up to Mach 7. However, UHTCs are found to degrade much more rapidly under arcjet conditions. The behavior under arcjet appears to be unique and different from other test methods and thus the least understood.

REFERENCES

1. Jackson TA, Eklund DR, Fink AJ. High speed propulsion: performance advantage of advanced materials. J Mater Sci 2004;39:5905–5913.
2. Voland RT, Huebner LD, McClinton CR. X-43A hypersonic vehicle technology development. Acta Astronaut 2006;59:181–191.
3. Cook S, Hueter U. NASA's integrated space transportation plan—3rd generation reusable launch vehicle technology update. Acta Astronaut 2003;53:719–728.
4. Norris G. High-speed strike weapon to build on X-51 flight. Aviation Week & Space Technology; May 20, 2013.
5. McClinton CR, Rausch VL, Shaw RJ, Metha U, Naftel C. Hyper-X: foundation for future hypersonic launch vehicles. Acta Astronaut 2005;57:614–622.
6. Kerans RJ. Concurrent material and structural design with innovative ceramic composites, Final Report for DARPA/DSO [AFRL Technical Report AFRL-ML-WP-TR-2006-4130] WPAFB, OH: Air Force Research Laboratory; 2005.

7. Ames Research Staff. Equations, tables and charts for compressible flow. National Advisory Committee for Aeronautics [Report 1135]. Washington, DC: NASA HQ; 1953.

8. Lees L. On the boundary layer equations in hypersonic flow and their approximate solutions. J Aero Sci 1953;20 (2):143–145.

9. Fay JA, Riddell FR. Theory of stagnation point heat transfer in dissociated air. J Aero Sci 1958;25:73–85.

10. Zoby EV. Empirical stagnation-point heat-transfer relation in several gas mixtures at high enthalpy levels. NASA tech note, NASA TN D-4799. Washington, DC: NASA HQ; 1968.

11. Scala SM, Gilbert LM. Theory of hypersonic laminar stagnation region heat transfer in dissociating gases. NASA tech note NAS 7-100 Accession No. N71-70918. Washington, DC: NASA HQ; 1963.

12. NASA. U. S. standard atmosphere. NOAA document ST 76-1562. Washington, DC: NASA HQ; 1976.

13. Clougherty EV, Pober RL, Kaufman L. Synthesis of oxidation resistant metal diboride composites. Trans Met Soc AIME 1968;242:1077–1082.

14. Kaufman L, Clougherty E. Technical Report RTD-TDR-63-4096: part 1. WPAFB, OH: Air Force Research Laboratory; December 1963.

15. Kaufman L, Clougherty E. Investigation of boride compounds for very high temperature applications. Technical Report RTD-TDR-63-4096: part 2. WPAFB, OH: Air Force Research Laboratory; February 1965.

16. Kaufman L, Clougherty EV, Berkowitz-Mattuck JB. Oxidation characteristics of hafnium and zirconium diboride. Trans Metall Soc AIME 1967;239:458–466.

17. Kaufman L, Nesor H. Stability characterization of refractory materials under high velocity atmospheric flight conditions, part III, vol. III: experimental results of high velocity hot gas/cold wall tests, AFML-TR-69–84; 1970.

18. Opeka MM, Talmy IG, Zaykoski JA. Oxidation-based materials selection for 2000°C+ hypersonic aerosurfaces: theoretical considerations and historical experience. J Mater Sci 2004; 39:5887–5904.

19. Montverde F, Savino R, Fumo M. Dynamic oxidation of ultra-high temperature ZrB_2–SiC under high enthalpy supersonic flows. Corros Sci 2011;53:922–929.

20. Levine SR, Opila EJ, Halbig MC, Kiser JD, Singh M, Salem JA. Evaluation of ultra-high temperature ceramics for aeropropulsion use. J Eur Ceram Soc 2002;22:2757–2767.

21. Marschall J, Pejakovic DA, Fahrenholtz WG, Hilmas GE, Zhu S, Ridge J, Fletcher DG, Asma CO, Thomel J. Oxidation of ZrB_2–SiC ultrahigh-temperature ceramic composites in dissociated air. J Thermophys Heat Transfer 2009;23 (2):267–278.

22. Tripp WC, Graham HC. Thermogravimetric study of the oxidation of ZrB_2 in the temperature range of 800 to 1500°C. J Electrochem Soc 1971;118 (7):1195–1199.

23. Carney CM, Mogilevsky P, Parthasarathy TA. Oxidation behavior of zirconium diboride silicon carbide produced by the spark plasma sintering method. J Am Ceram Soc 2009;92 (9):2046–2052.

24. Carney CM. Oxidation resistance of hafnium diboride-silicon carbide from 1400–2000°C. J Mater Sci 2009;44:5673–5681.

25. Carney CM, Parthasarathy TA, Cinibulk MK. Oxidation resistance of hafnium diboride ceramics with additions of silicon carbide and tungsten boride or tungsten carbide. J Am Ceram Soc 2011;94 (8):2600–2607.

26. Opeka MM, Talmy IG, Wuchina EJ, Zaykoski JA, Causey SJ. Mechanical, thermal and oxidation properties of refractory hafnium and zirconium compounds. J Eur Ceram Soc 1999;19:2405–2414.

27. Meng S, Chen H, Hu J, Wang Z. Radiative properties characterization of ZrB_2–SiC-based ultrahigh temperature ceramic at high temperature. Mater Design 2011;32:377–381.

28. Zhang X-H, Hu P, Han J-C. Structure evolution of ZrB_2–SiC during the oxidation in air. J Mater Res 2008;23 (7):1961–1972.

29. Han W-B, Hu P, Zhang X-H, Han J-C, Meng S-H. High-temperature oxidation at 1900°C of ZrB_2–xSiC ultrahigh-temperature ceramic composites. J Am Ceram Soc 2008;91 (10): 3328–3334.

30. Opila E, Levine S, Lorincz J. Oxidation of ZrB_2- and HfB_2-based ultra-high temperature ceramics: effect of Ta additions. J Mater Sci 2004;39:5969–5977.

31. Opila EJ, Halbig MC. Oxidation of ZrB_2–SiC. Ceram Eng Sci Proc 2001;22 (3):221–228.

32. Gasch M, Johnson S. Physical characterization and arcjet oxidation of hafnium-based ultra high temperature ceramics fabricated by hot pressing and field-assisted sintering. J Eur Ceram Soc 2010;30:2337–2344.

33. Savino R, Fumo MDS, Paterna D, Maso AD, Monteverde F. Arc-jet testing of ultra-high-temperature-ceramics. Aerosp Sci Technol 2010;14:178–187.

34. Jayaseelan DD, Jackson H, Eakins E, Brown P, Lee WE. Laser modified microstructures in ZrB_2, ZrB_2/SiC and ZrC. J Eur Ceram Soc 2010;30 (11):2279–2288.

35. Karlsdottir SN, Halloran JW. Oxidation of ZrB_2–SiC: influence of SiC content on solid and liquid oxide phase formation. J Am Ceram Soc 2009;92 (2):481–486.

36. Gangireddy S, Karlsdottir SN, Halloran JW. High-temperature oxidation at 1900°C of ZrB_2–xSiC ultrahigh-temperature ceramic composites. Key Eng Mater 2010;434–435:144–148.

37. Paul A, Jayaseelan DD, Venugopal S, Zapata-Solvas E, Binner J, Vaidhyanathan B, Heaton A, Brown P, Lee WE. UHTC composites for hypersonic applications. Bull Am Ceram Soc 2012;91 (1):1–8.

38. Gruber M, Donbar J, Jackson KJ, Mathur T, Baurle R, Eklund D, Smith C. Newly developed direct-connect high-enthalpy supersonic combustion research facility. J Propul Power 2001;17 (6):1296–1304.

39. Parthasarathy TA, Petry MD, Jefferson G, Cinibulk MK, Mathur T, Gruber MR. Development of a test to evaluate aerothermal response of materials to hypersonic flow using a scramjetwind tunnel. Int J Appl Ceram Tech 2011;8 (4):832–847.

40. Parthasarathy TA, Petry MD, Cinibulk MK, Mathur T, Gruber MR. Thermal and oxidation response of UHTC leading edge samples exposed to simulated hypersonic flight conditions. J Am Ceram Soc 2013;96 (3):907–915.

41. Wagner C. Theoretical analysis of the diffusion processes determining the oxidation rate of alloys. J Electrochem Soc 1952;99 (10):369–380.

42. Wagner C. Passivity and inhibition during the oxidation of metals at elevated temperatures. Corros Sci 1965;5:751–764.

43. Fenter JR. Refractory diborides as engineering materials. SAMPE Q 1971;2 (3):1–15.

44. Hill ML. Materials for small radius leading edges for hypersonic vehicles, AIAA/ASME Conference on Structures, Structural Dynamics and Materials; Palm Springs, CA, March 29-3 1967.

45. Monteverde F, Bellosi A. Oxidation of ZrB_2-based ceramics in dry air. J Electrochem Soc 2003;150 (11):B552–B559.

46. Fahrenholtz WG, Hilmas GE, Chamberlain AL, Zimmermann JW. Processing and characterization of ZrB_2-based ultra-high temperature monolithic and fibrous monolithic ceramics. J Mater Sci 2004;39 (19):5951–5957.

47. Wuchina E, Opeka M, Causey S, Buesking K, Spain J, Cull A, Routbort J, Guitierrez-Mora F. Designing for ultrahigh-temperature applications: the mechanical and thermal properties of HfB_2, HfCx, HfNx and a-Hf(N). J Mater Sci 2004;39:5939–5949.

48. Fahrenholtz WG. Thermodynamic analysis of ZrB_2–SiC oxidation: formation of a SiC-depleted region. J Am Ceram Soc 2007;90 (1):143–148.

49. Fahrenholtz WG. The ZrB_2 volatility diagram. J Am Ceram Soc 2005;88 (12):3509–3512.

50. Talmy IG, Zaykoski JA, Opeka MA. Properties of ceramics in the ZrB_2/ZrC/SiC system prepared by reactive processing. Ceram Eng Sci Proc 1998;19 (3):105–112.

51. Parthasarathy TA, Rapp RA, Opeka M, Kerans RJ. A model for transitions in oxidation regimes of ZrB_2. Mater Sci Forum 2008;595–598:823–832.

52. Parthasarathy TA, Rapp RA, Opeka M, Kerans RJ. A model for the oxidation of ZrB_2, HfB_2 and TiB_2. Acta Mater 2007;55:5999–6010.

53. Parthasarathy TA, Rapp RA, Opeka M, Kerans RJ. Effect of Phase change and oxygen permeability in oxide scales on oxidation kinetics of ZrB_2 and HfB_2. J Am Ceram Soc 2009;92 (5):1079–1086.

54. Eppler RA. Viscosity of molten B_2O_3. J Am Ceram Soc 1966;49 (12):679–680.

55. Patterson JW. Conduction domains for solid electrolytes. J Electrochem Soc 1971;118: 1033–1039.

56. Carney CM, Parthasarathy TA, Cinibulk MK. Separating test artifacts from material behavior in the oxidation studies of HfB_2–SiC at 2000°C and above. Int J Appl Ceram Technol 2013;10 (2).

57. Parthasarathy TA, Rapp RA, Opeka M, Cinibulk MK. Modeling oxidation kinetics of SiC-containing refractory diborides. J Am CeramSoc 2012;95 (1):338–349.

58. Holcomb GR, St.Pierre GR. Application of a counter-current gaseous diffusion model to the oxidation of Hafnium carbide at 1200°C to 1530°C. Oxid Met 1993;40 (1/2):109–118.

59. Mathur T, Gruber M, Jackson K, Donbar J, Donaldson W, Jackson T, Billig F. Supersonic combustion experiments with a cavity-based fuel injector. J. Propul Power 2001;17 (6):1305–1312.

60. Gruber M, Smith S, Mathur T. Experimental characterization of hydrocarbon-fueled axisymmetric scramjet combustor flowpaths. AIAA Paper 2011–2311; 2011.

61. Scatteia L, Alfano D, Cantoni S, Monteverde F, Fumo MDS, Maso AD. Plasma torch test of an ultra-high-temperature ceramics nose cone demonstrator. J Spacecraft Rockets 2010;47: 271–279.

62. Gardi R, Vecchio AD, Marino G, Russo G. CIRA activities on UHTC's: on-ground and in flight experimentations. Proceedings of the 17th AIAA International Space Planes and Hypersonic Systems and Technologies, Paper 2011-2303; San Francisco, CA, April 11–14, 2011.

63. Monteverde F, Alfano D, Savino R. Effects of LaB_6 addition on arc-jet convectively heated SiC-containing ZrB_2-based ultra-high temperature ceramics in high enthalpy supersonic air-flows. Corros Sci 2013;75:443–453.

64. Scatteia L, Alfano D, Monteverde F, Sans J-L, Balat-Pichelin M. Effect of the machining method on the catalycity and emissivity of ZrB_2 and ZrB_2–HfB_2-based ceramics. J Am Ceram Soc 2008;91:1461–1468.

65. Scatteia L, Borrelli R, Cosentino G, Beche E, Sans J-L, Balat-Pichelin M. Catalytic and radiative behaviors of ZrB_2–SiC ultrahigh temperature ceramic composites. J Spacecraft Rockets 2006;43:1004–1012.

66. Talmy I. Effect of SiC content on oxidation kinetics of SiC-ZrB_2, Unpublished work. Naval Surface Warfare Center: Carderock (MD); 2005.

67. Wang M, Wang C-A, Yu L, Huang Y, Zhang Z. Oxidation behavior of SiC platelet-reinforced ZrB_2 ceramic matrix composites. Int J App Ceram Tech 2012;9 (1):178–185.

12

TANTALUM CARBIDES: THEIR MICROSTRUCTURES AND DEFORMATION BEHAVIOR

Gregory B. Thompson[1] and Christopher R. Weinberger[2,3]

[1] *Department of Metallurgical and Materials Engineering, University of Alabama, Tuscaloosa, AL, USA*
[2] *Sandia National Laboratories, Albuquerque, NM, USA*
[3] *Department of Mechanical Engineering, Drexel University, Philadelphia, PA, USA*

Tantalum carbides comprise a class of ultra-high-temperature ceramics (UHTCs). These particular carbides, along with other carbides of the group V transition metals, exhibit a range of high to ultra-high melting temperatures, microstructures controlled by the co-precipitation of metal-rich carbide phases, and varied mechanical behaviors dependent on the phase content. Consequently, tantalum carbides and similar transitional metal carbides have been proposed for thermal heat protection, automotive wear resistant liners, and other types of thermomechanical loading applications [1–3]. In this chapter, the crystallography, microstructures, and mechanical attributes are reviewed for this specific material system.

12.1 CRYSTALLOGRAPHY OF TANTALUM CARBIDES

Arguably, Gusev *et al.* [4, 5] have provided the most thorough phase diagram for the tantalum-rich portion of the Ta–C system, as shown in Figure 12.1. For C/Ta ratios greater than 1, not shown in Figure 12.1, a two-phase mixture of the TaC phase and

Ultra-High Temperature Ceramics: Materials for Extreme Environment Applications, First Edition.
Edited by William G. Fahrenholtz, Eric J. Wuchina, William E. Lee, and Yanchun Zhou.
© 2014 The American Ceramic Society. Published 2014 by John Wiley & Sons, Inc.

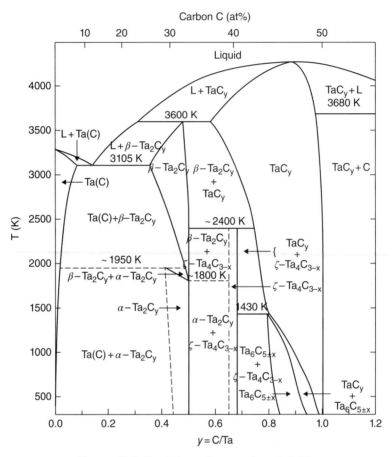

Figure 12.1. Ta–C phase diagram. From Ref. [5].

graphite has been reported with a eutectic point at approximately 3749 K and at a composition of 70 at %C [6]. To the authors' knowledge, little to no research has been done exploring or exploiting microstructures from this portion of the phase diagram. In contrast, several papers have been published on the precipitation and phase equilibrium of the monocarbide- and tantalum-rich carbide phases.

The monocarbide phase of TaC forms the B1, or rock salt, crystal structure (space group $Fm\bar{3}m$), with a cubic lattice parameter of 0.4457 nm [7]. From stoichiometry, the tantalum atoms form a face-centered cubic (FCC) sublattice with the carbon atoms occupying all of the octahedral interstitial sites (Fig. 12.2a) [8, 9]. Hence, the carbon atoms also form a FCC structure. Similar to other group V transition metal carbides, this structure is not a line compound and can accommodate the loss of some carbon, creating several site-occupancy defects [10]. The ability to accommodate this change in carbon content while maintaining the B1 phase offers one of the first qualitative clues that this phase is not dominated by ionic bonding, which is common in other ceramics that exhibit the B1 structure, such as MgO, FeO, and NiO. In these ionic crystals, the

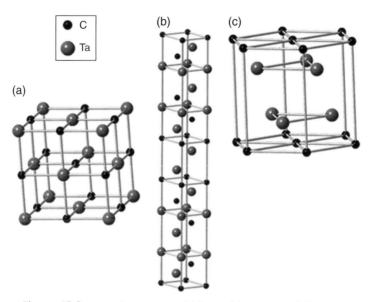

Figure 12.2. Crystal structures of (a) TaC, (b) ζ-Ta$_4$C$_3$, and (c) α-Ta$_2$C.

equivalent loss of cation and anion sites must occur to maintain charge neutrality. The monocarbide TaC phase exhibits a sharp maximum in the solidus near 4273 K for TaC$_{\sim0.89}$, which is one of the highest melting substances known at ambient pressure [8]. The high melting temperature also qualitatively indicates strong covalent bonding. Several computational reports on tantalum carbides [11–14] and experimental measurements on bonding energies [15, 16] have confirmed the dominant covalent bonding nature within this monocarbide. This covalent bonding is a result of strong sigma bonds that form between the overlaps of the $5d$ electrons in tantalum and the $2p$ electrons in carbon. Jhi *et al.* [17] reported density functional theory calculations for the density of states (DOS) for NbC, also a group V monocarbide. In their report, a minimum was observed in the DOS at the Fermi level for a slight substoichiometric NbC$_{0.875}$ composition. Such a DOS decrease is indicative of the loss of metallic-like bonding character and a shift toward a more covalent nature. Considering that the group V metals have a valence of 5 and carbon prefers 4, the loss of carbon in the B1 structure likely offsets these electronic differences, yielding the optimal bonding strength configuration. This change in DOS may explain why the melting temperature is higher for substoichiometric group V monocarbide compositions.

With the loss of carbon in TaC$_{1-x}$, a linear increase in lattice parameter with carbon content occurs [7] and provides an indirect means of determining chemical composition without direct chemical detection methods. The high electrical and thermal conductivity of TaC [18, 19] also supports the existence of metallic bonding within the compound. This is confirmed by numerous electronic structure theory calculations that show a continuity of the DOS through the Fermi level, an example of which is shown in Figure 12.3. The metallic bonding occurs through the sigma bonds of the valence d-electrons that form between neighboring tantalum atoms.

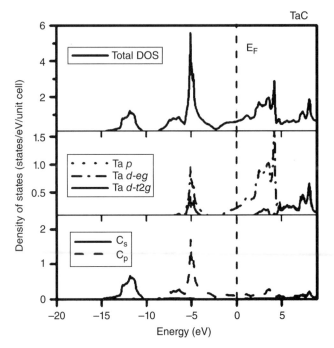

Figure 12.3. The total and partial density of states (DOS) for TaC computed from electronic structure density function theory using full-potential linearized augmented plane wave plus local orbitals and the local density approximation. The continuity of the DOS through the Fermi level (0 eV) demonstrates that TaC has no band gap and is an electrical conductor. The partial density of states (pDOS) shows that the DOS at the Fermi level is mainly due to tantalum d-electrons, which suggests that the metallic nature is associated with the bonding between neighboring tantalum atoms through overlap of the d-orbitals. Taken from Ref. [11].

Upon closer inspection of the phase diagram in Figure 12.1, at lower temperatures and with a loss of carbon, a Ta_6C_5 phase can precipitate. This phase, unlike the other substoichiometric phases of tantalum carbides reported in the following, is thought to be simply derived from cubic TaC_x where the vacancies create an ordered structure. In most of the experimental literature, this phase is not reported at the temperatures and compositions seen in the phase diagram of Figure 12.1. The Ta_6C_5 crystal symmetry is still under discussion as either monoclinic or trigonal [4]. In this same reference, the Nb_6C_5 and V_6C_5 compounds have been reported to form. The lack of consistent reporting of the Ta_6C_5 may be attributed to the larger scattering ratio of tantalum to carbon in comparison to these other group V transition metals, where this phase is more readily identified. Thus, it is reasonable to expect that this phase does form in the tantalum carbides, if present, but is difficult to detect. In addition, M_6C_5 phases require very slow cooling conditions to precipitate, ~0.5 K/min, indicative of some dependence on vacancy site ordering, which likely contributes to the uncertainty in which crystal structure forms. As a consequence of the very controlled manner by which this phase must form, the substoichiometric TaC_{1-x} phase is commonly quenched in and observed.

Unlike the Ta_6C_5 phase, the ζ-Ta_4C_3 phase has been observed by several groups [5, 8, 20, 21]. Similar to the Ta_6C_5 phase, some debate continues on the equilibrium stability. In the pioneering work of Rudy and Harmon [22], they noted that the Ta_4C_3 phase formed under compressive stresses by epitaxially growing off of the TaC_{1-x} phase when the temperature was too low to phase transform directly into Ta_2C. After 400 h at 1700°C, Ta_4C_3 decomposed suggesting to them the system was metastable. Arguably, no thorough experimental or computational equilibrium study of this phase has been conducted to definitively determine its stability, mechanical attributes, and range of carbon substoichiometry. Regardless of its stable or metastable thermodynamic state, the Ta_4C_3 phase is commonly observed. It has a rhombohedral structure ($R\bar{3}m$) with lattice parameters of $a = 0.3166$ nm and $c = 3.00$ nm [23] (Fig. 12.2b). The very large c-lattice parameter would make all but basal slip during deformation energetically unfavorable since a $<c+a>$ Burgers vector, **b**, would be very large because the elastic strain energy scales as $\mathbf{b}^2$ [24]. Consequently, by the nature of its structure, the Ta_4C_3 should be a relatively hard phase. Recent reports for tantalum carbides with a high volume fraction of Ta_4C_3 have shown significant increases in fracture toughness [25, 26] and will be further discussed in the mechanical behavior section of this chapter.

In the review article on metal-rich transition carbides and nitrides, Demyashev [27] described this Ta_4C_3 and similar structures as a sequence of cubic-based and hexagonal-based stacking faults. Using this perspective, the Ta_4C_3 phase forms from a TaC matrix by the removal of every fourth {111} carbon plane with a subsequent shear between the contacted Ta {111} planes (Fig. 12.4) [28]. As noted by Gusev *et al.* [4, 5], the tantalum carbides have an inherent vacancy ordering behavior, and the condensation of vacancies on every fourth {111} plane in TaC is necessary for the correct Ta_4C_3 composition but is not sufficient to form the observed crystal structure. The correct stacking sequence of Ta_4C_3 is obtained by passing a Shockley partial dislocation between each of the newly formed tantalum–tantalum layers. This transforms the FCC stacking sequence of the {111} planes of TaC, AγBαCβAγBαCβAγBαCβAγBαCβ, to the stacking sequence for

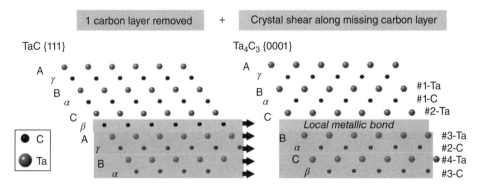

Figure 12.4. Left image is a series of {111} planes in TaC viewed along <110>. The right image is the removal of a {111} plane of carbon atoms and the subsequent shear along the now formed metallic Ta–Ta bonds. Stacking of three sequential highlighted boxes, one on top of another (or, alternatively, the loss of every fourth carbon plane with a shear), results in a unit cell of Ta_4C_3.

Ta_4C_3 as AγBAγBαCβACβAγBαCBαCβ. The Roman letters represent planes of tantalum atoms and the Greek letters planes of carbon atoms. Removal of carbon or, alternatively, diffusion of vacancies onto this plane to form this critical loop would imply that precipitation is by nucleation and growth rather than a simple martensitic shear reaction. This formation is analogous to the nucleation of Frank loops in FCC metals with the subsequent precipitation of a similar close-packed-based structure classically described in the Al–Ag system [29].

With further reduction in carbon, the Ta_2C phase is stable. Dependent on temperature, above or below approximately 1800 K, two similar but distinct structures exist for this composition [8, 9]. The low-temperature α phase is a CdI antitype structure, shown in Figure 12.2c, with the hexagonal close-packed (HCP) array of metal atoms alternatively separated by filled and unfilled planes of carbon. α-Ta_2C's lattice parameters are $a = 0.310$ nm and $c = 0.494$ nm [30]. Similar to Ta_4C_3, this structure allows for direct metal-to-metal atom bonding between close-packed planes, with the former Ta_4C_3 being every fourth plane and the latter α-Ta_2C on every other plane, and the resulting stacking sequence for α-Ta_2C is AγB. The contribution to the cohesive energy for α-Ta_2C is now equal between covalent and metallic bonds. Notably, this direct metallic bonding character is manifested with an order (α) to disorder (β) transition temperature at approximately 1800 K and melting temperature of approximately 3573 K for Ta_2C as compared to no disordering behavior and significantly higher melting temperature of approximately 4273 K for TaC. The increase in metallic bonding characteristics changes the mechanical properties of Ta_2C by reducing the brittle-to-ductile transition, reducing hardness, and increasing plastic flow, all of which are described in detail in the mechanical behavior section. The α-Ta_2C phase is trigonal, $P\overline{3}m1$, though it is often incorrectly referenced as hexagonal because the settings, that is, use of the same set of axes, are equivalent. A consequence of this loss in symmetry is that equivalent planes in HCP structures, such as $\{10\overline{1}1\}$ and $\{10\overline{1}\overline{1}\}$, are no longer equivalent in trigonal structures. This asymmetry will have consequences on the nonbasal slip behavior for this phase. The high-temperature β phase maintains the same close-packed structure as α phase but the carbon sublattice disorders yielding an **L'3** structure, which is now hexagonal. This phase melts at approximately 3573 K. Both phases of Ta_2C are able to exhibit some range of substoichiometric carbon loss while maintaining their structures, as shown in phase diagram in Figure 12.1.

Further loss in carbon does not result in any additional compounds. From Figure 12.1, only at high temperatures does tantalum indicate little solubility for carbon. At lower temperatures, the presence of carbon results in a two-phase equilibrium of Ta+Ta_2C microstructure. At C/Ta approximately 0.12 in Figure 12.1, a eutectic is present just above 3273 K. The development of an engineered eutectic microstructure processed above or below this temperature and composition has not been exhaustively studied to the authors' knowledge.

12.2 MICROSTRUCTURES OF TANTALUM CARBIDES

The crystallography of each tantalum carbide phase can and does have an impact upon the microstructure morphology. In single-phase regimes, such as TaC and Ta_2C, the microstructures are nominally equiaxed grains (Fig. 12.5). If precipitation occurs

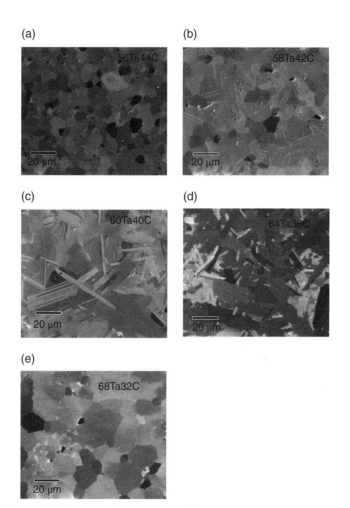

Figure 12.5. SEM backscattered micrographs of HIP processed tantalum carbides in at%: (a) 56Ta44C, (b) 58Ta42C, (c) 60Ta40C, (d) 64Ta36C, and (e) 68Ta32C. Images taken from Ref. [31].

because of a change in the C/Ta ratio, the microstructure can exhibit equiaxed grains with secondary-phase laths in a crisscross pattern or acicular grains with secondary-phase laths parallel to the major axis of the grains, again shown in Figure 12.5. The ability to tune a microstructure dependent on phase content provides ample engineering opportunities for tailored microstructures for regulating mechanical properties such as tensile strength, creep, and fracture toughness.

The processing used to fabricate a bulk ceramic can influence the microstructures the ceramic forms. In the context of this chapter, these bulk processing routes are divided into two generic categories: bottom-up, which refers to a solid-state reaction between powders, or a top-down method, which refers to a solidification of molten powders. In the bottom-up process, hot pressing (HP) [32–37], spark plasma sintering (SPS) [26, 38–41], and hot isostatic pressing (HIP) [31, 42] of constituent tantalum carbide powders result in relatively the same microstructures for similar compositions. The major difference between each process is usually related to the final grain size as the powders consolidate. The variation between process parameters such as applied pressure, heating and cooling rates, peak temperature, and time at peak temperature is critical in controlling densification and grain size. In the top-down method, such as arc melting or vacuum plasma spraying (VPS), the microstructure can initially appear very different than those in the bottom-up route. This is because the powder has undergone a liquid-to-solid transformation. Though SPS has localized melting, this melting is confined at powder contact points (surface) and the entire powder is not molten. Because of this distinction, it has been classified a bottom-up route. In contrast, arc melting or VPS causes the powders to completely liquefy and undergo solidification. In VPS, this rapid solidification results in a splat-like layered microstructure when it is deposited onto a substrate surface [2], as shown in Figure 12.6 [43]. Since carbon can be preferentially lost during melting because of the large vapor pressure differences between carbon and tantalum [44], the final composition may not be equivalent to the starting powder composition. Hence, blending in excess carbon with the starting powders helps to retain the desired C/Ta ratio. Upon subsequent post-processing of arc-melted or VPS materials using sintering or HIP to consolidate and homogenize the as-solidified microstructure, Morris *et al.* [43] reported that the splat-like VPS microstructure reverted to those microstructures observed in bottom-up processing.

The C/Ta composition ratio is dominant in determining the morphology of the phases within the processed grains using the synthesis routes described earlier. In single-phase regimes, such as TaC and Ta_2C, the microstructures are nominally equiaxed grains [31] (Fig. 12.5a and e). If precipitation occurs because of a change in the C/Ta ratio, the microstructure reveals either equiaxed grains with secondary-phase laths in a crisscross pattern (Fig. 12.5b) or acicular grains with secondary-phase laths parallel to the major axis of the grains (Fig. 12.5c and d). Elucidating how composition regulates

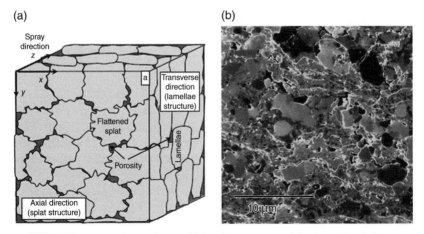

Figure 12.6. VPS processed tantalum carbide microstructure: (a) schematic of the as-sprayed microstructure taken from Ref. [2]. (b) SEM micrograph of an as-sprayed tantalum carbide microstructure.

the secondary-phase morphology within these grains is paramount to tailoring the microstructure.

Wiesenberger *et al.* [45], Morris *et al.* [31], and Santoro and Probst [9, 46] have reported diffusion-based studies to explain the types of microstructures that form as a function of C/Ta ratio. Wiesenberger *et al.* heated tantalum wire next to graphite powder, whereas Morris *et al.* studied hot isostatic pressed Ta powders next to TaC powders. Santoro and Probst investigated carburization of tantalum metal through metal–gas reactions. In Wiesenberger *et al.* and Morris *et al.*, other than the Ta_6C_5 phase in Figure 12.1, all other tantalum-rich carbide phases were observed. Morris *et al.* reported how the sequential precipitation sequence was paramount in controlling the microstructures. In this study, the carbon depleted from the TaC powder and reacted with the tantalum metal powders. The depletion of carbon from TaC resulted in the precipitation of Ta_4C_3 within the TaC_{1-x} grains. Transmission electron microscopy (TEM) revealed that the Ta_4C_3 phase precipitated from the TaC {111} planes with an orientation relationship of $\{0001\}Ta_4C_3$ // {111} TaC and $\langle 11\bar{2}0 \rangle$ Ta_4C_3 // $\langle 110 \rangle$ TaC. Since TaC has four variant {111} planes, the Ta_4C_3 forms a crisscross pattern (Figs. 12.5b and 12.7a) since each habit plane is equivalent. An interesting attribute of these Ta_4C_3 laths, and similar phases noted by Demyashev [27], is that they tend to be very thin and planar. Morris *et al.* [43] conducted electron tomography studies that revealed that these phases tended to be no more than a few nanometers thick and spanned large planar surfaces on the TaC {111} planes (Fig. 12.7a). Often, micrographs of these Ta_4C_3 phases, such as the one in Figure 12.5b, can lead one to believe that this phase has a range of thicknesses. In reality, one must be careful as the thickness of these laths is a two-dimensional viewing artifact of inclined laths to the surface. The presence of such a thin vertical direction and wide planar direction is suggestive of significant interfacial energy differences between the side and top surface energies of Ta_4C_3 in TaC in controlling the Ta_4C_3 morphology.

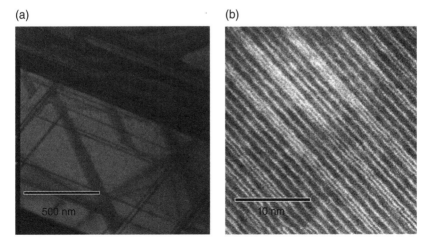

(a) (b)

500 nm 10 nm

Figure 12.7. (a) TEM bright-field micrograph showing the Ta_4C_3 crisscross lath pattern in TaC. (b) Scanning TEM high-angle annular dark-field (HAADF) micrograph revealing a series of parallel Ta_4C_3 laths (gray contrast) parallel to Ta_2C (bright contrast).

The acicular morphologies (Fig. 12.5c and d) were seen predominately when the composition was within the $Ta_2C + Ta_4C_3$ phase field. Similar to TaC, the Ta_4C_3 phase desires to maintain a close-packed plane and direction orientation relationship with the Ta_2C phase. This has been reported to be $\{0001\}\,Ta_4C_3\,//\,\{0001\}\,Ta_2C$ and $\langle 11\bar{2}0 \rangle\,Ta_4C_3\,//\,\langle 11\bar{2}0 \rangle\,Ta_2C$ [43], but unlike precipitation from TaC, which has four close-packed plane variants, the trigonal Ta_2C phase has only one close-packed plane, which is the basal plane. Consequently, the Ta_4C_3 and Ta_2C phases form a single, parallel planar arrangement (Fig. 12.7b). A consequence of the single variant orientation relationship contributes to acicular grains having the laths parallel to each other and along the major axis of the grain. The parallel orientation appears to influence the grain growth morphology to be acicular.

12.3 MECHANICAL PROPERTIES OF TANTALUM CARBIDES

12.3.1 Elastic Properties

In untextured polycrystalline form, TaC exhibits elastic isotropy at the macroscopic level and is characterized by two independent elastic constants, E (elastic modulus) and ν (Poisson's ratio). However, since the monocarbide has a range of substoichiometry (Fig. 12.1), this generates the possibility of a large number of atoms having a different chemical coordination environment. This alteration is manifested by the slight variations in the elastic properties as a function of carbon content within the monocarbide phase. Table 12.1 is the experimentally reported elastic constants for TaC_{1-x} determined using acoustical measurements at room temperature.

Besides carbon content, the porosity within an experimentally tested bar will also influence the elastic constant measurements. Jun *et al.* [47] reported the isotropic elastic constants of $TaC_{0.99}$ as a function of porosity and temperature. Under these controlled conditions, the elastic constants varied linearly for both porosity, changed from 6 to 12%, and temperature, from 300 to 1400 K.

In its single-crystal form, TaC is elastically anisotropic and is characterized by the three cubic elastic constants: c_{11}, c_{12}, and c_{44}. A limited number of experiments have determined these constants with reliability, and these values are listed in Table 12.2. However, a large number of theoretical studies have computed these constants using electronic structure density functional theory (Table 12.2). In order to facilitate comparisons between the single-crystal data in Table 12.2 and the polycrystal data in Table 12.1, the Voigt and Reuss average Young's modulus is listed with the single-crystal elastic

TABLE 12.1. Experimentally determined elastic constants of polycrystalline TaC

Comp	E	nu	Bulk modulus	Reference
TaC0.99	560	0.21	329	[47]
TaC0.90	477	0.152	229	[48]
TaC0.994	537	0.24	344	[49]
TaC0.98	567±68		332	[50]

TABLE 12.2. Cubic elastic constants and lattice constants of TaC single crystals obtained from experiments and simulation

Method	Comp	a_0	C_{11}	C_{12}	C_{44}	EV	ER	Reference
Phonon dispersion	TaC0.9		500	90	80	327	266	[51]
Neutron data	TaC		550	150	190	474	474	[51]
Ultrasonics	TaC		595		153			[52]
DFT/NCPP–LDA	TaC		740	165	176	549	523	[13]
DFT/USPP–LDA	TaC	4.424	611	243	160	440	439	[12]
DFT/USPP–PBE	TaC	4.564	466	224	120	320	319	[12]
DFT/HGHPP–GGA	TaC	4.525	621	155	167	480	470	[52]
DFT/USPP–PBE	TaC	4.39	704	110	170	538	498	[14]

All of the reported simulations are performed under the framework of electronic structure density functional theory (DFT) with different pseudopotentials (PP) and exchange-correlation (XC) functions used to approximate the nucleus–electron and electron–electron interactions.

DFT, density functional theory; NCPP, norm-conserving pseudopotential; USPP, ultrasoft pseudopotential; HGHPP, Hartwigsen–Goedecker–Hutter pseudopotential; LDA, local density approximation exchange-correlation function, GGA, generalized gradient approximation exchange-correlation function; PBE, Perdew–Burke–Ernzerhof exchange-correlation function.

constants. To the authors' knowledge, the elastic constants, whether experimental or simulated, have not been published for the metal-rich carbide phases.

12.3.2 Plastic Properties of TaC

The plastic deformation of tantalum carbides, and other transition metal carbides, may appear confusing given the small amount of literature and complicated bonding behavior previously alluded to in the crystallography section of the chapter. For example, the FCC symmetry of the B1 phase indicates two possibilities for slip, depending on the bonding characteristics. If the bonding is dominated by either covalent or metallic character, {111} slip planes in the ⟨110⟩ directions would be expected at low temperatures [53]. Only at elevated temperatures would slip on the {110} and {100} planes be suspected. In contrast, if the bonding has more ionicity, slip on the {110} would be probable at low temperatures as is observed for typical B1-structured ionic compounds [53].

Most of the plastic properties of TaC are attributed to dislocation motion, whether by dislocation glide or climb. Hardness anisotropy has shown that ⟨110⟩{111} slip occurs at room temperature [54, 55] confirmed using TEM dynamical electron diffraction [56] and etch pit analysis [57]. In this regime, the dislocations were straight screw segments, which are indicative that the motion is restricted by high lattice friction. These results support the notion that any ionic nature of the bonding is minimal. However, etch pit analysis of Rowcliffe and Warren [57] showed that dislocations may also move on planes other than {111}, which they interpreted to be {110} or {100}. Substantially, more experimental work is needed as these observations occurred near inclusions during cooling and the conditions for deformation are not readily established. Furthermore, the test temperature of 1573 K was estimated; this is the temperature at which the inclusions solidify, further complicating the behavior of the material. At intermediate temperatures (1473–1973 K), Kim *et al.* [56] reported the slip system to be $\langle 110 \rangle \{111\}$ and deformation comprised of curved dislocations. Thus, from these experiments, it is reasonable to state that slip in TaC occurs on $\{111\}$ planes with the potential of additional slip systems becoming active at higher temperatures.

While the prevalence of $\langle 110 \rangle \{111\}$ slip in TaC is generally agreed upon, no clarity exists in determining the temperature dependence of plastic flow and dislocation mobility. A number of theories have emerged to explain the temperature dependence of hardness measurements, bending stresses, and tensile stresses [56, 58–62]. If one assumes that TaC is a ceramic that behaves like a metal stabilized by the presence of carbon, then the natural assumption is to associate the strong temperature dependence of plastic flow on a Peierls mechanism. In this case, the strong metallic–covalent bonding in TaC creates an intrinsic lattice resistance that must be overcome by both stress and temperature for flow to occur [53]. If thermal activation is not sufficient to support bulk plasticity, brittle fracture is likely, which would result in a classic brittle-to-ductile transition. This transition is observed in TaC [63–65]. The argument for the importance of simple lattice resistance is perhaps best supported by the observations of Kim *et al.* [56] of long screw dislocations under room-temperature indents, which is indicative of strong lattice resistance similar to observations of low-temperature deformation in body-centered cubic metals [66].

While the Peierls resistance certainly must contribute to the resistance to plastic flow, a number of authors [59–62] have argued and presented evidence to support the notion that plasticity under noncreep conditions is governed by the diffusion of carbon. While this can be viewed as part of the Peierls resistance, this type of resistance is distinct from the Peierls resistance associated with direct slip between Ta and C bonds. This idea of carbon diffusion-limited plasticity starts with the premise that dislocations on $\{111\}$ planes in TaC split into partials, which has been reported in a limited number of experiments [67–69], and requires the diffusion of carbon through the dislocation core for dislocation glide [60, 70, 71]. To understand this idea, consider Figure 12.8, which shows a schematic of the stacking of the $\{111\}$ planes in TaC. The stacking sequence shown is $\ldots A\gamma B \ldots$ where the Latin letters represent the tantalum atoms and the Greek letters are the carbon atoms. The A atoms are gray, the B atoms are white, and the γ (carbon) atoms are black. If we consider the B layer to slip relative to the A layer, then motion associated with a perfect dislocation is shown by the black arrow in Figure 12.8a. Alternatively, the perfect dislocation could split into Shockley partials and the associated motion is shown by black dashed arrows in Figure 12.8a. Slip by either of these mechanisms would be limited by the Peierls resistance to the glide mechanism. If glide is assumed to be unfavorable such that metal atoms slide over one another, then slip of the B atoms must occur through the c sites (black dashed arrows in Fig. 12.8b), which are occupied by carbon atoms. This is presumably even more unfavorable than slip of metal atoms over metal atoms discussed previously. However, if carbon atoms move in a coordinated fashion, as shown in Figure 12.8b by gray dashed arrows, then glide would be easier. This can occur if carbon atoms diffuse out of these sites; hence, dislocation motion would be limited by the rate of carbon diffusion and the substoichiometric carbon content or, alternatively, the number of available site defects created on

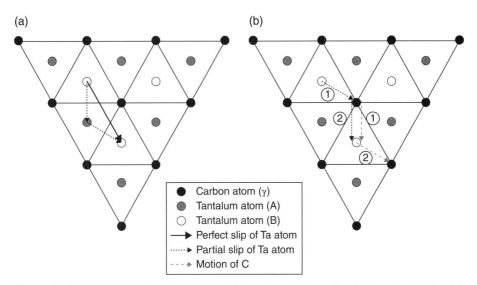

Figure 12.8. The potential mechanisms of dislocation glide on {111} planes in TaC. In this figure, the slip occurs between two tantalum layers, A and B, whose stacking is shown as white open circles and gray circles. The carbon atoms are shown as solid black circles and sit in the octahedral interstices between the tantalum layers. In this simple example, it is assumed that layer B will slip relative to layer A by a perfect dislocation, $a/2<110>$. (a) The large black arrow represents the slip of layer B relative to A along the $<110>$ direction, that is, a perfect dislocation. If it is assumed the dislocations split into Shockley partials, the atoms at B will move first from its original position over the atom at site A and then back to a position at B, which is illustrated by the shorter dashed black arrows. (b) Under the assumption that the motion of the B tantalum atom over the A tantalum atom is unfavorable, then it is possible for the B atom to move to the C position, which is occupied by the carbon atoms. This motion is certainly less favorable than even the motion to the A position. This can be favorable if the carbon atom moves in a coordinated fashion with the A atom. The black dashed arrows show the slip associated with these new partial dislocations, and the dashed gray arrows represent the coordinated motion of the carbon atoms. If the material is substoichiometric, then the carbon atoms in B may not be present, making slip even easier and explaining the loss of hardness with decreasing carbon content for single-phase TaC_{1-x} materials.

the carbon sublattice. This mechanism is commonly referred to as **synchroshear** [10]. However, if carbon atom motion actually occurs via another dislocation, then two sets of partial dislocations glide together, one between the A and γ layers and one between the γ and B layers. These dislocations are termed zonal dislocations [72] and have been proposed in the case of TaC by [73].

The evidence in support for carbon diffusion-controlled plasticity in TaC is presented in terms of the temperature/time dependence of hardness [58] and bending stresses [61]. Fitting of experimental data (hardness and yield stresses) to an appropriate Arrhenius law results in an activation energy that is argued to be the activation energy of carbon diffusion [58, 61, 71]. Specifically, Kumashiro *et al.* [58] determined the activation energy associated with indentation between 1473 and 1573 K as 4.8 eV (464 kJ/mol). This activation energy is not considerably different from Kim *et al.* [56] who found a value of 4.3 eV in creep studies at similar temperatures. Both of these values are much smaller than the values determined by Steinitz [64], who reported 7.4 eV. Martin *et al.* [61] analyzed their four-point bend data in temperature ranges from 1473 to 2473 K and found an activation energy of 5.8 and 6.2 eV with activation volumes of $48b^3$ and $52b^3$ depending on the fit used, where b is the magnitude of the Burgers vector. Numerous studies of the diffusion of carbon in TaC provide a range of activation energies between 3.7 and 4.3 eV [74–78]. The activation energies of most of the deformation experiments are too high to represent diffusion of carbon, especially if one considers the reduction in energy for diffusion near the dislocation core. The only exception to this is possibly the creep experiments of Kim *et al.* [56], which resides at the upper end of the diffusion data.

Alternatively, both Kim *et al.* [54] and Steinitz [62] argued that plasticity in creep conditions is controlled by dislocation climb, which would require diffusion of both carbon and tantalum. The activation energy for diffusion of Ta in TaC has not been measured, but is estimated to be 7 eV, which agrees with the measurements of Steinitz, but such conclusions are tentative given that estimated activation energy of TaC is based on analogies with properties of FCC metals. Clearly, discrepancies between measured activation energies are prevalent in the literature and may be a result in the experimental difficulties of consistent samples between different research groups in terms of carbon content, porosity, and testing environment where (de)carburization could occur from heating or the heating source itself, if graphite filaments are used. It is also worth pointing out that the activation energies associated with the Peierls mechanism have not been estimated and cannot be compared with these values. Thus, this collection of data provides some evidence for the resistance of dislocation glide by diffusion; however, it is not conclusive and further work is warranted.

12.3.3 Ductile-to-Brittle Transition

TaC exhibits a classic ductile-to-brittle transition (DBT) in its bulk polycrystalline form, as reported in several comprehensive studies [59, 63–65]. These different studies each used a different test method—Kelly and Rowcliffe [63] used four-point bending, Johansen and Cleary [65] used a cantilever bend test, and Steinitz [59, 64] used standard tensile bars. From the collective observations of these authors, it is reasonable to assign a ductile-to-brittle transition temperature (DBTT) of 2123 ± 100 K.

While the DBTT observed is around 2123 K, significant plastic flow is typically not observed until temperatures around 2373 K. While Kelly and Rowcliffe reported a DBT, they did not observe any measureable ductility in their specimens, but they did not test above 2273 K. Steinitz found appreciable ductility in TaC at temperatures at 2433 K and above with elongation to failure of 33%. From this, a tensile stress–strain curve was extracted, shown in Figure 12.9, which may be the only in existence.

The work of Johansen and Cleary [65] determined the effect of carbon content on the DBTT, which linearly decreased with carbon content. The carbon content of TaC_x included

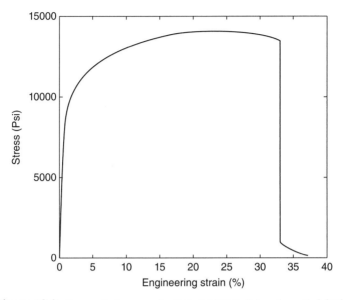

Figure 12.9. Stress–strain curve for TaC at 2433 K. Taken from Ref. [59].

$x = 1.26$, 0.98, 0.90, 0.84, and 0.73 with all the specimens containing the rock salt TaC structure except $x = 1.26$, which had graphite, and $x = 0.73$, which contained Ta$_2$C.

Despite the consistency of these studies regarding the DBT, it is important to note that plastic flow occurs at room temperature in TaC [54, 56] under microindents. Additionally, plastic flow also occurs at elevated temperatures below this measured DBTT in creep experiments [56]. Thus, the DBT may be dependent on many factors including the type of loading, surface conditions, and presence of flaws, as well as the number of grains in the volume being tested. It is important to bring into discussion the brittle-to-ductile nature of tungsten, the most refractory metal that is brittle in polycrystalline form but ductile in polished single crystals at low temperatures [79–83]. Thus, the agreement seen here may be largely related to the similarity of the test conditions: polycrystals largely subjected to tensile and compressive normal stresses.

12.3.4 Creep

Creep tests on TaC have been performed by both Steinitz [59] and Kim *et al.* [56]. Steinitz conducted tests at temperatures of 2233, 2313, and 2373 K with carbon content varying between 0.85 and 0.98. Kim *et al.* conducted creep tests at 1573, 1973, and 1773 K at stresses that ranged from 60 to 300 MPa. From these creep tests, Steinitz and Kim *et al.* were able to determine the stress exponent, n, for creep, $\dot{\varepsilon}$:

$$\dot{\varepsilon} = C\sigma^n \qquad (12.1)$$

Steinitz reported it to be between 2 and 4. The range is suggestive that no reliable number was established, likely because of potential carbon loss during the experiment.

Steinitz also reports a decrease in creep rate with decreasing carbon content, suggesting that decarburization improves creep resistance. Kim *et al.* determined the stress exponent at 1773°C and found a value of 2.05. Both authors reported activation energies, *Q*, using the Arrhenius relationship between the strain rate and temperature given as

$$\dot{\varepsilon} = A \exp\left(-\frac{Q}{kT}\right) \qquad (12.2)$$

where *k* is Boltzmann's constant, *T* is temperature, and *A* is a temperature-independent pre-exponential factor. Steinitz activation energy value was $7.37 \pm 1.3\,\text{eV}$ (711 kJ/mol), and Kim *et al.* reported an energy of 4.34 eV (420 kJ/mol). Steinitz interpreted his measured activation energy to be associated with diffusion of tantalum in the lattice. However, the values of Kim *et al.* are too low for tantalum diffusion (as estimated by Steinitz [59]) and are at the upper end of carbon diffusion (3.7–4.3 eV, 360–420 kJ/mol). Kim *et al.* suggest that at their intermediate temperatures, creep is accomplished by grain boundary sliding controlled by dislocation climb. If this is the case, dislocation climb would require both carbon and tantalum diffusion. Further work in the area of creep needs to be conducted to clarify these issues, and better estimates for the activation energy of Ta diffusion would further help establish the operant mechanisms.

12.3.5 Hardness of Tantalum Carbides

Hardness measurements are probably the most common mechanical measurements made on transition metal carbides and TaC in particular. Knoop hardness is the most common and potentially the most reliable measurements for these hard materials, though some values are reported with a Vickers indenter. These result in a large number of hardness measurements with values in the range of 150–5200 kg/mm² with dependence on carbon content, porosity, temperature, and the orientation of the surface indented. The highest hardness reported corresponds to indents on single-crystal faces of $TaC_{0.83}$ at room temperature [20], and the lowest hardness values correspond to indents in polycrystals at 2173 K [84].

As discussed in the crystallography section, we might expect that the hardness of tantalum carbides should continually decrease as a function of carbon content if one assumes that the covalent bonding character is decreasing. This is what is observed by Samsonov and Rukina [85], as shown in Figure 12.10a, which includes carbon contents that span compositions that include TaC, Ta_4C_3, and Ta_2C. However, Santoro [46] and Steinitz [59, 64] and Gusev *et al.* [5] have measured hardness values above $TaC_{0.75}$ and found that hardness increased nearly linearly with a decreasing carbon content until C/Ta = 0.8, which is in direct contrast with Samsonov and Rukina. The reason for this increase in hardness with a decrease in carbon content is difficult to explain, though it has been argued to be related to the optimal filling of the d-bands [57] or by dislocation–vacancy hardening [17, 65, 71] or may be related to the ordered Ta_6C_5 structure [86, 87]. To further complicate this issue, TiC, arguably the most widely studied group IV transition metal carbide with a rock salt structure, exhibits a continual decrease in hardness with decreasing carbon content. It can be safely stated that below C/Ta = 0.8, the hardness of tantalum carbides decreases with the trends above this threshold not well understood.

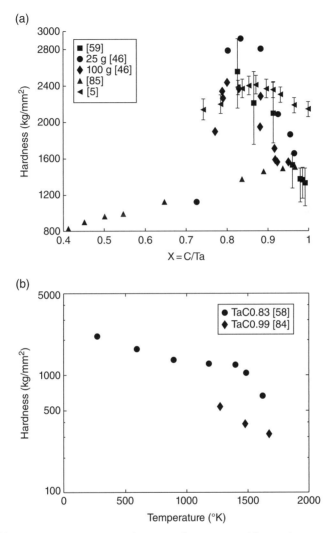

Figure 12.10. Hardness variation as a function of C/Ta composition and temperature. (a) The hardness at room temperature from various groups plotted as a function of carbon content. All hardness measurements are presumed to be made on polycrystals [5, 46, 59, 85]. (b) The hardness of TaC plotted as a function of temperature. The data of Kumashiro [58] is for indentation on a {001} surface, while the data of Koester and Moak [84] represent the hardness of a polycrystalline surface.

The hardness of TaC decreases with temperature, as shown in Figure 12.10b, as expected. The data of Kumashiro *et al.* [55, 58], which are Vickers hardness measurements on a {001} TaC surface, shows a knee in the hardness data near 1473 K. This may be indicative of a transition in deformation mechanisms. However, the data of Koester and Moak [84], which is a Knoop hardness on polycrystalline TaC, does not show such a transition.

12.3.6 Strength

The strength of tantalum carbide is difficult to establish because of its brittleness at low temperatures and its high DBTT. The strength of TaC below the DBTT should be statistical in nature with a strong dependence of flaw distributions, but no study to date has explored this issue. Above the DBTT, the high temperatures required for significant plasticity further limit the available data since few mechanical testing facilities are able to exceed 2373 K. Only a very small number of tensile tests have been conducted on TaC, one of which is shown in Figure 12.9, which reveals no well-defined yield point and only the ultimate tensile strengths (UTS) were reported by Steinitz [59]. The values of the UTS range from 109 MPa (15,800 psi) to 89 MPa (12,900 psi) for temperatures that range from 2153 to 2473 K [59]. In this temperature range, the ductility ranges from 2 to 39% and the carbon content was $C/Ta = 0.98$.

The strength in bending has also been reported as a function of temperature and carbon content. For $TaC_{0.98}$, Johansen and Cleary [65] found the strength to range between 89.6 MPa (13,000 psi) and 158.6 MPa (23,000 psi) for temperature between 2173 and 2423 K. As a function of carbon content, the authors showed that the bending stress at the BDTT has a minimum around $TaC_{0.9}$ with the strengths of $TaC_{0.8}$ and $TaC_{0.98}$ being similar. Santoro [46] performed experimental studies using TaC wires under three-point bending at room temperature and reported strengths between 360 MPa (52,000 psi) and 460 MPa (67,000 psi) for C/Ta from 0.70 to 0.95. The extrapolated strength to $TaC_{1.0}$ would be 520 MPa (75,000 psi). The minimum strength was at $C/Ta = 0.8$. Martin et al. [60, 61] measured bending stresses in polycrystalline $TaC_{0.98}$ formed through carburization of Ta sheets using four-point bending. They found that bend strengths depended on both strain rate and temperature, with strength values leveling off around 50 MPa (7250 psi) at high temperatures (typically above 2473 K) and rising to over 1780 MPa (25,000 psi) for temperatures around 1473 K. It is very interesting to note from these tests that bending strength has a local minimum either around $C/Ta = 0.8$ or 0.9. This stoichiometry is also where a maximum occurs in hardness (Fig. 12.10a). The creep resistance reported by Steinitz [59] also increases with decreasing carbon content through $C/Ta = 0.85$. Thus, one must be cautious in developing a consistent model that explains experimentally observed changes in hardness, creep resistance, or bending strength alone, assuming that the trends observed are accurate.

12.3.7 Fracture Toughness

By varying the C/Ta ratio, not only are hardness and strength varied but also fracture toughness. K_{IC} values in tantalum carbides have been reported with values that range from approximately 2 to 13 MPa√m [26–28]. The unusually high fracture toughness in some reports is associated with compositions near Ta_4C_3. Though no detailed study has been done on the mechanisms that lead to this increase in fracture toughness, it is most likely associated with the intrinsic microstructure and mechanical attributes of the phases present.

Additional toughness in these ceramics could be obtained from the increased metallicity of the bonds. As already discussed, in TaC, room-temperature dislocation activity has been reported around indents [56, 57]. This indicates some intrinsic strengthening behavior. Though no room-temperature dislocation activity has been reported in the Ta_4C_3 and Ta_2C phases, one could speculate that it would behave similarly to TaC, which is a much more carbon-rich phase, via dislocation motion on close-packed planes. These two metal-rich phases have basal planes made of metal-on-metal bonds (Fig. 12.2), which could accommodate dislocation motion much easier than that already observed for {111} planes of TaC, which has alternating metal and carbon close-packed layers.

The additional metallic bonds are not likely to provide the observed toughening in these ceramics, and the large improvement is likely to occur from the microstructure itself. As discussed in the section on microstructure, the microstructure of the high K_{IC} compositions would contain a crisscross pattern of Ta_4C_3 thin laths in a TaC matrix. Consequently, the multiple directional laths, with limited slip confinement to the basal plane, provide extrinsic barriers for crack propagation through crack deflection, contact shielding, and zone shielding [88]. The emission of dislocations in these materials would also provide additional intrinsic strengthening contributions. In general, ductile materials are intrinsically toughened by impeding crack propagation with obstacles in front of the crack, while brittle materials are extrinsically toughened by impeding crack propagation via resistance in the crack wake [89, 90]. The combined intrinsic and extrinsic characteristics of this multiphase material likely contribute to the high K_{IC}, which is one of the highest values reported for a monolithic ceramic [25]. Clearly, a more detailed investigation that understands the mechanical strength of these high K_{IC} properties is needed to elucidate the mechanisms by which these increases are achieved.

12.3.8 Plasticity of Ta_2C

Much of the discussion has focused on the mechanical behavior of the monocarbide. To date, little is known about the mechanical attributes of metal-rich carbides. Clearly, from the fracture toughness, these phases can exhibit interesting strengthening characteristics whether as a single phase or precipitated within a matrix phase. As discussed previously, the hardness of Ta_2C is about 33% lower than TaC [85]. This is to be expected because of the increase in localized metallic character of the bonding in Ta_2C. Four-point bend tests of DiPietro [91] showed that Ta_2C is brittle at room temperature but quite ductile at 2200 K, much more ductile than TaC. This result was confirmed by De Leon *et al.* [92] who provided dynamical electron diffraction quantifying the deformation mechanism of Ta_2C and found $\langle 11\bar{2}3\rangle\{10\bar{1}1\}$ pyramidal slip activity. Unfortunately, the presence of stacking faults caused by excess carbon uptake during the test prevented imaging of basal plane dislocations. However, theoretical studies of generalized stacking fault energy curves in Ta_2C by Wang *et al.* [93] demonstrated that basal slip should be favored. Wang *et al.* also showed that pyramidal slip, observed in De Leon *et al.*'s [92] experiments, would be the second most likely activated slip planes. The trigonal symmetry of Ta_2C also would regulate slip behavior between the different $\{10\bar{1}1\}$ and $\{10\bar{1}\bar{1}\}$ planes, and the interatomic spacing, not necessarily a local metal-on-metal bond, dictated the lowest stacking fault energy.

12.4 SUMMARY

Tantalum carbides comprise a class of UHTC with a range of melting temperatures and phases dependent on the C/Ta ratio. The phase structures and precipitation sequence of these materials are paramount in regulating the microstructure morphologies. Microstructures can range from equiaxed grains to equiaxed grains with secondary laths in a crisscross pattern to acicular grains with secondary laths all parallel to the major axis of the grain. These microstructures have been shown to regulate the mechanical responses, particularly increasing the fracture toughness to approximately $13\,\text{MPa}\sqrt{m}$ with a high volume fraction of Ta_4C_3. Most of the mechanical property work has been done on the monocarbide TaC phase. The mechanical behavior of this carbide shows a range of responses, dependent upon carbon content, porosity of the test specimen, and the testing means or environment. Because of these issues, reported mechanical data show variations. For example, agreement can be found in certain areas, such as decreasing hardness with decreasing carbon content below $C/Ta = 0.8$, whereas other tests show wide variations, including the reported activation energies in creep experiments. This demonstrates the challenges in studying these classes of materials. It is imperative that any conclusions based on literature reports carefully consider the intrinsic and extrinsic characteristics of the tantalum carbide sample tested. Regardless of these issues, the range of mechanical properties with corresponding varied microstructures offers several exciting opportunities to tailor phases together that couple ultra-high melting temperature capability with metallic-like and ceramic-like mechanical responses.

ACKNOWLEDGMENTS

GBT acknowledges the support of AFOSR-FA9550-12-1-0104 DEF (Dr. Ali Sayir, program manager) to write this review. In addition, GBT recognizes Dr. Lawrence Matson of AFRL, Dr. Daniel Butts of Plasma Processes, Inc., and Dr. Stephen DiPietro of Exothermics for technically stimulating discussions through the years on tantalum carbides. Finally, GBT thanks Drs. Robert Morris and Billie Wang and Mr. Nicholas De Leon and Bradford Schulz for their research efforts while at the University of Alabama. Sandia National Laboratories is a multiprogram laboratory managed and operated by Sandia Corporation, a wholly owned subsidiary of Lockheed Martin Corporation, for the U.S. Department of Energy's National Nuclear Security Administration under contract DE-AC04-94AL85000.

REFERENCES

1. Wang CR, Yang J, Hoffman W. Thermal stability of refractory carbide/boride composites. Mater Chem Phys 2002;74 (3):272–281.
2. Balani K, Gonzalez G, Agarwal A, Hickman R, O'Dell JS, Seal S. Synthesis, microstructural characterization and mechanical property evaluation of vacuum plasma sprayed tantalum carbide. J Am Ceram Soc 2006;89 (4):1419–1425.

3. Upadhyay K, Yang J, Hoffman W. Materials for ultrahigh temperature structural applications. Bull Am Ceram Soc 1997;76 (12):51–56.

4. Gusev AI. Sequence of phase transformations in the formation of superstructures of the M_6C_5 type in nonstoichiometric carbides. J Exp Theor Phys 2009;109 (3):417–433.

5. Gusev AI, Kurlov AS, Lipatnikov VN. Atomic and vacancy ordering in carbide Z-Ta_4C_3-X (0.28 < X < 0.40) and phase equilibria in the Ta-C system. J Solid State Chem 2007;180 (11):3234–3246.

6. Barabash OM, Koval YN. *A Handbook on the Structure and Properties of Metals and Alloys.* Kiev: Naukova Dumka; 1986.

7. Bowman AL. The variation of lattice parameter with carbon content of tantalum carbide. J Phys Chem 1961;65 (9):1596–1598.

8. Storms EK. The tantalum-tantalum carbide system. In: *The Refractory Carbides.* New York: Academic Press; 1967.

9. Santoro G, Probst HB. An explanation of microstructures in the tantalum-carbon system. Advances in X-ray Analysis: Proceedings of the 12th Annual Conference on Applications of X-ray Analysis, August 7–9, 1963; University of Denver. New York: Plenum Press.

10. Lewis MH, Billingham J, Bell PS. Non-stoichiometry in ceramic compounds. Electron Microscopy and Structure of Materials, Proceedings of the 5th International Materials Symposium, September 13–17, 1971; University of California Press, Berkley; 1972.

11. Sahnoun M, Daul C, Driz M, Parlebas JC, Demangeat C. Fp-Lapw investigation of electronic structure of TaN and TaC compounds. Comput Mater Sci 2005;33 (1):175–183.

12. Li H, Zhang L, Zeng Q, Guan K, Li K, Ren H, Lui S, Cheng L. Structural, elastic and electronic properties of transition metal carbides TmC (Tm = Ti, Zr, Hf and Ta) from first-principles calculations. Solid State Commun 2011;151 (8):602–606.

13. Wu Z, Chen XJ, Struzhkin VV, Cohen RE. Trends in elasticity and electronic structure of transition-metal nitrides and carbides from first principles. Phys Rev B 2005;71 (21):214103.

14. Yang J, Gao F. First principles calculations of mechanical properties of cubic 5d transition metal monocarbides. Phys B Condens Matter 2012;407 (17):3527–3534.

15. Khyzhun OY. XPS, XES, and XAS studies of the electronic structure of substoichiometric cubic TaC_x and hexagonal Ta_2C_y carbides. J Alloys Compd 1997;259 (1):47–58.

16. Khyzhun OY, Zhurakovsky EA, Sinelnichenko AK, Kolyagin VA. Electronic structure of tantalum subcarbides studied by XPS, XES, and XAS methods. J Electron Spectrosc Relat Phenomena 1996;82 (3):179–192.

17. Jhi SH, Louie SG, Cohen ML, Ihm H. Vacancy hardening and softening in transition metal carbides and nitrides. Phys Rev Lett 2001;86 (15):3348–3351.

18. Copper JR, Hansler RL. Variation of electrical resistivity of cubic tantalum carbide with composition. J Chem Phys 1963;39 (1):248–249.

19. Eckstein BH, Forman R. Preparation and some properties of tantalum carbides. J Appl Phys 1962;33 (1):82–87.

20. Brizes WF, Tobin JM. Isolation of the zeta phase in the system tantalum-carbon. J Am Ceram Soc 1967;50 (2):115–116.

21. Zaplatynsky I. Observations of the zeta phase in the system Ta-C. J Am Ceram Soc 1966;49 (2):109–110.

22. Rudy E, Harmon DP. *Afml-Tr-65-2,* Part I. Volume V, Wright-Patterson A.F.B. (OH): Air Force Materials Laboratory Research and Technology Division, Air Force Command; 1965.

23. Yvon K, Parthe E. On the crystal chemistry of the close-packed transition-metal carbides. I. The crystal structure of the Z-V, Nb and Ta carbides. Acta Crystallogr B 1970;26 (2):149–153.

24. Hull D, Bacon DJ. *Introduction to Dislocations*. 5th ed. Boston (MA): Butterworth-Heinemann; 2011.

25. Hackett K, Verhoef S, Cutler RA, Shetty DK. Phase constitution and mechanical properties of carbides in the Ta-C system. J Am Ceram Soc 2009;92 (10):2404–2408.

26. Limeng L, Feng Y, Yu Z. Microstructure and mechanical properties of spark plasma sintered $TaC_{0.7}$ ceramics. J Am Ceram Soc 2010;93 (10):2945–2947.

27. Demyashev GM. Review: transition metal-based nanolamellar phases. Prog Mater Sci 2010;55 (7):629–674.

28. Rowcliffe DJ, Thomas G. Structure of non-stoichiometric TaC. Mater Sci Eng 1975;18 (2):231–238.

29. Shewmon PG. *Transformations in Metals*. J. Williams Book Company: Tulsa (OK); 1983.

30. Bowman AL, Wallace TC, Yarnell JL, Wenzel RG, Storms EK. The crystal structures of V_2C and Ta_2C. Acta Crystallogr 1965;19 (1):6–9.

31. Morris RA, Wang B, Matson LE, Thompson GB. Microstructural formations and phase transformation pathways in hot isostatically pressed tantalum carbides. Acta Mater 2012;60 (1):139–148.

32. Zhang X, Hilmas GE, Fahrenholtz WG. Hot pressing of tantalum carbide with and without sintering additives. J Am Ceram Soc 2007;90 (2):393–401.

33. Liu J, Kan Y, Zhang G. Pressureless sintering of tantalum carbide ceramics without additives. J Am Ceram Soc 2010;93 (2):370–373.

34. Zhang X, Hilmas GE, Fahrenholtz WG. Densification and mechanical properties of TaC-based ceramics. Mater Sci Eng A 2009;501 (1):37–43.

35. Yohe WC, Ruoff AL. Ultrafine-grain tantalum carbide by high pressure hot pressing. Bull Am Ceram Soc 1978;57 (12):1123–1130.

36. Sciti D, Silvestroni L, Bellosi A. High-density pressureless-sintered HfC-based composites. J Am Ceram Soc 2006;89 (8):2668–2670.

37. Samsonov GV, Petrikina RY. *Sintering of Metals, Carbides, and Oxides by Hot-Pressing*. Institute for Development of Materials: Kiev; 1970.

38. Khaleghi E, Lin YS, Myers MA, Olevsky EA. Spark plasma sintering of tantalum carbide. Scr Metall 2010;16:557–580.

39. Bakshi SR, Musaramthota V, Lahiri D, Singh V, Seal S, Agarwal A. Spark plasma sintered tantalum carbide: effect of pressure and nano-boron carbide addition on microstructure and mechanical properties. Mater Sci Eng A 2011;528 (3):1287–1295.

40. Bakshi SR, Musaramthota V, Virzi DA, Keshri AK, Lahiri D, Singh V, Seal S, Agarwal A. Spark plasma sintered tantalum carbide–carbon nanotube composite: effect of pressure, carbon nanotube length and dispersion technique on microstructure and mechanical properties. Mater Sci Eng A 2011;528 (6):2538–2547.

41. Sciti D, Guicciardi S, Nygren M. Densification and mechanical behavior of HfC and HfB_2 fabricated by spark plasma sintering. J Am Ceram Soc 2008;91 (5):1433–1440.

42. Pierson HO. *Handbook of Refractory Carbides and Nitrides: Properties, Characteristics, Processing and Applications*. Saddle River (NJ): Noyes Publications; 1996.

43. Morris RA, Wang B, Thompson GB, Butts D. Variation in tantalum carbide microstructures with changing carbon content. Int J Appl Ceram Technol 2012;10 (3):540–551.

44. Nesmeianov A. *Vapor Pressure of the Chemical Elements*. Amsterdam: Elsevier; 1963.

45. Wiesenberger H, Lengauer W, Ettmayer P. Reactive diffusion and phase equilibria in the V-C, Nb-C, Ta-C and Ta-N systems. Acta Crystallogr 1998;46 (2):651–666.

46. Santoro G. Variation of some properties of tantalum carbides with carbon content. Trans Metall Soc AIME 1963;227:1361–1368.

47. Jun CK, Shaffer PPTB. Elastic moduli of niobium carbide and tantalum carbide at high temperature. J Less Common Met 1971;23 (4):367–373.

48. Bartlett RW, Smith CW. Elastic constants of tantalum monocarbide, $TaC_{0.90}$. J Appl Phys 1967;38 (13):5428–5429.

49. Brown HL, Armstrong PE, Kempter CP. Elastic properties of some polycrystalline transition-metal monocarbides. J Chem Phys 1966;45 (2):547–549.

50. Dodd SP, Cankurtaran M, James B. Ultrasonic determination of the elastic and nonlinear acoustic properties of transition-metal carbide ceramics: TiC and TaC. J Mater Sci 2003;38 (6):1107–1115.

51. Weber W. Lattice dynamics of transition metal carbides. Phys Rev B 1973;8 (11):5082–5092.

52. Lopez-de-la-Torre L, Winkler B, Schreur J, Knorr K, Avalos-Borja M. Elastic properties of tantalum carbide (TaC). Solid State Commun 2005;134 (4):245–250.

53. Hirth JP, Lothe J. *Theory of Dislocations*. Malabar (FL): Krieger; 1982.

54. Rowcliffe DJ, Hollox GE. Plastic flow and fracture of tantalum carbide and hafnium carbide at low temperatures. J Mater Sci 1971;6 (10):1261–1269.

55. Kumashiro Y, Nagai Y, Kato H, Sakuma E, Watanabe K, Misawa S. The preparation and characteristics of ZrC and TaC single crystals using an R.F. floating-zone process. J Mater Sci 1981;16 (10):2930–2933.

56. Kim C, Gottstein G, Grummon DS. Plastic flow and dislocation structures in tantalum carbide: deformation at low and intermediate homologous temperatures. Acta Metall Mater 1994;42 (7):2291–2301.

57. Rowcliffe DJ, Warren WJ. Structure and properties of tantalum carbide crystals. J Mater Sci 1970;5 (4):345–350.

58. Kumashiro Y, Nagai Y, Kato H. The vickers micro-hardness of NbC, ZrC, and TaC single crystals up to 1500°C. J Mater Sci Lett 1982;1(2):49–52.

59. Steinitz R. Mechanical properties of refractory carbides at high temperatures. In: Nucl Appl Nonfissional Ceram, edited by Boltax A, Handwerk JH. Am Nucl Soc 1966. p 75–100.

60. Martin JL, Lacour-Gayet P, Costa P. Variation de la contrainte avec la vitesse de deformation et la temperature pure le carbure de tantale entre 1200 et 2200°C. C R Acad Sci Paris 1971;272:2127–2130.

61. Martin JL, Lacour-Gayet P, Costa P. Plastic Deformation of Tantalum Carbide up to 2200°C. Research supported by the Delegation Generale a la Recherche Scientifique et Technique, ONERA; 1971.

62. Rowcliffe DL, Hollox GE. Hardness anisotropy, deformation mechanisms and brittle-to-ductile transition in carbides. J Mater Sci 1971;6 (10):1270–1276.

63. Kelly A, Rowcliffe DJ. Deformation of polycrystalline transition metal carbides. J Am Ceram Soc 1967;50 (5):253–256.

64. Steinitz R. Physical and mechanical properties of refractory compounds. In: Hausner HH, Bowman MG, editors. *Fundamentals of Refractory Compounds*. New York: Plenum Press; 1968. p 155–183.

65. Johansen HA, Cleary JG. The ductile-brittle transition in tantalum carbide. J Electrochem Soc 1966;133 (4):378–381.

66. Christian W. Some surprising features of the plastic deformation of body-centered cubic metals and alloys. Metall Trans A 1982;14 (7):1237–1256.

67. Allison C, Hoffman M, Williams WS. Electron energy loss spectroscopy of carbon in dissociated dislocations in tantalum carbide. J Appl Phys 1982;53 (10):6757–6761.

68. Martin JL, Jouffrey B. Dislocations partielles dans un carbure de tantale sous-stoechiometrique. J Phys 1968;29 (10):911–916.

69. Martin JL. Evidence of dislocation dissociation in nearly stoichiometric tantalum carbide using the weak-beam technique. J Microsc 1973;98 (2):209–213.

70. Hollox GE. Microstructure and mechanical behavior of carbides. Mater Sci Eng 1968;3 (3): 121–137.

71. Hoffman M, Williams WS. A simple model for the deformation behavior of tantalum carbide. J Am Ceram Soc 1986;69 (8):612–614.

72. Amelinckx S. Dissociations in particular structures. In: Nabarro FRN, editor. *Dislocations in Solids: Dislocations in Crystals*. Volume 2, Amsterdam: North-Holland; 1979. p 67–460.

73. Kelly A, Rowcliffe DJ. Slip in titanium carbide. Phys Status Solidi (B) 1966;14 (1):29–33.

74. Resnick R, Steinitz R, Seigle L. Determination of diffusivity of carbon in tantalum and columbium carbides by layer-growth measurements. Trans Metall Soc AIME 1965;233: 1915–1918.

75. Brizes WF. Diffusion of carbon in the carbides of tantalum. J Nucl Mater 1968;26 (2):227–231.

76. Fromm E, Gebhardt E, Roy U. Diffusion of carbon in the carbide phase of tantalum. Z Metall 1966;57:808–811.

77. Resnick R, Seigle L. Diffusion of tantalum in tantalum monocarbide. Trans Metall Soc AIME 1966;236:1732–1738.

78. Rafaja D, Lengauer W, Wiesenberger H. Non-metal diffusion coefficients for the Ta-C and Ta-N systems. Acta Mater 1998;46 (10):3477–3483.

79. Gumbsch P, Riedle J, Hartmaier A, Fischmeister HF. Controlling factors for the brittle-to-ductile transition in tungsten single crystals. Science 1998;282 (5392):1293–1295.

80. Gumbsch P. Brittle fracture and the brittle-to-ductile transition of tungsten. J Nucl Mater 2003;323 (2):304–312.

81. Argon AS, Maloof SR. Fracture of tungsten single crystals at low temperatures. Acta Metall 1966;14 (11):1463–1468.

82. Argon AS, Maloof SR. Plastic deformation of tungsten single crystals at low temperatures. Acta Metall 1966;14 (11):1449–1462.

83. Brunner D, Glebovsky V. The plastic properties of high-purity W single crystals. Mater Lett 2000;42 (5):290–296.

84. Koester RD, Moak DP. Hot hardness of selected borides, oxides and carbides to 1900°C. J Am Ceram Soc 1967;50 (6):290–296.

85. Samsonov GV, Rukina VB. Microhardness and electrical resistance of tantalum carbide in their homogeneous region. Dopovidi Akade Nauk Ukrain 1957;3:247–249.

86. Williams WS. Transition-metal carbides. Prog Solid State Chem 1971;6:57–118.

87. Hollox GE, Venables JD. The microstructure and mechanical properties of pure and boron-doped $VC_{0.85}$. Proceedings of the International Conference on Strength of Metals and Alloys; September 4–8, 1967; Tokyo, Japan

88. Ritchie RO, Lankford J. Small fatigue cracks: a statement of the problem and potential solutions. Mater Sci Eng 1986;84:11–16.

89. Ritchie RO. Mechanisms of fatigue-crack propagation in ductile and brittle solids. Int J Fract 1999;100 (1):55–83.

90. Ritchie RO, Gilbert CJ, McNaney JM. Mechanics and mechanisms of fatigue damage and crack growth in advanced materials. Int J Solids Struct 2000;37 (1):311–329.

91. DiPietro S, Wuchina E, Opeka MM, Buesking K, Matson LE, Spain J. Studies of phase stability and microstructure in the Ta-C system. ECS Trans 2007;3 (14):143–150.

92. De Leon N, Wang B, Weinberger CR, Matson LE, Thompson GB. Elevated-temperature deformation mechanisms in Ta_2C: an experimental study. Acta Mater 2013;61 (11):3905–3913.

93. Wang B, De Leon N, Weinberger CR, Thompson GB. A theoretical investigation of the slip systems of Ta_2C. Acta Mater 2013;61 (11):3914–3922.

13

TITANIUM DIBORIDE

Brahma Raju Golla[1], Twisampati Bhandari[2], Amartya
Mukhopadhyay[2], and Bikramjit Basu[3]

[1] *Metallurgical and Materials Engineering Department, National Institute of
Technology, Warangal, India*
[2] *High Temperature and Energy Materials Laboratory, Department of
Metallurgical Engineering and Material Science, IIT Bombay, Mumbai, India*
[3] *Materials Research Center, Indian Institute of Science, Bangalore, India*

13.1 INTRODUCTION

Ceramics with melting points in excess of 3000°C are classified as ultra-high-temperature ceramics (UHTCs) [1–3]. In the last two decades, different transition metal borides (especially those with melting points higher than 3000°C) have received wide attention for a range of technological applications including extreme environment aerospace applications such as wing leading edges and nose cones of hypersonic vehicles, scramjet propulsion, rocket propulsion, and atmospheric reentry vehicles [1–14]. More recently, TiB_2-based materials have been deemed to be promising UHTCs due to the attractive combination of mechanical, tribological, thermophysical, and chemical properties [1–4, 14–29]. The lower density and coefficient of thermal expansion (CTE) of TiB_2 are advantageous for applications demanding higher property to weight ratios. The properties of monolithic TiB_2 are summarized in Table 13.1 and compared to the

Ultra-High Temperature Ceramics: Materials for Extreme Environment Applications, First Edition.
Edited by William G. Fahrenholtz, Eric J. Wuchina, William E. Lee, and Yanchun Zhou.
© 2014 The American Ceramic Society. Published 2014 by John Wiley & Sons, Inc.

TABLE 13.1. Basic physical, mechanical, and oxidation properties of TiB$_2$, as well as the other popular Ultra-High-temperature transition metal borides [1–4]

Material	Crystal structure	Melting point (°C)	Density (g cm^{-3})	Co-efficient of thermal expansion (α; 10^{-6} K^{-1})	Thermal conductivity (W m^{-1} K^{-1})	Elastic modulus (GPa)	Hardness (GPa)	Fracture toughness (MPa m$^{1/2}$)	Flexural strength (MPa)	Oxidation resistance (°C)
TiB$_2$	Hex	3225	4.5	7.4	60–120	560	25–35	4–5	700–1000 (3-point)	<1200
HfB$_2$	Hex	3380	11.2	6.3	104	480	28	4	350–450 (4-point)	<1400
ZrB$_2$	Hex	3200	6.1	6.8	60–105	350	20–25	4	300–400	<1400
TaB$_2$	Hex	3040	12.5	8.2	16–35	550	25	4.5	555 (4-point)	<1400

corresponding properties of the other transition metal borides. The strength and directional nature of the primary covalent bonds are the basic factors responsible for such promising properties of the transition metal diborides. However, covalent bonding is also a cause for some of the shortcomings of these materials.

Although TiB_2 possesses excellent mechanical, physical, and chemical properties, it is difficult to densify [1–14]. Additionally, TiB_2 lacks fracture toughness [1–4, 15, 16]. Research on this material is complicated by the difficulty evaluating properties at extremely high temperatures and under harsh environments [1, 2, 5, 7, 11–13, 27–29, 20, 30]. For example, measurements of flexural creep are typically limited to less than equal to 1500°C, owing to reactions between fixtures and specimens. Recently, a novel noncontact method of mechanical testing at high temperatures was developed to study high-temperature creep and fatigue properties of electrically conducting materials [21]. Research dating from the 1960s, and continuing today, has failed to resolve issues limiting implementation of these materials [1–5].

In the recent past, considerable research has addressed the high sintering temperatures of TiB_2. Metallic and ceramic additives, new sintering techniques (conventional and advanced), and optimized processing parameters have been studied [1–4, 24–29, 20, 30–50]. However, incorporation of secondary phases can have serious drawbacks. For example, metallic additives degrade the high-temperature properties and corrosion resistance of diborides [4]. Ceramic additives like $MoSi_2$ and $TiSi_2$ have been introduced to improve densification and mechanical properties [4, 22–27]. Another important aspect is improving the fracture toughness of UHTCs by microstructural tailoring and incorporation of various reinforcements. However, composition and microstructural design should not have negative impacts over the high-temperature performance.

The present chapter starts with an introduction to processing and mechanical, physical, and chemical properties before ending with a description of applications of TiB_2. Research results concerned with some critical issues have been highlighted. Section 13.2 introduces the phase diagram, crystal structure, and bonding of transition metal diborides. Issues related to the synthesis and productions of nanopowders are presented in Section 13.3. Conventional and advanced sintering techniques used to produce dense borides are reviewed in Section 13.4 with the influence of ceramic additives on sintering temperature, densification, and properties highlighted along with the effects of processing conditions on microstructure and high-temperature mechanical properties. Section 13.5 describes the effect of temperature on hardness, elastic modulus, flexural strength, and thermal shock resistance (TSR). Thermophysical properties and oxidation behavior are reviewed in Sections 13.6 and 13.7, respectively. In Section 13.8, the tribological properties are discussed. The various applications and future prospects of TiB_2-based ceramics are discussed in Section 13.9. Finally, Section 13.10 concludes with a brief overview of the present status of TiB_2-based ceramics and raising important issues that remain unresolved despite the extensive research.

13.2 PHASE DIAGRAM, CRYSTAL STRUCTURE, AND BONDING

Attempts have been made to correlate transition metal diboride equilibrium phase diagrams to crystal chemistries [16]. The binary Ti–B phase diagram (Fig. 13.1a) shows three intermetallic phases: orthorhombic TiB, orthorhombic Ti_3B_4, and hexagonal TiB_2.

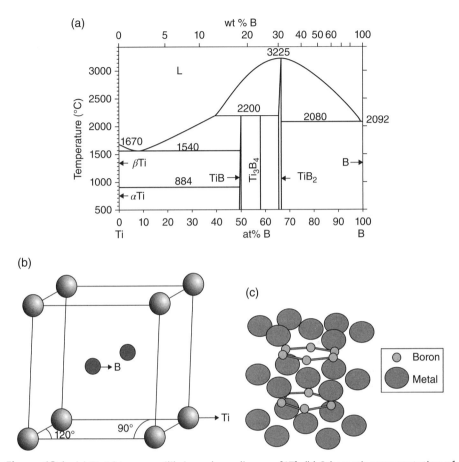

Figure 13.1. (a) Ti–B binary equilibrium phase diagram [17]. (b) Schematic representation of AlB_2 crystal structure of TiB_2 [76]. (c) Illustration of the hexagonal network of boron atoms, with metal atoms situated above and below the boron network [1].

Both TiB and TiB_2 have narrow homogeneity ranges, whereas Ti_3B_4 has a fixed stoichiometric composition. The TiB and Ti_3B_4 phases decompose peritectically at 2180 and 2200°C, respectively, whereas TiB_2 melts congruently at 3225°C. As evident from the phase diagram, the high melting point and stability make TiB_2 an important material for Ultra-High-temperature applications. Similarly, ZrB_2 has a high melting point of 3250°C, but unlike TiB_2, slight fluctuations in composition lead to the formation of liquid phase in the Zr–B system.

Transition metal diborides including TiB_2, ZrB_2, HfB_2, and TaB_2 possess primitive hexagonal crystal structures of the AlB_2 type (Fig. 13.1b and c). The hexagonal unit cell of TiB_2 belongs to the P6/mmm space group with lattice parameters: $a=b=3.029$Å, $c=3.229$Å ($\alpha=\beta=90°$, $\gamma=120°$). The layers of B atoms form two-dimensional (2D) graphite-like rings that alternate with the Ti layers (Fig. 13.1c) [1, 4]. In hexagonal close-packed (hcp) structures, the stacking sequence is ABABAB. The AA stacking sequence of Ti planes is represented in Figure 13.1b. The boron (B) atoms are located interstitially between the Ti layers, forming a strong covalently bonded hexagonal network structure. For the unit cell, a Ti atom is located at (0,0,0), while B atoms are at (1/3, 2/3, 1/2) and (2/3, 1/3, 1/2). Each Ti atom is surrounded by six equidistant Ti neighbors in its plane and 12 equidistant B neighbors (6 above and 6 below the Ti layer). Each B is surrounded by three B neighbors in its plane and by six Ti atoms (three above and three below the B layer). Bond strengths and the combination of bond types in the Ti–B system determine most of the properties. In other structures, B often forms octahedra or icosahedra consisting entirely of boron. Due to the high strength of the covalent B–B and Ti–B bonds, these intermetallic compounds are very hard and refractory. In addition to high melting points (~3225°C for TiB_2), room-temperature hardness, elastic modulus, and chemical resistance of diborides can be attributed to the crystal structure and atomic bonding. Hence, atomic bonding in TiB_2 is reflected in its high elastic modulus (~560 GPa), hardness (~32 GPa), and flexure strength (~700–1000 MPa). In addition to mechanical properties, the thermal conductivity can be of the same order of magnitude as highly conducting metallic materials, such as Cu or Al. The properties reported in papers and discussed in this chapter (Table 13.1) are for polycrystals, where the properties are isotropic, unlike those for single crystals.

13.3 SYNTHESIS OF TITANIUM DIBORIDE POWDERS

This section focuses on synthesis routes for fine TiB_2 powder, including reduction processes, reactive processes, and chemical routes. Limiting factors for commercial fabrication of TiB_2 powders are the relatively high costs of elemental boron and low production rates.

The most common synthesis route is reduction processes, which use reducing agents (carbon, boron, boron carbide or aluminum, silicon, magnesium) or a combination of them [1, 4, 16]. One solid-state reaction is borothermic reduction (Eq. 13.1), which can result in particle sizes as fine as approximately 1 µm (Fig. 13.2a); [4]

$$2TiO_2 + B_4C + 3C \rightarrow 2TiB_2 + 4CO \uparrow \qquad (13.1)$$

Nanocrystalline diboride powders (10–20 nm) have been synthesized by reacting anhydrous chlorides with sodium borohydride ($NaBH_4$) above 500°C under pressure (Fig. 13.2b): [53–56]

$$MCl_4 + 2NaBH_4 \rightarrow MB_2 + 2NaCl + 2HCl + 3H_2 \qquad (13.2)$$

Gu and coworkers [57] processed nanocrystalline TiB_2 (Fig. 13.2c) using a solvothermal reaction of metallic sodium with amorphous B and $TiCl_4$ at 400°C:

$$TiCl_4 + 2B + 4Na \rightarrow TiB_2 + 4NaCl \qquad (13.3)$$

Self-propagating high-temperature synthesis (SHS) has been used to produce nanosized TiB_2 using NaCl as a diluent [49]. As the amount of NaCl increased, the TiB_2 particle

Figure 13.2. Varying sizes and morphological features of TiB$_2$ powder, synthesized via different routes. (a) Nanocrystalline TiB$_2$ powder prepared by the reaction of TiCl$_4$ with NaBH$_4$ [55]. (b) TiB$_2$ powders synthesized via benzene–thermal reaction of metallic sodium with amorphous boron powder and TiCl$_4$ at 400°C [57]. (c) High-energy ball-milled and mechanical alloyed Ti–67B nanocrystalline TiB$_2$ powder [52]. (d and e) Sol–gel-synthesized nanocrystalline TiB$_2$ powder [115].

size decreased, reaching 26 nm for 20 wt% NaCl. The SHS method is advantageous for synthesizing TiB$_2$-based composites. High heating and cooling rates in SHS produce high defect concentrations in the resulting borides, which have been linked to improved densification. Submicron TiB$_2$ powders can also be produced by mechanical alloying of elemental Ti and B powders [44]. Similarly, high-energy ball milling (HEBM) has been used to produce nanocrystalline diboride powders (Fig. 13.2d) [51, 52]. However, the major problem associated with HEBM is contamination from the milling media and container. Furthermore, precise control and variation of stoichiometry are not possible with HEBM, unlike wet-chemical synthesis routes.

A number of laboratory-scale routes have been developed to synthesize submicron diboride powders. Researchers have also devised sol–gel routes, which are energy-efficient and cost-effective ways to produce high-purity nanocrystalline TiB_2 powder [115]. Temperatures above 1400°C produce highly faceted hexagonal TiB_2 crystals, whereas TiC is present in the form of spherical particles (Fig. 13.2e and f) [115]. Irrespective of the synthesis route, attention has to be paid to controlling powder characteristics such as initial particle size, particle size distribution, agglomeration, and purity as they influence densification.

13.4 DENSIFICATION OF TRANSITION METAL BORIDES

Monolithic transition metal diborides, such as TiB_2, are difficult to densify primarily due to a low self-diffusion coefficient, extremely high melting points, and high vapor pressures, which are due to the strong covalent bonding (Table 13.2). To obtain good densities (>95% relative density), sintering temperatures need to be at least approximately $0.7T_m$, where T_m is the melting point. Hence, extremely high sintering temperatures (≥ 2000°C) and long durations are necessary to obtain near-theoretical sinter densities for TiB_2 and other transition metal borides. However, high-temperature sintering is also associated with grain coarsening, including abnormal grain growth, which is detrimental to mechanical properties.

To improve densification of bulk TiB_2 for structural applications, sintering additives, external pressure, and advanced sintering techniques have been employed. Even though metallic additives are usually effective at reducing the sintering temperature of TiB_2, they have not been included in this chapter because their presence in as-sintered TiB_2 degrades the high-temperature properties and corrosion resistance.

13.4.1 Pressureless Sintering

Pressureless sintering (PS) is the most commonly used conventional sintering technique. PS can be used to consolidate ceramic powders to near-net shape with complex geometries of large size using standard methods [1, 48–50]. For PS, TiB_2 requires temperatures (>$0.7T_m$) on the order of 2000–2300°C to promote grain boundary and volume diffusion. Furthermore, in the case of monolithic TiB_2 and other diborides, the hexagonal crystal structure allows for anisotropic grain growth and concomitant retention of residual porosity after PS [1]. The major approaches used for improving the densities by PS of TiB_2 are the reduction of starting powders' particle size and the use of sintering additives. Sintering additives influence the microstructure of TiB_2, both in terms of phase evolution and grain size. However, among the possible additives, metallic additives are not preferred for high-temperature applications.

In addition to low self-diffusion coefficients and refractoriness, native oxide layers and other impurities are present on the surfaces of TiB_2 particles, which are introduced during synthesis or subsequent processing. Oxides and impurities increase surface diffusivity (not bulk), which promotes in particle/grain coarsening rather than densification [1, 3, 4, 61]. To achieve higher density and inhibit abnormal grain growth, oxygen content

TABLE 13.2. Summary of densification, microstructure, and mechanical properties of TiB_2 and other transition metal borides

Material composition	Sintering conditions	Sintered density (% pth)	Microstructural phases	Grain size (μm)	Hardness (GPa)	Fracture toughness (MPa m$^{1/2}$)	Flexural strength (MPa)	Ref.
Pressureless sintering								
TiB_2	1800–2275°C, 60 min	99	TiB_2	1–75	—	5	71–325	[81]
TiB_2–$MoSi_2$ (10–25 wt%)	1900°C, 120 min	83–91	TiB_2, $MoSi_2$, $TiSi_2$	2.5	—	—	—	[4]
ZrB_2	1500–1800°C, 60 min	94	ZrB_2	5–30	—	—	—	[84]
ZrB_2–$MoSi_2$ (0–10 vol%)	1800–1950°C, 60 min	86–99	ZrB_2, $MoSi_2$, (Zr, Mo)B	2.5	15–16	2.9–4.0	531–569	[85]
HfB_2–$MoSi_2$ (0–10 vol%)	1900–1950°C, 60 min	89–98	HfB_2, $MoSi_2$, HfO_2, HfC, MoB	1.6	18	4.0	405–472	[85]
Hot pressing								
TiB_2–$MoSi_2$ (0–10 wt%)	1700–1800°C, 60 min, 32 MPa	90–99	TiB_2, $MoSi_2$, Mo_5Si_3, $TiSi_2$	1.2–1.5	20–33 (HV$_{0.2}$), 20–30(HV$_5$)	4.1–5.7	268–390	[24]
TiB_2–$MoSi_2$ (0–10 wt%)	1650°C, 60 min, 30 MPa	94–99	TiB_2, $TiSi_2$, Ti_5Si_3	2.5–9	18–25	3.8–5.8	338–425	[37]
$TiBi_2$–AlN (0–10 wt%)	1800°C, 60 min, 30 MPa	87–98	TiB_2, AlN, BN, TiN, Al_2O_3	3.0	12–22	4.5–6.8	360–650	[32]
$TiBi_2$–Si_3N_4 (0–10 wt%)	1800°C, 60 min, 30 MPa	94–99	TiB_2, TiN, BN	3–7	21–27	4.4–5.8	400–810	[36]
ZrB_2–SiC (0–30 vol%)	1900°C, 45 min, 30 MPa	~100	ZrB_2, SiC, WC	3–5	23–24	3.5–5.3	565–1089	[87]
ZrB_2–15 vol% $MoSi_2$	1750°C, 20 min, 30 MPa	98	ZrB_2, $MoSi_2$, ZrO_2, SiO_2	1.8	15	—	704	[95]
TaB_2 10 vol% $MoSi_2$	1680°C, 8 min, 30 MPa	95	TaB_2, $MoSi_2$, $TaSi_2$, Ta_5Si_3, (Ta,Mo)Si_2, SiC	3.5	18	4.5	626	"

(Continued)

TABLE 13.2. (Continued)

Material composition	Sintering conditions	Sintered density (% ρth)	Microstructural phases	Grain size (μm)	Hardness (GPa)	Fracture toughness (MPa m$^{1/2}$)	Flexural strength (MPa)	Ref.
Reactive processing								
Ti–B$_4$C (3:1)	1600°C, 240 min, 41 MPa	92	TiB$_2$, TiC, Ti$_3$B$_4$	—	18–24	4–5	100	[88]
Ti–B$_4$C (4:1)	1600°C, 240 min, 41 MPa	99	TiB$_2$, TiC, Ti$_3$B$_4$	—	18–30	5.6	590	"
Ti–B$_4$C(3:1/4.8:1)	1800°C, 60 min, 35 MPa	100	TiB$_2$, TiC, TiC$_{0.5}$	—	—	8.4–12.2	454–680	[62]
TiH$_2$BN–B$_4$C (2 wt% Ni)	1850°C, 60 min, 35 MPa	97	TiB$_2$, TiCN	—	25	6.4	435	[63]
TiH$_2$–Al–BN (0–5 wt% Ni)	1850°C, 30 min, 25 MPa	98	TiB$_2$, AlN	—	18	5.1	539	[89]
TiH$_2$–SiC–Si–BN (2 wt% Ni)	2000°C, 60 min, 30 MPa	99	TiB$_2$, SiC	2	—	6.2	392	[90]
Zr–B$_4$C–Si	1900°C, 60 min, 30 MPa	98	ZrB$_2$, SiC	3–10	21	4.0	506	[91]
Zr–Si–B$_4$C	1800°C, 60 min, 20 MPa	97	ZrB$_2$, SiC, ZrC	3–10	17	5.1	—	[92]
ZrH$_2$–B$_4$C–Si	1890°C, 10 min, 30 MPa	98	ZrB$_2$, SiC	6	23	3.5	720	[93]
Hf–Si–B$_4$C	1900°C, 15 min, 50 MPa		HfB$_2$, SiC, HfC	3	19	—	770	[94]
Spark plasma sintering								
TiB$_2$	1400°C, 10 min, 30 MPa	97.6	TiB$_2$, TiB	2–3	18	5.8	—	[133]
TiB$_2$–5 wt% TiSi$_2$	1400°C, 10 min, 30 MPa (SSS)	92	TiB$_2$, TiSi$_2$, Ti$_5$Si$_3$	1–1.5	20	1.5	326	[70]

Material	Sintering conditions	Density (%)	Phases					Reference
TiB$_2$–5 wt% TiSi$_2$	1450°C, 10 min, 30 MPa (single stage sintering)	98	TiB$_2$, TiSi$_2$, Ti$_5$Si$_3$	1–1.5	25	4.3	533	[70]
TiB$_2$–5 wt% TiSi$_2$	1200°C 8 min, 30 MPa 1400°C, 2 min, 30 MPa (two stage sintering)	95	TiB$_2$, TiSi$_2$, Ti$_5$Si$_3$	1–1.5	22	2.9	468	[70]
TiB$_2$–5 wt% TiSi$_2$	1200°C, 3 min, 30 MPa 1400°C, 5 min, 30 MPa 1450°C, 2 min, 30 MPa (multi stage sintering)	98	TiB$_2$, TiSi$_2$, Ti$_5$Si$_3$	1–1.5	28	4 8	575	[70]
ZrB$_2$–15 vol%MoSi$_2$	1750°C, 7 min, 30 MPa	98	ZrB$_2$, MoSi$_2$, SiC	1.4	16	4.4	643	[96]
60 ZrB$_2$–30 ZrC–10 SiC	2100°C, 2 min, 30 MPa	99	ZrB$_2$, ZrC, SiC, ZrO$_2$	2	19	3.5	723	[96]
HfB$_2$	2220°C, 5 min, 65 MPa	80	HiB$_2$	10	7	—	—	[98]
HfB$_2$–9 vol% MoS1$_2$	1750°C, 3 min, 100 MPa	99	HfB$_2$, MoSi$_2$, HfO$_2$, SiC, SiO$_2$	1	21	5.0	690	[98]

of the starting powder must be less than 0.5 wt%, or reducing agents need to be incorporated to remove TiO_x below the coarsening temperature, which is approximately 1600°C.

Achieving high relative densities in TiB_2-based composites, such as TiB_2–TiC, by conventional PS is extremely difficult without additional powder processing steps [44]. HEBM can increase the powder reactivity, internal energy, and surface area, all of which favor densification. In the work of Wang et al., [134] nanosized TiB_2 and TiC powders obtained after 48 h of ball milling produced uniform microstructures and, more importantly, greater than 98% relative density by PS (Fig. 13.3). The pressureless sintered TiB_2 had moderate room-temperature mechanical properties, with hardness of 14–18 GPa, fracture toughness of 3–5 MPa·m$^{1/2}$, and flexural strength of 325–570 MPa (Table 13.2).

13.4.2 Hot Pressing

Application of pressure during sintering improves densities at lower temperatures compared to PS [1, 3, 48–50]. Simultaneous application of temperature and pressure (typically uniaxial from 20 to 50 MPa) is called hot pressing (HP). Hot-pressed diborides

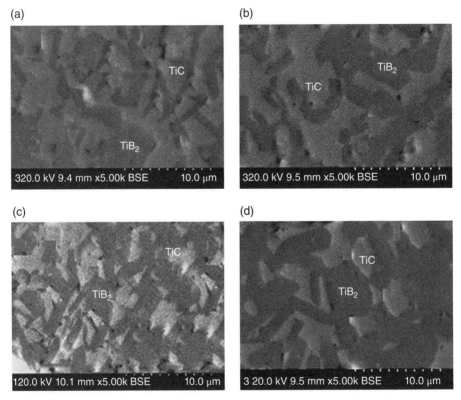

Figure 13.3. SEM images of pressureless sintered TiB_2–TiC composites sintered at (a) 1700°C without HEBM, (b) 1750°C without HEBM, (c) 1700°C with HEBM for 48 h, and (d) 1750°C with HEBM for 48 h [134].

exhibit better properties compared to PS diborides, even for similar relative densities. However, the major limitation of HP is the inability to produce complex shapes.

Fully dense TiB_2 ceramics can be obtained by HP between 1650 and 1900°C for 1 h, depending on the type and amount of additives (Table 13.2). Recent work revealed that dense (up to 98%) monolithic TiB_2 (i.e., without additives) was obtained by HP at 1800°C for 1 h [25]. The TiB_2 ceramics exhibited a maximum hardness of 32 GPa, fracture toughness of 6.8 MPa·m$^{1/2}$, and flexural strength of 810 MPa [26]. Higher relative density by HP is one of the notable advantages of TiB_2, with respect to the other transition metal diborides. For instance, HP at 1900°C only produced 87% relative density for ZrB_2 without additives [34].

Some additives react with oxides (MO_x or B_2O_3) that are present on the starting powders. Reactions can lead to liquid-phase sintering. For example, some of the reactions reported for sintering of TiB_2 with Si_3N_4 are noted below [35]:

$$3TiO_2 + Si_3N_4 \rightarrow 3TiN + 3SiO_2 + \frac{1}{2}N_2 \tag{13.4}$$

$$2B_2O_3 + Si_3N_4 \rightarrow 3BN + 3SiO_2 \tag{13.5}$$

$$TiB_2 + \frac{3}{2}N_2 \rightarrow TiN + 2BN \tag{13.6}$$

More recently, $MoSi_2$ and $TiSi_2$ additions can also enhance the sinterability of TiB_2 at lower HP temperatures of 1650–1700°C [24, 26, 36, 70, 133]. In the TiB_2– $MoSi_2$ system, sintering reactions take place during HP to produce Ti_5Si_3 and Mo_5Si_3 (Fig. 13.4). Detailed analysis revealed that surface oxides (TiO_2 and B_2O_3) reacted with $MoSi_2$ to form the secondary phases, as shown below [26]:

$$5TiO_2 + 5.714MoSi_2 \rightarrow 1.143Mo_5Si_3 + Ti_5Si_3 + 5SiO_2 \tag{13.7}$$

$$2.5Ti_3O_2 + 5MoSi_2 \rightarrow Mo_5Si_3 + 1.5Ti_5Si_3 + 2.5SiO_2 \tag{13.8}$$

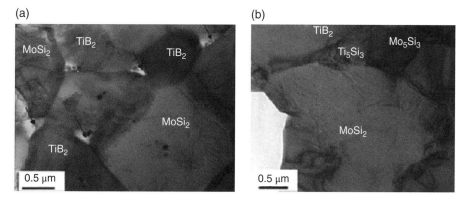

Figure 13.4. STEM bright-field images of hot-pressed TiB_2–$MoSi_2$, showing the phase assemblage (Ti_5Si_3, Mo_5Si_3) and grain structure [26].

Thermodynamic calculations revealed that Reactions 13.7 and 13.8 were feasible at typical HP temperatures (negative $\Delta G°$). Near-theoretical density (~99.6%) was obtained by HP at 1650°C for incorporation of $TiSi_2$, compared to 94.6% obtained for monolithic TiB_2 for HP under similar conditions. The beneficial effect of $TiSi_2$ was attributed to liquid-phase sintering due to its lower melting point (1540°C). Secondary phases, such as Ti_5Si_3, were observed, which was initially attributed to direct reaction (Reaction 13.9) between TiB_2 and $TiSi_2$ during HP [37];

$$3TiB_2 + 3TiSi_2 \rightarrow Ti_5Si_3 + SiB_6 \qquad (13.9)$$

However, thermodynamic analysis (Fig. 13.5) revealed that Reaction 13.9 was not thermodynamically feasible (positive $\Delta G°$) under typical HP conditions. Similar to other ceramic additives, reactions among $TiSi_2$, surface oxides (TiO_2), and carbon (present as contamination) were also explored (Reactions). Analysis revealed that only few Reactions were thermodynamically feasible (Fig. 13.5): [37]

$$2TiO_2 + TiB_2 + 2TiSi_2 \rightarrow Ti_5Si_3 + SiO_2 + B_2O_2 \qquad (13.10)$$

$$7TiO_2 + 8TiSi_2 \rightarrow 3Ti_5Si_3 + 7SiO_2 \qquad (13.11)$$

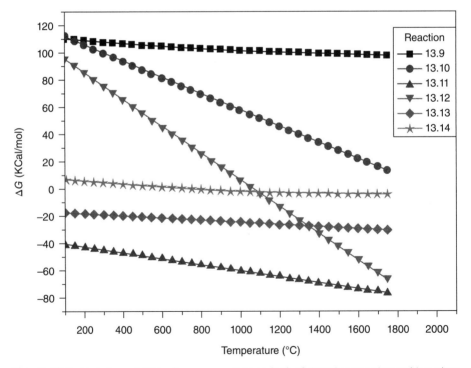

Figure 13.5. Variation of Gibbs free energy change (ΔG) of sintering reactions taking place during hot pressing of TiB_2 with $TiSi_2$ [37].

$$5TiO_2 + 5TiSi_2 + 2C \rightarrow 3Ti_5Si_3 + 7SiO_2 + 2CO(g) \qquad (13.12)$$

$$Ti_3O_2 + 2TiSi_2 \rightarrow Ti_5Si_3 + SiO_2 \qquad (13.13)$$

$$2.8TiO + 2.2TiSi_2 \rightarrow Ti_5Si_3 + 1.4SiO_2 \qquad (13.14)$$

It is plausible that ceramic additives can form secondary phases during HP due to reaction with surface oxides on TiB_2 particles or contamination and not due to direct reaction with TiB_2. Changes in microstructures, high relative densities, minimal grain growth, and formation of secondary phases during HP influence the mechanical properties including hardness (23–25 GPa), indentation toughness (4.2–5.9 MPa·m$^{1/2}$), 4-point flexural strength (338–426 MPa), and thermal conductivity (122.1 Wm^{-1}k^{-1}) of TiB_2–$TiSi_2$ composites [24, 37]. Overall, HP TiB_2-based composites exhibit better room-temperature mechanical properties and oxidation resistance compared to monolithic TiB_2 [25, 26, 36, 37].

13.4.3 Reactive Processing

Reactive processing (RP) is a two-step process during which reaction of precursor powders and densification occur *in situ*. RP is a potential route to develop second phases within a TiB_2 matrix. Compared to conventional processes, RP has advantages including lower impurity levels and higher density at lower sintering temperatures. It can also decrease formation of grain boundary phases [1, 3, 48]. Proper selection of precursors can lead to formation of transient liquid phases that facilitate densification. For instance, compositions such as 3Ti–B_4C and 4.8Ti–B_4C reach nearly full density between 1700°C and 1800°C due to liquid-phase formation but contain only TiB_2 and $TiC_{0.5}$ after processing (Fig. 13.6) [62]. Reaction-processed TiB_2 exhibited a good combination of room-temperature mechanical properties (hardness from 18 to 30 GPa, fracture toughness from 4 to 12 MPa·m$^{1/2}$, and flexural strength from 390 to 770 MPa), which was attributed to absence of secondary phases (Table 13.2). The absence of deleterious secondary phases, along with the platelet morphology of the TiB_2 grains, helped improve the high-temperature strength and fracture toughness of the TiB_2–TiC composite.

13.4.4 Spark Plasma Sintering

Spark plasma sintering (SPS) is an advanced sintering technique that utilizes simultaneous application of uniaxial pressure and pulsed direct current to densify materials more quickly than conventional processes [65–69]. SPS can produce nearly full density, even for the difficult to sinter materials such as monolithic transition metal borides, within 10 min at the sintering temperature. Enhanced densification is due to faster heating (up to 1000°C/min), breakdown of the surface films, possible formation of plasma between the powder particles, and electromigration-induced mass transport. Shorter sintering times also minimize grain growth. More detailed descriptions of the mechanistic aspects and various applications of SPS are available elsewhere [133].

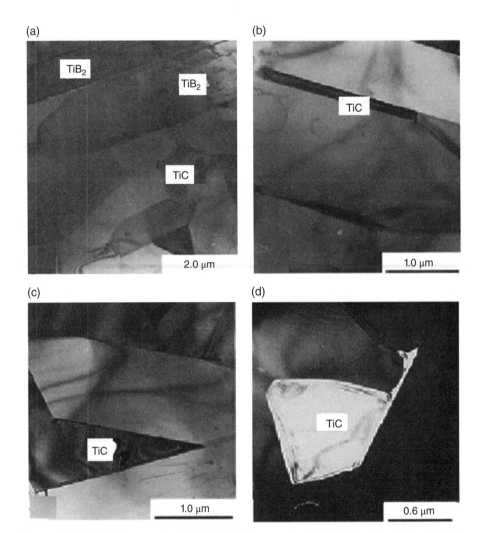

Figure 13.6. TEM micrographs of 4.8Ti–B$_4$C hot pressed at 1800°C, showing (a) well-distributed TiB$_2$ and TiC grains, (b) platelike TiC$_x$ grain, (c) triangular TiC$_x$, and (d) irregular TiC$_x$ [62].

The effects of sintering time and temperature on SPS densification of TiB$_2$–6 wt% Cu and ZrB$_2$–6 wt% Cu are shown in Figure 13.7. Relative densities of up to approximately 99% were obtained for TiB$_2$–6 wt% Cu at 1500°C for 15 min [34]. A significant increase in density occurred between 1200 and 1500°C. Microstructural features of SPS TiB$_2$–6 wt% Cu are presented in Figure 13.8. The densification rate was comparatively higher for the TiB$_2$-based composite than a ZrB$_2$-based composite, which reached approximately 95% density under identical conditions. This is in agreement with poorer densification obtained with ZrB$_2$ by HP. Similar to HP, ceramic additives such as MoSi$_2$ result in high relative density (~98%) at lower SPS temperatures such as 1400°C [34]. TiB$_2$-based materials produced by SPS were had finer

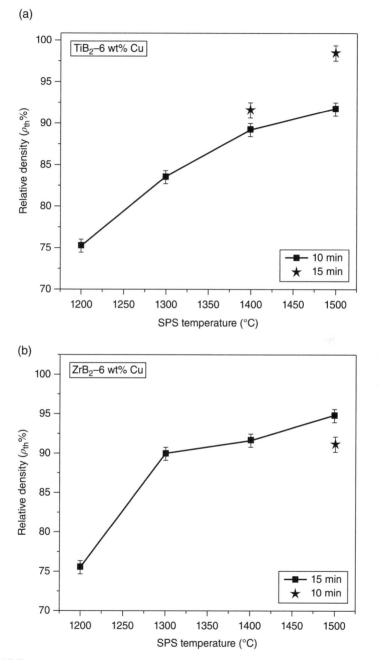

Figure 13.7. Variations of sinter densities of (a) TiB$_2$–6 wt% Cu and (b) ZrB$_2$–6 wt% Cu, processed using SPS [34].

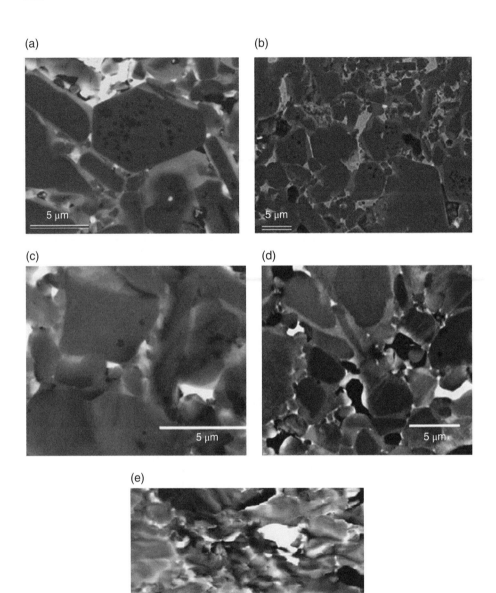

Figure 13.8. (a, b) SEM micrographs of TiB$_2$–6wt%Cu, processed using SPS (c) fracture surface showing a small amount of porosity for the same. By contrast, for ZrB$_2$–6wt% Cu composite, prepared under identical conditions, (d) irregular morphology and (e) (fractured surface) greater porosity can be observed [34].

microstructure scale and concomitantly excellent properties (Table 13.2). Interestingly, the fracture surface of the TiB$_2$-based composite showed predominantly intergranular fracture (Fig. 13.8), whereas ZrB$_2$ exhibited mixed mode fracture. Replacement of MoSi$_2$ with TiSi$_2$ resulted in improved densification behavior [70]. Also, multistage

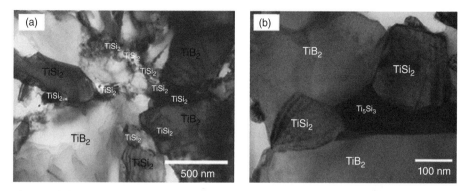

<u>Figure 13.9.</u> (a) Lower and (b) higher magnification TEM micrographs of multi-stage spark plasma sintered TiB_2–5%$TiSi_2$ (at 1450°C) [70].

SPS (MS-SPS, which involves constant temperature holds before reaching the final temperature) [34, 70] led to more uniform, finer microstructures for TiB_2–$TiSi_2$ (Fig. 13.9) [70].

Despite the advantages of SPS, recent work showed possibility of phase instability during SPS of monolithic TiB_2, which resulted in the formation of TiB that was not observed for HP of the same powder [133]. The presence of TiB after SPS resulted in inferior hardness, even with slightly higher relative density. Reactions of TiB_2 with either residual oxygen present in the SPS chamber or with a surface oxide layer (TiO_2) were thermodynamically feasible under the SPS conditions. The corresponding reactions are noted in the following text, and the change in Gibbs free energy change is presented in Figure 13.10. It must be noted here that even though the optical pyrometer shows 1400°C during SPS, the actual temperature within the die could be higher by 200–300°C, which renders Reaction 13.15 also feasible [133]:

$$4TiB_2 + 3O_2 \rightarrow 4TiB + 2B_2O_3 \tag{13.15}$$

$$TiB_2 + TiO_2 + 2C \rightarrow 2TiB + 2CO(g) \tag{13.16}$$

13.5 MECHANICAL PROPERTIES AT AMBIENT AND ELEVATED TEMPERATURES

Materials with favorable combinations of mechanical properties and good chemical stability are needed for Ultra-High-temperature structural applications. Properties, such as hardness and stiffness, are important for dimensional stability at elevated temperatures. TiB_2-based materials have superior room-temperature mechanical properties, such as Young's modulus of 560 GPa, hardness of up to 32 GPa, and flexural strengths as high as 1000 MPa (Table 13.1). However, the extent to which such properties are maintained at elevated temperatures is probably more important. The elevated-temperature properties of polycrystalline TiB_2 strongly depend on microstructural features such as grain size, composition, grain boundary segregation, and overall phase

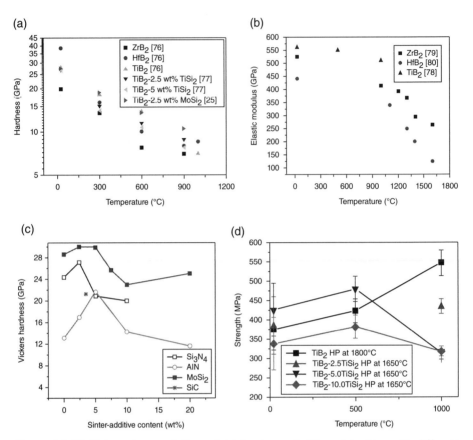

Figure 13.10. Temperature variations of (a) hardness of different borides with different additives, (b) elastic modulus, (c) room temperature hardness of TiB_2-based ceramics [24], and (d) flexural strength of monolithic TiB_2 and TiB_2–$TiSi_2$ [75].

assemblage, which are dependent on the processing route. The following discussion highlights results on elevated-temperature hardness and strength of monolithic TiB_2 and TiB_2-based composites.

13.5.1 Hardness

Hardness is a basic mechanical property, typically measured by indentation. Temperature dependence of hardness is a standard analytical tool to evaluate high-temperature mechanical behavior. The varying indentation behavior can be correlated to service temperature, microstructure, porosity, and density [71–76]. Hot hardness values for TiB_2 and other diborides are presented in Figure 13.11a. The hardness of diborides, like most materials, decreases with increasing temperature. At high temperatures, softening enhances deformation during indentation, which is responsible for decrease in hardness.

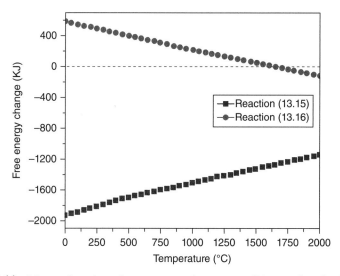

Figure 13.11. ΔG as a function of temperature for two possible reactions leading to the formation of TiB during SPS of TiB$_2$. The dotted line shows $\Delta G = 0$ [133].

The hardness of TiB$_2$-based materials can be improved with ceramic additives, such as TiSi$_2$ and MoSi$_2$, albeit only up to an optimal content. For example, the reported increase in hardness between monolithic TiB$_2$ (7.2 GPa) and TiB$_2$–TiSi$_2$ (8.9 GPa) at 900°C is primarily due to higher relative density. Interestingly, TiB$_2$ with 2.5 wt% TiSi$_2$ has higher hardness due to the formation of a smaller amount of soft second phase and higher relative density. Even larger improvements in room- and elevated-temperature hardness of TiB$_2$–MoSi$_2$/TiSi$_2$ ceramics can be obtained by MS-SPS due to finer grain sizes and enhanced homogenity [70]. The reduction in Vickers hardness of TiB$_2$–TiSi$_2$ composite with temperature is presented in Figure 13.11a [24]. Comparison of the hardness of TiB$_2$-based ceramics sintered with different additives is presented in Figure 13.11c. Among diboride ceramics, HfB$_2$ exhibits the highest room-temperature hardness of 38 GPa and retains a maximum hardness of 8.5 GPa up to 1000°C [74]. This further highlights the importance of characterization of high-temperature hardness, which might or might not vary with composition in line with room-temperature hardness. Table 13.3 summarizes the high-temperature hardness of the various TiB$_2$-based ceramics.

13.5.2 Elastic Modulus

Strain measurement, resonance, and ultrasound are three methods that are commonly used to determine elastic modulus [78], although indentation has also been used to compute elastic modulus from the initial slope of unloading curve [24]. Elastic modulus is an intrinsic material property but depends strongly on porosity and composition. Among the different transition metal diborides, TiB$_2$ exhibits the highest room-temperature elastic modulus (560 GPa), which only decreases slightly, approximately 510 GPa at 1000°C [113]. The typical trend for decreasing elastic modulus with increasing

TABLE 13.3. Summary of variation of hardness with respect to temperature for TiB$_2$-based ceramics with different compositions and processing conditions

Material composition by wt%	Processing route	Relative density (% ρth)	Hardness at GPa					Ref.
			25°C	200°C	300°C	600°C	900°C	
Monolithic TiB$_2$	HP at 1800°C	97.8	25.3	—	13.9	10.8	8.2	[75]
Monolithic TiB$_2$	HP at 1650°C	94.4	21.2	—	13.8	9.8	7.2	[75]
TiB$_2$–2.5TiSi$_2$	HP at 1650°C	98.8	27.1	—	15.0	11.5	8.9	[75]
TiB$_2$–5.0TiSi$_2$	"	99.6	26.8	—	14.1	10.6	7.8	[75]
TiB$_2$–10TiSi$_2$	"	99.6	23.7	—	13.1	9.8	8.0	[75]
TiB$_2$–2.5MoSi$_2$	HP at 1700°C	99.0	27.6	—	18.9	13.6	10.5	[26]
TiB$_2$–5.0TiSi$_2$	MSS sintering	98.9	31.4	—	—	—	—	[70]
TiB$_2$–20 vol% (Fe+Cr+Ni+Fe$_2$B)	HIP at 1500°C	99.2	20.4	15.6	10.3	—	6.4	[157]
TiB$_2$–7 wt% Ti-5 vol% (Fe+Cr+Ni)	HIP at 1500°C	98.8	16.1	11.3	7.3	—	4.6	[157]

temperature (above 900°C) for diborides is presented in Figure 13.11b [76]. Even though all of the diborides follow similar trends, the elastic modulus of TiB_2 is higher than ZrB_2 or HfB_2 at all temperatures. The elastic modulus of TiB_2–$MoSi_2$ (460 GPa) decreased with the addition of $MoSi_2$ due to the lower stiffness of $MoSi_2$ and the other secondary phases compared to TiB_2. Effects of $MoSi_2$ or other additives on elevated-temperature elastic modulus have not been investigated comprehensively.

13.5.3 Fracture Strength

For polycrystalline ceramics including transition metal diborides, strength typically refers to flexural (fracture) strength and is usually measured using 3-point or 4-point bending. Elevated-temperature strength is sensitive to microstructural phase assemblage, grain boundary phases, and additive content [1, 3, 4, 75]. Hence, for Ultra-High-temperature applications, it is necessary to ascertain whether additives are detrimental to elevated-temperature strength.

Table 13.4 compares the room- and elevated-temperature strengths of TiB_2 with other transition metal borides, in terms of composition (type and amount of additives), processing conditions, densification, and test conditions (type of test, sample dimensions, environment, etc.). The composition of TiB_2-based ceramics (i.e., appropriate additives in optimal amounts) influences elevated-temperature strength. For TiB_2, high relative density (>99%), high hardness (23–31 GPa), moderate fracture toughness (~4–6 MPa·m$^{1/2}$), and flexural strength (>500 GPa) can be obtained through small additions (1–2 wt%) of metallic binders, such as Fe, Cr, or Ni. However, metallic additives are detrimental to strength retention at elevated temperatures. For monolithic TiB_2, the strength of PS TiB_2 increases to a maximum value of approximately 470 MPa at 1200°C, whereas hot-pressed TiB_2 exhibits a higher strength of approximately 550 MPa at around 1000°C. In general, the trend for TiB_2-based systems (Table 13.4) is that flexural strength is higher at approximately 500°C compared to room temperature due to relief of internal stresses generated during cooling from the sintering temperatures [75]. The optimal amount of ceramic additives, about 2.5 wt% for $MoSi_2$ and $TiSi_2$ [24, 26, 27, 70], does not degrade the strength at elevated temperatures, in contrast to metallic additives. Greater amounts of the same additives lead to deterioration of strength compared to monolithic TiB_2 by more than a factor of 2. At 1000°C, degradation of strength occurs (Fig. 13.11c) for TiB_2-based materials containing 5 wt% or more $TiSi_2$ due to grain pullout, grain boundary sliding, and microcracking [75]. For other diborides, ZrB_2-based has better mechanical properties than TiB_2 at room temperature.

13.5.4 TSR

TSR must be considered during the design of TiB_2-based ceramics for applications where materials are exposed rapid temperature changes [80]. Rapid heating or cooling cycles generate transient tensile stresses due to instantaneous temperature gradients between the surface and inner parts of a component. Fracture occurs when thermal stresses exceed the strength. Thermal shock behavior is influenced by parameters like elastic modulus, flexural strength, fracture toughness, thermal expansion, thermal conductivity, and heat transfer rate [1, 80, 81].

TABLE 13.4. The effects of temperature on strength of TiB$_2$ and other transition metal diborides [4-point (4-P) and 3-point (3-P) flexural strength]

Material composition	Processing details °C, min, MPa	Relative density (% pth)	Flexural/bend test conditions	Flexural strength (MPa)				Ref.
				RT	1000°C	1200°C	1500°C	
TiB$_2$	PS, 1900, 60	>95	3-P, Argon 3.8×6.9×38 mm³	310	370	405	—	[83]
TiB$_2$	PS, 2100, 60	99	"	290	390	400	—	[76]
TiB$_2$	—	99.5	—	400	459	471	—	[25]
TiB$_2$–0 wt% MoSi$_2$	HP, 1800, 60, 30	98	4-P, Air, 3×4×40 mm³	387±52	546±33	—	—	"
TiB$_2$–2.5 wt% MoSi$_2$	HP, 1700, 60, 30	99	"	391±31	503±27	—	—	
TiB$_2$–10.0 wt% MoSi$_2$	HP, 1700, 60, 30	97	"	268±70	261±30	—	—	
TiB$_2$–2.5 wt% TiSi$_2$	HP, 1650, 60, 30	99	"	381±74	433±17	—	—	[75]
TaB$_2$–10.0 vol% MoSi$_2$	HP, 1680, 8, 30	90	"	626±11	—	219±20	114±5	[79]
ZrB$_2$	HP, 2150, 60, 32	99	4-P, Air, 3×4×45 mm³	381±41	399±37	392±37	—	[77]
ZrB$_2$–5.0 vol% MoSi$_2$	PS, 1800, 60	96	4-P, Air, 2×2.5×20 mm³	569±54	—	533±87	488±46	[99]
HfB$_2$–5.0 vol% MoSi$_2$	PS, 1950, 60	97	"	472±10	—	501±89	486±19	[99]

The common parameters used to describe TSR can be divided into two categories: (i) crack initiation (R, R′, and R″) and (ii) crack propagation (R‴ and R⁗). Resistance to crack initiation is maximized by increasing flexural strength and decreasing elastic modulus, whereas crack propagation resistance can be increased by increasing elastic modulus and decreasing flexural strength. TiB_2-based materials retain strengths up to high temperatures, possess appreciable Young's modulus (~560 GPa), and have high thermal conductivity ($k \sim 60\text{--}120\,Wm^{-1}K^{-1}$), which render them excellent materials for resisting thermal shock. TSR can be further improved by making composites, with tailored microstructures that can promote crack tip blunting, crack deflection, and crack bridging.

13.6 PHYSICAL PROPERTIES AND OXIDATION RESISTANCE

For a material to be useful under extreme conditions, physical properties and oxidation resistance are equally important to chemical and mechanical properties. In this respect, TiB_2 possesses attractive physical properties compared to the other transition metal diborides. For instance, TiB_2 has lower density ($4.53\,g/cm^3$) along with high thermal conductivity ($60\text{--}120\,Wm^{-1}K^{-1}$) and melting point ($3225°C$). By contrast, even though HfB_2 possesses a higher melting temperature of approximately $3380°C$, along with favorable mechanical, tribological, and oxidation properties, its higher density ($11.2\,g/cm^3$) may limit its usage. The following subsections discuss various physical properties and oxidation resistance of TiB_2 and compare them with other transition metal diborides.

13.6.1 CTE and Thermal Conductivity

Lower CTEs lead to less problems with dimensional changes and stress development during heating and cooling cycles. The CTE of TiB_2, as measured by dilatometry, is approximately $6.5 \times 10^{-6}/°C$ [109], which is less than those of other diborides, such as ZrB_2 (~$7 \times 10^{-6}/°C$) [109]. The temperature dependence of CTE is an important consideration (Fig. 13.12a) [37, 158], especially when materials are subjected to a wide temperature window. A good correlation between CTE and temperature has been reported for TiB_2 [109]. Incorporation of $TiSi_2$ into TiB_2 ceramics produces a secondary phase (Ti_5Si_3) that has a larger CTE (~$10.8 \times 10^{-6}/°C$) than TiB_2 [37]. This leads to development of compressive residual stress in the TiB_2 matrix, which can improve fracture toughness [37, 70]. Murthy *et al.* [158] reported that CTE varies with temperature and that TiB_2–$TiSi_2$ possesses a lower CTE (~$5.8 \times 10^{-6}/°C$) than monolithic TiB_2, possibly due to a new phase that hinders dimensional changes. During use, temperature gradients can develop between the surfaces and interior, which can lead to the development of residual stresses. Figure 13.12b and c show the position of peak tensile stress (as indicated with red color) and the stress gradient for a cross section of rocket nozzle during the early stages of firing, as determined using finite element modeling [107]. The surface temperature of the components can be

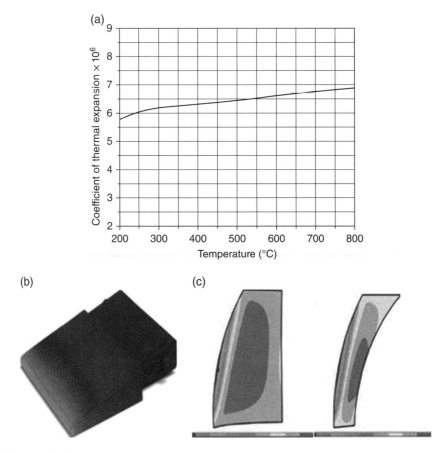

Figure 13.12. (a) The variation of CTE as a function of temperature for TiB$_2$–TiSi$_2$ [158]. (b) Schematic representation of sharp wing leading-edge component [110]. (c) Finite element modeling results showing the generation of stress gradient across the cross section in rocket motor nozzles during the early stages of firing [107]. Red (or darker shade) signifies position of the peak tensile stress.

balanced through heat conduction through the materials and radiation to the atmosphere (Fig. 13.13a), as expressed by Equation 13.17 [121]:

$$Q_{conv} = Q_{cond} + Q_{rad} \qquad (13.17)$$

where Q_{conv} is the convective heat flux to the surface and Q_{cond} and Q_{rad} are the conduction and radiation heat fluxes out of the surface.

The high thermal conductivity of TiB$_2$ (60–120 Wm^{-1}K^{-1}) helps transport heat to equilibrate temperature and concomitantly reduce thermal gradients and stresses. As compared to stoichiometric TiB$_2$, slight nonstoichiometry and the resulting defects (such as vacancies) in TaB$_2$ and ZrB$_2$ lead to lower thermal conductivities (16–35 Wm^{-1}K^{-1} and 60–110 Wm^{-1}K^{-1}, respectively) [1–4] due to phonon scattering [4, 104].

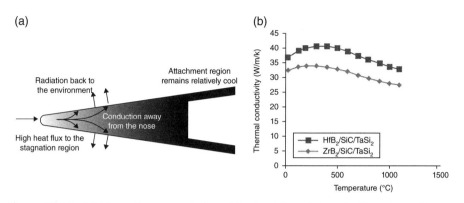

Figure 13.13. (a) Schematic representation of the heat flux of sharp leading edges of hypersonic vehicles [121]. (b) Plot showing dependence of thermal conductivity with temperature for different diboride-based composites [109].

The addition of 2.5 wt% $MoSi_2$ increases thermal conductivity from approximately $60\,Wm^{-1}K^{-1}$ to approximately $75\,Wm^{-1}K^{-1}$ (Table 13.5) [113]. Hence, $MoSi_2$ not only improves densification and mechanical properties but also physical properties of TiB_2. Since thermal transport depends on scattering mechanisms, the addition of 2.5 wt% $MoSi_2$ reduces phonon scattering compared to monolithic TiB_2 (Fig. 13.14). However, higher amounts of $MoSi_2$ (up to 10 wt%) reduce thermal conductivity due to lower thermal conductivity of $MoSi_2$ and higher thermal resistance at TiB_2–$MoSi_2$ grain boundaries. Additionally, higher volume fractions of secondary phases such as Ti_3Si_5 and Mo_5Si_3 also affect thermal conductivity for higher $MoSi_2$ contents. Even though advanced sintering techniques, such as SPS, can improve densification and reduce grain size, which are beneficial for mechanical properties, the presence of more grain boundaries might negatively affect thermal conductivity, irrespective of composition [47, 120]. The negative effect of increased grain boundary area would compete with enhanced density, and the combined effect would dictate thermal conductivity.

13.6.2 Effects of Physical Properties on TSR

This subsection attempts to establish a correlation between TSR and physical properties. Thermal shock arises during applications involving cycling and sudden variations in temperature and loading. Rather than only depicting the effects of mechanical properties on TSR (Eq. 13.18), the effects of physical parameters, such as thermal conductivity and CTE (α) on the TSR, can be depicted using the following relation [122]:

$$\text{TSR} = \frac{\sigma_f k}{E\alpha} \qquad (13.18)$$

where σ_f is fracture strength and E is Young's modulus. As described previously, the high thermal conductivity of TiB_2 (up to $120\,Wm^{-1}K^{-1}$) can readily dissipate heat from hotter to cooler regions while at the same time the low CTE would minimize dimensional

TABLE 13.5. Summary of thermal conductivity as a function of temperature for TiB_2 ceramics

Materials composition	Processing route	Thermal conductivity(TC) at Wm^{-1} K^{-1}						Ref.
		25°C	200°C	400°C	600°C	800°C		
TiB_2–0% $MoSi_2$	HP at 1700°C	68.2	68.7	67.2	67.5	67.3		[113]
TiB_2–2.5% $MoSi_2$	HP at 1700°C	77.0	79.1	76.7	76.2	75.9		[113]
TiB_2–10% $MoSi_2$	HP at 1700°C	64.9	66.4	65.5	65.7	65.9		[113]
TiB_2	—	96.0	—	82.9	—	79.8		[76]
TiB_2–10% $TiSi_2$	HP at 1650°C	122.1	108.6	99.8	96.7	95.2		[155]
TiB_2–Cu	Self propagating reaction(1700°C)	131.6	85.9	81.8	79.3	76.9		[128]
ZrB_2–0% SiC	HP at 1800°C	—	55.1	53.0	—	50.2		[3]
ZrB_2–5% SiC	HP at 1800°C	—	100.0	95.1	—	87.4		[3]
ZrB_2–10% SiC*	HP at 1800°C	—	105.1	104.0	—	102.1		[3]
HfB_2–0 SiC	HP at 2000°C	107.0	100.0	96.1	79.0	—		[112]
HfB_2–20 SiC	HP at 2000°C	78.0	74.0	70.1	69	—		[112]

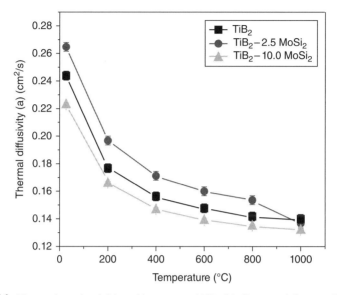

Figure 13.14. Thermal conductivities of hot-pressed TiB$_2$–MoSi$_2$, containing varying amounts of MoSi$_2$ [113].

changes during thermal cycling. The combined effect reduces the mismatch strain and, therefore, improves resistance to damage under sudden temperature fluctuations [121]. Introduction of Cu in particulate form and SiC in the form of whiskers has been reported to further improve the TSR of TiB$_2$-based composites [132].

13.7 OXIDATION RESISTANCE

The oxidation rate of TiB$_2$ depends on oxygen partial pressure, temperature, and composition (i.e., amount of additives/reinforcements). For monolithic TiB$_2$, oxidation begins below 400°C by initially forming TiBO$_3$, which subsequently oxidized to form B$_2$O$_3$ by Reactions 13.19 and 13.20 [123]:

$$4TiB_2 + 9O_2 \rightarrow 4TiBO_3 + 2B_2O_3 \quad \text{at} \sim 400°C \text{ and } 0.05 \text{ ppm } O_2 \quad (13.19)$$

$$4TiBO_3 + O_2 \rightarrow 4TiO_2 + 2B_2O_3 \quad \text{at } 400 - 900°C \text{ and } 10 \text{ ppm } O_2 \quad (13.20)$$

Even though B$_2$O$_3$ initially forms a protective layer, it evaporates above 900°C, leaving behind porous TiO$_2$. Thus, a compact, protective oxide scale is not present on unreinforced TiB$_2$ under application conditions.

Poor oxidation resistance of monolithic TiB$_2$, especially above 900°C, has been confirmed by observing the surface morphology of oxide layers formed at different temperatures (Figs. 13.15 and 13.16). Crystalline TiO$_2$ and amorphous B$_2$O$_3$ are present after exposure at temperatures between 800 and 900°C (Fig. 13.16a and b). However, above

(a)

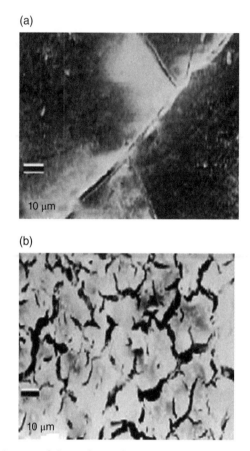

(b)

Figure 13.15. SEM images of the surfaces of monolithic TiB_2 after oxidation at 850°C for (a) 1 h and (b) 4 h [4].

900°C, B_2O_3 evaporates and the oxide scale becomes highly textured crystalline rutile (TiO_2) at long exposures (~30 h) at 1100°C (Figs. 13.17a and b, and 13.18). Upon further exposure, the TiO_2 layer thickens, which leads to intergranular cracking due to volume expansion, making the oxide layer nonprotective [4]. The mass change during oxidation of monolithic TiB_2 is usually parabolic up to 900°C since B_2O_3 acts as a protective layer to limit diffusion of oxygen to the boride surface. At higher temperatures, the oxide scale becomes nonprotective and oxidation kinetics become linear (Fig. 13.18) [4, 50].

The poor oxidation resistance of monolithic TiB_2 at elevated temperatures can be improved by adding silica-forming agents such as Si or $MoSi_2$, which can also act as sintering aids. These additives lead to formation of borosilicate glasses, which are more stable at elevated temperatures and form more effective diffusion barriers. For example, adding 2.5 wt% Si_3N_4 to TiB_2 maintains parabolic oxidation kinetics up to 1200°C.

Another route to improve oxidation resistance of TiB_2 is to add other transition metal borides or silicides, which form complex immiscible glasses that reduce the

(a)   (b)

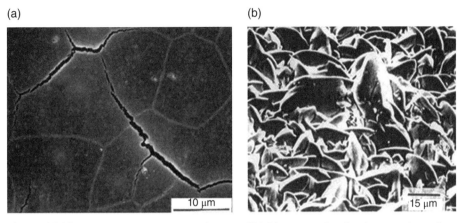

Figure 13.16. Internal cracking in oxide layer and highly textured rutile phase on the surface of monolithic TiB_2 oxidized at 800°C for (a) 10h and 1000°C for (b) 30h [4].

Figure 13.17. Cross-sectional SEM images showing the surface and subsurface layers after oxidation at (a) 800°C for 10h and (b) 1000°C for 2h. Note the absence of B_2O_3 at high temperature [128].

oxygen diffusion rates. Such additions increase the service life by up to a factor of 10 [126, 127]. With increasing WSi_2 content, rate of weight gain due to oxidation decreases (Fig. 13.18). Mass gain kinetics is still linear for additions up to 5 wt% WSi_2 but becomes parabolic with additions of 10 wt% or more.

One of the more common additives that improves oxidation resistance of borides is SiC [127]. Incorporation of SiC in TiB_2 reduces mass gain significantly compared to monolithic TiB_2, at least up to 1200°C [131]. By comparison, the linear kinetics of pure ZrB_2 shifted to parabolic for ZrB_2–SiC, even at temperatures as high as approximately 1700°C [126–130]. Similar to WSi_2, SiC content also influences oxidation behavior. Contents of SiC in the range of 20–25 vol% appear to be optimum up to approximately 2000°C [78]. The overall reactions taking place during oxidation of transition metal diborides (MB_2) and SiC-reinforced MB_2 can be summarized as follows [126]:

$$MB_2 + 2.5O_2 \rightarrow MO_2 + B_2O_3(l) \qquad (13.21)$$

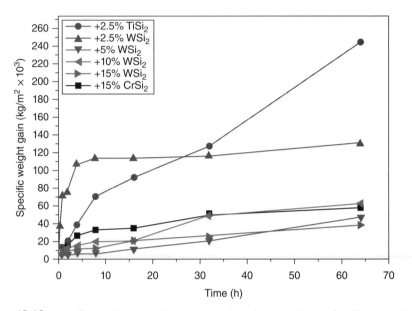

Figure 13.18. Specific weight gain with respect to time during isothermal oxidation at 850°C for TiB_2–WSi_2 composites with different WSi_2 loadings [131].

$$B_2O_3(l) \rightarrow B_2O_3(g) \text{ at above } 1100°C \tag{13.22}$$

$$SiC + 1.5O_2 \rightarrow SiO_2(l) + CO(g) \tag{13.23}$$

$$B_2O_3(l) + SiO_2 \rightarrow B_2SiO_3 + O_2 \text{ above } 1100°C \tag{13.24}$$

The addition of silica-forming agents such as SiC or WSi_2 is more beneficial for the oxidation resistance of ZrB_2 than TiB_2. This may be due to disintegration and cracking of the surface oxide layer, which is more prevalent for TiB_2. For ZrB_2–SiC, a borosilicate glass forms that effectively covers and passivates the surface. The overall oxide scale is comprised of different layers on top of the ZrB_2–SiC bulk including (i) an outer borosilicate glassy layer and (ii) an inner layer composed of ZrO_2 [128]. Presence of ZrO_2 in the surface layer is believed to add to the mechanical integrity and stability of the overall scale on ZrB_2–SiC, but this is less effective in TiB_2–SiC. However, the presence of SiO_2 improves the oxidation resistance of diborides (Fig. 13.19a and b).

13.8 TRIBOLOGICAL PROPERTIES

TiB_2-based materials are candidates for tribological applications because of high hardness, elastic modulus, and wear resistance [4, 135]. Tribology involves interacting surfaces in relative motion, commonly known as **friction**, **wear**, and **lubrication** science [136]. Studies on wear behavior of hard materials like ceramics, cermets, and

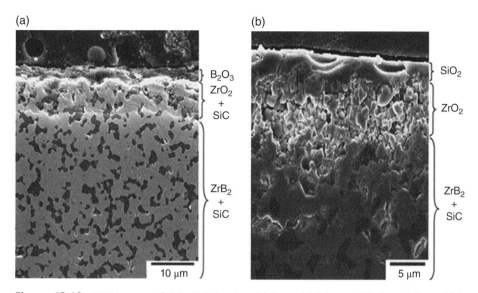

Figure 13.19. SEM images of ZrB$_2$–SiC showing (a) 2 μm thick layer of B$_2$O$_3$ and 6 μm thick ZrO$_2$-rich layer formed on the surface of ZrB$_2$–SiC after exposure to air at 1000°C for 30 min and (b) SiO$_2$ outer layer and second layer composed of ZrO$_2$ above of ZrB$_2$–SiC layer after exposure to air at 1200°C for 30 min [128].

ceramic composites have been widely reported, but limited research has been reported on wear behavior of transition metal borides [4, 135–142]. Tribological performance depends on parameters including fracture toughness, hardness, microstructural characteristics, coefficient of friction (COF), and the test environment. In general, the wear of brittle ceramics involves fracture or grain pullout along with tribochemical reactions, which dominate tribological interactions in sliding couples involving nonoxide ceramics. The tribolayer chemistry and its protectiveness under dynamic sliding conditions are important factors in the wear of ceramics. In particular, TiB$_2$-containing materials have great potential for triboapplications for two reasons: high hardness and the lubrication efficiency of the boric acid film produced at the tribological interface [4].

13.8.1 Wear Properties of Bulk TiB$_2$-Based Ceramics

The friction and wear behavior of monolithic TiB$_2$ and TiB$_2$ containing cermets and ceramic composites were studied by fretting [22, 23, 143–148]. Fretting is small-amplitude (below 300 μm) relative reciprocal displacement and sliding motion. Basically, fretting is a form of adhesive or abrasive wear, where the normal load causes adhesion between asperities and oscillatory movement causes rupture, resulting in the formation of wear debris [23, 146, 148, 149]. The other damaging aspect is fretting fatigue. These forms of damage can arise in any assembly of engineering components/machinery, if a source of vibration is present. Among the various modes, mode I fretting wear has been studied widely for various engineering materials [23, 139, 141]. Basu and coworkers performed fretting tests on TiB$_2$ with 5 vol% SiC and TiB$_2$ with 16 vol% Ni$_3$(Al, Ti) binder against different counterbody materials (alumina, steel, tungsten carbide, cobalt, and SiAlON) under identical fretting conditions (Fig. 13.20a) [135, 143, 144]. Depending on the counterfaces, the COF varies from 0.3 to 0.7. An interesting observation is that a combination of lower COF (up to 0.5) and lower wear rate (in the order of 10^{-5} mm³/N·m) can be obtained during unlubricated fretting of TiB_2 against alumina or SiAlON counterfaces. Also, both COF and wear rate are lower in the case of water or oil lubrication, as compared with unlubricated fretting (Fig. 13.20a). Fretting in water resulted tribochemical reactions coupled with mild abrasion, which played a major role in the wear behavior. In the case of the fretting tests in paraffin oil, the worn surfaces on all of the flats and balls were covered by an adherent carbon-rich (graphite) tribochemical lubricating layer. Tribodegradation of the paraffin oil was the major source for the carbon deposition.

The wear behavior of monolithic TiB_2 and other structural ceramics against different counterbodies is summarized in Figure 13.20b. Wear data were based on unlubricated sliding tests (plate-on-plate configuration) at rotation speed of 220 revolutions/min with sliding speeds of 0.2 m/s and a load of 214 N [147]. The material removal rate of the tribocouples falls under the mild wear regime and is approximately 10^{-5} mm³/N·m. A comparison of the wear data of different ceramics reveals that wear resistance of mullite is relatively high when compared to TiB_2, SiC, or Al_2O_3 ceramics [147]. Such differences in the wear rates are attributed to hardness and the nature of the tribofilm formed during the test. TiB_2 ceramics exhibit good wear resistance against TiB_2, SiC, or Al_2O_3 counterbodies. Wasche *et al.* [148] investigated the influence of relative humidity on friction and wear behavior of PS TiB_2 against SiC and Al_2O_3 balls under unlubricated conditions at room temperature. TiB_2 exhibited low wear rates ($<1 \times 10^{-6}$ mm³/N·m), that is, high wear resistance.

As a case study, we present work on fretting wear of TiB_2 and TiB_2–5 wt% $TiSi_2$ against bearing grade steel and WC–6 wt% Co [23]. The TiB_2–5 wt% $TiSi_2$ ceramic showed better wear properties compared to monolithic TiB_2 as it exhibited very good mechanical properties and high density. This study also revealed that TiB_2 ceramics exhibited negative wear behavior when fretted against steel but normal wear against WC–Co. The wear volumes of TiB_2 ceramics were estimated using laser surface profilometry. A typical 3D profile of a worn surface of TiB_2–5 wt% $TiSi_2$ is shown in Figure 13.21, revealing peaks instead of troughs. This indicates negative wear for TiB_2, when fretted against steel, indicating that 2D profiles, if taken at various places on worn surfaces, showed hills instead of valleys that are commonly observed for worn surfaces. In contrast, the fretted surfaces on TiB_2 ceramics, after testing against WC–Co, show regular 3D or 2D profiles, as typically observed for worn material (Fig. 13.21c and d).

Based on the SEM–EDS of fretted TiB_2 surfaces, the oxide layer formed mainly from steel and it transformed to TiB_2 flat (Fig. 13.22a and b). Possible tribochemical reactions involved for TiB_2–steel system are as follows:

$$2TiB_2 + 5O_2 \rightarrow 2TiO_2 + 2B_2O_3 \tag{13.25}$$

$$Fe + (1/2)O_2 \rightarrow FeO \tag{13.26}$$

$$2Fe + (3/2)O_2 \rightarrow Fe_2O_3 \tag{13.27}$$

$$3Fe + 2O_2 \rightarrow Fe_3O_4 \tag{13.28}$$

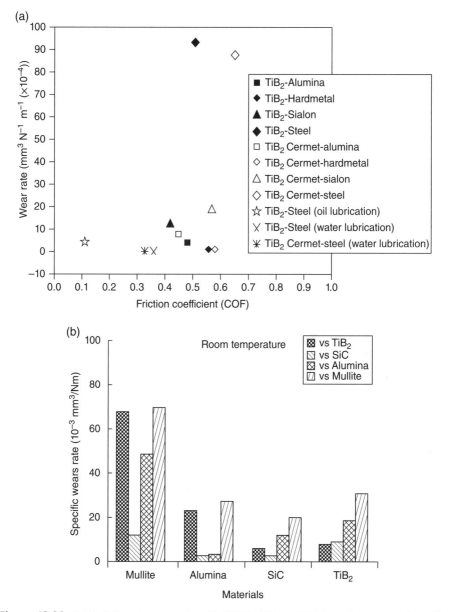

Figure 13.20. (a) Variations in wear rate with COF for TiB$_2$ and TiB$_2$-based cermet against different counter bodies. (b)The TiB$_2$-cermet has a composition of TiB$_2$-16vol% Ni$_3$(Al,Ti); while that of monolithic TiB$_2$ has TiB$_2$-5 vol% SiC. Fretting parameters include: load of 8N for 100,000 cycles with a frequency of 10 Hz and a displacement of 200 μm [139, 147, 148]. Wear of different ceramics after sliding against each other. The sliding wear test conditions : (plate-on-plate configuration) at rotation speed of 220 revolutions/min with sliding speed of 0.2 m/s and a load of 214N under unlubricated conditions.

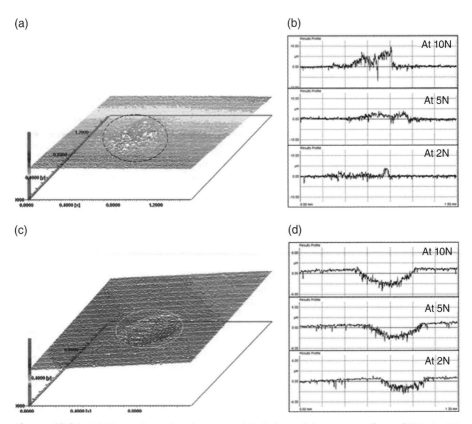

Figure 13.21. (a) Three dimensional topographical view of the worn surface of TiB$_2$-5 wt% TiSi$_2$/steel at 10 N load, (b) the maximum wear peak profiles obtained at various loading conditions of the TiB$_2$ composite after fretting against steel counterbody, (c) three dimensional topographical view of worn surface of the TiB$_2$-5 wt% TiSi$_2$/ WC-6 wt% Co at 10 N load and (d) the maximum wear depth profiles obtained at various loading conditions of the TiB$_2$ composite after fretting against WC-6 wt% Co counterbody. Fretting conditions: 4 Hz frequency, 100,000 cycles and 100 μm stroke length.

In the case of TiB$_2$/WC–Co, SEM images revealed that worn surfaces had mild abrasive scratches, grain pullout, and polishing (Fig. 13.22c and d). For TiB$_2$, adhesion, abrasion, and three-body wear mechanisms were observed for both TiB$_2$–steel and TiB$_2$/ WC–Co. The third body was predominantly a tribochemical layer for TiB$_2$–steel and loose wear debris particles for TiB$_2$/WC–Co [23].

13.8.2 Tribological Properties of TiB$_2$ Coatings

The tribological properties of TiB$_2$ coatings have also been investigated [151–156]. The tribological behaviors of TiB$_2$-coated cemented carbide (CC) cylinders were evaluated in dry sliding contact against an Al cylinder and compared with uncoated and

(a)   (b)

(c)   (d)

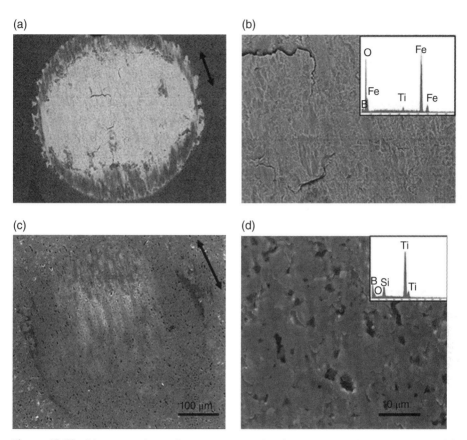

Figure 13.22. (a) Worn surfaces of TiB$_2$-5wt% TiSi$_2$ after fretting against bearing steel ball, (b) the magnified image of the worn surface along with the EDS reveals the tribochemical layer. (c) The worn surfaces of TiB$_2$-5wt% TiSi$_2$ after fretting against WC-6wt% Co ball and (d) the magnified image of the worn surface along with the EDS. Fretting conditions: 10 N load, 4Hz frequency, 100μm stroke length and 100,000 cycles. (Arrow marks indicate the fretting direction.)

TiN-coated CC [153]. No wear, adhesive detachment, or chemical dissolution was evident for TiB$_2$ coatings under low stress conditions, whereas the reference TiN coating suffered wear loss due to both chemical degeneration and mechanical decohesion/detachment. The TiB$_2$ coating had a reduced tendency to pick up Al from the counter surface compared to TiN and uncoated CC, which correlated to a lower COF. Another study compared the tribological properties and counter material pickup for physical vapor deposited (PVD) TiB$_2$ coatings in dry sliding with commercial PVD titanium nitride (TiN), titanium aluminum nitride (TiAlN), and titanium carbonitride (TiCN) [152]. The adhesion of counter material to the coating was initially observed for all combinations. The TiB$_2$ coating outperformed the others in sliding contact against Al, with respect to friction, resistance against pickup of counter material, and smoothness in the contact area. This was attributed to the chemical stability of TiB$_2$ against Al.

Magnetron-sputtered TiB_2 showed lower abrasive wear rates than TiN [154]. The use of a hard TiB_2 coating was reported to increase the wear resistance by a factor of 10 compared to bulk TiB_2 [154]. The low wear rate of the TiB_2 coatings is due to extremely high hardness (45–53 GPa).

13.9 APPLICATIONS OF TIB_2

The preceding sections highlight the suitability of TiB_2 and its composites for extreme environments. One such application, which has led to widespread research on the borides as structural materials, is thermal protection system (TPS) of space vehicles [1–7, 11–15]. The thermophysical and thermomechanical properties of TiB_2, as compared to the other transition metal borides, render it suitable for this application. Furthermore, the lower density of TiB_2 also lends additional support toward use of this material for aerospace applications. On a different note, the low density and high thermal conductivity of TiB_2 also make it attractive for engine components in airborne structures. TiB_2-based materials are also attractive for defense applications such as armor.

TiB_2-based materials are used as cutting tools and high-temperature wear-resistant parts. Furthermore, the chemical inertness and low electrical resistivity of TiB_2 render it suitable as an electrode material for aluminum electro-smelting and vacuum metal deposition equipment [1, 2, 4, 7, 8]. TiB_2-based materials also have potential applications in renewable energy. Their low neutron absorption cross sections make them suitable for control rods in nuclear reactors.

13.10 CONCLUSIONS

This chapter focused on the processing, thermophysical, thermomechanical, and oxidation properties of TiB_2-based ceramics. The combination of properties including refractoriness, high hardness, elastic modulus, and chemical resistance is attributed to the crystal structure and atomic bonding. The low self-diffusion coefficient and high melting points of monolithic TiB_2 necessitate sintering at high temperatures for long periods, which can deteriorate the mechanical properties due to grain growth. To mitigate these issues, advanced sintering techniques and additives have been employed to attain high relative densities at lower temperatures and shorter times. Ceramic additives such as SiC, Si_3N_4, $MoSi_2$, and $TiSi_2$ are more suitable than conventional metallic additives for retention of mechanical properties at higher temperatures. Among the advanced sintering techniques, SPS limits grain coarsening, resulting in superior mechanical properties. The effects of different processing routes and additives on the microstructure and properties of TiB_2 were described. In addition to the mechanical properties, thermal conductivity, TSR, oxidation behavior, and wear properties were reviewed. TiB_2 is suitable for a variety of applications in extreme environments. To realize the suitability of these diborides for specific Ultra-High-temperature applications, the relevant properties must be further characterized at elevated temperatures and under harsh conditions. Comprehensive investigation is still lacking, and simulation of response to such conditions is still needed.

REFERENCES

1. Fahrenholtz WG, Hilmas GE, Talmy IG, Zaykoski JA. Refractory diborides of zirconium and hafnium. J Am Ceram Soc 2007;90 (5):1347–1364.

2. Fahrenholtz WG, Hilmas GE. Oxidation of ultra high temperature transition metal diboride ceramics. Int Mater Rev 2012;57 (1):61–72.

3. Guo SQ. Densification of ZrB_2-based composites and their mechanical and physical properties: a review. J Eur Ceram Soc 2009;29:995–1105.

4. Basu B, Raju GB, Suri AK. Processing and properties of monolithic TiB_2-based materials. Int Mater Rev 2006;51:352–374.

5. Monteverde F, Savino R. ZrB_2–SiC sharp leading edges in high enthalpy supersonic flows. J Am Ceram Soc 2012;95 (7):2282–2289.

6. Pierson HO. *Handbook of Refractory Carbides and Nitrides: Properties, Characteristics, Processing, and Applications*. Park Ridge (NJ): Noyes Publications; 1996.

7. Cotton J. Ultra-high-temperature ceramics. Adv Mater Processes 2010;168:26–28.

8. Sciti D, Silvestroni L, Guicciardi S, Fabbriche DD, Bellosi A. Processing, mechanical properties and oxidation behavior of TaC and HfC composites containing 15 vol% $TaSi_2$ or $MoSi_2$. J Mater Res 2009;24:2056–2065.

9. Silvestroni L, Sciti D. Densification of ZrB_2–$TaSi_2$ and HfB_2–$TaSi_2$ Ultra-High-temperature ceramic composites. J Am Ceram Soc 2011;94 (6):1920–1930.

10. Sciti D, Guicciardi S, Nygren M. Spark plasma sintering and mechanical behaviour of ZrC-based composites. Scr Mater 2008;59:638–641.

11. Blum YD, Marschall J, Hui D, Young S. Thick protective UHTC coatings for SiC-based structures: process establishment. J Am Ceram Soc 2008;91 (5):1453–1460.

12. Corral EL, Loehman RE. Ultra-High-temperature ceramic coatings for oxidation protection of carbon–carbon composites. J Am Ceram Soc 2008;91 (5):1495–1502.

13. Levinea SR, Opila EJ, Halbig MC, Kisera JD, Singh M, Salem JA. Evaluation of Ultra-High temperature ceramics for aero propulsion use. J Eur Ceram Soc 2002;22:2757–2767.

14. Bellosi A, Monteverde F. Ultra-high temperature ceramics: microstructure control and Properties improvement related to materials design and processing procedures. Proceedings 5th European Workshop on Thermal Protection Systems and Hot Structures; May 17–19, 2006; Noordwijk, the Netherlands. ESA SP-631. Noordwijk: ESA Publications Division; August 2006. p 17–19.

15. Brewer L, Sawyer DL, Templeton DH, Dauben CH. A study of the refractory borides. J Am Ceram Soc 1951;34:173–179.

16. Cutler RA. Engineering properties of borides. In: *Engineered Materials Handbook*. Volume 4, Metals Park (OH): ASM International, The Materials Information Society; 1991. Ceramics and Glasses; p 787–801.

17. Telle R. Boride and carbide ceramics. In: Cahn RW, Haasen P, Kramer EJ, editors. *Materials Science and Technology*. Volume 11, Structure and Properties of Ceramics. Swain M Weinheim: VCH; 1994. p 175.

18. Clougherty EV, Pober RL, Kaufman L. Synthesis of oxidation resistant metal diboride composites. Trans Metal Soc AIME 1968;242:1077–1082.

19. Peng F, Speyer RF. Oxidation resistance of fully dense ZrB_2 with SiC, TaB_2, and $TaSi_2$ additives. J Am Ceram Soc 2008;91 (5):1489–1494.

20. Gasch M, Ellerby D, Irby E, Beckman S, Gusman M, Johnson S. Processing, properties and arc jet oxidation of hafnium diboride/silicon carbide ultra-high temperature ceramics. J Mater Sci 2004;39:5925–5937.

21. Gangireddy S, Halloran JW, Wing ZN. Non-contact mechanical property measurements at ultrahigh temperatures. J Eur Ceram Soc 2010;30:2183–2189.

22. Mukhopadhyay A, Raju GB, Basu B. Understanding influence of $MoSi_2$ addition (5 wt. %) on tribological properties of TiB_2. Metal Mater Trans A 2008;39:2998–3013.

23. Raju GB, Basu B. Wear mechanisms of TiB_2 and TiB_2-$TiSi_2$ ceramics at fretting contacts with steel and WC-6wt. %Co. Int J App Ceram Tech 2010;7 (1):89–103.

24. Mukhopadhyay A, Raju GB, Basu B, Suri AK. Correlation between phase evolution, mechanical properties and instrumented indentation response of TiB_2-based ceramics. J Eur Ceram Soc 2009;29:505–516.

25. Raju GB, Basu B, Suri AK. Oxidation kinetics and mechanisms of hot pressed TiB_2-$MoSi_2$ composites. J Am Ceram Soc 2008;91:3320–3327.

26. Raju GB, Mukhopadhyay A, Biswas K, Basu B. Densification and high temperature mechanical properties of hot pressed TiB_2-(0-10 wt.%) $MoSi_2$ composites. Scr Mater 2009;61:674–677.

27. Murthy TSRC, Basu B, Balasubramaniam R, Suri AK, Subramonian C, Fotedar RK. Processing and properties of TiB_2 with $MoSi_2$ sinter-additive: a first report. J Am Ceram Soc 2006;89:131–138.

28. Tang S, Deng J, Wang S, Liu W, Yang K. Ablation behaviors of ultra-high temperature ceramic composites. Mater Sci Eng A 2007;465:1–7.

29. Savino R, Fumo MDS, Silvestroni L, Sciti D. Arc-jet testing on HfB_2 and HfC-based ultra-high temperature ceramic materials. J Eur Ceram Soc 2008;28:1899–1907.

30. Marschall J, Chamberlain A, Crunkleton D, Rogers B. Catalytic atom recombination on ZrB_2/SiC and HfB_2/SiC ultrahigh-temperature ceramic composites. J Spacecraft Rockets 2004;41 (4):576–581.

31. Telle R, Petzow G. Strengthening and toughening of boride and carbide hard material composites. Mater Sci Eng A 1988;105/106:97–104.

32. Li LH, Kim HE, Kang ES. Sintering and mechanical properties of titanium diboride with aluminum nitride as a sintering aid. J Eur Ceram Soc 2002;22:973–977.

33. Telle R, Meyer S, Petzow G, Franz ED. Sintering behavior and phase reactions of TiB_2 with ZrO_2 additives. Mater Sci Eng A 1988;105/106:125–129.

34. Venkateswaran T, Basu B, Raju GB, Kim D-Y. Densification and properties of transition metal borides-based cermets via spark plasma sintering. J Eur Ceram Soc 2006;26:2431–2440.

35. Park JH, Koh Y, Kim H, Hwang C, Kong E. Densification and mechanical properties of titanium diboride with silicon nitride as sintering aid. J Am Ceram Soc 1999;82 (11):3037–3042.

36. Biswas K, Basu B, Suri AK, Chattopadhyay K. A TEM study on TiB_2 –20% $MoSi_2$ composite: microstructure development and densification mechanisms. Scr Mater 2006;54:1363–1368.

37. Raju GB, Basu B. Densification, sintering reactions and properties of titanium diboride with titanium disilicide as a sintering aid. J Am Ceram Soc 2007;90 (11):3415–3423.

38. Monteverde F, Bellosi A. Efficacy of HfN as sintering aid in the manufacture of ultrahigh-temperature metal diborides-matrix ceramics. J Mater Res 2004;19:3576–3585.

39. Monteverde F, Bellosi A. Development and characterization of metal-diboride-based composites toughened with ultra-fine SiC particulates. Solid State Sci 2005;7:622–630.

40. Wang SS, Vasiliev AL, Padture MP. Improved processing and oxidation resistance of ZrB_2 ultra high temperature ceramics containing SiC nano-dispersoids. Mater Sci Eng A 2007;464:216–224.

41. Monteverde F, Bellosi A. Effect of the addition of silicon nitride on sintering behavior and microstructure of zirconium diboride. Scr Mater 2002;46:223–228.

42. Monteverde F, Bellosi A. Beneficial effects of AlN as sintering aid on microstructure and mechanical properties of hot-pressed ZrB_2. Adv Eng Mater 2003;5:508–512.

43. Sciti D, Silvestroni L, Celotti G, Melandri C, Guicciardi S. Sintering and mechanical properties of ZrB_2–$TaSi_2$ and HfB_2–$TaSi_2$ ceramic composites. J Am Ceram Soc 2008;91 (10):3285–3291.

44. Wang H, Wang CA. Processing and mechanical properties of zirconium diboride-based ceramics prepared by spark plasma sintering. J Am Ceram Soc 2007;90 (7):1992–1997.

45. Guo SQ, Nishimura T, Kagawa Y, Yang JM. Spark plasma sintering of zirconium diborides. J Am Ceram Soc 2008;91 (9):2848–2855.

46. Hu C, Sakka Y, Gao J, Tanaka H, Grasso S. Microstructure characterization of ZrB_2–SiC composite fabricated by spark plasma sintering with $TaSi_2$ additive. J Eur Ceram Soc 2012;32:1441–1446.

47. Mallik M, Kailath AJ, Ray KK, Mitra R. Electrical and thermophysical properties of ZrB_2 and HfB_2 based composites. J Eur Ceram Soc 2012;32:2553–2563.

48. Rangaraj L, Divakar C, Jayaram V. Processing of refractory metal borides, carbides and nitrides. Key Eng Mater 2009;395:69–88.

49. Mishra SK, Pathak LC. Self-propagating high-temperature synthesis (SHS) of advanced high-temperature ceramics. Key Eng Mater 2009;395:15–38.

50. Raju GB, Basu B. Development of high temperature TiB_2-based ceramics. Key Eng Mater 2009;395:89–124.

51. Zamora V, Ortiz AL, Guiberteau F, Nygren M, Shaw LL. On the crystallite size refinement of ZrB_2 by high-energy ball-milling in the presence of SiC. J Eur Ceram Soc 2011;31: 2407–2414.

52. Tang WM, Zheng ZX, Wu YC, Wang J-M, Lu J, Liu JW. Synthesis of TiB_2 nanocrystalline powder by mechanical alloying. Trans Nonferrous Met Soc China 2006;16 (3):613–617.

53. Chen L, Gu Y, Shi L, Yang Z, Ma J, Qian Y. Synthesis and oxidation of nanocrystalline HfB_2. J Alloys Compd 2004;368 (1–2):353–356.

54. Chen L, Gu Y, Yang Z, Shi L, Ma J, Qian Y. Preparation and some properties of nanocrystalline ZrB_2 powders. Scr Mater 2004;50 (7):959–961.

55. Chen L, Gu Y, Qian Y, Shi L, Yang Z, Ma J. A facile one-step route to nanocrystalline TiB_2 powders. Mater Res Bull 2004;39:609–613.

56. Bates SE, Buhro WE, Frey CA, Sastry SML, Kelton KF. Synthesis of titanium boride (TiB_2) nanocrystallites by solution-phase processing. J Mater Res 1995;10:2599–2612.

57. Gu Y, Qian Y, Chen L, Zhou F. A mild solvothermal route to nanocrystalline titanium diboride. J Alloy Comp 2003;352:325–327.

58. Forsthoefel K, Sneddon LG. Precursor routes to Group 4 metal borides, and metal boride/carbide and metal boride/nitride composites. J Mater Sci 2004;39:6043–6049.

59. Schwab ST, Stewart CA, Dudeck KW, Kozmina SM. Polymeric precursors to refractory metal borides. J Mater Sci 2004;39:6051–6055.

60. Guron MM, Kim MJ, Sneddon LG. A simple polymeric precursor strategy for the syntheses of complex zirconium and hafnium-based ultra high-temperature silicon-carbide composite ceramics. J Am Ceram Soc 2008;91 (5):1412–1415.

61. Baik S, Becher PF. Effect of oxygen contamination on densification of TiB_2. J Am Ceram Soc 1987;70 (8):527–530.

62. Wen G, Li SB, Zhang BS, Guo ZX. Reaction synthesis of TiB_2–TiC composites with enhanced toughness. Acta Mater 2001;49:1463–1470.

63. Zhang GJ. Preparation of TiB_2-$TiC_{0.5}N_{0.5}$ ceramic composite by reactive hot-pressing and its microstructure. Ceram Int 1995;21 (1):29–31.

64. Chamberlain AL, Fahrenholtz WG, Hilmas GE. Low-temperature densification of zirconium diboride ceramics by reactive hot pressing. J Am Ceram Soc 2006;89 (12):3638–3645.

65. Mukhopadhyay A, Basu B. Consolidation-microstructure-property relationships in bulk nanoceramics and ceramic nanocomposites: a review. Int Mater Rev 2007;52 (5):257–288.

66. Nishimura T, Xu X, Kimotoa K, Hirosakia N, Tanaka H. Fabrication of silicon nitride nanoceramics powder preparation and sintering: a review. Sci Tech Adv Mater 2007;8:635–643.

67. Orru R, Licheri R, Locci AM, Cincotti A, Cao G. Consolidation/synthesis of materials by electric current activated/assisted sintering. Mater Sci Eng R 2009;63:127–287.

68. Munir ZA, Quach DV, Ohyanagi M. Electric current activation of sintering: a review of the pulsed electric current sintering process. J Am Ceram Soc 2011;94 (1):1–19.

69. Raj R, Cologna M, Francis JSC. Influence of externally imposed and internally generated electrical fields on grain growth, diffusional creep, sintering and related phenomena in ceramics. J Am Ceram Soc 2011;94 (7):1941–1965.

70. Jain D, Reddy KM, Mukhopadhyay A, Basu B. Achieving uniform microstructure and superior mechanical properties in ultrafine grained TiB_2–$TiSi_2$ composites using innovative multi stage spark plasma sintering. Mater Sci Eng A 2010;528 (1):200–207.

71. Lankford J. Comparative study of the temperature dependence of hardness and compressive strength in ceramics. J Mater Sci 1983;18:1666–1674.

72. Wang HL, Hon MH. Temperature dependence of ceramics hardness. Ceram Int 1999;25: 267–271.

73. Kutty TRG, Ganguly C, Sastry DH. Development of creep curves from hot indentation hardness data. Scr Mater 1996;34:1833–1838.

74. Koester RD, Moak DP. Hot hardness of selected borides, oxides and carbides to 1900°C. J Am Ceram Soc 1967;50:290–296.

75. Raju GB, Basu B, Tak NH, Cho SJ. Temperature dependent hardness and strength properties of TiB_2 with $TiSi_2$ sinter-aid. J Eur Ceram Soc 2009;29:2119–2128.

76. Munro RG. Material properties of titanium diboride. J Res Natl Inst Stand Technol 2000;105: 709–720.

77. Neuman EW, Hilmas GE, Fahrenholtz WG. Strength of zirconium diborides to 2300°C. J Am Ceram Soc 2013;96 (1):47–50.

78. Opeka MM, Talmy IG, Wuchina EJ, Zaykoski JA, Causey SJ. Mechanical, thermal, and oxidation properties of refractory hafnium and zirconium compounds. J Eur Ceram Soc 1999;19: 2405–2414.

79. Silvestroni L, Guicciardi S, Melandri C, Sciti D. TaB_2-based ceramics: microstructure, mechanical properties and oxidation resistance. J Eur Ceram Soc 2012;32:97–105.

80. Zimmermann JW, Hilmas GE, Fahrenholtz WG. Thermal shock resistance of ZrB_2 and ZrB_2-30% SiC. Mater Chem Phys 2008;112:140–145.

81. Monteverde F, Scatteia L. Resistance to thermal shock and to oxidation of metal diborides-SiC ceramics for aerospace application. J Am Ceram Soc 2007;90:1130–1138.

82. Thomson R. Production, fabrication, and uses of borides. In: Freer R, editor. *The Physics and Chemistry of Carbides, Nitrides and Borides*. Dordrecht: Kluwer Academic; 1990. p 113–120.

83. Baumgartner HR, Steiger RA. Sintering and properties of TiB_2 made from powder synthesized in a plasma-arc heater. J Am Ceram Soc 1984;67 (3):207–212.

84. Mishra SK, Das S, Das SK, Ramachandrarao P. Sintering studies on ultrafine ZrB_2 powder produced by a self-propagating high-temperature synthesis process. J Mater Res 2000;15 (11):2499–2504.

85. Silverstroni L, Sciti D. Effects of $MoSi_2$ additions on the properties of Hf– and ZrB_2 composites produced by pressureless sintering. Scr Mater 2007;57:165–168.

86. Silverstroni L, Sciti D, Bellosi A. Microstructure and properties of pressureless sintered HfB_2-based composites with addition of ZrB_2 or HfC. Adv Eng Mater 2007;9 (10):915–920.

87. Chamberlain AL, Fahrenholtz WG, Hilmas GE. High-strength zirconium diboride-based ceramics. J Am Ceram Soc 2004;87:1170–1172.

88. Barsoum MW, Houng B. Transient plastic phase processing of titanium-boron-carbon composites. J Am Ceram Soc 1993;76 (6):1445–1451.

89. Zhang GJ, Jin ZZ. Reactive synthesis of AlN/TiB_2 composite. Ceram Int 1996;22:143–147.

90. Zhang GJ, Yue XM, Jin ZZ, Dai JY. In-situ synthesized TiB_2 toughened SiC. J Eur Ceram Soc 1996;16 (9):409–412.

91. Zhang GJ, Deng ZY, Kondo N, Yang JF, Ohji T. Reactive hot pressing of ZrB_2-SiC composites. J Am Ceram Soc 2000;83 (9):2330–2332.

92. Wu WW, Zhang GJ, Kan YM, Wang PL. Reactive hot pressing of ZrB_2-SiC-ZrC ultra high-temperature ceramics at 1800 degrees C. J Am Ceram Soc 2006;89 (9):2967–2969.

93. Zimmermann JW, Hilmas GE, Fahrenholtz WG, Monteverde F, Bellosi A. Fabrication and properties of reactively hot pressed ZrB_2-SiC ceramics. J Eur Ceram Soc 2007;27:2729–2736.

94. Monteverde F. Progress in the fabrication of ultra-high-temperature ceramics: "*in-situ*" synthesis, microstructure and properties of a reactive hot-pressed HfB_2-SiC composite. Compos Sci Tech 2005;65:1869–1879.

95. Sciti D, Monteverde F, Guicciardi S, Pezzotti G, Bellosi A. Microstructure and mechanical properties of ZrB_2-$MoSi_2$ ceramic composites produced by different sintering techniques. Mater Sci Eng A 2006;434:303–309.

96. Bellosi A, Monteverde F, Sciti D. Fast densification of ultra-high-temperature ceramics by spark plasma sintering. Int J App Ceram Tech 2006;3 (1):32–40.

97. Monteverde F. Ultra-high-temperature HfB_2-SiC ceramics consolidated by hot-pressing and spark plasma sintering. J Alloy Compd 2007;428:197–205.

98. Sciti D, Guicciardi S, Nygren M. Densification and mechanical behaviour of HfC and HfB_2 fabricated by spark plasma sintering. J Am Ceram Soc 2008;91 (5):1433–1440.

99. Silvestroni L, Sciti D. Effect of $MoSi_2$ additions on the properties of Hf- and Zr-B_2 composites produced by pressureless sintering. Scr Mater 2007;57:165–168.

100. Monteverde F, Guicciardi S, Bellosi A. Advances in microstructure and mechanical properties of zirconium diboride based ceramics. Mater Sci Eng A 2003;346:310–319.

101. Zimmermann JW, Hilmas GE, Fahrenholtz WG. Thermal shock resistance and fracture behaviour of ZrB_2-based fibrous monolith ceramics. J Am Ceram Soc 2009;92:161–166.

102. Monteverde F, Melandri C, Guicciardi S. Microstructure and mechanical properties of an HfB_2 + 30 vol% SiC composite consolidated by spark plasma sintering. Mater Chem Phys 2006;100:513–519.

103. Zamora V, Ortiz AL, Guiberteau F, Nygren M. Spark-plasma sintering of ZrB_2 ultra-high-temperature ceramics at lower temperature via nanoscale crystal refinement. J Eur Ceram Soc 2012;32:2529–2536.

104. Courtright EL, Graham HC, Katz AP, Kerans RJ. Ultra high temperature assessment study-ceramic matrix composites. WL-TR-91-4061. Wright-Patterson Air Force Base (OH); 1991.

105. Wuchina E, Opila E, Opeka M, Fahrenholtz W, Talmy I. UHTCs: ultra-high temperature ceramic materials for extreme environment applications. Electrochem Soc Interface 2007;16 (Winter):30–36.

106. Campbell IE, Sherwood EM. *High Temperature Materials and Technology*. New York: John Wiley & Sons, Inc.; 1967.

107. Pluscauskis J, Dion M, Buesking K. Development and preliminary TGA correlation results of a finite element based durability model for CMC materials. Proceedings of the 32nd Conference on Composites, Materials, and Structures; 2008; Daytona Beach (FL); January 2008.

108. Weng L, Han W, Li X, Hong C. High temperature thermo-physical properties and thermal shock behavior of metal–diborides-based composites. Int J Refract Metals Hard Mater 2010;28:459–465.

109. Justin JF, Jankowiak A. Ultra high temperature ceramics: densification, properties and thermal stability. Onera J Aerosp Lab 2011;3:1–11.

110. Squire TH, Marschall J. Material property requirements for analysis and design of UHTC components in hypersonic applications. J Eur Ceram Soc 2010;30:2239–2251.

111. Monteverde1 F, Savino R. Stability of ultra-high-temperature ZrB_2–SiC ceramics under simulated atmospheric re-entry conditions. J Eur Ceram Soc 2007;27:4797–4805.

112. Gasch M, Johnson S. Physical characterization and arc jet oxidation of hafnium-based ultra-high temperature ceramics. J Eur Ceram Soc 2010;30:2338–2342.

113. Raju GB, Basu B, Suri AK. Thermal and electrical properties of TiB_2–$MoSi_2$. Int J Refract Metals Hard Mater 2010;28:174–179.

114. Loehman R, Corral E, Dumm H-P, Kotula P, Tandon R. Ultra high temperature ceramics for hypersonic vehicle applications. *SAND* 2006-2925. Albuquerque (NM); June 2006.

115. Baca L, Stelzer N. Adapting of sol–gel process for preparation of TiB_2 powder from low-cost precursors. J Eur Ceram Soc 2008;28:907–911.

116. Zhang L, Pejakovic̆ı DA, Marschall J, Gasch M. Thermal and electrical transport properties of spark plasma sintered HfB_2 and ZrB_2ceramics. J Am Ceram Soc 2011;94 (8):2562–2570.

117. Zimmermann JW, Hilmas GE, Fahrenholtz WG, Dinwiddie RB, Porter WD, Wang H. Thermophysical properties of ZrB_2 and ZrB_2-SiC ceramics. J Am Ceram Soc 2008;91 (5):1405–1411.

118. Fan Yang L, Fang D. Thermal shock modelling of ultra-high temperature ceramics under active cooling. J Comput Math Appl 2009;58:2373–2378.

119. Basu B. Toughening of yttria-stabilised tetragonal zirconia ceramics. Int Mater Rev 2005; 50:239–255.

120. Smith DS, Fayette S., Grandjean S, Martin C. Thermal resistance of grain boundaries in alumina ceramics and refractories. J Am Ceram Soc 2003;86 (1):105–111.

121. Johnson SM. Ultra high temperature ceramics: application, issues and prospects. Proceedings of the 2nd Ceramic Leadership Summit; Baltimore, MD; August 3, 2011.

122. Callister WD Jr. *Callister's Materials Science and Engineering*. New Delhi: Wiley India; 2008.

123. Kulpa A, Troczynski T. Oxidation of TiB_2 powders below 900°C. J Am Ceram Soc 1996;79 (2):518–520.

124. Li W, Zhang Y, Zhang X, Hong C, Han W. Thermal shock behavior of ZrB_2–SiC ultra-high temperature ceramics with addition of zirconia. J Alloy Compd 2009;478:386–391.

125. Solomah AC, Reichert W, Rondinella V, Esposito L, Toscano E. Mechanical properties, thermal shock resistance, and thermal stability of zirconia-toughened alumina-10 vol% silicon carbide whisker ceramic matrix composite. J Am Ceram Soc 1990;73:740–743.

126. Guo W-M, Zhang GJ. Oxidation resistance and strength retention of ZrB_2–SiC ceramics. J Eur Ceram Soc 2010;30:2387–2395.

127. Talmy I, Zaykoski J, Opeka M, Dallek S. In: McNallan M, Opila E, editors. *High Temperature Corrosion and Materials Chemistry III: The Electrochemical Society Proceedings Series. PV 2001-12*. Pennington (NJ): Electrochemical Society; 2001. p 144.

128. Rezaie A, Fahrenholtz WG, Hilmas GE. Evolution of structure during the oxidation of zirconium diboride-silicon carbide in air up to 1500°C. J Eur Ceram Soc 2007;27:2495–501.

129. Fahrenholtz WG. Thermodynamic analysis of ZrB_2–SiC oxidation: formation of a SiC-depleted region. J Am Ceram Soc 2007;90:143–8.

130. Momozawa A, Tu R, Goto T, Kubota Y, Hatta H, Komurasaki K. Quantitative evaluation of the oxidation behavior of ZrB_2-15 vol.% SiC at a low oxygen partial pressure. Vacuum 2013;88:98–102.

131. Murthy TSRC, Sonber JK, Subramanian C, Hubli RC, Suri AK. Densification, characterization and oxidation studies of TiB_2–WSi_2 composite. Int J Refract Metals Hard Mater 2012;33:10–21.

132. Han J, Hong C, Zhang X, Wang B. Thermal shock resistance of TiB_2–Cu interpenetrating phase composites. Compos Sci Tech 2005;65:1711–1718.

133. Mukhopadhyay A, Venkateswaran T, Basu B. Spark plasma sintering may lead to phase instability and inferior mechanical properties: a case study with TiB_2. Scr Mater 2013;69:159–164.

134. Wang H, Sun S, Wang D, Tu G. Characterization of the structure of TiB_2/TiC composites prepared via mechanical alloying and subsequent pressureless sintering. Powder Tech 2012;217:340–346.

135. Basu B. Fretting wear behavior of advanced ceramics and cermet against alumina. Metals Mater Processes 2002;14 (2):133–144.

136. Bhusan B, Gupta BK. *Handbook of Tribology: Materials, Coatings, and Surface Treatments*. New York: McGraw-Hill; 1991.

137. Kato K, Adachi K. Wear of advanced ceramics. Wear 2002;253:1097–1104.

138. Engqvist H, Axen N, Hogmark S. Tribological properties of a binderless carbide. Wear 1999;232:157–162.

139. Basu B, Vitchev RG, Vleugels J, Celis JP, VanDerBiest O. Influence of humidity on the fretting wear of self-mated tetragonal zirconia ceramics. Acta Mater 2000;48:2461–2471.

140. Jones AH, Dobedoe RS, Lewis MH. Mechanical properties and tribology of Si_3N_4–TiB_2 ceramic composites produced by hot pressing and hot isostatic pressing. J Eur Ceram Soc 2001;21:969–980.

141. Basu B, Vleugels J, Van Der Beist O. Unlubricated tribological performance of advanced ceramics and composites at fretting contacts with alumina. J Mater Res 2003;18 (6):1314–1324.

142. Basu B, Sarkar D, Venkateswaran T. Pressureless sintering and tribological properties of WC-ZrO_2 composites. J Eur Ceram Soc 2005;25:1603–1610.

143. Vleugels J, Basu B, Hari Kumar KC, Vitchev RG, VanDerBiest O. Unlubricated fretting wear of TiB_2 containing composites against bearing steel. Metal Mater Trans A 2002;33:3847–3859.

144. Basu B, Vleugels J, Van Der Biest O. Influence of lubrication on the fretting wear performance of TiB_2 based materials. Wear 2001;250:631–641.

145. Murthy TSRC, Basu B, Srivastava A, Balasubramaniam R, Suri AK. Tribological properties of TiB_2 and TiB_2–$MoSi_2$ ceramic composites. J Eur Ceram Soc 2006;26:1293–1300.

146. Raju GB, Basu B. Influence of MoSi$_2$ addition on load dependent fretting wear properties of TiB$_2$ against cemented carbide. J Am Ceram Soc 2009;92 (9):2059–2066.

147. Yang Q, Senda T, Kotani N, Hirose A. Sliding wear behavior and tribofilm formation of ceramics at high temperatures. Surf Coat Technol 2004;184:270–277.

148. Wasche R, Klaffke D, Troczynski T. Tribological performance of SiC and TiB$_2$ against SiC and Al$_2$O$_3$ at low sliding speeds. Wear 2004;256:695–704.

149. Waterhouse RB. Fretting wear. Wear 1984;100:107–118.

150. Klaffke D. Fretting wear of ceramics. Trib Int 1989;22 (2):89–101.

151. Prakash B, Ftikos C, Celis JP. Fretting wear behavior of PVD TiB$_2$ coatings. Surf Coat Technol 2000;124:253–261.

152. Burger M, Hogmark S. Tribological properties of selected PVD coatings when slid against ductile materials. Wear 2002;252:557–565.

153. Burger M, Hogmark S. Evaluation of TiB$_2$ coatings in sliding contact against aluminium. Surf Coat Technol 2002;149:14–20.

154. Burger M, Larsson M, Hogmark S. Evaluation of magnetron-sputtered TiB$_2$ intended for tribological applications. Surf Coat Technol 2000;124:253–261.

155. Berger M, Karlsson L, Larsson M, Hogmark S. Low stress coatings with improved tribological properties. Thin Solid Films 2001;401:179–186.

156. Yang Y, Zheng Z, Wang X, Liu X, Han JG, Yoon JS. Microstructure and tribology of TiB$_2$ and TiB$_2$/TiN double-layer coatings. Surf Coat Technol 1996;84:404–408.

157. Jungling T, Sigl LS, Oberacker R, Thummler F, Schwetz KA. Fabrication of TiB$_2$-Fe-Al cermet alloys synthesized by pulsed current process. Int J Refract Metals Hard Mater 1993; 12:71–88.

158. Murthy TSRC, Subramanian C, Fotedar RK, Gonal MR, Sengupta P, Kumar S, Suri AK. Preparation and property evaluation of TiB$_2$ + TiSi$_2$ composite. Int J Refract Metals Hard Mater 2009;27:629–636.

159. Golla BR, Basu B. Hot pressed TiB$_2$-10 wt. % TiSi$_2$ ceramic with extremely good thermal transport properties at elevated temperatures (up to 1273K). Scr Mater 2013;68:79–82.

THE GROUP IV CARBIDES AND NITRIDES

Eric J. Wuchina and Mark Opeka

Naval Surface Warfare Center, West Bethesda, MD, USA

14.1 BACKGROUND

Ultra-high-temperature ceramics (UHTCs) are a family of materials that display a unique set of properties including extremely high melting points (>3000°C), low vapor pressures, chemical stability, high hardness, and high-temperature strength and are typically the group IVB and VB carbides, nitrides, and borides. This combination of properties postures these materials as candidates for high-temperature applications including aerosurfaces for hypersonic vehicles, propulsion system components, plasma arc electrodes, furnace elements, high-temperature shielding, and cutting tools [1]. The hypersonic vehicles and propulsion systems (such as for missiles) are the primary applications in view at this time.

Current interest in hypersonic flight provides a strong impetus for research on UHTCs. At the velocities associated with hypersonic flight (>Mach 5), the deceleration of the air at the vehicle surface produces a very high-temperature flow field, which yields stringent requirements on the thermal protection system. The highest heating rates are found on the nose tips and leading edges of the wings and engine inlets, and as a result, these locations will experience the highest surface temperatures on the vehicle. Vehicle performance is maximized with sharp nose tips and leading edges, but the

Ultra-High Temperature Ceramics: Materials for Extreme Environment Applications, First Edition.
Edited by William G. Fahrenholtz, Eric J. Wuchina, William E. Lee, and Yanchun Zhou.
© 2014 The American Ceramic Society. Published 2014 by John Wiley & Sons, Inc.

heating rates and resulting surface temperatures also increase steeply with decreasing radii. Thus, a major material selection criterion for nose tip and leading-edge aerodynamic control materials (ACMs) is the melting point. Since these components operate in air, oxidation resistance and ablation resistance at high temperatures are secondary material selection criterion to sustain optimum vehicle aerodynamic performance [2, 3].

At the present time, structural materials and coatings for use in high-temperature oxidizing environments are based almost exclusively on SiC. For example, the US Space Shuttle nose tip and wing leading edges were constructed from carbon–carbon (C–C) composite with a SiC coating. Silicon-containing materials such as SiC, Si_3N_4, and $MoSi_2$ are highly oxidation resistant because they form a SiO_2-based coating that is protective (known as passive oxidation) due to its very low O_2 diffusivity [4]. However, the SiO_2-forming materials have a temperature limit of approximately 1700°C. This limit is due to the onset of active oxidation, a condition where the protective SiO_2 coating scale is unable to form and instead SiO vapor is formed on the surface with resultant high ablation (thermally induced mass loss) rates [5]. The transition temperature from passive to active oxidation is further reduced as ambient pressure decreases, therefore stringently limiting the hypersonic vehicle trajectory envelope. Further, oxidation rates of SiO_2-forming materials increase in moisture-containing environments [6]. Since the 1700°C limit holds for all SiO_2-forming materials, higher-temperature-capable materials (e.g., 2000°C or higher) are desired for next-generation hypersonic vehicle ACMs. The benefit and incentive for research on the Zr- and Hf-based UHTCs is the formation of a solid and ablation-resistant high-temperature oxide scale. As will be discussed later, these materials can be used at temperatures above which active oxidation occurs with SiO_2 formers such as SiC. As seen in other chapters in this book, one of the primary goals of current research is to optimize UHTC compositions to minimize oxidation rates at relevant temperatures.

Missile propulsion combustion temperatures with solid propellants are typically in the range of 2200–3500°C depending on propellant chemistry. Since these rocket motors operate uncooled, very stringent demands are placed on motor components. Current materials such as graphite, carbon–carbon composites (C–C), and phenolic composites (carbon phenolic, silica phenolic, etc.) exhibit significant ablation at these temperatures, which decrease the rocket motor performance [7]. UHTCs could provide a significant improvement in missile performance, if ablation-resistant materials are used.

The significant difference exists between very short-duration, single heating exposure experienced by missile propulsion system components and much longer-duration, cyclic heating exposure experienced by hypersonic vehicle components. This distinction could result in different material selection paths among the family of UHTCs, which are listed in Table 14.1 [3, 8–10].

Other compounds may also be members of this set since refractory nitrides have pressure-dependent melting points. For example, the melting point of HfN is 3390°C at 0.1 MPa and 3810°C at 8.0 MPa. Similar pressure dependence is known for TiN and ZrN [11–13]. Rare earth metal nitrides (GdN, $T_m > 2900$°C at 0.1 MPa) [14] are also likely to have highly pressure-dependent melting points (which may exceed 3000°C), but this has not been verified experimentally. Rocket motor combustion pressures are very high (10–20 MPa), so these rare earth nitrides may be candidates for such components.

TABLE 14.1. Materials with melting temperatures above 3000°C

Group IVB	Group VB	Elements	Other
		C	BN (high P_{N2})
TiC, TiN, TiB$_2$			
ZrC, ZrN, ZrB$_2$	NbC, NbB$_2$		
HfC, HfN, HfB$_2$	Ta$_2$C, TaC, TaN, TaB, TaB$_2$	W, Re, Os, Ta	ThO$_2$, Re$_2$Hf

TABLE 14.2. Selected eutectic temperatures (°C)[a]

	ZrB$_2$	ZrC	ZrN	HfB$_2$	HfC	HfN
C	2390	2910	Reaction	2515	3180	Reaction
SiC	2270	2410	Reaction	2350	2640	Reaction
ZrC	2830	—	SS	>2830	SS	SS
ZrN	2780	SS	—	>2780	SS	SS
W	2250	2800	<2800	2280	2890	2800
Re	2100	2670	2500?	2140	2720	>2500

[a]From Refs. [15–21].

Hypersonic ACM and rocket motor propulsion materials are exposed to oxidizing environments at high temperatures and pressures. Because of this, the refractoriness of oxidation products is a key consideration for material selection. Elements such as Ti, Nb, and Ta, and their compounds, are not candidates since the condensed oxides that form *in situ* have melting temperatures that are too low (Ta$_2$O$_5$ has the highest of these at 1887°C) to be used in high-shear environments. Liquid oxidation products would be removed from the surface by shear, resulting in unacceptable ablation rates. For solid rocket propellant environments, refractory metals such as Ta, W, Re, and Os, along with their carbides and nitrides, may be candidate materials, but these materials are addressed elsewhere and are excluded from this chapter [8]. Boron nitride forms the very low melting point oxide B$_2$O$_3$ (T_m, 450°C) and, therefore, is an ablator similar to carbon (graphite, C–C, etc.). Thorium oxide is perceived to have an unacceptable radioactivity risk, which removes it from serious consideration. Thus, UHTCs based on Zr and Hf (borides, carbides, and nitrides) are considered the most promising candidate materials for propulsion and hypersonic leading-edge components due to the refractoriness of the base compounds and their respective oxides ZrO$_2$ (2700°C) and HfO$_2$ (2800°C). While oxides may also be considered candidates for these applications, since they are stable in oxidizing environments, poor thermal shock resistance (due to high thermal expansion and Young's modulus and low thermal conductivity) typically introduces significant design problems that eliminate them from consideration for most aerospace applications.

Recent research on UHTCs has been substantially focused on ceramics based on the ZrB$_2$–SiC and HfB$_2$–SiC systems due to their potential for use on hypersonic vehicles. However, they have relatively low eutectic temperatures with many adjoining structural materials of interest, such as C–C and W. As shown in Table 14.2, the eutectic temperatures of boride–SiC ceramics are relatively low (2270 and 2350°C, respectively).

Pure diborides also have similarly low eutectic temperatures in contact with C–C and W, introducing significant design complexities for rocket motor components. However, Zr and Hf carbides and nitrides have high eutectic temperatures in contact with these structural materials and may be more suitable for the highest temperature applications.

Other physical, thermal, and mechanical properties such as density, thermal expansion coefficient, thermal conductivity/diffusivity, stiffness, strength, creep, and so on, at ambient and elevated temperatures must be also considered during the material selection process. These properties, however, can be modified to varying degrees with composition variations and fabrication processes. Many properties can be suitably tailored for specific hypersonic vehicle and rocket propulsion system designs so that acceptable performance can be achieved.

In summary, the Zr and Hf carbides and nitrides are good candidates for missile propulsion and hypersonic applications. By meeting three major material selection criteria of high melting point, high oxide melting point, and high eutectic temperature in contact with C–C and W-based structural materials, these UHTCs are of high interest for the hypersonic and propulsion communities.

14.2 GROUP IV CARBIDES

Group IV metal carbides have been known for over 150 years. For example, metal carbides were known to steelmakers in the mid-1800s. They identified and extracted Fe_3C, TiC, and WC from steels as strength-limiting brittle phases. By 1900, Moissan had synthesized refractory carbides in an arc furnace, later focusing his research on SiC [22]. Westbrook and Stover [23] investigated TaC while working on light emitters with longer life and higher brightness than W. The high hot hardness of ZrC and HfC motivated investigations for cutting tool applications. As early as 1950, ZrC was explored for potential gas turbine applications [24]. Zirconium carbide coatings on graphite were investigated for nuclear applications due to the very high stability of ZrC in high vacuum [25]. Hafnium carbide and ZrC cermets (with W) were evaluated as bearing materials in liquid Li environments also for nuclear power generation [26]. Regarding nuclear applications, one of the most striking differences between Zr and Hf is the neutron transparency of Zr and the high neutron absorptivity of Hf. The corrosion of graphite by H_2 at high velocities has led to the study of ZrC coatings for space solar propulsion [27]. Hafnium carbide has also been investigated for electrodes in arcjet thrusters [28]. Large programs to identify oxidation-resistant coatings for C–C composites, including HfC, were conducted in the mid-1980s to early 1990s by various defense agencies. Hafnium carbide was proposed as an oxidation-resistant coating for 2000°C carbon–carbon composite gas turbine components due to its low-carbon transport properties and its *in situ* formation of a protective refractory oxide (R.A. Rapp, Personal communications).

The properties of transition metal carbides including Zr and Hf have been reviewed extensively. Perry [29–31] and others [14, 22, 25, 32–37] compiled summaries of the literature on HfC. In addition, the binary and ternary phase equilibria associated with these carbides have been compiled [15–17].

The transition metal carbides are materials of interest due to their extremely high melting points, hardness, and stiffness (e.g., Young's modulus). The monocarbides of Ta and Hf are of particular interest because they possess the highest melting temperatures of any known compounds (3980 and 3930°C, respectively). A melting point maximum between the two was reported in the initial melting point determinations by Agte [38], but later work by Rudy [39] did not confirm this and determined that HfC and TaC form a continuous solid solution. The extremely high melting point, hardness, and stiffness of the Hf and Zr carbides (and nitrides) are due to their complex combination of metallic, covalent, and ionic bonding. While the most significant component to the interatomic forces is speculated to be covalent bonding, the precise nature of bonding in these compounds is not well understood, as discussed by Davis [40].

The group IVB and VB monocarbides all exist in the NaCl-type face-centered cubic (fcc) crystal structure. While the chemical formula is generally written as MeC, a wide homogeneity region exists for the monocarbides, as shown in the Hf–C phase diagram in Figure 14.1 [41]. Due to this wide homogeneity region, the formula MeC_x is more appropriate where x is the C/Me ratio. Many transition metal carbides have potentially tailorable thermomechanical and thermophysical properties due to the large, substoichiometric phase stability field associated with vacancies in the C sublattice.

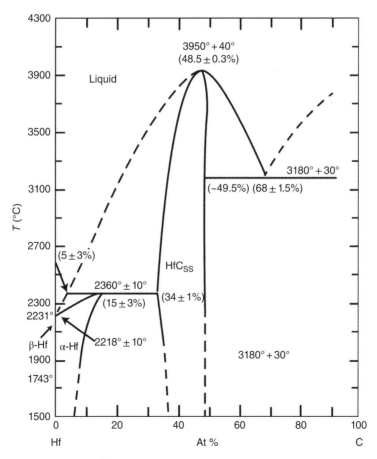

Figure 14.1. Hf–C phase diagram [41].

14.3 PREPARATION AND PROCESSING

The carbides may be prepared by the direct carburization of the metal:

$$Hf + C \rightarrow HfC \qquad (14.1)$$

However, carbides are typically produced by carbothermal reduction of the corresponding oxide, according to the reaction

$$HfO_2 + 3C \leftrightarrow HfC + 2CO \uparrow \qquad (14.2)$$

The P_{CO} has been measured in the temperature range of 1300–1650°C [42]. Using these reactions, carbide powders are obtained by grinding and milling the reaction product using traditional techniques. High-purity materials are also produced using gas-phase synthesis (condensation) and self-propagating high-temperature synthesis (SHS). High-purity films and coatings can be prepared by direct carburization of metals, physical vapor deposition, molten salt synthesis (electrolysis), and chemical vapor deposition (CVD). For CVD, an *in situ* chlorination of hafnium sponge to produce hafnium tetra-chloride, followed by a subsequent reaction with methane, is typical:

$$Hf + 2Cl_2 \rightarrow HCl_4 \qquad (14.3)$$

$$HfCl_4 + CH_4 \rightarrow HfC + 4HCl \qquad (14.4)$$

Densification of powders is typically carried out at high temperatures by sintering, hot pressing, hot isostatic pressing (HIP), spark plasma sintering, or vacuum plasma spray-ing. Air plasma spraying cannot be utilized due to the high affinity of these compounds to oxygen. Vacuum plasma spraying has the advantage of being able to produce near-net-shape articles, but stoichiometry control is difficult due to preferential loss of C at extremely high melt temperatures. The HIP process is used when control of carbon content is of high importance. A metallic encapsulation provides an effective barrier for the inward diffusion of C from the heating elements and outward diffusion of C and N from the compounds being densified, allowing the control of stoichiometry and purity during the pressure-aided sintering process.

14.4 MECHANICAL AND PHYSICAL PROPERTIES

Measurement of mechanical properties of these extremely high-melting carbides requires the fabrication of dense samples, which is difficult due to the extremely high sintering temperatures. Chiotti [43] cold pressed Zr or ZrO_2 and C powders and sintered at 2550°C, but did not report density, final composition, or microstructure characteris-tics. He described the ceramics as dense and having appreciable strength, but did not quantify these. Gangler [24] prepared ZrC by hot pressing, but did not describe the starting powders. He reported a density of 97.8% and a free carbon content of 4% residing at the grain boundaries, but no grain size was given. Tensile strengths at

1000 and 1200°C were 99.6 and 109.3 MPa, respectively. Leipold *et al.* [44] also prepared ZrC by hot pressing and reported 1–2% of free C located at grain boundaries. Tensile strengths at 20, 1600, 2000, 2200, 2400, and 2600°C were measured. The strength at 20°C was over 207 MPa, while at 2600°C it was 18.5 MPa. Creep measurements revealed over 30% elongation at 2600°C. Based on these data, Leipold concluded that ZrC had **little structural usefulness** as a high-temperature material. However, he recognized that impurities such as N, O, and free C played a significant role in the high-temperature properties.

Sanders [45] prepared HfC ceramics from powders (3.1 μm average particle size) that were 95–97% dense with very low O, N, and free C (0.03 wt%) contents. He measured flexural strengths of 235 MPa at room temperature and 109 MPa 1315°C for ceramics having an average grain size of 30 μm. A second powder with a smaller 1.3 μm starting particle size, yet larger O, N, and free C (0.3 wt%) impurity contents, yielded HfC ceramics with a 12.7 μm average grain size. The resulting relative densities were 91–96%, but strength was lower. Sanders compared HfC flexural strength with TiC data from Glaser [46], who measured TiC strength as a function of density. With relative densities of 95, 99.8, and 100%, Glaser obtained room-temperature strengths of 393, 634, and 855 MPa, respectively.

Kendall [47, 48] prepared hypo- and hypereutectic arc-cast ZrC–C and HfC–C compositions. The ceramics contained excess carbon by design and had excellent thermal shock resistance in arc-heater testing. While flexural strengths were not reported for the ZrC–C and HfC–C materials, it was inferred that the strengths would be similar—a strength of 240 MPa measured for TiC–C. Gangler [24] reported strengths of nearly 700 MPa for HfC and ZrC cermet compositions, which contained 8–17 at% W and Mo.

Plastic deformation was observed at elevated temperatures for these carbides, as was cited previously for ZrC [44]. The ductile-to-brittle transition temperatures (DBTTs) have been identified for stoichiometric and substoichiometric compositions using flexural strength testing. For the $HfC_{0.98}$ composition, the DBTT was approximately 1900°C but dropped to 1000°C for $HfC_{0.67}$. Above these temperatures, significant plastic deformation was observed but reached the deformation limit of the apparatus, and the strength could not be measured [49]. Hypereutectic HfC compositions exhibited ductility only above 2100°C [50]. Thus, these carbides behave as metals with very high DBTTs, and the DBTTs decrease with decreasing C content.

In the 1960s, research on single crystals was conducted to understand the nature of the bonding and deformation in NaCl-type cubic carbides. Most of the research was conducted on TiC and only limited work on ZrC and HfC. However, no difference in the fundamental physics was identified. Williams and Schaal [51, 52] and Hollox and Smallman [53] recognized that the Ti–Ti interatomic spacing in TiC is only 2% larger than that in Ti metal, and both teams concluded that the deformation mechanisms were likely governed by the metal lattice. This was found to be true, yet only at temperatures high enough that the Peierls stress from the ionic and covalent bonds would be sufficiently relaxed and dislocation mobility increased. In single crystals tested in flexure and compression, plasticity was observed at temperatures as low as 800°C. Lower C contents did reduce this onset further. A flexural strength of 5.5 GPa has been reported for

TiC [52]—this astonishing value was measured by Williams at room temperature using an electropolished single-crystal sample of stoichiometric TiC. Williams [54] recognized the exponential drop in the critical resolved shear stress (CRSS) with increasing temperature and found a significant retention of the CRSS with a boron addition of a few tenths of a percent. Extremely fine TiB_2 platelets developed coherently in the TiC matrix parallel to {111} planes since the close-packed Ti planes of both phases have the interatomic spacing difference of only 1%. The result was a fivefold increase in the CRSS at 1600°C. An extensive review of the physical properties and solid-state physics of transition metal carbides was compiled by Davis *et al.* in 1985 [55]. Williams [56] also published a summary of his work on solid-state physics of transition metal carbides in 1988.

DePoorter and Wallace [57] provided an extensive review of the literature up to 1970 on the self-diffusion rates of C in refractory carbides including ZrC and HfC and also on the chemical diffusion constants of the binary carbides. X-ray diffraction (XRD) studies of mixed HfC–TaC ceramics sintered at 1900C for up to 200h showed a higher intrinsic diffusivity of Hf versus Ta [58].

14.5 OXIDATION OF THE UHTC CARBIDES AND NITRIDES

At temperatures up to 2000°C, especially for long-duration applications in oxidizing environments, pure HfO_2-forming materials (or Zr-based) such as elemental Hf [59] and Hf–Ta alloys [60] exhibit oxidation rates lower than carbide, nitride, or boride UHTCs. The oxidation rates of these pure HfO_2-forming materials are governed by diffusion of O^{2-} through the dense HfO_2 scale. The UHTCs form gaseous phases in the oxidation process, such as B_2O_3, CO, CO_2, or N_2, resulting in a nondense network of interconnected porosity from gas outward diffusion from the interface through the crystalline HfO_2 scale. As has been implied previously, even relatively rapid oxidation is acceptable in rocket propulsion applications, if the oxide scale remains intact and ablation does not occur. Thus, the primary requirement for the rocket nozzle application is that the HfO_2 or ZrO_2 scale forms at a sufficiently low rate and remains adherent such that ablation does not occur. While the oxidation behavior of HfB_2 (and ZrB_2) has been extensively researched, the oxidation of the Hf and Zr carbide and nitrides has been much less studied.

Four considerations regarding the oxidation of materials should be noted. First, reference to the Pilling–Bedworth ratio (the ratio of oxide volume to the substrate material, or matrix volume), or R_{PB}, has often been made relative to protective versus nonprotective oxidation. The material scope for which this ratio is relevant is not clear since it was based on the protective or nonprotective behavior of a limited number of pure metals. Nevertheless, it may be useful to consider it for the behavior of UHTCs. A R_{PB} ratio between 1 and 2 is considered favorable for forming a protective oxide scale [61]; a ratio above 2 leads to spalling and a nonprotective scale [62], while ratios below 1 are considered to form a scale that do not provide sufficient surface coverage [63]. The R_{PB} can be related to the formation and development of stresses in the oxide scale—a R_{PB} greater than unity indicates a lower-density oxide scale relative to the matrix density; hence, in-plane compressive stresses develop in the scale. For the oxidation of HfC and ZrC, the R_{PB} is 1.44 and 1.43, respectively, based on the density of the low-temperature monoclinic phases of the HfO_2 and ZrO_2. It is also important to note that hafnium oxide has three polymorphs, which are α-HfO_2 (monoclinic $P2_1/c$), β-HfO_2 (tetragonal $P4_2/nmc$) above approximately 1600°C, and δ-HfO_2 (cubic $Fm\overline{3}m$) above approximately 2700°C [64]. The same polymorphs exist for zirconium oxide, with transformation temperatures of 1050 and 2370°C, respectively. These phases yield different values for R_{PB}; thus, in principle, compressive stress states will be different depending on the oxide polymorph formed. Transformations also confound the understanding of microstructures since phase transformations occur upon cooling and may introduce features (e.g., microcracking) not present at test temperatures.

Second, these carbides and nitrides form CO and/or CO_2 and N_2, which must **vent** in some way through the oxide scale since the solubility of these gaseous oxides in the crystalline oxide structure is extremely low or negligible. Thus, the oxidation process may include generation of porosity.

Third, these oxides are n-type semiconductors, and O^{2-} ions are formed at the external surface of the oxide and diffuse inward from the oxide surface to the matrix/oxide interface. This process sustains very strongly bonds between the oxide scale and the matrix material.

Fourth, since the crystal structures of ZrC, HfC, ZrN, and HfN are the same (cubic NaCl type) and the structures of the oxides are also the same, very strong similarities should be observed in the oxidation behavior of the two carbides and the two nitrides. One of the significant differences between the Zr and Hf compounds is that while the oxide phases and transformations are the same, the transformation temperatures are different, with those of HfO_2 occurring at higher temperatures than those of ZrO_2.

14.6 OXIDATION OF THE UHTC CARBIDES

The initial oxidation study on Hf and Zr carbides was conducted by Bartlett [65] in 1963 in the temperature range of 450–580°C with an O_2 pressure range of 6.5 Pa up to 100 kPa. Zirconium carbide powder was utilized, but the purity was only described as Hf-free and having 1% total impurities and "less than 0.1% on any single impurity constituent". Free carbon was present in the initial powders but was removed by a flotation technique. The oxidation kinetics were initially diffusion controlled followed by reaction rate-controlled process. He observed elemental C remaining in the oxidation product and a rapid oxidation rate above 500°C. He speculated that a ZrC_xO_y phase may play a role but lacked evidence for the phase. He also noted an increased oxidation rate when water vapor was introduced into the O_2–He gas mixture used for the experiments. Kuriakose [66] used electron-beam melted ZrC samples and found rapid, linear oxidation in the range of 554 to 654°C. Shimada [67] also investigated the low-temperature oxidation of ZrC using well-characterized powders and observed parabolic oxidation from 380 to 550°C at O_2 pressures of 1.3, 2.6, and 7.9 kPa and a total pressure of 39.5 kPa. He speculated that formation of ZrC_xO_y was initially rapid followed by diffusion-controlled formation of an amorphous ZrO_2 scale with embedded hexagonal nuclei. Shimada [68] obtained similar results in his oxidation study of HfC powders from 480 to 600°C. He also speculated that the initial step was uptake of oxygen to form a HfC_xO_y phase and observed a two-layer oxide scale composed of a dense inner layer and a cracked outer layer, both containing ZrO_2 and amorphous C.

Using single crystals of HfC, Shimada [69] reported linear oxidation kinetics at 600°C but parabolic kinetics from 700 to 900°C in O_2–Ar with a P_{O2} of 8 kPa. Samples were oxidized to a total extent of 20%. In this temperature range, a two-layer oxidation zone was again observed. Wavelength-dispersive spectroscopy (WDS), X-ray micro-analysis (XMA), and Raman spectroscopy were used to quantify the elemental compositions of the scale layers. Both the inner and outer layers were HfO_2 with inclusions of carbon. In the dense and crack-free inner layer, very steep carbon and oxygen gradients were identified. During oxidation, the inner layer increased in thickness but then reached an equilibrium value—an additional scale growth occurred in the outer layer. The cracked outer layer of HfO_2 contained 7–14% of elemental C (amorphous). Shimada claimed that cracks did not form *in situ* but rather from the thermal expansion mismatch of HfO_2 with HfC during cooling. The presence of HfC_xO_y was inferred, but was not clearly identified. While not explicitly stated, Hf was preferentially oxidized at the HfC–HfO_2 interface, resulting in C remaining in the scale. Even in the outer scale, P_{O2} was not high enough to fully oxidize the C, but no thermodynamic analysis was provided [70]. Using ZrC samples hot pressed at 2500°C, Shevchenko [71] observed oxidation rate kinetics with an exponent of 1.4–1.6 in the temperature range of 500–700°C. Free carbon was identified in the scale and attributed to preferential oxidation of the Zr. At 750 and 800°C, linear kinetics were observed and attributed to the significantly accelerated C combustion rate. The powder composition included 0.3 wt% free carbon, and the mean grain size of the hot-pressed specimens was approximately 10 μm.

Berkowitz-Mattuck [72] studied the oxidation of ZrC and HfC from 1130 to 2160 K at P_{O2} from 0.5 and 2.6 kPa. Samples of the ZrC were cut from zone-refined bars, which contained 11.2 wt% C, while the HfC was arc melted and a composition of $HfC_{0.952}$ was stated. No grain sizes were identified. Linear kinetics were observed throughout the entire temperature range with preferential oxidation occurring along grain boundaries. Below 1560 K, intergranular fracture was observed. Similar grain boundary disintegration has also been observed in arc-melted Hf–N ceramics (Talmy I. Unpublished data, 2012). Above this temperature, stresses developed during oxidation were sufficiently relieved and the samples did not fracture. Voitovich [73] found that hot-pressed ZrC and HfC oxidized differently in the range 500–1200°C, with HfC showing almost no oxide scale at lower temperatures. This was attributed to the rise in the vacancy concentration in the metallic sublattice of hafnium oxycarbide compared to that in the zirconium analog. He reported that both ZrC and HfC form oxide and free carbon and the C then oxidizes to CO and CO_2.

Shimada [69] used zone-refined HfC single crystals and determined the oxidation kinetics from 700 to 1500°C with P_{O2} values of 0.08 and 80 kPa for 4 h exposures. Linear kinetics were reported for all temperatures, but at longer hold times, the rates at 1000 to 1500°C decreased (no rate exponent provided). This was attributed to the closing of pores by sintering of the oxide scale. At 1500°C, the uptake of oxygen in the HfC lattice was only 1.4%, and the composition of the oxycarbide phase was approximately $HfC_{0.99}O_{0.01}$. He observed a two-layer oxide scale with 23–25 at% C in the inner layer and 6–11 at% C in the outer layer, both as amorphous carbons. The layer junctions exhibited steep oxygen and carbon gradients. The HfC-scale interface probably acted as

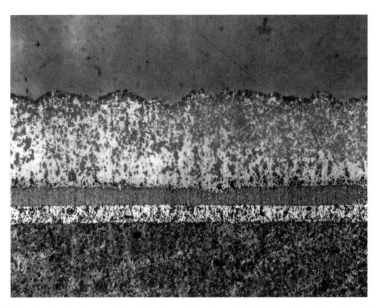

Figure 14.2. Optical micrograph of HfC$_{0.5}$ oxidized at 1865°C for 600 s. Showing the bilayer carbide–oxide scale with sharp interface boundary [74].

the O and C barrier during oxidation. The residual C in both layers points to preferential oxidation of Hf. At the P$_{O2}$ of 80 kPa, oxidation was so rapid that analysis could not be performed. Similar results were observed for ZrC [67].

Bargeron [74, 75] determined oxidation kinetics of HfC prepared by CVD at 1400 and 2060°C and a P$_{O2}$ of 7.2 kPa in Ar with a total pressure of 100 kPa. Based on lattice constants determined by XRD, a composition of HfC$_{0.5}$ was calculated. Parabolic rate kinetics were reported for both temperatures. Although parabolic kinetics transitioned to linear for long-term 2060°C exposure, insufficient data were collected to conclude the same at 1400°C. Using SEM with energy-dispersive X-ray spectroscopy (EDS), a two-layer scale was observed (Fig. 14.2). The outer layer was identified by XRD as HfO$_2$. The inner layer was stated to be HfC$_x$O$_y$ with NaCl structure. Although the two layers exhibited the same O/Hf ratio, the sharp interface was interpreted as a difference in crystal structure.

Holcomb [76] identified the oxidation kinetics of HfC, HfC–25 wt%TaC, and HfC–7 wt%PrC$_2$ from 1200 to 2200°C at O$_2$ pressures of 2, 18, and 100 kPa for exposures up to 75 min. The samples were prepared by hot pressing mixtures of the respective powders, with phase identification by XRD. The TaC addition was intended to mimic Hf–Ta alloy compositions known for good oxidation resistance [60]. The PrC$_2$-containing composition was an attempt to form the pyrochlore-phase Hf$_2$Pr$_2$O$_7$ in situ, which has an anionic mobility and an order of magnitude lower than HfO$_2$. However, the kinetics for the HfC–PrC$_2$ composition could not be obtained due to catastrophic (or runaway) oxidation. The HfC and HfC–TaC compositions exhibited parabolic rate kinetics over the entire temperature range, and rate constants for the HfC–TaC composition were higher than for HfC. Rate constants decreased from 1200 to 1800°C

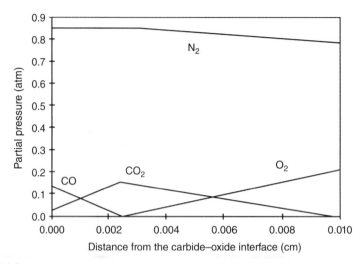

Figure 14.3. Partial pressures across a porous HfO_2 scale at 1400°C calculated using the Courtright–Holcomb gas shuttle model [76, 77].

and then increased to 2200°C. A two-layer oxide scale structure was observed throughout the temperature range. The inner scale was black, relatively pore-free, and dense with cracks oriented normal to the oxide growth direction. The inner layer was proposed to be O-deficient HfO_2; however, it was not verified. A white, porous outer layer was confirmed by XRD to be pure HfO_2. Below 1800°C, a gas transport model was used to explain the kinetics. The formation of CO would occur at the carbide–oxide interface but would be oxidized by incoming O_2 to form CO_2 in the scale where the morphology abruptly changed. The CO_2 acted as a shuttle for oxygen transport through the inner layer of the scale, as shown schematically in Figure 14.3. Courtright [77] correlated oxidation kinetics to anionic diffusion rates in HfO_2 above 1800°C. The scale substantially densified, yet remained sufficiently porous to allow escape of CO. This raises questions about the relative magnitudes of gaseous transport versus ionic diffusion.

As shown in the Hf–C phase diagram [41], hafnium carbide is not a line compound, but has a range of carbon stoichiometries (from 37.5 to 50 at% carbon or from $HfC_{0.6}$ to $HfC_{1.0}$). If it is assumed that the activity of hafnium approaches one near the left side of the HfC_x stability region in the phase diagram and the activity of carbon approaches one toward the right side of that region, then the equilibrium P_{CO} at the oxide/carbide interface can be calculated for each condition, creating upper and lower bounds for the substoichiometric carbides.

Wuchina [78] conducted furnace oxidation testing of near-stoichiometric $HfC_{0.98}$ and substoichiometric compositions of $HfC_{0.82}$ and $HfC_{0.67}$ in air at 1500°C and found significant differences in oxidation rates based on weight gains measured after 15 min. Weight gains of 21, 14, and 1 mg/cm², respectively, were measured for these three compositions as shown in Figure 14.4. Posttest optical photographs also revealed the significant difference in the scale thicknesses as shown in Figure 14.5. Since CO generation at the HfC_x–HfO_2 interface was believed to be a significant factor, the CO pressure

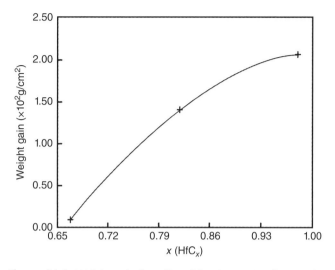

Figure 14.4. Weight gain for HfC$_x$ oxidized at 1500°C for 15 min.

Figure 14.5. Optical micrographs of HfC$_{0.67}$ (left) and HfC$_{0.98}$ (right) samples oxidized at 1500°C for 15 min showing oxide scale thickness [78].

at the interface for the limiting conditions of unit Hf and C activities was calculated and shown in Figure 14.6a. For the case of unit C activity, Reaction 14.5

$$HfC + 2CO \leftrightarrow HfO_2 + 3C \tag{14.5}$$

yielded a P_{CO} of 101 kPa at 1580°C. For the case of unit Hf activity, Reaction 14.6

$$3Hf + 2CO \leftrightarrow HfO_2 + 2HfC \tag{14.6}$$

yielded a P_{CO} of 101 kPa at 2850°C. (These equations were not provided in the original reference.) The P_{CO} values seem to follow the trend in oxide thickness shown

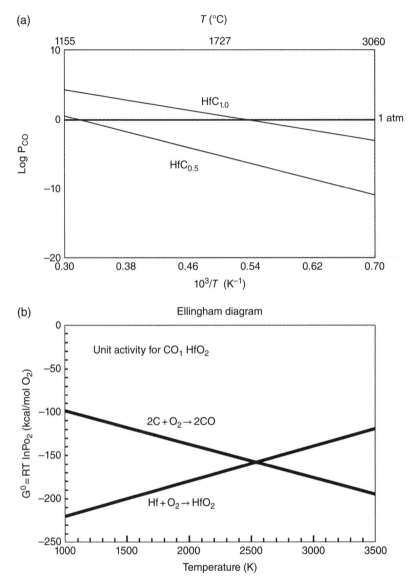

Figure 14.6. (a) Calculated CO pressure at the HfC_x–HfO_2 interface for $x = 1.0$ and 0.5 [78]. (b) Free energies of formation for CO and HfO_2 as a function of temperature (Opeka M. Unpublished data).

in Figure 14.5 for the different carbon stoichiometries. It was also recognized that C could be formed at the HfC–HfO_2 interface through Reaction 14.7:

$$HfC + O_2 \leftrightarrow HfO_2 + C \tag{14.7}$$

and both HfO_2 and C were identified in the scale by WDS and FTIR as found by Shimada [67–70] and Wuchina [78]. Thermochemically, the presence of C at the interface can be shown graphically (Fig. 14.6b) from the free energy versus temperature Ellingham-type diagrams of the hafnium and carbon oxidation reactions (Opeka M. Unpublished data). Additions of TaC, TaB$_2$, WB, and WC, in amounts of 1, 2.5, 5, 10, and 25 mol%, which largely resided at grain boundaries, did not enhance the oxidation resistance of any HfC compositions at 1500°C. For samples tested in the arc heater with surface temperatures in the range of 2000–2400°C and Po$_2$ of approximately 10 kPa (total pressure was 58 kPa), an interlayer of Hf C$_x$O$_y$ was proposed as the composition between the HfC and HfO$_2$ layers. The oxycarbide phase was identified with SEM–WDS, and the absence of electroluminescence and elemental carbon was cited as evidence that the interlayer was not HfO$_2$. The precise composition of the oxycarbide was not identified [79].

14.7 UHTC NITRIDES

The UHTC nitrides ZrN, HfN, and TaN are less well known than the respective carbides and borides, but they do have some advantageous properties. Chiotti [43] cites six reviews on these compounds dating back to 1933. Blocher [80] cites significant research on these transition metal nitrides conducted in the 1940s for the Manhattan Project. Air- and moisture-stable crucible materials were sought, and successfully identified, for the melting of metals including Ce, Fe, and Be. Additional uses for TiN and ZrN coatings include inert surfaces for melting of Zr, Al, and radioactive metals. The most significant body of reported research and development on transition metal nitrides has focused on hard coatings for cutting tool and machining applications. While TiN hard coatings for tool steels became commercially available in the early 1980s, ZrN was also investigated for its higher oxidation resistance [81]. Leverenz reported the benefit of using HfN coatings on WC–Co tools over TiC or TiN due to a higher hot hardness and oxidation resistance, a closer match in thermal expansion coefficient, and chemical inertness [82].

At least two significant literature reviews of the properties of the nitrides have been published [83, 84]. The Hf–N phase diagram [85] is shown in Figure 14.7. Pure Hf exists in the hexagonal close-packed (hcp) structure (α-Hf, $P6_3/mmc$) from room temperature up to 1743°C and transforms to the body-centered cubic (bcc) structure (β-Hf) phase, which exists up to the 2230°C melting point. While the β-Hf phase has only a 3 at% solubility of N, the α-Hf phase is strongly stabilized by N with a solubility up to 29 at% and increases the melting point up to the peritectic temperature of 2910°C. At room temperature, from 46 to slightly above 50 at% N, the mononitride δ-HfN exists with the B1 or cubic $Fm\overline{3}m$ NaCl-type structure, and the substoichiometry range is smaller than for the corresponding NaCl-type monocarbides. Under a N$_2$ pressure of 100 kPa, HfN has a melting temperature of 3387°C. The melting point dramatically increases with N$_2$ pressure, and at 6.0 MPa, the melting point is 3800°C. Because of the preferential loss of nitrogen, the use temperatures of the nitrides are lower than the corresponding carbides. TaN is considerably less stable than HfN, with a significantly lower melting point and decomposition temperature.

From 29 to 46 at% N, the rhombohedral phases ε-Hf$_3$N$_2$ and ζ-Hf$_4$N$_3$ are stable, with each phase having the $R\overline{3}m$ point group symmetry. They were initially identified as a single phase with the hcp Hf$_2$N structure [86].

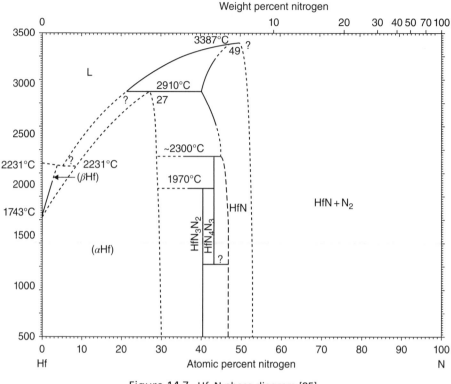

Figure 14.7. Hf–N phase diagram [85].

The ordered phases in the Hf–N system were studied by Rudy [87] who found that Hf_2N did not exist, but identified Hf_3N_2 and Hf_4N_3 as distinct phases. The particular stoichiometries of the two ordered phases can be understood from their structural similarities to the hcp M_2X and fcc (NaCl-type) δ-HfN structures [87, 88]. This becomes evident from the crystallographic (110) sectional cut of both unit cells. The ζ-Hf_4N_3 structure is formed by adding a cubic metal layer (c) after every third layer in the ϵ-Hf_3N_2 structure, leading to the sequence $(hhcc)_3$ in ζ-Hf_4N_3 from ϵ-Hf_3N_2's $(hhc)_3$ [87]. Both phases are rhombohedral and structurally similar to the hcp M_2X phase in the group VB carbides, which led to uncertainty and confusion about the existence of Hf_2N [87–90].

14.8 PREPARATION, DIFFUSION, AND PHASE FORMATION

Lengauer [91] heated Hf sheet and wedges at 1160–1800°C at N_2 pressures from 0.1 to 3.0 MPa to study phase formation in the Hf–N system. He found that during diffusion of N into hafnium metal, an outer layer of HfN_{1-x} was formed, followed inward by the ζHf_4N_{3-x} phase, with ηHf_3N_{2-x} above the nitrogen-saturated alpha-hafnium (α-Hf(N)) and the base hafnium. He also measured diffusion activation energies [92]. Edwards and Malloy [93, 94] found that the reaction rate of nitrogen with hafnium metal follows a simple parabolic law over the range of 876–1034°C within a N_2 pressure range of 5,000–53,000 kPa. An activation energy of 57 kcal/mol was identified.

Lyutaya and Poroskaya [95] used 30 μm hafnium powder to study phase formation at 600–1200°C in nitrogen and ammonia streams. They reported the formation of α-Hf(N) and HfN, as well as hafnium nitrohydrides when using ammonia. The rate of nitridation was lower in ammonia. Dating back to the 1960s in the Soviet Union, SHS [96] was developed and used to prepare a wide variety of refractory materials. Borovinskaya and Pityulin [97] studied combustion of Hf and Zr in nitrogen. They found differences, with Zr having a combustion front that propagated by formation of a solid solution of N in Zr, while the main stage of combustion for Hf was direct formation of HfN_x, making it a slower combustion process. Desmaison-Brut [98] densified HfN powder by HIP at 1950°C and 195 MPa for 1 h, with the resultant ceramics having a Vickers hardness of 14.5 GPa, a bend strength of 350 MPa, and fracture toughness of 4.5 MPa/m$^{1/2}$. Alexandre *et al.* [99] studied the densification behavior of ZrN powder by hot pressing and found that its rate depended on the surface area and shape of starting powders. They also measured mechanical properties as a function of porosity.

Pshenichnaya [100] studied densification of ZrN for use in heating elements and resistors. The starting powder had surface area of 1.35 m^2/g (<1 μm particle size), and she showed the loss of N as sintering temperature increased and that isothermal holds not only increased densification but also contributed to grain growth. In vacuum sintering studies, a postsinter annealing in N_2 was used to return to the original composition. An increase in sintering density in vacuum was attributed to loss of N.

14.9 MECHANICAL AND PHYSICAL PROPERTIES

Few data have been published on the properties for ceramics in the Hf–N system. Opeka *et al.* [49] reported a thermal expansion coefficient for $HfN_{0.92}$ that was almost identical to those of HfB_2, $HfC_{0.98}$, and $HfC_{0.67}$, but deviated to higher values above 1250°C. Room-temperature thermal conductivity for $HfN_{0.92}$ was 15 W/m·K and increased slowly with temperature—the data very closely matches those of the respective carbide. A value of 28 W/m·K was determined for stoichiometric HfN (Opeka M. Unpublished data).

Desmaison-Brut *et al.* [98] obtained a flexural strength of 350 MPa and Young's modulus of 420 GPa at room temperature for HfN with 5% porosity. Room-temperature strength of 290 MPa and modulus of 420 GPa were also obtained for stoichiometric HfN of unknown porosity. In the same study, elevated-temperature properties were acquired. The onset of plastic behavior was just under 1600°C. At the highest test temperature of 2315°C, strengths of 60 MPa and Young's modulus of 55 GPa were recorded. For ZrN with 2% porosity, Alexandre *et al.* [97] measured a room-temperature flexural strength of 330 MPa and Young's modulus of 380 GPa.

Johannson [101] studied the hardness and electrical resistivity of reactively sputtered Hf–N films. Both increase with N content. The highest hardness value for the HfN films was 35 GPa, much higher than that reported for bulk material. While film thickness or substrate data were not reported, the authors claimed that the high hardness was due to a high defect density, primarily dislocations. Straumanis [102] studied the dissolution of nitrides in HF with characterization of structures by XRD and electrical conductivity measurements.

While the thermomechanical properties of alpha-Hf have been described [101], relatively little has been published on this phase with N additions. As was shown in the Hf–N phase diagram in Figure 14.7, α-Hf can dissolve up to 29 at% N. The addition of N to pure Hf significantly increases the melting point, but also embrittles it. Only one study has reported ductility and strength of α-Hf(N) ceramics [103]. While the N-free Hf exhibited a yield point of ≈395 MPa at 20°C (without fracture in compression testing), the addition of 5 at% N (Hf–5N) resulted in a fracture stress of 535 MPa at 20°C and was independent of strain rates between 10^{-3} and $10^{-6} \cdot s^{-1}$. For this composition, the brittle-to-ductile transition temperature was ≈500°C. The Hf–5N also exhibited a temperature-dependent yield stress of 463 MPa at 600°C, 422 MPa at 700°C, and 376 MPa at 800°C. The stress exponent was equal to ≈3.2, indicative of a dislocation deformation mechanism. The addition of greater percentages of N increased the DBTT—yielding occurred at 800°C and 1000°C for Hf–10N and Hf–30N compositions, respectively.

14.10 OXIDATION OF NITRIDES

Relatively little research has been published on the oxidation of ceramics in the Zr–N and Hf–N systems compared to the number for the corresponding carbides and borides. Suni and coworkers [104] reported the oxidation of 2500Å thickness sputtered films of HfN in dry and wet conditions at 450–800°C. The oxidant was described as O_2, but the P_{O_2} was not reported. The wet condition was produced by bubbling the O_2 through hot deionized water. The oxide film thickness was measured using the backscattering yield spectra from 2 MeV $^4He^+$ ions. Parabolic rate kinetics were observed for dry condition throughout the temperature range, and the oxide layer composition was identified as monoclinic HfO_2 by XRD. Rate constants at test temperatures followed a simple Arrhenius relationship, which indicated a diffusion-controlled process throughout this temperature range. A complex oxidation response was seen for wet conditions with **severe pitting** observed in the oxide film.

Caillet and coworkers [105] used 30 μm thick ZrN coatings produced by nitriding Zr samples in ammonia and evaluated oxidation kinetics from 550 to 700°C with a P_{O_2} range of 50–500 torr. Linear rates were observed at all temperatures and pressures and rates increased with P_{O_2}. However, significant, initial mass losses were observed at 550 and 576°C before proceeding to linear kinetics. The scale included the hcp α-Zr phase, as well as the monoclinic and cubic ZrO_2 phases. A mechanism was proposed in which oxidation began with rejection of nitrogen from ZrN, resulting in an oxygen-saturated α-Zr layer that formed prior to ZrO_2.

Desmaison and coworkers [106] prepared HfN by HIP with a relative density of 96%. They evaluated oxidation kinetics by thermal gravimetric analysis. Testing was performed from 755 to 950°C in dry O_2 with a pressure range of 20–100 kPa. Oxidation initiated at 800°C, and a linear, nonprotective response was observed under all test conditions. Rates increased with increasing P_{O_2}. At lower temperatures, a powdered oxide product was formed, while at higher temperatures, an attached, but porous, **Maltese cross** scale morphology resulted. The scale product was identified as monoclinic HfO_2.

Voitovich *et al.* [107] conducted a comprehensive study of the oxidation kinetics of ZrN and HfN from 400 to 1200°C. However, the study was marred by a lack of description of the sample fabrication process and details of test procedures. Since hardness was measured across the scale, it seems that solid samples were used, rather than powders. A test gas mixture of O_2 and N_2 suggests that oxidation experiments were conducted in air. Significant quantitative and qualitative differences were found for the responses of the two compositions. Significant oxidation began for ZrN above 500°C and above 600°C for HfN. For ZrN, the oxidation rates increased significantly up to 700°C yet could be construed as semiprotective since rates decreased with time. From 800 to 1000°C, weight losses were observed, as was found by Caillet [105] at 550 and 576°C, yet it was not clear from the published plots whether Voitovich also observed this between 500 and 800°C. A rapid, linear weight gain was exhibited at test temperatures of 1100 and 1200°C. The oxidation scale products included α-Zr, monoclinic ZrO_2, and cubic ZrO_2, which were claimed to have been stabilized by nitrogen. The HfN exhibited no weight reductions and very low oxidation rates below 700°C, and after an initial rise, the rates increased negligibly at 1000–1200°C. The 700 and 900°C tests revealed high and linear oxidation rates. A significant difference in the stoichiometry ranges of the ZrN_xO_y and HfN_xO_y phases was explained by the very different oxidation responses, and these phases were proposed as the source of the rate-limiting diffusion steps in the oxidation process. Hardness measurements were used to support this claim. No micrographs of samples or oxidation products were provided to support the conclusions, nor was the method used to quantify O and N contents described.

In oxidation experiments described by Wuchina and Opeka [108], furnace oxidation testing of HfN at 1500°C was conducted using the same procedure as was described for the HfC materials. The oxidation of HfN proceeded according to the overall reaction

$$HfN + O_2 \rightarrow HfO_2 + N_2 \qquad (14.8)$$

Similar to the oxidation of carbides, a gaseous product is expected to form at the base material/oxide scale interface. Furnace oxidation testing revealed similarities in oxidation between the HfN_x and HfC_x ceramics. In furnace testing at 1100–1500°C, all nitride compositions formed a porous scale and had a temperature region in which pesting was observed (Wuchina E, Opeka M. Unpublished data). Whiles nitrides did not show significant pesting, the scale was not protective and showed evidence of porosity and cracking (Wuchina E, Opeka M. Unpublished data).

High-temperature arc-heater testing was conducted with these materials at NASA Ames. A nominal 2000°C surface temperature was obtained with a 58 kPa stagnation pressure and test duration of 3 min. The samples were 1.9 cm in diameter and 6.3 mm thick. Three Hf–N compositions were included in the test series: Hf(33N), $HfN_{0.75}$, and $HfN_{0.95}$. The latter two compositions were within the HfN_x phase stability field, while the Hf(33N) was α-Hf near the limit of N solubility in the α-Hf region.

The change of stoichiometry in HfN_x yielded significant differences in oxidation response. The multilayer scale that formed on the near-stoichiometric composition

Figure 14.8. Optical photograph of HfN$_{0.95}$ after 3 min arcjet exposure at 2000°C showing an adherent oxide scale [78].

(HfN$_{0.95}$) was porous and very similar to that observed on the corresponding carbide as shown in Figure 14.8 (Wuchina E, Opeka M. Unpublished data).

A dense oxide scale was formed on the Hf(33N) and HfN$_{0.75}$ specimens. For both compositions, the scale was a wavy bubble across the entire surface and was only attached at the sample periphery. The dense scale allowed accumulation of gas beneath it, and surface temperatures exceeded the yield point of the oxide, resulting in bubble formation. For Hf(33N), the bubble remained intact as shown in Figure 14.9. For the HfN$_{0.75}$ composition, the bubble bust during the test as shown in Figure 14.10. As also shown in Figures 14.9 and 14.10, the scale exhibited plastic deformation in the substoichiometric samples. It was deduced that lower N contents yielded lower N$_2$ gas volumes and/or gas generation rates at the base of the oxide scale, resulting in fewer and smaller pore channels.

This behavior was not observed in arc-heater testing of the respective substoichiometric carbide compositions despite the similarity of forming a gaseous product at the interface between the base compounds and their respective oxide scale. This revealed a qualitative difference in the oxidation of both carbides and nitrides at this test condition, yet the mechanisms are not known.

Rocket nozzle tests of HfN$_{0.95}$ were conducted with a 2000°C nonaluminized propellant at relatively high (vs. furnace and arc-heater testing) total pressures of 3.43 and 10.3 MPa with exposure times of 5, 10, and 20 s. For the 20-s exposures, 25 and 55 μm-thick scales were formed at the 3.43 and 10.3 MPa conditions, respectively, as shown in Figure 14.11 [109]. The nitride ceramics were considered to have the better performance compared to HfC$_{0.67}$, HfC$_{0.98}$, and HfB$_2$ compositions.

(a)

(b)

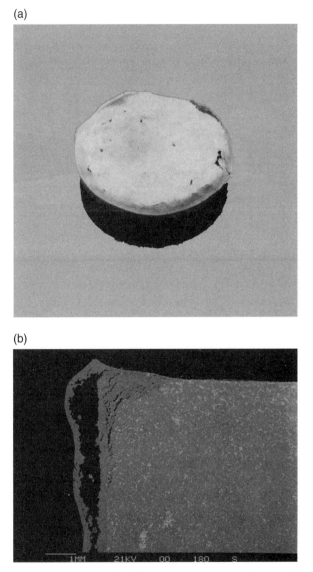

Figure 14.9. (a) Posttest micrograph of Hf(33N) showing wavy oxide scale (Wuchina E, Opeka M. Unpublished data). (b) SEM micrograph of Hf(33N) showing a cavity within the oxide scale (Wuchina E, Opeka M. Unpublished data).

14.11 CONCLUSIONS AND FUTURE RESEARCH

While a substantial body of research has been conducted with group IVB carbide and nitride UHTCs, multiple gaps remain in both the theoretical and practical arenas that need to be addressed. Significant research was dedicated to the nature of interatomic bonding and associated properties of carbides during the 1960s, especially with TiC.

Figure 14.10. Posttest micrograph of $HfN_{0.75}$ showing oxide scale disruption (Wuchina E, Opeka M. Unpublished data).

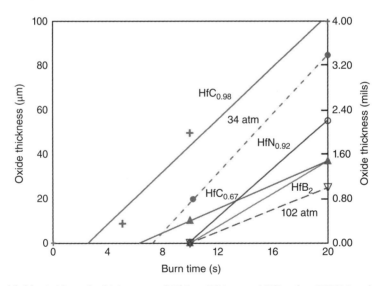

Figure 14.11. Oxide scale thicknesses of $HfC_{0.98}$, $HfN_{0.92}$, and HfB_2 after 2000°C rocket motor firings at 3.43 and 10.3 MPa total pressure for 5, 10, and 20 s.

Very high flexural strength (85.5 GPa) at room temperature was observed in TiC single crystals. The elevated-temperature strength was substantially improved by very small additions (tenths of a percent) of boron, which yielded coherent precipitates on grain boundaries. Plasticity in single crystals tested in flexure and compression was observed at temperatures as low as 800°C. In comparison, plasticity in polycrystalline samples was observed at approximately 1700°C, which decreased with substoichiometric compositions. While less research has been conducted on ZrC and HfC, no differences in the fundamental physics were found or expected. Significantly less research has been carried out on the group IV nitrides than the carbides. Some properties of nitrides, such as mechanical, may be expected to be similar to carbides since both possess the same NaCl structure. The differing electronic structure should result in differences in associated properties.

Relatively modest room-temperature strengths were measured for polycrystalline HfC and ZrC ceramics (maximum of ~250 MPa) and for HfN and ZrN ceramics (maximum of 350 MPa). Composition, purity, density, and grain size all play a major role in governing the mechanical properties of ceramics, but these have not yet been optimized for carbides or nitrides. In addition, very little data exist for these ceramics with second phase additions. Finally, elevated-temperature flexural strength data are insufficient for design of components that operate above the onset of plasticity. Stress–strain data gained from tensile and compressive testing are required.

Conclusions regarding the oxidation of the group IV UHTC carbides and nitrides may be made although the behavior is complex and significant questions remain. The *in situ* formation of gaseous products CO, CO_2, and N_2 yields a porous oxide scale under virtually all testing conditions. For carbides, the kinetics were typically linear followed by parabolic rates yet with large rate constants (e.g., $k_p = 10^{-7} \cdot g^2 cm^4/s$ for HfO_2 on HfC at 1500°C [76, 77]) relative to ceramics forming a slow-growing scale ($k_p = 10^{-11} \cdot g^2 cm^4/s$ for SiO_2 on SiC, and extrapolated to 1500°C [3]). A multilayer scale was observed with significant porosity and cracking found in the outermost layer. A poorly adherent or **pesting** scale was observed for HfC oxidation, but the data of Shimada and others indicate that this behavior is observed at temperatures only below 1200°C and that at higher temperatures the oxide densified and remained adherent. The innermost layer typically consisted of amorphous or graphitic carbon indicating preferential oxidation of the Zr or Hf. The outer layer (or layers) revealed the presence of C within the oxide, yet it was not always clear that an oxycarbide phase had formed. An oxycarbide layer was reported by both Courtright [77] and Wuchina [79] at the oxide/carbide interface after exposures to higher temperatures, although the exact composition was not confirmed. The combustion model by Holcomb is persuasive in understanding the multilayer scale. The CO formed at the surface of the carbide diffuses outward to form CO_2 at the midscale location, where O_2 diffusion inward yields the combustion reaction. Wuchina observed lower oxidation rates in substoichiometric carbide compositions at 1500°C, but this distinction vanished in arc-heater testing at 2000–2400°C. It is speculated here that the higher diffusivity rates of C and O at very high temperatures negated the potential benefits of the substoichiometric compositions. For nitrides, linear oxidation kinetics were always observed, except for the apparent surface sintering observed in arc-heater testing for the substoichiometric compositions. This sintering behavior was not observed for the substoichiometric carbides under the same test conditions.

The oxidation of the group IV UHTC ceramics is still not well understood, and additional investigations of a number of aspects are warranted. First, a rigorous analysis of the relative thermodynamic stability of carbide, nitride, oxide, gas-saturated metal, and C condensed phases and the gaseous phases CO, CO_2, N_2, O_2, and O could provide insight into the oxide scale morphology and attendant oxidation rate. The solid-state and gaseous diffusion rate contributions have not been fully addressed. For example, Courtright correlated HfC oxidation kinetic rates above 1800°C with oxygen ion diffusion rates in the HfO_2 scale. This implied that the ordinarily much more rapid gaseous transport was not the rate-limiting step, but that the scale was dense, which then raised questions about the mechanism for C transport outward. While formation of the bilayer (or trilayer) scale has been observed from under 600 up to 2700°C, the composition and structure of the layers were not fully characterized. Only limited comparisons with the significant body of Hf and Hf–Ta oxidation research, especially considering time-dependent changes in the rate constants, have been made. The oxidation rate reduction associated with substoichiometric carbide compositions has been observed in furnace testing, but not in arc-heater testing, and is not understood mechanistically. The apparent surface densification of the substoichiometric nitrides in arc-heater testing, but not in furnace testing, is also not understood. Finally, the scale morphology development of both the carbides and the nitrides formed under very high pressure and oxidizing rocket motor environments is not understood relative to lower pressure furnace and arc-heater oxidizing environments.

Special attention in future research has to be paid to an increase in the oxidation resistance of these ceramics by the introduction of additives, which can modify the composition of the oxide scale leading to its densification and adhesion at use temperatures. Especially important is optimization of processing of carbide and nitride ceramics with special emphasis on composition and particle size of the starting powders, development of pressureless sintering, and the use of HIP for densification. The latter is particularly promising for significantly reducing densification temperatures, compared to conventional hot pressing, and control of the composition, which is essential in processing nitrides with their tendency to lose nitrogen at high temperatures. The smaller grain size would lead to a significant increase in strength of the final ceramics. HIP, as well as pressureless sintering, presents an opportunity to process ceramic articles in near net shape. Important is also the use of carbides and nitrides in multiphase materials including cermets with Mo or W as a metallic component. Research in these areas should significantly increase the potential of carbides and nitrides in high-temperature applications.

ACKNOWLEDGMENTS

The authors would like to thank Dr. Inna Talmy for her constructive comments and editing.

REFERENCES

1. Wuchina E, Opila E, Opeka M, Fahrenholtz W, Talmy I. UHTCs: Ultra-High temperature ceramic materials for extreme environment applications. Interface 2007;16 (4):30–36.
2. Hubbard LR. Hypersonic materials summary; Requirements and properties. Report number DDRE-WS-2008-TR-04. The Washington Center for Internships and Academic Seminars; July 2008.

3. Opeka MM, Talmy IG, Zaykoski JA. Oxidation-based materials selection for 2000°C+ hypersonic aerosurfaces: theoretical considerations and historical experience. J Mater Sci 2004;39 (19):5887–5904.

4. Deal BE, Grove AS. General relationship for the thermal oxidation of silicon. J Appl Phys 1965;36 (12):3770–3778.

5. Jacobson N. Corrosion of silicon-based ceramics in combustion environments. J Am Ceram Soc 1993;76 (1):3–28.

6. Opila EJ, Hann RE. Paralinear oxidation of CVD SiC in water vapor. J Am Ceram Soc 1997;80 (1):197–205.

7. Sutton GP. *Rocket Propulsion Elements*. 6th ed. New York: John Wiley & Sons, Inc; 1992.

8. Opeka M. Thermodynamics-based materials selection for corrosion-resistant performance. In: Opila E, et al., editors. High temperature missile propulsion systems. Electrochemical Society Proceedings, Volume 2004–16, Pennington, NJ, The Electrochemical Society; 2005. p 253–267.

9. Kaufman L, Nesor H. Stability characterization of refractory materials under high velocity atmospheric flight conditions: part 1, Volume 1, Summary of Results. Report Number AFML-TR-69-84. Wright-Patterson Air Force Base (OH): Air Force Materials Laboratory; 1970.

10. Samsonov GV, editor. *Protective Coatings on Metals (English Translation)*. Volume 1, New York: Consultants Bureau; 1969.

11. Eronyan MA, Avarbe RG, Danisina TN. Effect of equilibrium nitrogen pressure on the melting points of TiN and HfNn (Engl. Transl.). High Temp 1976;14 (2):359–360.

12. Eronyan MA, Avarbe RG, Nikolskaya TA. Determination of the congruent melting point of zirconium nitride (Engl. Transl.). J Appl Chem 1973;46 (2):440.

13. Levinskii YV. Pressure-temperature projection of the phase diagram of the titanium-nitrogen system. Izv Akad Nauk SSSR 1974;10 (9):1628–1631.

14. Kosolapova TY. *Handbook of High Temperature Compounds: Properties, Production, Applications*. New York: Hemisphere Pub. Corp.; 1990.

15. Rudy E. Ternary phase equilibria in transition metal-boron-carbon-silicon systems: part V. Compendium of Phase Diagram Data. Technical Report Number AFML-TR-65-2. Wright-Patterson Air Force Base (OH): Air Force Materials Laboratory; May 1969.

16. McHale A, editor. *Phase Equilibria Diagrams: Phase Diagrams for Ceramists*. Volume X, Westerville (OH): The American Ceramic Society; 1994.

17. Holleck H. Binare und Ternare Carbide und Nitride der Ubergangsmetalle und ihre Phasenbeziehungen. Report KfK 3087-B. Karlsruhe: Kernforschungszentrum Karlsruhe G.m.b.H; January 1981.

18. Kaufman L. Calculation of multicomponent refractory composite phase diagrams. Report Number NSWC TR 86-242. Dahlgren (VA): Naval Surface Weapons Center; 1986.

19. Ordanyan SS. Interaction of rhenium and other refractory metals with certain metal-like compounds. Poroshkovaya Metallurgiya 1975;2 (146):48–53.

20. Ordanyan SS. Interaction in the hafnium boride-tungsten system. Izv Akad Nauk SSSR 1980;16 (5):839–841.

21. Ordanyan SS. Reaction of hafnium diboride with rhenium and chromium. Poroshkovaya Metallurgiya 1980;4 (208):73–78.

22. Schwartzkopf P, Kieffer R. Introduction and History. *Refractory Hard Metals: Borides, Carbides, Nitrides, and Silicides*. New York: MacMillan Company; 1953. p 5.

23. Schwartzkopf P, Kieffer R. Tantalum Carbides. In: *Refractory Hard Metals: Borides, Carbides, Nitrides, and Silicides*. New York: MacMillan Company; 1953.p 121.

24. Gangler J. Some physical properties of eight refractory oxides and carbides. J Am Ceram Soc 1950;33 (12):367–375.

25. Westbrook J, Stover E. Carbides for high-temperature applications. In: Campbell I, Sherwood E, editors. *High Temperature Materials and Technology*. New York: John Wiley & Sons, Inc; 1967. p 312–348.

26. Gangler J. NASA research on refractory compounds. High Temp 1971;3 (5):487–502.

27. Rothman A. Aerospace nuclear propulsion technology. In: Hove J, Riley W, editors. *Ceramics for Advanced Technologies*. New York: John Wiley and Sons, Inc; 1965. p 306–337.

28. Hardy TL. Electrode emission in arc discharges at atmospheric pressure. NASA Technical Report Number NASA-TM-87087. Cleveland (OH): NASA Lewis Research Center; 1985.

29. Perry A. Refractories HfC and HfN-A survey. Powder Metall Int 1987;19 (1):29–35.

30. Perry A. Refractories HfC and HfN-A survey: II, phase relationships. Powder Metall Int 1987;19 (2):32–36.

31. Perry A. Refractories HfC and HfN-A survey: III, cemented carbides and coatings. Powder Metall Int 1987;19 (6):17–19.

32. Smiley WD, Sobon LE, Hruz FM. Mechanical property survey of refractory nonmetallic crystalline materials and intermetallic compounds. Technical Report WADC TR 59-448. Wright-Patterson Air Force Base (OH): Air Force Materials Laboratory; 1960.

33. Battelle Memorial Institute; Metals and Ceramics Information Center. Engineering Property Data on Selected Ceramics, Volume 2, Carbides. Report MCIC-HB-07-VOL. II. Columbus (OH): Metals and Ceramics Information Center; August 1979.

34. Baum PJ. *Group IV Metal Carbides: Processing and Engineering Properties: Critical Review and Technology Assessment*. Rome (NY): Advanced Materials and Processes Technology Information Analysis Center, IIT Research Institute; 1999.

35. Samsonov GV, Vinitskii IM. *Handbook of Refractory Compounds, translated by K. Shaw*. New York: IFI Plenum; 1980.

36. Storms EK. *The Refractory Carbides*. New York: Academic Press; 1967.

37. Toth LE. *Transition Metal Carbides and Nitrides*. New York: Academic Press; 1971.

38. Agte C, Moers K. Methoden zur Reindarstellung hochschmelzender Carbide, Nitride und Boride und Beschreibung einiger ihrer Eigenschaften. Z Anorg Allg Chem 1931;198 (1):233–275.

39. Rudy E. Ternary phase equilibria in transition metal-boron-carbon,-silicon systems: part II. Ternary Systems, Vol I. Ta-Hf-C System. Technical Report Number AFML-TR-65-2 Part II. Wright-Patterson Air Force Base (OH): Air Force Materials Laboratory; 1965.

40. Davis R. Advances in Ceramics. In: Catlow C, Mackrodt W, editors. *Nonstoichiometric Compounds*. Volume 23, Westerville (OH): The American Ceramic Society; 1987. p 529–557.

41. Okamoto H. The C-Hf system. Bull Alloy Phase Diagrams 1990;11 (4):396–403.

42. Achour M, Pialoux A, Dode M. Study of the progressive carboreduction of hafnium dioxide-II: pressure determination for a monovariant equilibrium between 300°C and 650°C. Revue Internationale des Hautes Temperatures et des Réfractaires 1975;12 (3):281–287.

43. Chiotti P. Experimental refractory bodies of high-melting nitrides, carbides, and uranium dioxide. J Am Ceram Soc 1952;35 (5):123–130.

44. Leipold MH, Nielsen TH. Mechanical properties of hot-pressed zirconium carbide tested to 2600°C. J Am Ceram Soc 1964;47 (9):419–424.

45. Sanders WA, Creagh JR, Zalabak C, Gangler JJ. High temperature materials II: Proceedings of a Technical Conference. In: Ault GM, Barclay WF, Munger HP, editors. Metallurgical Society Conferences, Volume 18; Cleveland, Ohio; April 26–27, 1961. p 469–483.

46. Glaser FW, Moskowitz D, Post B. A study of some binary hafnium compounds. Trans Metallur Soc AIME 1953;197 (9):1119–1120.

47. Kendall EG, Riley WC, Slaughter JI. A new class of hypereutectic carbide composites. AIAA J 1966;4 (5):900–905.

48. Kendall EG, Riley WC, Slaughter JI. A new class of hypereutectic carbide composites. Air Force SSD-TR-65-78, Report Number TDR-469 (5250-10)-11. Wright-Patterson Air Force Base (OH): Air Force Materials Laboratory; June 1965.

49. Opeka M, Talmy I, Wuchina E, Zaykoski J, Causey S. Mechanical, thermal, and oxidation properties of refractory hafnium and zirconium compounds. J Eur Ceram Soc 1999;19: 2405–2414.

50. Adams RP, Copeland MI, Deardorff DK, Lincoln RL. Cast hafnium carbide-carbon alloys: preparation, evaluation, and properties. U.S. Bureau of Mines Report Number 7137. Albany (OR): U.S. Bureau of Mines; 1968.

51. Williams WS, Schall R. Elastic deformation, plastic flow, and dislocations in single crystals of titanium carbide. J Appl Phys 1962;33:955–959.

52. Williams WS. Influence of temperature, strain rate, surface condition, and composition on the plasticity of transition metal carbide crystals. J Appl Phys 1964;35 (4):1329–1334.

53. Hollox GE, Smallman R. Plastic behavior of titanium carbide. J Appl Phys 1966;37 (2):818–823.

54. Williams WS. Dispersion hardening of titanium carbide by boron doping. Trans Metallur Soc AIME 1966;236 (2):211–216.

55. Davis RF, Carter CH Jr, Chevacharoenkul S, Bentley J. The occurrence and behavior of dislocations during plastic deformation of selected transition metal and silicon carbides. In: Tressler RE, Bradt RC, editors. *Deformation of Ceramic Materials II*. New York: Springer; 1984. p 97–124.

56. Williams WS. Physics of transition metal carbides. Mater Sci Eng A 1988;105:1–10.

57. DePoorter GL, Wallace TC. Diffusion in binary carbides. In: Eyring L, editor. *Advances in High Temperature Chemistry*. New York: Academic Press; 1971. p 107–136.

58. Valvoda V, Dobiasova L, Karen P. X-Ray diffraction analysis of diffusional alloying of HfC and TaC. J Mater Sci 1985;20:3605–3609.

59. Kofstad P, Espevik S. Kinetic study of high-temperature oxidation of hafnium. J Less Common Metals 1967;12 (5):382–394.

60. Marnoch K. High-temperature oxidation-resistant hafnium-tantalum alloys. J Metals 1965;17 (11):1225–1231.

61. Pilling NB, Bedworth RE. The oxidation of metals at high temperatures. J Instit Metals 1923;29:529–591.

62. Alexandre J, Desmaison A. Comparison of the oxidation behavior of two dense hot isostatically pressed tantalum carbide (TaC and Ta2C) materials. J Eur Ceram Soc 1997;17:1325–1344.

63. American Society for Metals. *ASM Handbook Vol. 13 Corrosion*. Materials Park (OH): *ASM International*; 1987.

64. Shin D, Arroyave R, Liu Z-K. Thermodynamic modelling of the Hf-Si-O system. Calphad 2006;30 (4):375–386.

65. Barlett R, Wadsworth M, Cutler I. The oxidation kinetics of zirconium carbide. Trans Metallur Soc AIME 1963;227:467–472.

66. Kuriakose A, Margrave J. the oxidation kinetics of zirconium diboride and zirconium carbide at high temperatures. J Electrochem Soc 1964;111 (7):827–831.

67. Shimada S, Ishil T. Oxidation kinetics of zirconium carbide at relatively low temperatures. J Am Ceram Soc 1990;73 (10):2804–2808.

68. Shimada S, Inagaki M, Matsui M. Oxidation kinetics of hafnium carbide in the temperature range of 480 to 600°C. J Am Ceram Soc 1992;75 (10):2671–2678.

69. Shimada S, Nakajima K, Inagaki M. Oxidation of single crystals of hafnium carbide in a temperature range of 600° to 900°C. J Am Ceram Soc 1997;80 (7):1749–1756.

70. Shimada S, Yunazar F, Otani S. Oxidation of hafnium carbide and titanium carbide single crystals with the formation of carbon at high temperatures and low oxygen pressures. J Am Ceram Soc 2000;83 (4):721–728.

71. Shevchenko AS, Lyutikov RA, Andrievskii RA, Terekhova VA. Oxidation of zirconium and niobium carbides. Powder Metall Metal Ceram 1980;19 (1):48–52.

72. Berkowitz-Mattuck J. High temperature oxidation IV. Zirconium and hafnium carbides. J Electrochem Soc 1967;114 (10):1030–1033.

73. Voitovich RF, Pugach EA. High temperature oxidation of ZrC and HfC. Soviet Powder Metall Metal Ceram 1973;12 (11):916–921.

74. Bargeron CB, Benson RC, Jette AN, Phillips TE. Oxidation of hafnium carbide in the temperature range 1400°C to 2060°C. J Am Ceram Soc. 1993;76 (4):1040–1046.

75. Bargeron CB, Benson RC. X-ray microanalysis of a hafnium carbide film oxidized at high temperature. Surf Coat Technol 1988;36 (1):111–115.

76. Holcomb GR. The high temperature oxidation of hafnium carbide [PhD dissertation]. Columbus (OH): Ohio State University; 1988.

77. Courtright EL, Prater JT, Holcomb GR, Pierre GS, Rapp RA. oxidation of hafnium carbide and hafnium carbide with additions of tantalum and praseodymium. Oxidation Metals 1991;36 (5-6):423–437.

78. Wuchina E, Opeka M. Oxidation of Hf-based ceramics. Electrochem Soc Proc 1999; 99-38:477–488.

79. Wuchina E, Opeka M. The oxidation behavior of HfC, HfN, and HfB$_2$. Electrochem Soc Proc 2001;2001-12:136–143.

80. Blocher J. Carbide for high temperature applications. In: Campbell IE, Sherwood EM, editors. *High Temperature Materials and Technology*. New York: John Wiley & Sons, Inc; 1967. p 312–348.

81. Korhonen AS, Molarius JM, Penttinen I, Harju E. Hard transition metal nitride films deposited by triode ion plating. Mater Sci Eng A 1988;105:497–501.

82. Leverenz RV. Look at the hafnium nitride coatings. Manuf Eng 1977;79 (1):38–39.

83. Battelle Memorial Institute; Metals and Ceramics Information Center. Engineering Property Data on Selected Ceramics, Volume 1, Nitrides, Metals and Ceramics Information Center, Report Number MCIC-HB-07. Columbus (OH): Metals and Ceramics Information Center; March 1976.

84. Samsonov GV. Nitrides. Report Number FTD-MT-24-62-70, Foreign Technology Division, Wright-Patterson Air Force Base, OH, July 1970.

85. Okamoto H. The Hf-N (hafnium-nitrogen) system. Bull Alloy Phase Diagrams 1990;11 (2):146–149.

86. Zerr A, Miehe G, Riedel R. Synthesis of cubic zirconium and hafnium nitride having the Th3N4 structure. Nat Mater 2003;2 (3):185–198.

87. Rudy E. The crystal structures of H3N2 and Hf4N3. Metallur Mater Trans 1970;1 (5): 1249–1252.

88. Rudy E, Nowotny H. Untersuchungen im System hafnium-tantal-kohlenstoff. Monatshefte Chem Ver Teile Anderer Wissen 1963;94 (3):507–517.

89. Nowotny H, Rudy E, Benesovsky F. Untersuchungen in den Systemen: Hafnium Bor Kohlenstoff und Zirkonium Bor Kohlenstoff. Monatshefte Chem Ver Teile Anderer Wissen 1961;92 (2):393–402.

90. Rudy E, Benesovsky F. Untersuchungen in den Systemen: Hafnium Bor Stickstoff und Zirkonium Bor Stickstoff. Monatshefte Chem Ver Teile Anderer Wissen 1961;92 (2): 415–441.

91. Lengauer W, Rafaja D, Zehetner G, Ettmayer P. The hafnium-nitrogen system: phase equilibria and nitrogen diffusivities obtained from diffusion couples. Acta Mater 1996;44 (8):3331–3338.

92. Lengauer W, Ettmayer P, Rafaja D. Diffusion in the Hf-N system. Mater Sci Forum 1994;155:549–552.

93. Edwards RK, Malloy GT. The rate of reaction of nitrogen with hafnium metal. ONR Contract NONR 1406, Task Order II, Project 051-070. Chicago, IL: Illinois Institute of Technology; December 1956.

94. Edwards RK, Malloy GT. The rate of reaction of nitrogen with hafnium metal. J Phys Chem 1958;62 (1):45–47.

95. Lyutaya MD, Kulik OP, Timofeeva II. Soviet Powder Metall Metal Ceram 1974;13 (9):695–698.

96. Merzhanov, SHS technology AG. Adv Mater 1992;4 (4):294–295.

97. Borovinskaya IP, Pityulin AN. Combustion of hafnium in nitrogen. Combust Explos Shock Waves 1978;14 (1):111–114.

98. Desmaison-Brut M, Montintin J, Valin F, Boncoeur M. Mechanical properties and oxidation behaviour of HIPed hafnium nitride ceramics. J Eur Ceram Soc 1994;13 (4):379–386.

99. Alexandre N, Desmaison-Brut A, Valin FM, Boncoeur M. Mechanical properties of hot isostatically pressed zirconium nitride materials. J Mater Sci 1993;28 (9):2385–2390.

100. Pshenichnaya OV, Kuzenkova MA, Kislyi PS. Sintering of ZrN in vacuum and N2. Powder Metall Metal Ceram 1975;14 (12):986–989.

101. Johannson BO, Sundren JE, Helmerson U. Reactively magnetron sputtered Hf-N films. II. hardness and electrical resistivity. J Appl Phys 1985;58 (8):3112–3117.

102. Straumanis M. Bonding, imperfect structure, and properties of the refractory nitrides of titanium, zirconium, and hafnium. In: Vahldiek FW, Mersol SA, editors. *Anisotropy in Single Crystal Refractory Compounds*. NY: Plenum Press; 1968. p 121–138.

103. Wuchina E, Opeka M, Gutierrez-Mora F, Koritala RE, Goretta KC, Routbort JL. Processing and mechanical properties of materials in the Hf-N system. J Eur Ceram Soc 2002;22: 2571–2576.

104. Suni I, Sigurd D, Ho K, Nicolet M-A. Thermal oxidation of reactively sputtered titanium nitride and hafnium nitride films. J Electrochem Soc 1983;130 (5):1210–1214.

105. Caillet M, Ayedi H, Besson J. Etude de la Corrosion de Revetements Refractaires Sur le Zirconium I. Oxydation par l'oxygene de revetements de nitrure de zirconium. J Less Common Metals 1977;51 (2):305–322.

106. Desmaison-Brut M, Montintin J, Valin F, Boncoeur M. Mechanical properties and oxidation behavior of HIPed hafnium nitride ceramics. J Eur Ceram Soc 1994;13 (4):379–386.

107. Voitovich RF, Pugach EA. High-temperature oxidation of the nitrides of the group IV transition metals II. Oxidation of zirconium and hafnium nitrides. Soviet Powder Metall Metal Ceram 1975;9 (153):63–68.

108. Wuchina E, Opeka M, Causey S, Buesking K, Spain J, Cull A, Routbort J, Guitierrez-Mora F. Designing for Ultra-High temperature applications: the mechanical and thermal properties of HfB_2, HfC_x, and α-Hf(N). J Mater Sci 2004;39:5939–5949.

109. Talmy IG, Zaykoski JA, Opeka MM, Dallek S. Oxidation of ZrB_2 ceramics modified with SiC and group IV–VI transition metal borides. In: McNallan M, Opila E, editors. *High Temperature Corrosion and Materials Chemistry III*. Pennington (NJ): The Electrochemical Society; 2001. p 144–153.

15

NUCLEAR APPLICATIONS FOR ULTRA-HIGH TEMPERATURE CERAMICS AND MAX PHASES

William E. Lee[1], Edoardo Giorgi[1], Robert Harrison[1], Alexandre Maître[2], and Olivier Rapaud[2]

[1] *Centre for Advanced Structural Ceramics, Department of Materials, Imperial College London, London, UK*
[2] *Centre Européen de la Céramique, University of Limoges, Limoges Cedex, France*

This chapter begins with a brief presentation of future nuclear reactors. Next, current uses of ceramics in the nuclear industry are reviewed, which, with the exception of graphite moderators and B_4C control rods, rely largely on oxides or silicates. Then, the potential for using non-oxide Ultra-High Temperature Ceramics (UHTCs) and MAX phases[1] in future (so-called Generation IV) fission and fusion reactors is discussed.

15.1 FUTURE NUCLEAR REACTORS

The feasibility and performance capabilities of next generation nuclear energy systems have been established via the Generation IV International Forum (GIF). GIF, which is a cooperative international endeavor, has identified four main nuclear energy systems that employ a variety of reactor, energy conversion, and fuel cycle technologies. It had to

---

[1]Class of layered ternary compounds with the general formula $M_{n+1}AX_n$ (M is an early transition metal, A is an A group element $n = 1$–3, and X is C and/or N).

Ultra-High Temperature Ceramics: Materials for Extreme Environment Applications, First Edition.
Edited by William G. Fahrenholtz, Eric J. Wuchina, William E. Lee, and Yanchun Zhou.
© 2014 The American Ceramic Society. Published 2014 by John Wiley & Sons, Inc.

allow sustainable development of the nuclear sector [1], to increase its economic competitiveness and to reinforce the reliability and safety of future nuclear reactors. These concepts are described in detail in this section.

The Very High Temperature Reactor (VHTR) concept is a graphite-moderated reactor with a once-through uranium fuel cycle. The VHTR is a type of High-Temperature Reactor (HTR) that can conceptually have an outlet temperature of 1000°C. The reactor core can be either a "prismatic block" or a "pebble-bed" core. The high temperatures enable applications such as process heat or hydrogen production via the thermochemical sulfur–iodine cycle. Helium is the coolant because it is an inert gas, so it will generally not chemically react with any nuclear fuel or cladding material. Additionally, exposing helium to neutron radiation does not make it radioactive, unlike most other possible coolants. Super Critical Water Reactors (SCWRs) are one of the three types of Light Water Reactor (LWR), the other types being Boiling Water Reactors (BWRs) and Pressurized Water Reactors (PWRs). A SCWR usually operates at higher pressure and temperature, with a direct once-through cycle like a BWR, and the water is always in a single, liquid state as in the PWR. The SCWR is a promising advanced nuclear system because of its high thermal efficiency (~45% vs. ~33% for current LWRs) and simple design. A Molten Salt Reactor (MSR) is a class of nuclear fission reactor in which the primary coolant, or even the fuel itself, is a molten salt mixture. MSRs run at higher temperatures than water-cooled reactors for higher thermodynamic efficiency, while staying at low vapor pressure. In many designs, the nuclear fuel is dissolved in the molten fluoride salt coolant as uranium tetrafluoride (UF_4). The fluid becomes critical in a graphite core, which serves as the moderator. Solid fuel designs rely on ceramic fuel dispersed in a graphite matrix, with the molten salt providing low-pressure, high-temperature cooling.

Fast Neutron Reactors (FNRs) are a technological step beyond conventional power reactors. They offer the prospect of vastly more efficient use of uranium resources and the ability to burn actinides, which are otherwise the long-lived component of high-level nuclear wastes. The fast reactor has no moderator and relies on fast neutrons alone to cause fission, which for uranium is less efficient than using slow neutrons. Hence, a fast reactor usually uses plutonium as its basic fuel, since it fissions sufficiently with fast neutrons to sustain the reaction. The coolant is a liquid metal (normally sodium or lead for future reactors) or gas (helium) to avoid any neutron moderation and provide a very efficient heat transfer medium.

In this class, the Gas-cooled Fast Reactor (GFR) (i.e., FNR using helium as coolant) is a nuclear reactor design that is currently under development. It features a fast-neutron spectrum and closed fuel cycle for efficient conversion of fertile uranium and management of actinides. The reference reactor design is a helium-cooled system operating with an outlet temperature of 850°C. Several fuel forms are being considered for their potential to operate at very high temperatures and to ensure retention of fission products (FP): composite ceramic fuel, advanced fuel particles, or ceramic clad elements of actinide compounds.

The International Thermonuclear Experimental Reactor (ITER) program is the next step in studying the fusion reaction, which will be achieved in a tokamak device. Tokamaks use magnetic fields to contain and control the hot plasma in which the fusion reactions occur. The fusion between deuterium and tritium will produce one helium nuclei, one neutron, and energy. The helium nucleus carries an electric charge, which will respond to the magnetic fields of the tokamak and remain confined within the plasma. However, some 80% of the energy produced is carried away from the plasma by the neutron that has no electrical charge and is, therefore, unaffected by magnetic fields. The neutrons will be absorbed by the surrounding walls of the tokamak, transferring their energy to the walls as heat. Consequently, the wall materials will be submitted to extreme working conditions with the temperature reaching 1300°C (under normal conditions) and the fast neutron energy being 14 MeV.

Nuclear fuel and cladding materials that are of interest in these different nuclear reactor concepts are given in Table 15.1 which also summarizes the usual reactor working conditions and the corresponding nuclear fuel materials.

15.2 CURRENT NUCLEAR CERAMICS

Ceramics have properties that make them particularly useful to the nuclear industry, where they are currently used predominantly as fuels and waste forms [2]. They may contain fissionable species and have sufficient thermal conductivity, refractoriness, phase stability, and radiation resistance in fuels. In waste forms, they have the ability to accommodate radionuclides and other waste species, are relatively easy to process, and are durable including against radiation damage. Table 15.2 lists some current nuclear ceramics and their key properties.

Uranium dioxide ceramic is the main fuel used in nuclear reactors. UO_2 is highly stable, the high symmetry fluorite structure demonstrating good corrosion resistance to water and steam making the fuel pellets free from anisotropic effects, with no phase transformations occurring up to a melting point of 2865°C. Fluorite structured UO_2 (space group $Fm\bar{3}m$) exhibits a broad range of stoichiometry ($UO_{2\pm x}$) facilitated by oxygen defects (vacancies or interstitials), where the O/U ratio can reach 1.7 at approximately 2700 K on the oxygen-deficit side and over 2.2 by approximately 2000 K on the oxygen-excess side (see Fig. 15.1).

The major disadvantage of uranium dioxide as a fuel is its poor thermal conductivity (Table 15.2), which is exacerbated by the presence of oxygen defects. Due to the low thermal conductivity of UO_2, the interior of the pellets may reach much higher temperatures during reactor operation than the edges (center line temperatures are often 1100°C, while a few millimeters away at the edge of the pellet, the temperature will be approximately 380°C). To try to account for this, pellets are "dished" at each end to accommodate the increased thermal expansion of the pellet interior relative to that of the edges (Fig. 15.2).

In addition, like most ceramics, the vapor pressure of uranium dioxide increases rapidly with increasing temperature and it displays poor thermal shock resistance at low temperatures, behaving in a brittle manner at temperatures less than 1200°C. However, at temperatures above 1200°C, UO_2 is no longer brittle, which leads to further discrepancy between the behavior of the pellet interior and edge, although only during abnormal

TABLE 15.1. The GIF reactor designs

Acronym	Coolant	Neutron energy	Temperature (°C)	Fuel design	Cladding/structural materials
GFR	Helium	Fast	850	(U, Pu)C (+liner)	SiC
LFR	Pb-Bi or Pb	Fast	480–800	Metal alloy or nitride	Ferritic steel or ceramic (SiC, ZrN)
MSR	Molten fluoride salt	Fast/thermal	700–800	Th-U or U-Pu	Ni-based alloys
SFR	Liquid sodium	Fast	550	(U, Pu)O$_2$ + MA/Na in gap U-Pu-MA-Zr metal alloy	Ferritic or ODS
SCWR	Supercritical water	Thermal/fast	510–625	UO$_2$	Austenitic, ferritic-martensitic stainless steel, Ni-alloy cladding
VHTR	Helium	Thermal	900–1000	TRISO UO$_2$(+UC)	Graphite prismatic-type, graphite pebbles

LFR, Lead-cooled Fast Reactor; ODS, Oxide-Dispersion Strengthened ferritic stainless steels; SFR, Sodium-cooled Fast Reactor; TRISO, TRistructural ISOtropic.

TABLE 15.2. Typical properties of some current nuclear ceramics

Ceramic properties	UO_2	PuO_2	$(U_{0.8}Pu_{0.2})O_2$	Boro-silicate glass	Synroc
Theoretical density (g/cm^3)	10.96 [3]	11.44 [4]	11.04	2.23 [5]	4.35 [5]
Melting point (°C)	3120±20 [4]	2674±20 [4]	3256	1100 [6]	1370 [6]
Thermal conductivity (W/m·K)					
1273°C	2.9 [7]	2.6 [6]	2.6	1.0 (RT) [6]	3.0 (RT) [6]
2273°C	1.8 [7]		2.4		
Crystal structure	Fluorite	Fluorite	Fluorite	—	Hollandite-zirconolite-perovskite
Handling	In air	In air	In air	In air	In air
Processing capability	Demonstrated on industrial scale for aqueous and pilot scale for pyro-processes	Demonstrated on industrial scale for aqueous and pilot scale for pyro-processes	Demonstrated on industrial scale for aqueous and pilot scale for pyro-processes	Decades of industrial	Demonstrated on pilot scale
Fabrication/irradiation experience	Large	Moderate/high	Large	Large	Moderate/high

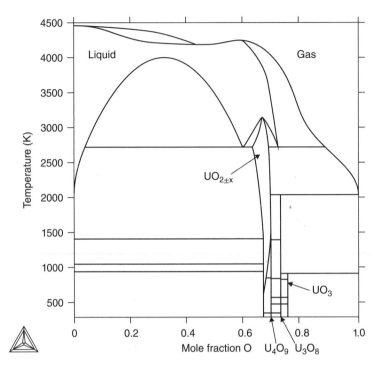

Figure 15.1. Phase diagram of the uranium–oxygen system. Reproduced from Guéneau *et al.* [8].

or transient temperature circumstances. This leads to cracking in UO_2 fuel pellets; the cracks form on initial ramp to power but will evolve further as fuel burnup proceeds.

Mixed Oxide (MOX) fuel was first developed in the late 1950s through research programs in reprocessing, as a mechanism to separate Pu from spent fuel, and in the development of fast reactor systems, to burn Pu as fuel. With the shutdown of the majority of fast reactor programs over the last 30 years, the only remaining option for the recycling of Pu was based upon the burning of Pu in LWRs. Drawing on the fuel research undertaken for fast reactor systems, this has led to the use of MOX fuel assemblies in LWR systems. In general, MOX fabrication involves blending UO_2 and PuO_2 powders with a higher content of plutonium (around 30%). Then the final plutonium enrichment (3–10%) is reached by dilution with uranium dioxide. This step is followed by pressing and sintering at temperatures of approximately 1700°C to produce MOX fuel pellets. These pellets are then loaded into fuel rods, which are then grouped together to form a MOX fuel assembly, in much the same way as for conventional UO_2 fuel (Fig. 15.2). After sintering, the final relative density of the pellets approaches 95%. This process is commonly called MIcronization MASter blend (MIMAS) [9]. In the future, the fabrication process may involve co-milling UO_2 and PuO_2 powders to directly provide the targeted Pu enrichment, ensuring better homogeneity in the nuclear fuel. Whatever the elaboration process, diffusion data are needed to achieve optimized heat treatment and develop a homogeneous Pu distribution within MOX pellets [10].

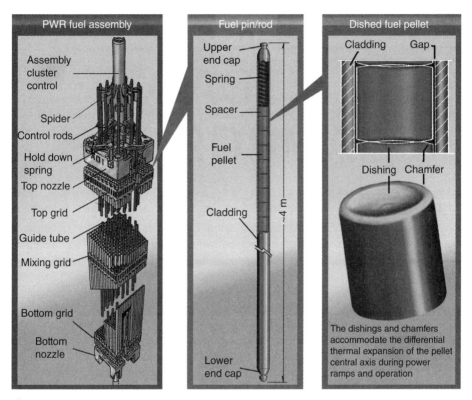

Figure 15.2. Schematic of a 17 × 17 PWR Fuel assembly with inserted control cluster, fuel pin, and dished pellet.

Choosing a suitable waste form to use for radioactive waste immobilization is not easy and durability is not the sole criterion [11]. A variety of matrix materials and techniques are available for immobilization. The main immobilization technologies that are available commercially and that have been demonstrated to be viable are cementation and vitrification, the latter using borosilicate glasses capable of hosting a range of radionuclides in the open glass structure. Ceramication, in which a range of crystals capable of hosting radionuclides in their structures, such as zirconolite, perovskite, and zircon, is close to commercial application [12]. Single phase systems have been examined predominantly for Pu immobilization while more complex waste streams need multiphase systems such as the Synroc ceramics developed in Australia.

While oxides/silicates are the predominant ceramics used by the nuclear industry, currently several non-oxides find key applications. Graphite is used as a moderator in many fission reactors operating today, although most are in the United Kingdom (Advanced Gas-cooled Reactors, AGR) and the countries of the former Soviet Union (water-cooled Reaktor Bolshoy Moshchnosti Kanalniy, RBMK, high power channel-type reactor). The role of graphite, in addition to mechanically supporting the fuel, is to facilitate the nuclear chain reaction by moderation of the high energy (~2 MeV) neutrons. Graphite is well suited to this role as it efficiently slows the neutrons, while being inexpensive, easily fabricated in large quantities, and compatible with the other core materials. Nuclear graphite consists of a classic grain-and-bond microstructure with a filler, binder phase, and controlled porosity. Overall porosity ranges between 14 and 21% with a few large pores constituting the majority of the overall porosity, which also has preferred orientation. Graphite remains the subject of extensive study due to its potential use as a moderator as well as a structural material in the VHTR. Boron carbide (B_4C) control rods are used in all current BWRs and fast reactors to slow neutrons, enabling control of fission reactions. B_4C is used because it is refractory, light weight, and chemically stabile to 2400°C, along with a large neutron absorption cross-section.

15.3 FUTURE NUCLEAR CERAMICS

While non-oxides are inherently more difficult to fabricate than oxide ceramics due to the need for atmosphere control, they find niche applications when they provide property benefits. UHTC properties of interest to the nuclear industry include that they may be fissile, and that they have high thermal conductivity, refractoriness, and phase stability. As a result, future nuclear ceramics will potentially include UHTCs, for example as non-oxide fuels (U/Pu carbides and nitrides) and fuel cladding (TaC, ZrC, HfC). The family of ternary compounds $M_{n+1}AX_n$, where M is an early transition metal, A is an A group element, X is either C or N, and n is from 1 to 3, are known as MAX phases. MAX phases are also likely to find applications in fuel cladding, so we include them here, even though they are strictly not UHTCs. ZrB_2 has already been examined as a potential neutron poison [13]. The main physical properties of these UHTCs are reported in Table 15.3.

MAX phases are attractive in nuclear structural applications in Generation IV fission and fusion reactor applications including for the former in near core fuel coating and cladding uses, so-called Accident Tolerant Fuels (ATFs). Such applications require materials able to function in extreme environments of high temperature, oxidation, stress, irradiation flux, and energy including, for example, above 1000°C in fast neutron environments. While such conditions are not those in which UHTCs are typically designed to operate, we include MAX phases here since they are advanced ceramics that will find application in future nuclear systems. They also, due to their unique crystal and microstructures, have excellent mechanical properties including high strength and toughness at elevated temperatures (Table 15.4).

15.4 NON-OXIDE NUCLEAR FUELS

The high melting temperatures of Pu and U carbides and nitrides make them *de facto* UHTCs. Recent developments of non-oxide nuclear fuels have been reviewed for carbides by Sengupta *et al.* [26] and for nitrides by Arai [27]. Fabrication of non-oxide ceramics is inherently more difficult than oxides since it cannot be carried out in air and the products are often susceptible to oxidation. Add the additional complication of radioactivity and in some cases pyrophoric behavior and the complexity of non-oxide

TABLE 15.3. Typical properties of UHTCs

Ceramic properties	ZrB$_2$	HfB$_2$	TaC	HfC	ZrC
Theoretical density (g/cm^3)	6.09 [14]	11.20 [14]	14.5 [14]	12.67 [14]	6.59 [14]
Melting point (°C)	3040 [14]	3250 [14]	3950 [14]	3928 [14]	3420 [14]
Thermal conductivity (W/m·K)					
25°C	56 [15]	100 [17]	22.1	20.0	25
1000°C	55 [16]	80	—	25 [17]	35
Crystal structure	Hexagonal	Hexagonal	Rock salt	Rock salt	Rock salt
Handling	Inert	Inert	Inert/low moisture	Inert/low moisture	Inert/low moisture
Processing capability	Demonstrated on lab scale	Demonstrated on lab scale	Lab scale	Lab scale	Stoichiometry control issues
Fabrication/irradiation experience	Low	Low	Low	Low	Low

TABLE 15.4. Typical properties of MAX phases

Ceramic properties	Ti_3SiC_2	Ti_3AlC_2	Ti_2AlC
Theoretical density (g/cm³)	4.48 [18]	4.25 [19]	4.42
Melting point (°C)	3000	Decomposes at 1460 [20]	Decomposes at 1400 [21]
Thermal conductivity (W/m·K) 25°C	43	40 [22]	46
Strength (MPa)			
RT	580	760 [23]	763
1200 °C	260	195	270 [24]
Toughness (MPa·m$^{1/2}$)			
RT	10 [25]	9.1	6.5
1200°C	5		
Crystal structure	Hexagonal	Hexagonal	Hexagonal
Handling	Air	Air	Air
Processing capability	Moderate, research scale	Moderate, research scale	Moderate, research scale
Fabrication/irradiation experience	Low	Low	Low

nuclear fuel fabrication become apparent. Nonetheless, carbide and nitride fuels are ideally suited for use in FNRs and GFRs due to their improved performance at high temperatures relative to oxide and metal fuels. Their higher fissile density gives them superior breeding characteristics, which, together with their high specific power operation, provides increased Pu production and reduced fuel cycle and power costs, with a gain in power levels over oxide fuels by a factor of 3 for carbide fuels and a factor of 4 for nitride fuels. Both fuel types can be fabricated either with a small He-bonded fuel/cladding gap or with a large Na-bonded fuel/cladding gap since interactions between fuel and cladding are problematic. These allow the fabricated fuel to be tailored to specific reactor/fuel temperature regimes to manage temperature gradients, fission gas retention, and fuel swelling in a way that effectively combines the most desirable characteristics of both oxide and metal fuels.

Thermodynamic and thermophysical properties of actinide carbides and nitrides have been recently reviewed [26, 27]. Carbide and nitride fuels exhibit higher melting points and thermal conductivity leading to lower temperature gradients within the fuel compared to oxide or metal fuels. This, in turn, leads to a reduction in the migration of fuel constituents and FPs through and out of the fuel. As a result they have been considered for higher linear power applications (up to 70 kW/m). Table 15.5 compares and summarizes the properties of carbide and nitride fuels.

Nitride fuel refers to a solid solution of UN and PuN, namely (U, Pu)N, in which the Pu/(U + Pu) molar ratio ranges from approximately 0.15 to −0.25. In addition, UN was developed as a potential fuel for space reactors in the United States. While interest in such fuels subsided with the demise of the global fast breeder reactor (FBR) programs in the 1980s, the solid solution of UN, PuN, and minor actinide (MA = Np, Am, and Cm)

TABLE 15.5. Properties of carbide and nitride fuels (Adapted from Sengupta *et al.* [26])

Properties	$(U_{0.8}Pu_{0.2})C$	$(U_{0.8}Pu_{0.2})N$
Theoretical density (g/cm³)	13.58	14.32
Melting point (°C)	2477	2797
Thermal conductivity (W/m·K)		
727°C	18.8	15.8
1727°C	21.2	20.1
Crystal structure	Rock salt	Rock salt
Handling	Inert atmosphere	Inert atmosphere
Dissolution and reprocessing capability	Dissolution not simple. Not yet demonstrated on industrial scale	Dissolution easy but risk of ¹⁴C in waste management
Fabrication/irradiation experience	Limited	Very little

mononitride (U, Pu, MA)N has been proposed as a candidate fuel for Generation IV fast reactors. In addition, U-free nitride fuel (such as (Pu, MA)N diluted by ZrN) for MA transmutation has been extensively studied in Japan.

Compared to oxide fuels, nitride fuels, in particular, have only one moderating atom per molecule and are compatible with existing established oxide fuel fabrication and reprocessing methods, with easier dissolution in the Plutonium and URanium Extraction (PUREX) process compared to carbide fuels.

Although the nitride fuels show a higher gain in power levels, they have some disadvantages over carbides. First, they sublime or dissociate into liquid U and N_2 gas at temperatures below their melting point if N_2 overpressure is not maintained. Increased pressure from the build up of N_2 gas can cause the fuel cell to swell and liquid U is highly corrosive to the fuel cell cladding; both factors can cause fuel assemblies to fail and have led to the use of additives such as Zr, Ti, and W to form a more stable fuel [28]. Second, the thermal neutron absorption cross-section of ¹⁴N (resulting in ¹⁴C) is high enough to reduce the breeder ratio of the fuel, although this can be addressed by enriching the ¹⁵N content of the natural N_2 used in the fabrication, albeit at additional cost. The reprocessing of nitride fuels fabricated using natural N_2 is also a concern as it yields significant volumes of ¹⁴C, which is biologically hazardous, although again this could be addressed by enriching in ¹⁵N. Nitride fuel is less hygroscopic than carbide fuel making processing easier, and its better dissolution in nitric acid without any Pu oxalate formation makes it compatible with hydrochemical reprocessing technology such as the established PUREX process.

Initially, melting and casting techniques or hydration processes were used for the fabrication of carbide fuels, but they ultimately proved too expensive and produced coarse-grained pellets, not as desirable as the fine-grained pellets resulting from powder-metallurgy processes. Later, at the end of 1960s, carbothermic reduction and internal gelation routes were examined. In the carbothermic method, uranyl and plutonium oxalate feedstocks (or uranium dioxide that has been previously reduced) are converted to UO_2 and PuO_2 by heat treatment to decompose the oxalate and thereby remove water and carbon oxides. UO_2 and PuO_2 powders are then blended to an enrichment level of typically 20% Pu and mixed with carbon in the form of either carbon black or graphite. To encourage formation of a homogeneous product, the mixed powder is consolidated by briquetting before undergoing carbothermic reduction, typically by heating for several hours between 1400 and 1700°C under vacuum or inert gas because of the high carbide reactivity in air (formation of oxides) or in nitrogen (formation of nitrides) atmospheres. The (U, Pu)C product is mechanically crushed and milled to produce a powder of particle size less than 44 µm, which is pressed under approximately 410 MPa. Then, pellets are sintered in Ar/H$_2$ at temperatures higher than 1650–1700°C to reach relative density values between 80 and 90% [29].

The method of internal gelation uses colloidal carbon in a co-conversion process to produce spherical carbide particles from nitrate feedstocks, thereby bypassing the mechanical milling, crushing, and blending processes employed in carbothermic reduction [30]. A homogeneous solution of the uranyl and plutonium nitrates is mixed with hexamethylenetetramine, urea and dispersed carbon black, and dropped into hot silicone oil to decompose the hexamethylenetetramine to ammonia. This precipitates ammonium diurinate (or plutonate) as microspheres within droplets that are washed, dried, and calcined to remove the silicone oil, solvents, and any volatiles. Sintering for 8 h at 1950°C under Ar produces carbide microspheres of greater than 95% theoretical density, which can be loaded directly into fuel pins.

Carbide fuel is a multiphase mixture of (U,Pu)C and (U,Pu)$_2$C$_3$ to avoid metal phase formation and to improve in-pile behavior. Indeed, a (U,Pu)C$_{1.00}$ single phase cannot be elaborated from the carboreduction route because the kinetics of carbon monoxide release during the carburizing treatment are dependent on those of the densification. Furthermore, effective oxygen removal from the carbide structure requires an increase in the carboreduction temperature or the use of severe atmospheres (vacuum). These conditions promote the volatilization of species such as metallic plutonium and make compositional control difficult [31, 32]. Finally, production of biphasic carbides with very low oxygen contents (<100 ppm) is possible for low Pu/(U + Pu) atomic ratios (around 20 wt%).

Carbide fuels have been used in test reactors in, for example, the United States (EBR-II), Russia (BOR60), Japan (JFR 2), and France (RAPSODIE). The Indian FBR program started with Pu-rich mixed U,Pu carbide as the driver fuel in its loop-type fast breeder test reactor that went critical in 1985 and remains the only reactor operating on a full core of carbide fuel.

As with carbides, nitride fuels can be fabricated via a carbothermic reduction process. In the fabrication of nitride fuels, following the carbothermic reduction of the oxide to the carbide, a further reduction step is performed by soaking the carbide in N$_2$-6% H$_2$ gas to form the nitride and liberate the carbon in the form of methane gas [33, 34]. This nitride powder is then mechanically crushed and milled to form the feed powder for pressing as in carbide fuel fabrication. Nitride fuels have also been fabricated through a hydride-dehydride-nitride process and other sol-gel and pyrochemical processes described by Arai [27]. Before non-oxide fuels can be used in commercial reactors, further research is needed, in particular, on both hydrochemical and pyrochemical reprocessing methods, fuel-clad interaction, effects of burnup, and irradiation performance.

15.4.1 Composite Fuels

Composite fuels consisting of a fissile phase dispersed in an inert, nonfuel matrix have been developed for high temperature and/or high power density applications such as material test reactors, isotope producing reactors, and reactors developed for both power generation and propulsion in space. The two main composite fuels are CERamic METal composites (CERMETS) consisting of ceramic fuel particles dispersed in a metal matrix and CERamic CERamic composites (CERCERS) consisting of ceramic fuel particles dispersed in a ceramic matrix. In particular, a composite with a Mo metal matrix and $(Pu, Np, Am, Cm)O_{2-x}$ fuel particles represents a promising CERMET fuel candidate dedicated to transmutation of MA accumulated in spent fuel from LWRs [35]. Indeed, this composite fuel reduces the reactivity swing and increases the energetic efficiency of MA transmutation.

CERCER fuels in the form of pellets are often considered as replacement materials for conventional uranium dioxide pellets. Their robust nature and high burn-up performance has encouraged research on composite fuels since the 1950s, but the largest current use of composite fuels is in research and test reactors where high power density is required, often at high burnup. Composite fuels are distinguished from the usual fuel types by the localization of fuel material within an inert matrix (as opposed to a solid solution of fissile material in an inert matrix). Indeed, an advantage of the composite design is that radiation damage is primarily localized in the dispersed phase, in which case the inclusions are greater than 100 µm in size [36]. A composite fuel is shown schematically in Figure 15.3.

Fuel particles are, in effect, individually clad in the matrix material (Fig. 15.3a), or the particles can be coated with a buffer layer (Fig. 15.3b). Encapsulation of fuel particles in a matrix prevents transport of fission gas and solid FPs outside of the local environment. Suitable matrix/buffer materials provide mechanical resistance to radiation-induced swelling. Particle dispersions are further classified as macro- and microdispersions. The particle size distribution in a macrodispersion (typically 50–200 µm) is selected to retain most of the fission fragments inside the fuel particles, leaving the majority of the matrix undamaged. Microdispersions of particles with diameters of a few microns have been considered for some concepts because of the convenience of fabrication or prevention of matrix cracking in CERCER fuels during irradiation. Macrodispersed particles may be fabricated with or without a coating layer. The simplest concept consists of a fuel particle embedded in an inert matrix. The particle may incorporate a thin barrier coating to prevent fuel-matrix chemical interaction on processing and irradiation. Such dispersions have proven effective in combination with ductile matrices such as steel or niobium. A buffer layer can be used to more effectively isolate the fuel particles from the matrix (Fig. 15.3b). The buffer layer may provide free volume for fission gas accumulation or fuel particle swelling without stressing the matrix. It may be free space, a coating, or a series of coatings. The most highly developed variants of coated particle fuel are the TRistructural ISOtropic (TRISO)[2]-coated fuel particles, which evolved from Bistructural ISOtropic (BISO) pellets and are discussed later. However,

---

[2]The fuel kernel is coated with four layers of three isotropic materials.

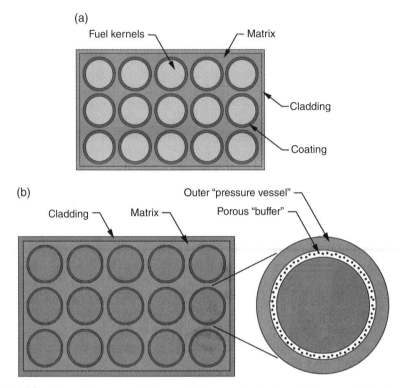

<u>Figure 15.3.</u> Schematic representations of (a) composite fuel and (b) composite fuel using a multilayer-coated fuel particle in which the coating acts as a buffer layer between the fuel particle and the matrix material. Reproduced from Meyer [37].

variants such as the QUADRISO (QUADRuple ISOtropic) concept have been suggested that incorporate an additional layer of burnable poison such as Eu_2O_3 [38].

TRISO fuel particles are envisaged for VHTRs considering two main fuel element concepts [39]: (1) the spherical fuel element used in the pebble-bed concept in Germany, Russia, China, and South Africa and (2) the block-type fuel element applied in the prismatic core in the United States, the United Kingdom, and Japan.

As reviewed by Sawa [40] and Petti *et al.* [41] the spherical fuel element is a graphitic sphere with a diameter of 60 mm composed of a fuel zone of 50 mm in diameter with around 10^4 coated ~0.5 mm diameter particles. The particles are over-coated with a matrix graphite layer with a thickness of around 200 μm to prevent direct contact of the particles and then dispersed uniformly in the same matrix graphite material. The outermost 5 mm of the fuel sphere is a shell of matrix graphite without any particles. The block-type fuel element in the U.S. design is a hexagonal graphite block (793 mm in length and 360 mm wide across the flat surface) containing 102 coolant channels and 210 fuel holes which are filled with fuel compacts and sealed. The fuel compacts are a mixture of TRISO-coated fissile and fertile particles and graphite shim particles bonded by a carbonaceous matrix. A fuel compact made

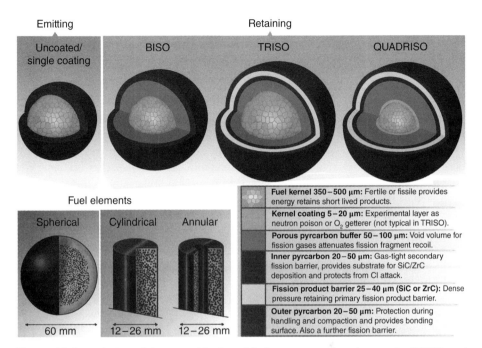

Figure 15.4. Evolution of the coated fuel particle from uncoated to the standard TRISO and proposed QUADRISO as well as common fuel element compacts.

of graphite matrix powder with the shape of an annular cylinder contains 13,500 TRISO-coated fissile particles. For operational and economic reasons, the preferred fuel kernel is low-enriched uranium dioxide for the pebble bed reactors and uranium oxycarbide for the prismatic designs.

Fuel designs for the VHTR, even though they are being developed in various countries, use the same generic refractory coated fuel particles. So, in TRISO, the actinide fuel is distributed in thousands of tiny spherical particles (Fig. 15.4).

In TRISO, the fuel kernel is a 300–500 µm diameter sphere of UO_2, UC_2, or a mixture of the two which has been termed uranium oxycarbide or UCO. Kernels based on Pu or Th are also possible. The kernel is the first barrier to FP release, as FPs must first diffuse to its outer surface. Kernels are coated by chemical vapor deposition in a fluidized bed reactor. The first coating is a 60–100 µm thick buffer layer of porous pyrolytic carbon, with acetylene as the carbon source and argon as the fluidizing gas. The buffer accommodates fission gases and allows for thermal expansion mismatch between the kernel and dense coatings. The key barrier to FP diffusion is a 35–40 µm thick layer of silicon carbide, sandwiched between two dense pyrocarbon layers, each 40 µm thick. The dense pyrocarbon layers are derived from propylene in argon, while SiC is obtained from methyltrichlorosilane in hydrogen and argon. The coated TRISO particle is 0.7–0.9 mm in diameter. The SiC layer acts as a load-bearing pressure vessel and is impermeable to FP release. The inner pyrocarbon layer protects the kernel from chlorine attack during SiC deposition and the SiC from carbon monoxide generated in the kernel upon fission, while the outer pyrocarbon layer adds structural support and is a substrate for bonding to the matrix when compacted into fuel elements.

TRISO fuel fabrication routes have been reviewed by Sawa [40]. Much research is being performed on TRISO fuels for Generation IV reactors focused on extending the fuel capabilities to higher burnups (10–20%) and operating temperatures. The higher temperatures envisaged in the VHTR (fuel temperature of 1250°C) compared to the Pebble Bed Modular Reactors mean that the SiC retention layer will be inefficient and alternatives (such as ZrC) are being examined. Minato and Ogawa [42] describe advanced concepts in TRISO fuel including developments in ZrC and TiN coatings while potential fuel recycling and MA burning in these fuels are also being examined.

15.4.2 Inert Matrix Fuels

At present, the only method for the commercial recycling of Pu is as MOX fuel. Current strategies for burning MOX in LWRs are not fully optimized in terms of Pu burnup. One option to improve Pu consumption rates is to optimize reactor design to accommodate a 100% MOX core. An alternative strategy is to replace the UO_2 component of the fuel with a neutron inactive or neutron transparent material—the "inert matrix"—thereby substantially reducing the capacity of the fuel for Pu production.

Strictly, an inert matrix fuel (IMF) refers to any nuclear fuel that does not contain a "fertile isotope" so that no conversion of fertile to fissile material takes place and only the initially present fissile atoms are burnt. This concept was originally aimed at improving fuel properties or to save U resources [43]. Thus, in the plutonium cycle, IMFs are those that are free from ^{238}U, and an inert material that does not contain a fissionable or fertile isotope is used as the matrix with a plutonium-bearing material. Spent IMF is much less radiotoxic than spent MOX and so disposal is easier. IMF is also used in the context of U-free fuels for transmutation of MA, although in many cases this is inappropriate as the fissile content is inadequate for fuel purposes and it is more correct to call these systems transmutation targets. Actinide-bearing fuels and transmutation targets, including those made by adding MAs to $(U,Pu)O_2$ at a low enough content that the performance of the fuel and core is unaffected, were discussed by Pillon [44]. The MAs form solid solutions with each other over large compositional ranges so that they are generally introduced in this form. However, the properties of actinide carbides and nitrides may make them more suitable than oxides as transmutation targets. Possible ceramic or metallic candidate materials can be chosen as IMFs using the following criteria [45]: (i) the inert matrix must be transparent to neutrons, its neutron cross-section must, therefore, be low and no FPs should be generated; (ii) specific physical properties are required such as high melting temperature. In the case of a phase transformation, the temperature must be at a value well above the operating temperature of the fuel; (iii) compatibility with neighboring materials is also a concern (cladding, coolant, fissile phase, and FPs). Indeed, new phases formed as a result of chemical reactions between the matrix and the fissile phase may lower the fuel melting temperature; (iv) the last criterion concerns the behavior under irradiation. Three main areas are of concern: the behavior of materials with respect to irradiation damage, and the thermomechanical and thermochemical behaviors. Considering these different criteria, IMFs fall into three categories [46]:

- Homogeneous fuels, in which the Pu forms a solid solution with the inert matrix. Principle candidate materials of this type of inert matrix are ceramic oxides such as ZrO_2, $(Y,Zr)O_2$ (YSZ), and CeO_2. These ceramics were chosen because of the small neutron cross-section of their components, high melting points, and low reactivities with the Zircaloy cladding and reactor water [47]. A major disadvantage of these matrices is their relatively low thermal conductivities compared to UO_2 at room temperature, which could induce large thermal gradients within the fuel.
- Heterogeneous particle fuels, in which the Pu particles are embedded in an inert matrix, such as MgO.
- Hybrid or composite fuels, which can either be CERCERS such as a (Y,Pu,Zr) O_{2-x} solid solution embedded in a YSZ matrix, or CERMETS, where the metal could be Mo, Cr, Al or certain ferritic steels.

Much of the key work in this area has been performed by the CEA in France and at the European Institute for TransUranium elements (ITU) in Karlsruhe, Germany. The primary concern for any potential inert matrix material is its neutronic properties; to be consistent with the definition of an IMF, it must remain neutron transparent, typically with a thermal neutron absorption cross-section smaller than 0.2×10^{-24} cm^{-2}. Beyond this, typical properties for an IMF are the same as those for any fuel: suitable thermal conductivity and acceptable thermal expansion coefficient, phase stability over large temperature ranges and high irradiation doses, and inactivity with respect to FPs, reactor water, and cladding. Much like composite fuels, fabrication of IMF pellets may be achieved through one of the three routes: coprecipitation of the oxy-hydroxides from concentrated nitrate solutions of all components, a dry powder mixing/milling route, or via a sol-gel process [48].

15.4.3 Other Fuel Cladding Applications

SiC$_f$/SiC composite ceramics (which like MAX phases are not UHTCs) are being examined as refractory cladding systems (e.g., in future GFRs as ATFs) to avoid the sort of problem that occurred at Fukushima where melting of the zircalloy cladding led to hydrogen generation and the explosions that destroyed the reactor buildings [49]. Indeed, SiC$_f$/SiC composites show toughness six times as high as the corresponding silicon carbide monoliths. Nevertheless, the significant porosity of typical SiC$_f$/SiC composite ceramics leads to a decrease in the thermal conductivity and to permeability to the FPs [50]. To enhance the density of composites, promising methods of synthesis being investigated include the Nanopowder Infiltration and Transient Eutectoid (NITE) process. In the same manner, replacement of SiC-based composites by high thermal conductivity ceramics such as TiC or ZrC is being examined. These latter materials show an excellent ability to retain the FPs under GFR severe accident scenarios [51].

The cladding material in GFRs is made of non-oxide ceramics but also of metallic liners (W-Re, Mo-Re, or Nb-Zr) that surround the ceramics. The inner liner is located between the nuclear fuels and ceramics and is dedicated to fission production retention

[52]. The outer liner avoids migration of the helium coolant in the cladding ceramics. Increasing the number of interfaces in the fuel assembly with additional liners may lead to damaging reactions between the liner and (U, Pu)C, leading to liquid phases from 1880°C or to reactions between the SiC-based cladding materials and liner (Mo, W, Re) causing formation of low melting mixed phases at approximately 1200°C.

Hoffmann [25] examined the potential of the MAX phases Ti_3SiC_2, Ti_3AlC_2, and Ti_2AlC for such applications. Many Al-containing MAX phases are stable in inert atmospheres up to >1500°C. Ti_3SiC_2 is stable up to 2200°C. However, on exposure to oxidizing atmospheres at high temperature, MAX phases form oxide scales, the nature of which varies with the composition of the MAX phase. In Ti_2AlC, the scales are protective to 1350°C for 8000 cycles, enabling their commercial application in furnace heating elements. The oxide scales of dense polycrystalline Ti_2AlC and Ti_3AlC_2 in air are comprised of Al_2O_3 and TiO_2 layers at temperatures below 1200°C. At higher temperatures, Al_2TiO_5 forms, which has a high thermal expansion coefficient and causes cracking of the protective oxide, limiting oxidation resistance above 1400°C.

A key factor in using materials in nuclear applications is activation during use, making waste disposal challenging due to transmutation reactions leading to formation of long-lived and highly radioactive species. MAX phases were predicted and empirically observed to behave well when exposed to idealized fast and thermal reactor neutron spectra, behaving similarly to SiC and much better than potential metal alloys as they are composed of low Z elements that exhibit no long-term activation. Ti_3SiC_2 and Ti_3AlC_2 have also been observed to have high tolerances to radiation damage due to the nature of their bonding and structural ability to recover from atomic displacement damage. The combination of oxidation resistance to 1400°C, tolerance to radiation damage, good mechanical properties, ease of fabrication, and lack of activation make MAX phases strong candidates for future reactor structural applications including fuel cladding.

15.5 OTHER POSSIBLE FUTURE FISSION AND FUSION APPLICATIONS

Advances in waste separation technology [53] afford the opportunity to develop ceramic targets containing separated wastes (e.g., Minor Actinides, MA), which can then be bombarded with neutrons in a reactor or accelerator to induce transmutation and remove difficult radionuclides. International partitioning and transmutation (P&T) programs have examined ceramic transmutation targets. The differences between IMFs for burning in a reactor and P&T targets are somewhat diffuse. P&T aims at separating long-lived isotopes, followed by their nuclear transformation into shorter-lived nuclides—based in part upon the precept that this has a safety benefit with respect to waste management. Reprocessing of spent nuclear fuel when the aim is to recycle actinides as fuel (e.g., as MOX) is effectively a form of P&T (in that it reduces the proportion of radionuclides with long half-lives in the resulting waste stream). Transmutation targets come in two main forms: *homogeneous* actinide-containing solid solutions (e.g., $(Y,Zr,Cm)O_{2-x}$ and $(Am,Y)N$ made by sol gel routes) or *heterogeneous* composites of sol-gel infiltrated actinide-containing particles (e.g., $(Y,Zr,Cm)O_{2-x}$, $(Pu,MA,Zr)O_2$ and $(Pu,MA,Zr)N$ mixed with inert matrix phases such as MgO, $MgAl_2O_4$, TiN, and ZrN, then pressed and sintered). Heterogeneous fuels have the benefit of minimizing radiation-induced property changes by localizing fission heavy ion damage to isolated regions containing the MA.

While graphite moderators have been used in some reactor designs (e.g., UK AGR) and B_4C control rods in others (e.g., BWRs), other large volume applications of non-oxide ceramics in current fission reactors have been limited. However, due to the far higher temperatures that will occur in, for example, the VHTR, large volumes of non-oxide, particularly graphitic and carbon-fiber reinforced carbon matrix composite (C/C) ceramics will be needed. For example, C/C composites are being examined for control rod elements in the VHTR with a focus on the impact of temperature and irradiation damage on properties and dimensional change. It has been shown that graphite material databases can be used to support evaluation of graphitised C/C [54]. In addition, SiC/SiC and C/C composites are potential candidates for blanket structural applications in Magnetic Confinement Fusion (MCF) tokamaks. The possibility of using monolithic and composite SiC thermal insulation for both fission and fusion systems is under investigation, although clearly oxidation behavior in service and in accident scenarios must be well understood [55].

Such applications require materials able to function in extreme environments including of high temperature, oxidation, stress, and irradiation flux and energy including, for example, above 1000°C in fast neutron environments. Hoffman *et al.* [25] examined the potential of Ti_3SiC_2, Ti_3AlC_2, and Ti_2AlC for such applications. UHTCs are typically defined as ceramics designed to operate at temperatures above 2000°C, and since the U and Pu carbides and nitrides illustrated in Table 15.5 typically melt at temperatures above 3000°C they are *de facto* UHTCs. Significant challenges remain including understanding neutron irradiation damage and its impact on dimensions and thermal conductivity, physical sputtering, chemical erosion, radiation-enhanced sublimation, and joining of carbon fiber composites to heat sink materials. Radiation damage is a key issue in all materials that will be used in fusion reactors.

Uses of ceramics in fusion reactor systems will be both functional (such as the ceramic superconductors in the magnet systems for controlling the plasma) and structural in various locations outside of the first wall in MCF including use of single-crystal sapphire in diagnostic windows and ports. A recent review [56] emphasized that the plasma facing components, first wall, and blanket systems of tokamak-based fusion power plants arguably represent the single greatest materials engineering challenge of all time. Wirth *et al.* [56] highlighted the numerous multiscale modeling grand challenges, in particular the plasma materials interactions, the extreme heat and particle flux environments and complexity of extracting the tritium from breeder blankets, and the large and time-varying thermomechanical stresses in structural materials all subject to 14.1 MeV neutrons causing extensive irradiation damage, which has significant effects on thermal, mechanical, and electrical properties. Snead and Ferraris [57] highlight the current use and future potential for carbon as a plasma facing material in tokamaks. Potential fusion applications of UHTCs include tokamak diverters and plasma facing materials.

15.6 THERMODYNAMICS OF NUCLEAR SYSTEMS

Determination of accurate phase equilibria and thermochemical behavior in such fuels is challenging for several reasons: (i) 60 elements are generated by actinide fission, (ii) complex metals and oxide solid solutions are formed, (iii) thermally and compositionally driven transport processes result in compositional inhomogeneities, and (iv) radiation effects in such complex phases are not well understood. Relationships among components in these complex and radioactive phases are defined only for a few systems. To begin to address such problems, an assessed database of phase equilibria and thermochemical values is required, as have been developed via the European Actinet and F-Bridge programs leading to the FUELBASE database for advanced nuclear fuel materials systems such as U-Pu-O-C, U-Pu-Am-Np-O, and U-Pu-Si-C [58].

The FUELBASE thermodynamic database is based on critical assessments and on the modeling of binary and ternary systems using the calculation of phase diagrams (CalPhaD) method, allowing further developments of higher order systems. This method relies on the Gibbs energy evaluation of each phase, as simple in appearance as pure elements [59], stoichiometric compounds (SSUB, SGTE) [60], and solid solutions. The Gibbs energy function of a specific phase is evaluated as polynomial expressions of the temperature. Various models are used for the chemical description of the considered phase (mainly related to crystallographic and ordering considerations) [61].

This database could be used for equilibrium calculations, representative of different steps in fuel processing from the fabrication to the behavior in operation, the interactions between the fuel and the selected cladding material, and also for different scenarios (nominal and off-normal operating conditions). It also allows selection of different materials based on possible interactions (solubility, formation of an undesired compound or a liquid phase). Besides the systems concerning U- and Pu-based kernels themselves, this database includes ceramic systems for cladding or structural materials and takes into account possible interactions between all these constituents with the fuel kernel.

This thermodynamic database represents an important modeling effort that allows the selection of materials for specific operating conditions without dedicated hazardous and difficult experiments (radioactivity, use of shielded cells, etc.). Currently, fundamental understanding of irradiated fuel chemistry as a function of temperature and burnup is not possible. FPs are not included, and minor actinide (Am, Np, Cm) introduction to the database is on-going.

The CalPhaD method reveals the stability domains of GFR reactor fuels. For example, the multiphase carbide fuel $(U,Pu)C+(U,Pu)_2C_3$ appears to be the most stable with no trace of free carbon and liquid phase under argon up to 1973 K. This last temperature could be correlated to that of the incidental conditions or could be related to the presence of heat points. From this temperature, a liquid phase could be evidenced in equilibrium with the two carbides (Fig. 15.5).

From thermodynamic calculations, the interactions between fuel and cladding materials in the GFR context could be studied. For a UC fuel embedded in a SiC inert matrix, phase diagram modeling has confirmed that no interaction occurs at the usual operating temperature, or at a higher temperature of 1600°C for a C-rich UC fuel (Fig. 15.6).

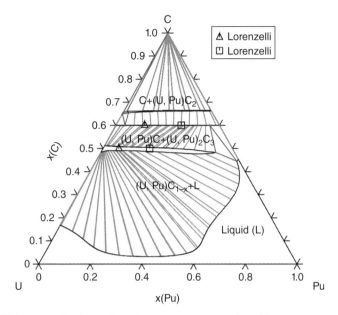

Figure 15.5. C-Pu-U isothermal section at 1973 K. Reproduced from Guéneau *et al.* [8].

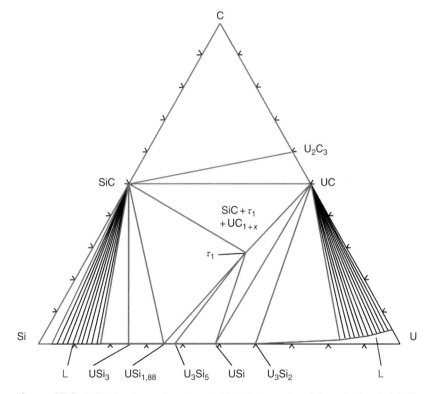

Figure 15.6. C-Si-U isothermal section at 1673 K. Reproduced from Rado *et al.* [62].

In contrast, for a nitride-based fuel like UN and a SiC cladding material, the thermodynamic approach shows that a U-Si intermetallic phase appears at 1723 K. At higher temperature, a liquid phase appears, thus prohibiting the use of this UN/SiC cladding couple [63].

15.7 CONCLUSIONS

The melting temperatures of current oxide and future non-oxide nuclear fuels make them *de facto* UHTCs. Other UHTCs as well as non-UHTCs such as MAX phases and SiC/SiC will undoubtedly find application in future fission and fusion reactor systems due to their ability to survive severe conditions of temperature, atmosphere, and radiation damage. Development of complex composite ceramics and careful consideration of thermodynamic aspects will be key to their development.

REFERENCES

1. Carre F, Yvon P, Anzieu P, Chauvin N, Malo J-Y. Update of the French R&D strategy on gas-cooled reactors. Nucl Eng Des 2010;240:2401–2408.
2. Lee WE, Gilbert M, Murphy ST, Grimes RW. Opportunities for advanced ceramics and composites in the nuclear sector. J Am Ceram Soc 2013;96:2005–2030.
3. Fink JK. Thermophysical properties of uranium dioxide. J Nucl Mater 2000;279:1–18.
4. Guéneau C, Chartier A, Brutzel LV. Thermodynamic and thermophysical properties of the actinide oxides. In: Konings JM, editor. *Comprehensive Nuclear Materials*. Volume 2, Oxford: Elsevier; 2012. p 21, 59.
5. Donald IW. Mechanical properties. In: *Waste Immobilization in Glass and Ceramic Based Hosts*. Chichester: John Wiley & Sons Ltd; 2010. p 331.
6. Ringwood AE, Oversby V, Kesson S, Sinclair W, Ware N, Hibberson W, Major A. Immobilization of high-level nuclear reactor wastes in SYNROC: a current appraisal. Nucl Chem Waste Manage 1981;2:287–305.
7. Pillai CGS, George AM. Thermal conductivity of uranium dioxide. J Nucl Mater 1993;200:78–81.
8. Guéneau C, Dupin N, Sundman B, Martial C, Dumas J-C, Gossé S, Chatain S, De Bruycker F, Manara D, Konings RJM. Thermodynamic modelling of advanced oxide and carbide nuclear fuels: description of the U-Pu-O-C system. J Nucl Mater 2011;419:145–167.
9. Anderson HH, Asprey LB. US Patent 2,924,506. February 9, 1960.
10. Noyau S, Garcia P, Pasquet B, Roure I, Audubert F, Maitre A. Towards measuring the Pu self-diffusion coefficient in polycrystalline $U_{0.55}Pu_{0.45}O_{2\pm x}$. Defect Diffus Forum 2012;323–325:203–208.
11. Ojovan MI, Lee WE. *An Introduction to Nuclear Waste Immobilisation*. Oxford: Elsevier; 2nd edn. 2014.
12. Burakov BE, Ojovan MI, Lee WE. *Crystalline Materials for Actinide Immobilisation*. Volume 1, London: Imperial College Press; 2011.
13. Middleburgh SC, Parfitt DC, Blair PR, Grimes RW. Atomic scale modeling of point defects in zirconium diboride. J Am Ceram Soc 2011;94:2225–2229.

14. Pierson HO. *Handbook of Refractory Carbides and Nitrides*. Westwood: William Andrew; 1996.

15. Guo S-Q. Densification of ZrB_2-based composites and their mechanical and physical properties: a review. J Eur Ceram Soc 2009;29:995–1011.

16. Thompson MJ, Fahrenholtz WG, Hilmas GE. Elevated temperature thermal properties of ZrB_2 with carbon additions. J Am Ceram Soc 2012;95:1077–1085.

17. Opeka MM, Talmy IG, Wuchina EJ, Zaykoski JA, Causey SJ. Mechanical, thermal, and oxidation properties of refractory hafnium and zirconium compounds. J Eur Ceram Soc 1999;19:2405–2414.

18. Barsoum MW, El-Raghy T. Synthesis and characterization of a remarkable ceramic: Ti_3SiC_2. J Am Ceram Soc 1996;79:1953–1956.

19. Zhou A, Wang C-A, Hunag Y. Synthesis and mechanical properties of Ti_3AlC_2 by spark plasma sintering. J Mater Sci 2003;38:3111–3115.

20. Chen JX, Zhou YC, Zhang HB, Wan DT, Liu MY. Thermal stability of Ti_3AlC_2/Al_2O_3 composites in high vacuum. Mater Chem Phys 2007;104:109–112.

21. Pang WK, Low IM, O'Connor BH, Peterson VK, Studer AJ, Palmquist JP. In situ diffraction study of thermal decomposition in Maxthal Ti_2AlC. J Alloy Compd 2011;509:172–176.

22. Wang XH, Zhou YC. Layered machinable and electrically conductive Ti_2AlC and Ti_3AlC_2 ceramics: a review. J Mater Sci Technol 2010;26:385–416.

23. Bao Y, Wang X, Zhang H, Zhou Y. Thermal shock behaviour of Ti_3AlC_2 from between 200°C and 1300°C. J Eur Ceram Soc 2005;25:3367–3374.

24. Spencer CB, Córdoba JM, Obando N, Sakulich A, Radovic M, Odén M, Hultman L, Barsoum MW. Phase evaluation in Al_2O_3 fiber-reinforced Ti_2AlC during sintering in the 1300°C–1500°C temperature range. J Am Ceram Soc 2011;94:3327–3334.

25. Hoffman EN, Vinson DW, Sindelar RL, Tallman DJ, Kohse G, Barsoum MW. MAX phase carbides and nitrides: properties for future nuclear power plant in-core applications and neutron transmutation analysis. Nucl Eng Des 2012;244:17–24.

26. Sengupta AK, Agarwal R, Kamath HS. Carbide fuel. In: Konings RJM, editor. *Comprehensive Nuclear Materials*. Volume 3, Oxford: Elsevier; 2012. p 55, 86.

27. Arai Y. Nitride fuel. In: Konings R, editor. *Comprehensive Nuclear Materials*. Volume 3, Oxford: Elsevier; 2012. p 41, 54.

28. Arai Y, Nakajima K. Preparation and characterization of PuN pellets containing ZrN and TiN. J Nucl Mater 2000;281:244–247.

29. Matthews RB, Harbst RJ. Uranium-plutonium carbide fuel for fast breeder reactors. Nucl Technol 1983;63:9–22.

30. Matthews RB, Hart PE. Nuclear fuel pellets fabricated from gel-derived microspheres. J Nucl Mater 1980;92:207–216.

31. Potter P. The volatility of plutonium carbides. J Nucl Mater 1964;12:345–348.

32. Anselin F, Dean G, Lorenzelli R, Pascard R. In: Russell LE, editor. *Carbides in Nuclear Energy*. Volume 1, London: Macmillan; 1964. p 162, 163.

33. Muromura T, Tagawa H. Formation of uranium mononitride by the reaction of uranium dioxide with carbon in ammonia and a mixture of hydrogen and nitrogen: I synthesis of high purity UN. J Nucl Mater 1977;71:65–72.

34. Muromura T, Tagawa H. Formation of uranium mononitride by the reaction of uranium dioxide with carbon in ammonia and a mixture of hydrogen and nitrogen: II. Reaction rates. J Nucl Mater 1979;80:330–338.

35. Uyttenhove W, Sobolev V, Maschek W. Optimisation of composite metallic fuel for minor actinide transmutation in an accelerator-driven system. J Nucl Mater 2011;416:192–199.

36. Chauvin N, Konings RJM, Matzke H. Optimisation of inert matrix fuel concepts for americium transmutation. J Nucl Mater 1999;274:105–111.

37. Meyer MK. Composite fuel (CERMET, CERCER). In: Konings JM, editor. *Comprehensive Nuclear Materials*. Volume 3, Oxford: Elsevier; 2012. p 257–273.

38. Talamo A, Pouchon MA, Venneri F. Alternative configurations for the QUADRISO fuel design concept. J Nucl Mater 2009;383:264–266.

39. Zhou XW, Tang CH. Current status and future development of coated fuel particles for high temperature gas-cooled reactors. Prog Nucl Energy 2011;53:182–188.

40. Sawa K. TRISO fuel production. In: Konings RJM, editor. *Comprehensive Nuclear Materials*. Volume 3, Oxford: Elsevier; 2012. p 143, 149.

41. Petti DA, Demkowicz PA, Maki JT, Hobbins RR. TRISO-coated particle fuel performance. In: Konings RJM, editor. *Comprehensive Nuclear Materials*. Volume 3, Oxford: Elsevier; 2012. p 151, 213.

42. Minato K, Ogawa T. Advanced concepts in TRISO fuel. In: Konings JM, editor. *Comprehensive Nuclear Materials*. Volume 3, Oxford: Elsevier; 2012. p 215, 236.

43. Pöml P, Konings RJM, Somers J, Wiss T, de Haas GJLM, Klaassen FC. Inert matrix fuel. In: Konings RJM, editor. *Comprehensive Nuclear Materials*. Volume 3, Oxford: Elsevier; 2012. p 237, 256.

44. Pillon S. Actinide-bearing fuels and transmutation targets. In: Konings RJM, editor. *Comprehensive Nuclear Materials*. Volume 3, Oxford: Elsevier; 2012. p 109, 141.

45. Chauvin N, Pelletier M. Inert matrix fuels. In: Buschow KHJ, editor. *Encyclopedia of Materials: Science and Technology*. Oxford: Elsevier; 2001. p 4066, 4068.

46. Schram RPC, van der Laan RR, Klaassen FC, Bakker K, Yamashita T, Ingold F. The fabrication and irradiation of plutonium-containing inert matrix fuels for the "Once through then out" experiment. J Nucl Mater 2003;319:118–125.

47. Ledergerber G, Degueldre C, Heimgartner P, Pouchon MA, Kasemeyer U. Inert matrix fuel for the utilisation of plutonium. Prog Nucl Energy 2001;38:301–308.

48. Konings RJM, Bakker K, Boshoven JG, Hein H, Huntelaar ME, van der Laan RR. Transmutation of actinides in inert-matrix fuels: fabrication studies and modelling of fuel behaviour. J Nucl Mater 1999;274:84–90.

49. Farid O, Shih K, Lee W, Yamana H. Fukushima: the current situation and future plans. In: Lee WE et al., editors. *Radioactive Waste Management and Contaminated Site Clean-up: Processes, Technologies and International Experience*. Oxford: Woodhead Pub Ltd.; 2013.

50. Cabrero J, Audubert F, Pailler R, Kusiak A, Battaglia JL, Weisbecker P. Thermal conductivity of SiC after heavy ion irradiation. J Nucl Mater 2010;396:202–207.

51. Gutierrez G, Toulhoat N, Moncoffre N, Pipon Y, Maître A, Gendre M, Perrat-Mabilon A. Thermal behaviour of xenon in zirconium carbide at high temperature: role of residual zirconia and free carbon. J Nucl Mater 2011;416:94–98.

52. Viaud C, Maillard S, Carlot G, Valot C, Gilabert E, Sauvage T, Peaucelle C, Moncoffre N. Behaviour of helium after implantation in molybdenum. J Nucl Mater 2009;385:294–298.

53. Nash KL, Lumetta GJ. *Advanced Separation Techniques for Nuclear Fuel Reprocessing and Radioactive Waste Treatment*. Oxford: Woodhead Publishing; 2011. Series in Energy No. 2.

54. Shibata T, Sumita J, Sawa K, Takagi T, Makita T, Kunimoto E. Irradiation-induced property change of C/C composite for application of control rod elements of Very High Temperature

Reactor (VHTR). Proceedings of the Structural Materials for Innovative Nuclear Systems (SMINS-2), OECD Nuclear Energy Agency (NEA Report n. 6896); August 31–September 3, 2010; Daejon, Republic of Korea; 2010.

55. Snead LL, Katoh Y, Nozawa T. Radiation effects in SiC and SiC–SiC. In: Konings RM, editor. *Comprehensive Nuclear Materials*. Volume 4, Oxford: Elsevier; 2012. p 215, 240.

56. Wirth B, Nordlund K, Whyte D, Xu D. Fusion materials modeling: challenges and opportunities. Mater Res Soc Bull 2011;36:216–222.

57. Snead LL, Ferraris M. Carbon as a fusion plasma-facing material. In: Konings RJM, editor. *Comprehensive Nuclear Materials*. Volume 4, Oxford: Elsevier; 2012. p 583, 620.

58. Guéneau C, Gossé S, Chatain S, Utton C, Dupin N, Sundman B, Martial C, Dumas J-C, Rado C. 2008. FUELBASE: a thermodynamic database for advanced nuclear fuels. Best poster award, Calphad Meeting 2008. Available at http://www.calphad.org/awards/2008-Best-Poster.pdf. Accessed July 1, 2014.

59. Dinsdale AT. SGTE data for pure elements. Calphad 1991;15:317–425.

60. SGTE Substances Database. 2013. Available at http://www.thermocalc.com/TCDATA.htm. Accessed July 01, 2014.

61. Lukas HL, Fries SG, Sundman B. *Computational Thermodynamics: The Calphad Method*. Cambridge: Cambridge University Press; 2007.

62. Rado C, Rapaud O, Chatain C, Guéneau C, Guyadec F, Deschamps B. Développement du combustible des futurs réacteurs rapides à gaz. etude de la compatibilité entre composé fissile et matrice inerte. Proceedings of the Materiaux 2006, Dijon, November 13–17, 2006.

63. Guéneau C, Chatain S, Gossé S, Rado C, Rapaud O, Lechelle J, Dumas JC, Chatillon C. A thermodynamic approach for advanced fuels of gas-cooled reactors. J Nucl Mater 2005; 344:191–197.

16

UHTC-BASED HOT STRUCTURES: CHARACTERIZATION, DESIGN, AND ON-GROUND/IN-FLIGHT TESTING

Davide Alfano[1], Roberto Gardi[1], Luigi Scatteia[2], and Antonio Del Vecchio[1]

[1] *Italian Aerospace Research Centre (CIRA), Capua (CE), Italy*
[2] *Booz & Company B.V., Amsterdam, the Netherlands*

16.1 INTRODUCTION

Passive thermal protection systems (TPS) for hypersonic vehicles are essential to protect cold structures from high heat fluxes during the reentry phase into the Earth's atmosphere. TPS materials fall into two broad categories: ablative and reusable.

Ablative materials as TPS were employed in the Apollo program [1–3]. Thermal protection is achieved through endothermic chemical processes such as melting, sublimation, or pyrolysis of ablative materials, which are usually low thermal conductivity insulators such as carbon phenolic composites. Reusable TPS must withstand aerodynamic loads at very high operational temperatures reached during reentry. Among reusable TPS are monolithic ceramics and ceramic matrix composites such as carbon fiber reinforced silicon carbide composites (C/SiC) [4, 5]. Ceramic compounds based on the metal borides, such as zirconium diboride (ZrB_2) and hafnium diboride (HfB_2), are

Ultra-High Temperature Ceramics: Materials for Extreme Environment Applications, First Edition.
Edited by William G. Fahrenholtz, Eric J. Wuchina, William E. Lee, and Yanchun Zhou.
© 2014 The American Ceramic Society. Published 2014 by John Wiley & Sons, Inc.

commonly referred to as Ultra-High-Temperature Ceramics (UHTCs). This class of materials is promising for use in extreme environments, such as sharp leading edges hot structures on future-generation slender-shaped reentry vehicles, because of their high melting points (ZrB$_2$ 3040°C, HfB$_2$ 3250°C) and relatively good oxidation resistance in reentry conditions [6–10].

In the framework of the unmanned space vehicle (USV) project funded by the Italian Aerospace Program (PRORA) and within various other European programs, since 2000 the Italian Aerospace Research Centre (CIRA) has studied, developed, and tested monolithic UHTCs to employ as passive structural TPS [11–14], henceforth referred to as hot structures. In Figure 16.1 CIRA's roadmap of experimental activities performed on hot structures in the last 10 years is summarized.

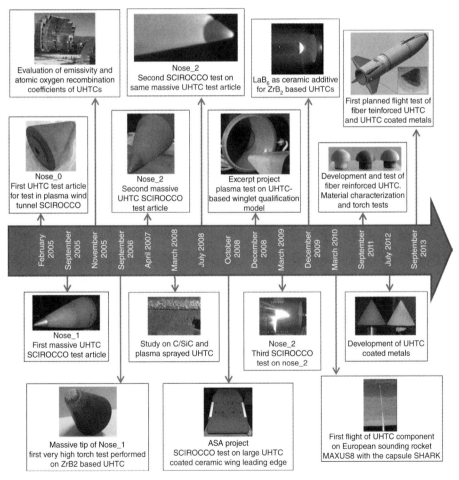

Figure 16.1. CIRA activities on UHTCs for thermal protection systems.

16.2 TPS: TEST ARTICLES AND PROTOTYPES

In the framework of the internal CIRA project named sharp hot structure (SHS), three full scale nose caps characterized by small curvature radii for vehicle reentry from LEO (low Earth orbit) were designed, manufactured, and then tested at high stagnation point heat flux. The nose caps were identified as Nose_0, 1, and 2.

The test article Nose_0 was a cone relatively small in size (height 174 mm and diameter 127 mm) constructed with a graphite core and an external layer of C/SiC (thickness 5 mm) applied by chemical vapor infiltration. The oxidation resistance of the structure was improved by a plasma-sprayed ZrB_2-based UHTC coating. Plasma deposition was carried out by an 80 kW plasma torch, which allows depositions in the pressure range 10–400 kPa under air or inert atmosphere. The protective coating was obtained starting from a mixture of ZrB_2 and SiC powders with 75 and 25 wt%, respectively (more details on deposition process are described in Refs. [15–17]). That composition was selected to obtain a coating with good oxidation resistance. Previous studies have described how the addition of SiC, used also as sintering aid, provides significant improvement to the oxidation resistance of monolithic ZrB_2-based UHTCs [18]. Although Nose_0 only employs UHTC materials in form of a coating, its manufacturing allowed CIRA to familiarize with the fabrication of C/SiC components, the deposition of UHTC by plasma spraying on axially symmetric structures, the design and manufacturing of a mechanical interface with a cold structure, and then with the testing of a SHS in the plasma wind tunnel (PWT) "Scirocco" (see Section 16.3).

Noses_1 and 2 were manufactured to obtain and then study more complex test articles wherein topics such as the attachment of massive UHTCs and C/SiC composites to substructures or the deposition of UHTC-based coatings on relatively large and complex shaped structures were faced.

Nose_1 and Nose_2 consisted of a monolithic ZrB_2-based UHTC conical nose tip with height of 100 mm attached to a truncated cone base (height 404 mm and diameter 304 mm) with a graphite core and an external layer of C/SiC (Fig. 16.2). The same UHTC coating (ZrB_2 and SiC with 75 and 25 wt%) was deposited by plasma spray on the truncated cone bases of both prototypes in order to improve their oxidation resistance. In the case of Nose_1, coating adhesion to the C/SiC substrate was not uniform due to the relatively large coating thickness of about 1 mm. These adhesion problems were not observed on the C/SiC base of Nose_2 where the coating thickness was reduced to about 300–500 μm.

The chemical composition of the monolithic nose tips is the main difference between Nose_1 and Nose_2: $ZrB_2 + 15$ vol% $SiC + 2$ vol% $MoSi_2$ (ZS) in the former and $ZrB_2 + 10$ vol% $HfB_2 + 15$ vol% $SiC + 2$ vol% $MoSi_2$ (ZHS) in the latter.

SiC addition was found to improve the processing by lowering sintering temperatures of diborides-based UHTCs [19–21] and improving the oxidation resistance. In fact the oxidation of pure ZrB_2 above 1200°C produces B_2O_3 which, being characterized by intensive volatility, is not able to form a stable and protective oxide scale

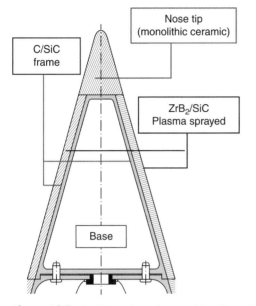
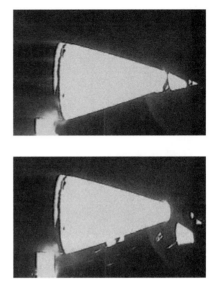

Figure 16.2. Configuration of test articles. Nose_1: breaking of nose tip during the plasma test.

[22–28]. The addition of SiC enhances resistance to the oxidation of diborides-based monolithic ceramics up to 1600°C through the formation of borosilicate glass [18, 24–26, 29–43].

Nose cone tips were manufactured by uniaxial hot-pressing in vacuum starting from powder mixtures containing $MoSi_2$ as a sintering aid. Peak temperatures, dwell times, and applied pressures were 1820°C/15 min/30 MPa for material ZS, and 1900–1940°C/45 min/40 MPa for material ZHS, with an average heating rate of 20°C/min. The relatively high electrical conductivity of ZS and ZHS compositions allows the final shapes in both nose tips to be fabricated from sintered billets by electrical discharge machining (EDM). Although EDMed surfaces had a higher roughness with respect to diamond-loaded tooled (DLTed) ones [44], EDM has proven to be an effective and flexible technique to machine UHTC sintered pieces into complex components. Thermomechanical and physical properties of ZS and ZHS formulations obtained by CIRA are summarized in Table 16.1.

For use as hot structures, parameters such as emissivity (ratio of surface radiance to blackbody radiance at the same temperature and wavelength) and catalicity need to be evaluated in addition to characterizing the conventional thermomechanical properties measurements. Emissivity can become the main mechanism for heat dissipation at high altitudes during reentry phase, when convection is low. Catalicity is also important since while crossing the atmosphere, a space vehicle creates a shock wave leading to very high temperatures from recombination of excited and ionized species (atoms, molecules, ions, and electrons).

TABLE 16.1. Thermomechanical and physical properties of $ZrB_2 + 15\,vol\%\,SiC + 2\,vol\%$ $MoSi_2$ (ZS) and $ZrB_2 + 10\,vol\%\,HfB_2 + 15\,vol\%\,SiC + 2\,vol\%\,MoSi_2$ (ZHS) formulations [45]

Property	ZS		ZHS	
Density (kg/m³)	5610 ± 10		6060 ± 15	
Porosity (%)	<1		<1	
Specific heat	**T (K)**	**C_P (J/g·K)**	**T (K)**	**C_P (J/g·K)**
	298	0.46 ± 0.01	298	0.52 ± 0.02
	500	0.60 ± 0.02	500	0.67 ± 0.02
	1000	0.70 ± 0.02	1000	0.76 ± 0.02
	1500	0.75 ± 0.02	1500	0.81 ± 0.02
	2000	0.79 ± 0.02	2000	0.84 ± 0.02
Total hemispherical emissivity[a] (measurement performed on EDMed sample at 200 Pa)	**T (K)**	**Emissivity**	**T (K)**	**Emissivity**
	1037	0.77 ± 0.04	1144	0.56 ± 0.10
	1169	0.71 ± 0.03	1434	0.73 ± 0.07
	1250	0.72 ± 0.04	1733	0.72 ± 0.10
	1319	0.73 ± 0.04		
	1485	0.75 ± 0.04		
	1681	0.72 ± 0.04		
Total hemispherical emissivity[a,b] (measurement performed on EDMed samples at 10 Pa)	**T (K)**	**Emissivity**	**T (K)**	**Emissivity**
	1099	0.53 ± 0.03	1171	0.50 ± 0.03
	1170	0.66 ± 0.03	1366	0.74 ± 0.04
	1418	0.81 ± 0.04	1533	0.81 ± 0.04
	1609	0.76 ± 0.04	1720	0.75 ± 0.04
	1865	0.66 ± 0.03		
Atomic oxygen recombination coefficient[b,c] (γ) (measurement performed on EDMed sample at 200 Pa)	**T (K)**	**$\gamma \times 10^{-2}$**	**T (K)**	**$\gamma \times 10^{-2}$**
	850–1200	0.8 ± 0.2	1000	3.0 ± 0.9
	1400	1.6 ± 0.5	1200	5.4 ± 1.6
	1600	1.7 ± 0.5	1400	6.5 ± 1.9
	1800	1.9 ± 0.6	1600	9.2 ± 2.8
	2000	2.9 ± 0.9	1800	9.7 ± 2.9
Total hemispherical emissivity[a,b] (measurement performed on DLTed sample at 10 Pa)	**T (K)**	**Emissivity**	**T (K)**	**Emissivity**
	1188	0.53 ± 0.03	1168	0.49 ± 0.02
	1397	0.71 ± 0.04	1422	0.62 ± 0.03
	1595	0.76 ± 0.04	1600	0.70 ± 0.04
	1720	0.73 ± 0.04	1701	0.71 ± 0.04
Atomic oxygen recombination coefficient[b,c] (γ) (measurement performed on DLTed sample at 200 Pa)	**T (K)**	**$\gamma \times 10^{-2}$**	**T (K)**	**$\gamma \times 10^{-2}$**
	1000	2.4 ± 0.7	1000	1.6 ± 0.5
	1200	4.4 ± 1.2	1200	4.0 ± 1.2
	1400	5.6 ± 1.7	1400	6.9 ± 2.1
	1600	7.7 ± 2.3	1600	10.5 ± 3.1
	1800	9.5 ± 2.8	1800	11.5 ± 3.4
Thermal conductivity (uncertain 7%)	**T (K)**	**λ (W/m·K)**	**T (K)**	**λ (W/m·K)**
	303	62.47	303	79.87
	523	64.02	523	83.14
	1023	64.83	1023	83.88
	1473	65.18	1473	85.03

TABLE 16.1. (*Continued*)

Property	ZS		ZHS	
Vickers hardness HV$_{1.0}$ (GPa)	17.7±0.4		18.2±0.5	
Fracture toughness K_{IC} (MPa·m$^{1/2}$)	4.07±0.03 at 298 K 2.5±0.2 at 1773 K		4.1±0.7 at 298 K	
Young modulus (GPa)	480±4		506±4	
Flexural strength (MPa)	8.9±1.2 E+02 at 298 K 2.5±0.2 E+02 at 1773 K		7.6±0.7 E+02 at 298 K 2.4±0.2 E+02 at 1773 K	
Poisson's ratio	0.13		0.13	
CTE (cm/cm·K)	**T (K)**	**CTE ($\times10^{-6}$)**	**T (K)**	**CTE ($\times10^{-6}$)**
	1073	6.67	1073	6.74
	1173	6.79	1173	6.85
	1273	6.95	1273	7.00
	1373	7.02	1373	7.09
	1473	7.01	1473	7.20
	1573	7.11	1573	7.24

[a]Total hemispherical emissivity: values obtained by angular integration (0°–80° by 10° step) of the directional emissivity measured in the wavelength range of 0.6–40 μm (for more details see References [48, 49]).

[b]Data from Reference [44].

[c]Atomic oxygen recombination coefficient: method and experimental set-up are described in Refs. [50, 51].

The key parameter for describing surface catalicity is the recombination coefficient, γ, which is defined as:

$$\gamma = \frac{N_r}{N}$$

where N_r is the number of atoms that recombine on a surface per unit area and time, and N is the number of atoms striking the surface per unit area and time.

Since 2005, hemispheric emissivity and atomic oxygen recombination coefficients of UHTCs were obtained through different experimental campaigns carried out at PROMES-CNRS laboratories [44, 46, 47] using MEDIASE (*Moyen d'Essai et de DIagnostic en Ambiance Spatiale Extrême*) [48, 49] and MESOX (*Moyen d'Essai Solaire d'Oxydation*) [50, 51]. Nose_1 and Nose_2 prototypes were tested in PWT "Scirocco" wherein heat flux up to 4.5 MW/m² was reached for a total duration of about 4 min. Details and results from these plasma tests are described in Section 16.3.

From the perspective of chemical reactions and thermodynamics, on-ground qualification and characterization of aerospace materials cannot reproduce all aspects of the hypersonic flight environment. Hence, to close the loop between scientific knowledge and the mutual relationship among theoretical, on-ground, and flight research, materials must be tested in real flight conditions. For this purpose, the Nose_1 and Nose_2 formulations were also used to manufacture test articles for the technological projects EXPERT (European eXPErimental Reentry Test-bed) and SHARK (Sounding Hypersonic

Atmospheric Reentering "Kapsule"). These two projects provided a means of evaluating UHTC-based thermostructural components under real flight conditions.

In the frame of the European Space Agency (ESA) Program EXPERT, CIRA developed a scientific payload to test a UHTC structure in flight. The objective was to design, manufacture, and test a structural subcomponent. Two small winglets were made of ZS (same composition of Nose_1 tip). They were in a dimension of 5×10 cm and mounted on the external surface of the capsule where they would be exposed to the high energy air flow of the suborbital reentry trajectory [52], which is described in more detail in Section 16.4.

SHARK is a small capsule designed and built at CIRA under an ESA contract. The aim of the project was to prove the viability of a low-cost experimental space platform and execute a reentry test flight by dropping a capsule from a sounding rocket. The capsule comprised of a stainless steel frontal shield with a UHTC nose tip machined from the Nose_2 (ZHS) prototype after its ground test in the PWT "Scirocco." The technical details of this flight experiment and the experimental findings are described in Section 16.5.

In the frame of the Advanced Structural Assembly (ASA) project, funded by Agential Spaziale Italiana (ASI), a full scale technological demonstrator (the forepart of a wing leading edge) was designed, manufactured, and then tested in the PWT. The base formulation was a ceramic material lighter than the UHTCs used in nose prototypes. The composition had a higher damage tolerance (a requirement for large-scale monolithic ceramic components), but with lower chemical stability at high temperatures, which was improved by a UHTC coating.

A wide characterization campaign was performed on different ceramic compositions to select the formulation with higher values of fracture toughness and thermal shock resistance. These properties were evaluated through the measurement of four-point residual flexure strength [53] after quenching in water (thermal shock up to $900°C$). Based on materials requirements defined by ASA project, the compositions listed in Table 16.2 were selected and then characterized.

Processing conditions for pressureless sintering were flowing inert gas, a heating rate of $10°C/\text{min}$, and temperatures up to $1900°C$. For hot pressing, the heating rate was about $20°C/\text{min}$ up to $1900°C$ at a pressure of $30\,\text{MPa}$.

The selected composition for the bulk ceramic part of the leading edge was $Si_3N_4 + 35\,\text{vol}\%\ MoSi_2 + 2.5\,\text{vol}\%\ Y_2O_3 + 1\,\text{vol}\%\ Al_2O_3$ (RAY2535M). Compared to the other formulations, this composition had the highest flexure strength values that were

TABLE 16.2. Ceramic formulations and related densification processes: hot-pressing (HP) and pressureless sintering (PLS)

ID	Basic composition (vol%)	Densification process
Z3M	$ZrB_2 + 3\ MoSi_2$	HP
RAY2535M	$Si_3N_4 + 35\ MoSi_2$	HP
Z20M	$ZrB_2 + 20\ MoSi_2$	PLS
ASM30	$AlN + 30\ MoSi_2 + 15\ SiC$	PLS

TABLE 16.3. Thermomechanical properties of massive $Si_3N_4 + 35\,vol\%\ MoSi_2 + 2.5\ Y_2O_3 + 1\ Al_2O_3$ (RAY2535)

Properties	$Si_3N_4 + 35\,vol\%\ MoSi_2 + 2.5\ Y_2O_3 + 1\ Al_2O_3$	
Density (kg/m³)	4200 at 298 K	
	4216 at 873 K	
	4235 at 1573 K	
Specific heat	T (K)	C_P (J/g·K)
	298	0.567
	573	0.717
	1073	0.787
	1573	0.832
Thermal diffusivity	T (K)	D_{TH} (cm²/s)
	298	0.131
	573	0.073
	1073	0.048
	1573	0.039
Thermal conductivity	T (K)	K_{TH} (W/m·K)
	298	31.2
	573	22.0
	1073	15.8
	1573	13.7
Elastic modulus (GPa)	328	
Microhardness (GPa)	15.8 ± 0.2	
Fracture toughness K_{IC} (MPa·m$^{1/2}$)	6.0 ± 0.2 at 298 K	
CTE (10^{-6}/K)	6.5; 298–1573 K	
Flexural strength (MPa)	915 ± 65 at 298 K	
	755 ± 90 at 1273 K	
	250 ± 28 at 1573 K	
	148 ± 7 at 1773 K	

retained even after a thermal shock temperature drop of 650°C. In Table 16.3, the values of some thermomechanical properties of RAY2535M are reported.

The leading edge was shaped by EDM starting from a monolithic ceramic billet. To increase oxidation resistance of the monolithic ceramic, a coating of $ZrB_2 + 30\,vol\%$ $SiC + 10\,vol\%\ MoSi_2$ with a thickness of about 150 μm was applied by plasma spray on the leading edge surface. That composition was selected among different coating formulations after a wide experimental characterization campaign in which morphological structure, adhesion to the selected massive ceramic substrate, and oxidation resistance (assessed by plasma torch test under high enthalpy flow conditions—maximum temperatures reached were about 1750°C) were evaluated.

At the same time as technology research in prototype manufacturing, base material research was conducted on the reference UHTC composition (ZrB_2-SiC), with the aim to improve its sinterability and oxidation resistance. In this study, LaB_6 was considered as a ceramic additive. The presence of LaB_6 could potentially increase the oxidation resistance ZrB_2 through the formation of a refractory oxidation by-product, lanthanum

TABLE 16.4. Thermomechanical properties of massive $ZrB_2 + 15\,vol\%$ $SiC + 10\,vol\%$ LaB_6 [59]

Properties	$ZrB_2 + 15\,vol\%$ $SiC + 10\,vol\%$ LaB_6	
Density (kg/m³)	5500 at 298 K	
Porosity (%)	<0.1	
Thermal conductivity	**T (K)**	**K_{TH} (W/m·K)**
	300	129.0
	1073	106.7
	1773	97.8
Elastic modulus (GPa)	465	
Fracture toughness K_{IC} (MPa·m$^{1/2}$)	2.8 ± 0.4 at 293 K	
CTE (10^{-6}/K)	6.95; 293–1573 K	
Flexural strength (MPa)	608 ± 103 at 293 K	
	405 ± 30 at 1573 K	

zirconate ($La_2Zr_2O_7$, LZ), to limit the inward diffusion of oxygen [54–56]. This compound has cubic pyrochlore structure [57] and is thermodynamically stable up to the melting point (at about 2300°C) [58]. Monolithic samples with 15 vol% SiC and different loadings of LaB_6 (from 10 to 20 vol%) were sintered by hot pressing under a rough vacuum of 0.2–1 mbar. The hot pressing cycle included heating up to 1930°C with an applied pressure of 30 MPa and a heating rate of 20°C/min, followed by an iso-thermal hold at 1930°C for 15 min under a pressure of 40 MPa. Oxidation tests on ZrB_2-LaB_6-SiC based UHTC formulations were carried out by arc-jet experiments on EDMed hemispherical-shaped samples for at least 6 min at a specific total enthalpy of about 10 MJ/kg to reproduce the typical experimental conditions during the hypersonic reentry phase. Microstructural characterization showed that surface damage increased with increasing ratios of LaB_6/ZrB_2, making the formulation with 10 vol% of LaB_6 the most resistant to oxidation. In any case, ZrB_2-LaB_6-SiC based ceramic formulations are char-acterized by lower oxidation resistance than ZS UHTCs, because a protective lanthanum zirconate layer did not form [59]. Thermomechanical properties of $ZrB_2 + 15\,vol\%$ $SiC + 10\,vol\%$ LaB_6 are summarized in Table 16.4.

16.3 PLASMA TESTS OF NOSE TEST ARTICLES

The test article Nose_0 withstood three sequential runs in the PWT "Scirocco," with a heat flux of about 1100 kW/m² on the stagnation point and an overall exposure time of about 140 s. After the tests, negligible changes to the external surfaces were observed. The test article Nose_1 was tested at a heat flux of 1200 kW/m² on the stagnation point and was intended to last for 72 s. The coating sustained the thermal load, while an unex-pected failure occurred to the monolithic tip. After 28 s, the monolithic tip broke into two large parts and produced several pieces of debris (Fig. 16.2). After the test, the causes of the failure were studied to identify corrective actions aimed at reproducing a second test article that would be able to sustain the thermal loads. In particular, attention was focused on the threaded hole manufactured by EDM in the monolithic nose tip to attach the nose tip to the C/SiC base using a titanium screw. Optical and electron microscopic analyses performed on the broken fragments revealed the presence of microcracks localized on the surface of the thread due to the EDM process. The low temperature at the moment of failure and the microscopic analyses led to the conclusion that the failure was caused by the concentration of micromechanical loads on microcracks present inside the tip that were in contact with the titanium screw.

As a result of the Nose_1 test results, Nose_2 fabrication was delayed and the attachment concept between tip and base was redesigned. In particular, the mechanical interface between the monolithic tip and the C/SiC base was changed to a metallic fastener preloaded with a spring, which avoided the use of any threaded pivot or hole [60]. The concept of the coupling pin is as follows: the pin is introduced into the hole and then rotated by 90° to ensure contact. The spring allows control of the applied preload even in presence of thermal expansion, minimizing stresses induced by heating. Moreover, the possibility of removing the tip allowed carrying out subsequent machining of the test article: the monolithic tip was retro-fitted with two temperature sensors and one deformation transducer based on fiber optics, without inducing stress on the ceramic tip.

Three plasma tests were performed on Nose_2. The first one was executed at the same conditions used for the test on Nose_1 and the test article sustained the heating load for the duration of the test. The second consecutive test was successfully carried out increasing the heat flux up to 2.1 MW/m² for 108 s. The measured temperature on the stagnation point of the tip was about 1800°C; no damage was detected at the end of the test. Finally, the third run pushed the heat flux up to 4.5 MW/m². After 30 s of exposure the test article failed dramatically.

16.4 EXPERT PROJECT: COMPUTATIONAL FLUID DYNAMICS COMPUTATIONS AND PLASMA TESTS

The ceramic monolith winglet from the EXPERT capsule is shown in Figure 16.3.

Computational fluid dynamics (CFD) simulations of a winglet on a flat surface similar to the capsule were carried out for the maximum peak heating foreseen during the reentry phase (altitude = 34 km, Mach = 14). The simulations used the conservative assumption of an isothermal 3D winglet ($T_{wall} = 300$ K) and imposed the appropriate inlet boundary conditions and axis-symmetric field computations at the winglet locations. Figure 16.4 shows the distribution of heat flux and pressure values at the maximum heat flux conditions under the assumption of a fully catalytic wall. Evaluation of the winglet temperature was performed combining CFD analysis with a thermal model of the structure. The maximum value of temperature reached on the leading edge surface, evaluated at time of 78 s starting from an initial altitude of 100 km, was about 2130 K for a fully catalytic wall and about 1700 K for an inert wall (Fig. 16.4). In any case, the temperature of the metallic support did not exceed its maximum allowable value of about 1500 K.

To evaluate the interaction of the shock wave from the winglet with the flaps, the pressure distributions on two planes (located at 1.5 cm and at 4.5 cm from the capsule

(a)

(b)

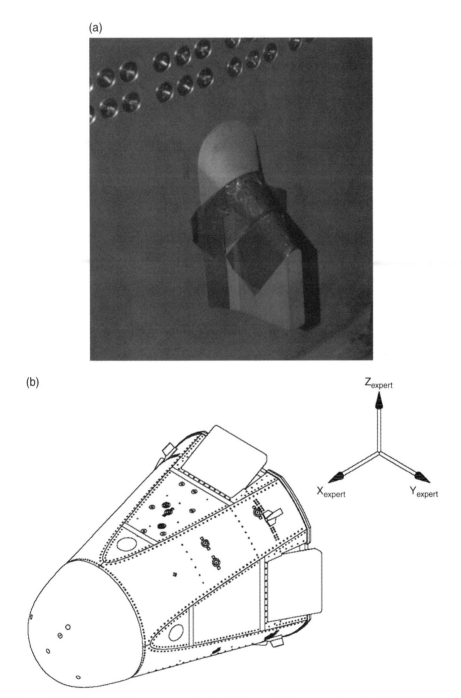

Figure 16.3. (a) Monolithic ceramic winglet flight models mounted on the EXPERT TPS and the (b) EXPERT capsule.

surface) were calculated. Also, the pressure distributions on the capsule wall and on the plane located 10 cm from the winglet front region were evaluated. Disturbances were found to be relatively low in pressure (15 kPa) in a limited region very close to the capsule (Fig. 16.4).

One of the most challenging aspects of a flight experiment is to produce a mechanical interface that allows coupling the UHTC winglet, which is subjected to large heat fluxes and mechanical loads, to the metallic structure of the capsule. Mechanical loads are generated from the thermal expansion of the subcomponents that are due to the hot environment and the load factor during reentry deceleration. Although UHTCs have a high coefficient of thermal expansion (CTE) compared to some ceramic materials, these

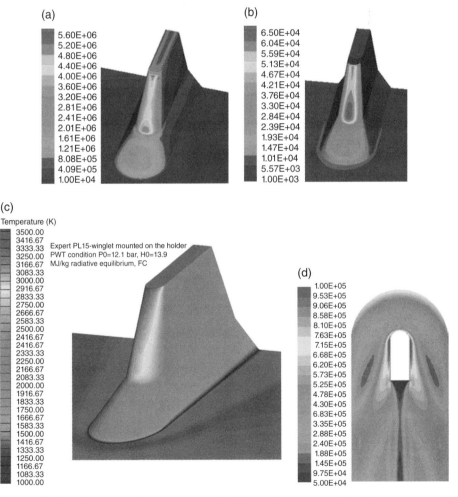

Figure 16.4. CFD analysis: (a) surface heat flux (W/m²), (b) pressure (Pa), (c) temperature (K) distributions in fully catalytic conditions, (d) heat fluxes (W/m²), and

(e)

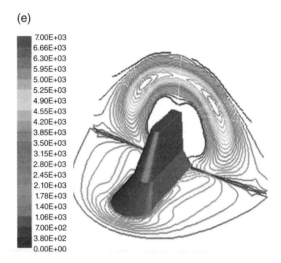

7.00E+03	
6.66E+03	
6.30E+03	
5.95E+03	
5.00E+03	
5.25E+03	
4.90E+03	
4.55E+03	
4.20E+03	
3.85E+03	
3.50E+03	
3.15E+03	
2.80E+03	
2.45E+03	
2.10E+03	
1.78E+03	
1.40E+03	
1.06E+03	
7.00E+02	
3.80E+02	
0.00E+00	

Figure 16.4. (*Continued*) (e) pressure (Pa) disturbances around the winglet.

values are still very different from the CTE of the metallic materials of capsule surface. To address this issue, the parts of the winglet directly interfaced with the capsule surface were made of the same metal, that is, PM 1000 ODS Nickel Alloy. The 2D thermal expansion analysis and 3D FEM models showed how the expansion of the metallic structures was isolated so that no themomechanical load was produced on the ceramic components. The winglets were equipped with five C-type thermocouples and four pressure lines to measure, during plasma tests, temperature and pressures, respectively. The required passing holes were the most critical issues tackled during the manufacturing phase.

Qualification tests for all the equipments that will fly on the capsule were required by ESA. In addition to the standard mechanical tests (vibration and shock), a plasma test was performed in the CIRA PWT "Scirocco" to reproduce the flight thermal loads to which the winglet will be subjected. Reproduction of the flight conditions needed a long design process that assessed the possibility of reproducing the thermal load, heat flux, and temperature on the UHTC within a 10–20% margin of accuracy. The design started from the consideration that the winglet could not be tested by itself, but an instrumented model holder was required. This model holder supported the winglet qualification model in the same way in which the flight models will be supported on the capsule and had to provide an aerothermodynamic field around the winglet able to generate the needed heat flux values and distributions. To have meaningful indications about temperature and heat flux levels reached on the winglet surface and about the reproducibility with respect to the one obtained in flight, preliminary two-dimensional (2D) analysis was matched with 2D computations on the winglet both in flight and in PWT conditions. Then, 3D CFD simulations on the complete model holder and winglet assembly were also conducted in PWT conditions.

During the test performed in December 2008 on the qualification model named as EXPERT PL15, wall pressures were measured at four points of the winglet leading

TABLE 16.5. Experimental flow conditions during plasma test on the qualification model EXPERT PL15

Test phase	Reservoir conditions		Calibration probe	
	Total pressure P_0 (bar)	Total enthalpy H_0 (MJ/kg)	Stagnation pressure P_s (mbar)	Stagnation heat flux Q_s (kW/m²)
Plateau (0–15 s)	9.6	8.8	94.5 ± 1.1	1776 ± 90
Heating ramp (instant of fracture at $t = 35$ s)	10.7	13.4	104.6 ± 1.1	2435 ± 90

edge by an electronic pressure scanning transducer (32HD type) while two Infrared thermo-cameras monitored the test article (TA) both from the top (emissivity 0.8, temperature range 250–3000°C, spatial resolution 5 mm/pixel) and from the side (emissivity 0.72, temperature range 250–3000°C, spatial resolution 1.65 mm/pixel). The test was executed but, after 35 s under the plasma flow, the holder fractured and the run was stopped. Nevertheless, flow total pressure, total enthalpy, stagnation pressure, and heat flux measured on the PWT hemispherical and cooled calibration probe were collected both during the first 15 s at constant enthalpy (plateau phase) and during the ramp at increasing enthalpy just before the facility shut down (Table 16.5). IR thermography images recorded 32 s into the test are shown in Figure 16.5.

The first European UHTC payload named EXPERT PL15 was on-ground qualified for the flight on a scientific space capsule by passing all of the demanding mechanical qualification tests—sinusoidal, random, and shock—based on increased test requirements and levels with respect to those expected in the real flight. The same qualification model was tested in the "Scirocco" PWT and the winglet performed as expected, matching CFD predictions until the failure of the model holder. PL15 has also proven that a small UHTC component can survive the foreseen conservative ground qualification tests, if great attention is used in the design of the mechanical interface with metallic structures of the capsule.

16.5 IN-FLING TESTING OF THE CAPSULE "SHARK"

SHARK is a small capsule (Fig. 16.6) designed and realized at CIRA under ESA contract and mounted on the European sounding rocket MAXUS 8 launched on March 26, 2010, from the Swedish space base ESRANGE, near Kiruna, in the far north of the country. Ninety seconds after launch and at 192 km of altitude, SHARK was released when the vehicle was flying at 3 km/s with an 88° flight path angle. The ballistic flight reached over 700 km altitude. Data acquisition was performed with an onboard computer, and the survival of the data in the memory unit was successful due to the design of the hull that protected the internal systems during all flight phases. The system was

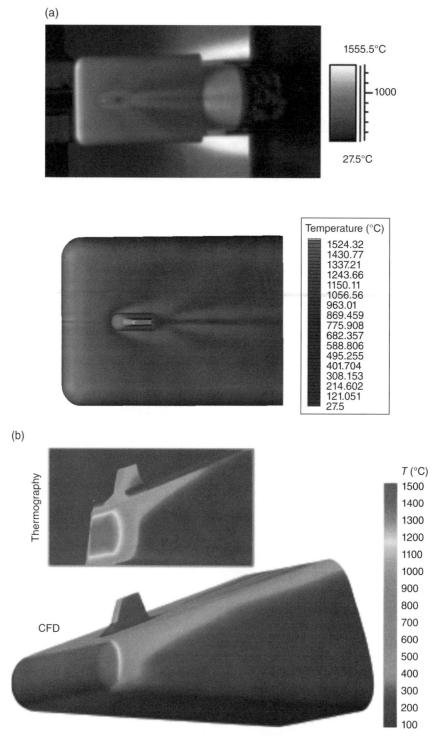

(a)

1555.5°C

1000

27.5°C

Temperature (°C)

1524.32
1430.77
1337.21
1243.66
1150.11
1056.56
963.01
869.459
775.908
682.357
588.806
495.255
401.704
308.153
214.602
121.051
27.5

(b)

Thermography

CFD

T (°C)

1500
1400
1300
1200
1100
1000
900
800
700
600
500
400
300
200
100

Figure 16.5. IR thermography images recorded during the test and before the holder fracture (t=32 s) (a) test article in centerline and (b) predicted CFD wall temperature in test conditions.

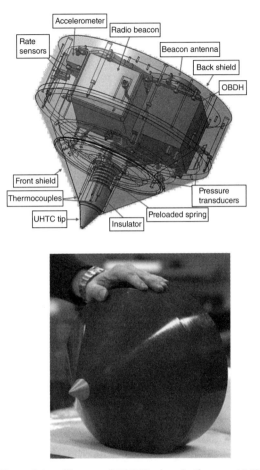

Figure 16.6. 3D model and image of SHARK wherein the gray UHTC tip is visible.

able to acquire data from 15 thermocouples and 16 analogical channels. Each temperature channel was connected to type-K thermocouple: three were installed inside the UHTC tip, some in the fore region close to the external surface aiming to measure the effect of the aerothermal heating, and some inside the vehicle to evaluate the effects of heating on the internal systems.

The accuracy of the predicted trajectory was improved by utilizing a landing point. The retrieval of the capsule on July 1, 2010, was carried out through a satellite emergency locator system, operating at 406 MHz. The metallic structure was found in good condition while the UHTC tip was broken into four parts, of which only three were recovered. The paint on the front stainless steel shield was totally removed by the aerodynamic heating, while it was intact on the backside, proving that the reentry attitude was nominal (i.e., it did not tumble).

Post-flight analysis showed that the UHTC tip ruptured during flight. The thermocouple inserted in the tip measured an abrupt temperature increment and then the signal was lost. At the same time, pressure transducers just downstream the UHTC tips measured a pressure variation, due to the different aerodynamic characteristics of the capsule after fracture of the tip. The mechanical interface was designed to crush inside the capsule allowing part of ceramic tip to survive the impact and offering the possibility of post-flight analyses on the UHTC.

Accelerometers sensed the reduction of the drag coefficient that occurred during the transition from the supersonic to the subsonic regime. During the reentry phase, the UHTC tip was exposed to about 9 MW/m² heat flux and the whole capsule sustained more than 40 g deceleration. The fracture of UHTC tip was triggered by a small defect introduced during machining of the component or during the last ground tests. On the basis of the recovered UHTC parts, extensive analyses of the mechanism of failure and its possible causes were performed. Two kinds of fractures were identified: the main one caused the loss of extreme tip and the secondary one resulted in a division of the recovered ceramic piece into three separated fragments. Both fractures seem to have occurred at the bottom hole, manufactured by EDM, in which the thermocouples were located. In Figure 16.7, the fracture surface of the nose tip with all fractographic findings and the hole for thermocouples are shown.

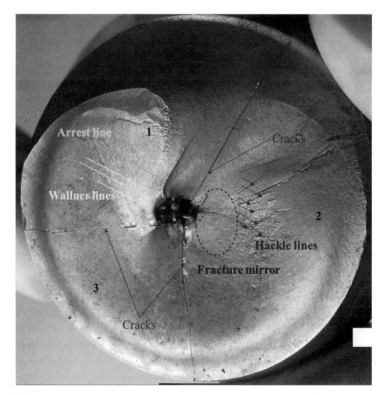

Figure 16.7. Fracture surface of SHARK Nose tip wherein the three separated fragments are also indicated.

16.6 FUTURE WORK

The most CIRA research activities on UHTCs were focused on increasing fracture resistance. This goal will be achieved by employing ZrB_2-based composites reinforced by discontinuous SiC whiskers or SiC fibers. Starting from laboratory scale, addition of up to 15 vol% fibers to ZrB_2 is being studied to evaluate the impacts on sintering process, mechanical properties, and oxidation resistance. Interesting preliminary results drove CIRA to use reinforced ZrB_2-SiC chopped fibers composites in the manufacturing of subsequent prototypes and, following the experience gained within the ESA project EXPERT, CIRA has designed, manufactured, and installed two UHTC winglets as passenger payload onto a free-flying scramjet (SCRAMSPACE), a project led by the University of Queensland (Australia), which flew in September 2013.

In parallel, CIRA is also developing and improving the features and the capabilities of UHTC coatings produced by plasma spray. Coatings are deposited on metals using metallic interlayers with the aim to permit use of metal alloys in the most severe thermodynamic conditions and at higher temperatures.

REFERENCES

1. Pavlosky JE, St. Leger LG. Apollo experience report—thermal protection subsystem. NASA technical note TN D-7564 January 1974. Washington (DC): National Aeronautics and Space Administration; 1974.

2. Sutton GW. The initial development of ablation heat protection, an historical perspective. J Spacecraft Rockets 1982;19:3–11.

3. Schneider PJ, Dolton TA, Reed GW. Mechanical erosion of charring ablators in ground-test and re-entry environments. AIAA J 1968;6:64–72.

4. Laux T, Ullmann T, Auweter-Kurtz M, Hald H, Kurz A. Investigation of thermal protection materials along an X-38 re-entry trajectory by plasma wind tunnel simulations. Proceedings of the 2nd International Symposium on Atmospheric Reentry Vehicles and Systems; March 26–29, 2001; Arcachon, France; 2001, AAAF, paper 1–9.

5. Glass DE. Ceramic Matrix Composite (CMC) Thermal Protection Systems (TPS) and hot structures for hypersonic vehicles. Proceedings of the 15th Space Planes and Hypersonic Systems and Technologies Conference; April 28–May 1, 2008; Dayton (OH); 2008, AIAA-2008-2682.

6. Paul A, Jayaseelan DD, Venugopal S, Zapata-Solvas E, Binner J, Vaidhyananthan B, Heaton A, Brown P, Lee WE. UHTC composites for hypersonic applications. Am Ceram Soc Bull 2012;91:22–28.

7. Squire T, Marschall J. Material property requirements for analysis and design of UHTC components in hypersonic applications. J Eur Ceram Soc 2010;30:2239–2251.

8. Ushakov SV, Navrotsky A. Experimental approaches to the thermodynamics of ceramics above 1500°C. J Am Ceram Soc 2012;95:1463–1482.

9. Johnson SM, Gasch MJ, Squire TH, Lawson JW. Ultra high temperature ceramics: issues and prospects. In: Krenkel W, Lamon J. 7th International Conference on High Temperature Ceramic Matrix Composites (HT- CMC 7); September 20-22, 2010; Bayreuth, Germany. p 818–831.

10. Justin JF, Jankowiak A. Ultra high temperature ceramics: densification, properties and thermal stability. Onera J Aerosp Lab 2011;3:1–11.

11. Russo G, Marino G. The USV program & UHTC development. Proceedings of the 4th European Workshop on Thermal Protection Systems for Space Vehicles; November 26–29, 2002; Palermo, Italy. Paris: European Space Agency; 2002, paper 157–163.

12. Savino R, De Stefano Fumo M, Paterna D, Serpico M. Aerothermodynamic study of UHTC-based thermal protection systems. Aerosp Sci Technol 2005;9:151–160.

13. Scatteia L, Del Vecchio A, De Filippis F., Marino G, Savino R. PRORA-USV SHS: Development of sharp hot structures based on ultrahigh temperature metal diborides current status. Proceedings of the 56th International Astronautical Congress; November 17–21, 2005; Fukuoka, Japan. West Chester (OH): Curran Associates Inc; 2005, paper IAC-05-C2.3.05.

14. Del Vecchio A, Di Clemente M, Ferraiuolo M, Gardi R, Marino G, Rufolo G, Scatteia L. Sharp hot structures project current status. Proceedings of the 57th International Astronautical Congress; Valencia, Spain; October 2–6, 2006, paper IAC-06-C2.4.05.

15. Bartuli C, Valente T, Tului M. Plasma spray deposition and high temperature characterization of ZrB_2-SiC protective coatings. Surf Coating Technol 2002;155:260–273.

16. Tului M, Ruffini F, Arezzo F, Lasisz S, Znamirowski Z, Pawlowski L. Some properties of atmospheric air and inert gas high-pressure plasma sprayed ZrB_2 coatings. Surf Coating Technol 2002;151–152:483–489.

17. Tului M, Marino G, Valente T. Plasma spray deposition of ultra high temperature ceramics. Surf Coating Technol 2006;201:2103–2108.

18. Tripp WC, Davis HH, Graham HC. Effect of SiC addition on the oxidation of ZrB_2. Am Ceram Soc Bull 1973;52:612–616.

19. Monteverde F, Guicciardi S, Bellosi A. Advances in microstructure and mechanical properties of zirconium diboride based ceramics. Mater Sci Eng 2003;346:310–319.

20. Chamberlain A, Fahrenholtz W, Hilmas G. High-strength zirconium diboride-based ceramics. J Am Ceram Soc 2004;87:1170–1172.

21. Monteverde F. Beneficial effects of an ultra-fine α-SiC incorporation on the sinterability and mechanical properties of ZrB_2. Appl Phys A Mater Sci Process 2006;82:329–337.

22. Kuriakose AK, Magrave JL. The oxidation kinetics of zirconium diboride and zirconium carbide at high temperatures. J Electrochem Soc 1964;111:827–831.

23. Tripp WC, Graham HC. Thermogravimetric study of the oxidation of ZrB_2 in the temperature range of 800°C to 1500°C. J Electrochem Soc 1971;118:1195–1199.

24. Opeka MM, Talmy IG, Wuchina EJ, Zaykoski JA, Causey SJ. Mechanical, thermal, and oxidation properties of refractory hafnium and zirconium compounds. J Eur Ceram Soc 1999;19:2405–2414.

25. Ban'kovskaya IB, Zhabrev VA. Kinetic analysis of the heat resistance of ZrB_2–SiC composites. Glass Phys Chem 2005;31:482–488.

26. Chamberlain A, Fahrenholtz W, Hilmas G, Ellerby D. Oxidation of ZrB_2-SiC ceramics under atmospheric and reentry conditions. Refract Appl Trans 2005;1:1–8.

27. Fahrenholtz WG. The ZrB_2 volatility diagram. J Am Ceram Soc 2005;88:3509–3512.

28. Fahrenholtz G, Hilmas GE, Talmy IG, Zaykoski JA. Refractory diborides of zirconium and hafnium. J Am Ceram Soc 2007;90:1347–1364.

29. Fahrenholtz WG. Thermodynamic analysis of ZrB_2-SiC oxidation: formation of a SiC-depleted region. J Am Ceram Soc 2007;90:143–148.

30. Levine SR, Opila EJ, Halbig MC, Kiser JD, Singh M, Salem JA. Evaluation of ultra-high temperature ceramics for aeropropulsion use. J Eur Ceram Soc 2002;22:2757–2767.

31. Fahrenholtz WG, Hilmas GE, Chamberlain AL, Zimmermann JW. Processing and characterization of ZrB_2-based ultra-high temperature monolithic and fibrous monolithic ceramics. J Mater Sci 2004;39:5951–5957.

32. Gasch M, Ellerby D, Irby E, Beckman S, Gusman M, Johnson S. Processing, properties and arc jet oxidation of hafnium diboride/silicon carbide ultra high temperature ceramics. J Mater Sci 2004;39:5925–5937.

33. Opeka MM, Talmy IG, Zaykoski JA. Oxidation-based materials selection for 2000°C + hypersonic aerosurfaces: theoretical considerations and historical experience. J Mater Sci 2004;39:5887–5904.

34. Opila E, Levine S, Lorincz J. Oxidation of ZrB_2- and HfB_2-based ultra-high temperature ceramics: effect of Ta additions. J Mater Sci 2004;39:5969–5977.

35. Monteverde F, Bollosi A. The resistance to oxidation of an HfB_2–SiC composite. J Eur Ceram Soc 2005;25:1025–1031.

36. Rezaire A, Fahrenholtz WG, Hilmas GE. Evolution of structure during the oxidation of zirconium diboride–silicon carbide in air up to 1500°C. J Eur Ceram Soc 2007;27:2495–2501.

37. Rezaire A, Fahrenholtz WG, Hilmas GE. Oxidation of zirconium diboride silicon carbide at 1500°C at a low partial pressure of oxygen. J Am Ceram Soc 2006;89:3240–3245.

38. Han J, Hu P, Zhang X, Meng S, Han W. Oxidation-resistant ZrB_2-SiC composites at 2200°C. Compos Sci Technol 2008;68:799–806.

39. Zhang X, Hu P, Han J, Meng S. Ablation behaviour of ZrB_2–SiC ultra high temperature ceramics under simulated atmospheric re-entry conditions. Compos Sci Technol 2008;68:1718–1726.

40. Carney CM, Mogilvesky P, Parthasarathy TA. Oxidation behaviour of zirconium diboride silicon carbide produced by the spark plasma sintering method. J Am Ceram Soc 2009;92:2046–2052.

41. Hu P, Goulin W, Wang Z. Oxidation mechanism and resistance of ZrB_2-SiC composites. Corros Sci 2009;51:2724–2732.

42. Karlsdottir SN, Halloran JW. Oxidation of ZrB_2-SiC: influence of SiC content on solid and liquid oxide phase formation. J Am Ceram Soc 2009;92:481–486.

43. Eakins E, Jayaseelan DD, Lee WD. Toward oxidation-resistant ZrB_2-SiC ultra high temperature ceramics. Metall Mater Trans A 2010;42:878–887.

44. Scatteia L, Alfano D, Monteverde F, Sans J-L, Balat-Pichelin M. Effect of the machining method on the catalycity and emissivity of ZrB_2 and ZrB_2–HfB_2-based ceramics. J Am Ceram Soc 2008;91:1461–1468.

45. Monteverde F, Bellosi A, Scatteia L. Processing and properties of ultra-high temperature ceramics for space applications. Mater Sci Eng A 2008;485:415–421.

46. Scatteia L, Borrelli R, Cosentino G, Bêche E, Sans J-L, Balat-Pichelin M. Catalytic and radiative behaviors of ZrB_2–SiC ultra high temperature ceramic composites. J Spacecraft Rockets 2006;43:1004–1012.

47. Alfano D, Scatteia L, Monteverde F, Bêche E, Balat-Pichelin M. Microstructural characterization of ZrB_2–SiC based UHTC tested in the MESOX plasma facility. J Eur Ceram Soc 2010;30:2345–2355.

48. Charpentier L, Dawi K, Eck J, Pierrat B, Sans J-L, Balat-Pichelin M. Concentrated solar energy to study high temperature materials for space and energy. J Sol Energy Eng 2011; 133:031005.

49. Eck J, Sans J-L, Balat-Pichelin M. Experimental study of carbon materials behavior under high temperature and VUV radiation: application to Solar Probe + heat shield. Appl Surf Sci 2011;257:3196–3204.

50. Balat-Pichelin M, Badie J-M, Berjoan R, Boubert P. Recombination coefficient of atomic oxygen on ceramic materials under earth re-entry conditions by optical emission spectroscopy. Chem Phys 2003;291:181–194.

51. Balat-Pichelin M, Vesel A. Neutral oxygen atom density in the MESOX air plasma solar furnace facility. Chem Phys 2006;327:112–118.

52. Reibaldi G, Gavira J, Ratti F, de Mey S, Muylaert J-M, Provera R, Massobrio F. EXPERT: the Esa eExperimental re-entry test-bed programme. Proceedings of the 59th International Astronautical Congress; Glasgow, Scotland; September 29 to October 3, 2008, IAC-08.D2.6.3.

53. BS EN 843-1:2006 Advanced technical ceramics. Mechanical properties of monolithic ceramics at room temperature. Determination of flexural strength. BS EN 820-1:2002 Advanced technical ceramics. Methods of testing monolithic ceramics. Thermo-mechanical properties. Determination of flexural strength at elevated temperatures.

54. Zhang XH, Hu P, Han J, Xu L, Meng S. The addition of lanthanum hexaboride to zirconium diboride for improved oxidation resistance. Scripta Mater 2007;57:1036–1039.

55. Hu P, Zhang XG, Han JC, Luo XG, Du SY. Effect of various additives on the oxidation behavior of ZrB_2-based ultra high temperature ceramics at 1800°C. J Am Ceram Soc 2010; 93:345–349.

56. Jayaseelan DD, Zapata-Solvas E, Brown P, Lee WE. In situ formation of oxidation resistant refractory coatings on SiC-reinforced ZrB_2 ultra high temperature ceramics. J Am Ceram Soc 2012;95:1247–1254.

57. Subramanian MA, Aravamudan G, Subba Rao GV. Oxide pyrochlores—a review. Prog Solid State Chem 1983;15:55–143.

58. Cao XQ, Vassen R, Stöver D. Ceramic materials for thermal barrier coatings. J Eur Ceram Soc 2004;24:1–10.

59. Monteverde F, Alfano D, Savino R. Effects of LaB_6 addition on arc-jet convectively heated SiC-containing ZrB_2-based UHTC in high enthalpy supersonic airflows. Corros Sci 2013;75:443–453.

60. Borrelli R, Riccio A, Tescione D, Gardi R, Marino G. Numerical/experimental correlation of a plasma wind tunnel test on a UHTC-made nose cap of reentry vehicle. J Aerospace Eng 2010;23:309–316.

INDEX

Ultra-High Temperature Ceramics: Materials for Extreme Environment Applications, First Edition.
Edited by William G. Fahrenholtz, Eric J. Wuchina, William E. Lee, and Yanchun Zhou.
© 2014 The American Ceramic Society. Published 2014 by John Wiley & Sons, Inc.